Fungal Biology

Series Editors

Fungal biology has an integral role to play in the development of the biotechnology and biomedical sectors. It has become a subject of increasing importance as new fungi and their associated biomolecules are identified. The interaction between fungi and their environment is central to many natural processes that occur in the biosphere. The hosts and habitats of these eukaryotic microorganisms are very diverse; fungi are present in every ecosystem on Earth. The fungal kingdom is equally diverse, consisting of seven different known phyla. Yet detailed knowledge is limited to relatively few species. The relationship between fungi and humans has been characterized by the juxtaposed viewpoints of fungi as infectious agents of much dread and their exploitation as highly versatile systems for a range of economically important biotechnological applications. Understanding the biology of different fungi in diverse ecosystems as well as their interactions with living and non-living is essential to underpin effective and innovative technological developments.This series will provide a detailed compendium of methods and information used to investigate different aspects of mycology, including fungal biology and biochemistry, genetics, phylogenetics, genomics, proteomics, molecular enzymology, and biotechnological applications in a manner that reflects the many recent developments of relevance to researchers and scientists investigating the Kingdom Fungi. Rapid screening techniques based on screening specific regions in the DNA of fungi have been used in species comparison and identification, and are now being extended across fungal phyla. The majorities of fungi are multicellular eukaryotic systems and therefore may be excellent model systems by which to answer fundamental biological questions.A greater understanding of the cell biology of these versatile eukaryotes will underpin efforts to engineer certain fungal species to provide novel cell factories for production of proteins for pharmaceutical applications. Renewed interest in all aspects of the biology and biotechnology of fungi may also enable the development of "one pot" microbial cell factories to meet consumer energy needs in the 21st century. To realize this potential and to truly understand the diversity and biology of these eukaryotes, continued development of scientific tools and techniques is essential. As a professional reference, this series will be very helpful to all people who work with fungi and should be useful both to academic institutions and research teams, as well as to teachers, and graduate and postgraduate students with its information on the continuous developments in fungal biology with the publication of each volume.

Ahmed M. Abdel-Azeem • Francesca Degola
Editors

Advances in Lichens and Non-Flowering Plants Fungal Endophytes Research

Volume 1: Biodiversity and Ecological Perspective

Springer

Editors
Ahmed M. Abdel-Azeem
Botany and Microbiology Department
Faculty of Science
Suez Canal University
Ismaliia, Egypt

Francesca Degola
Department of Chemistry, Life Sciences
and Environmental Sustainability
University of Parma
Parma, Italy

ISSN 2198-7777 ISSN 2198-7785 (electronic)
Fungal Biology
ISBN 978-3-032-01505-1 ISBN 978-3-032-01506-8 (eBook)
https://doi.org/10.1007/978-3-032-01506-8

This Springer imprint is published by the registered company Springer Nature Switzerland AG
The registered company address is: Gewerbestrasse 11, 6330 Cham, Switzerland

Foreword by Vijai Kumar Gupta

Advances in Lichens and Non-Flowering Plants Fungal Endophytes Research is a comprehensive and balanced introduction to fungal cell factories' biology, biotechnological applications, and environmental significance. This book provides a detailed exploration of fungal diversity and the mapping of fungal barcoding for biotechnological potential. Since the introduction of technology that optimises these organisms, fungi's practical applications have increased dramatically. The current climate offers an ideal environment for broad fungal diversity. Fungal species are predominantly unidentified and remain a largely untapped resource. At the same time, while research communities and industries lack a source of suitably adapted fungal species, scientists find themselves without a depository for fungal identification and culturing services; thus, our knowledge of fungal biodiversity is incomplete. To provide the necessary foundation for fungal research, this book will serve as a valuable resource for the multidisciplinary characterisation of species. The information presented in this book will further facilitate mining for potential fungal candidates for biotechnological product development. Consequently, innovative and highly specific molecules will be identified. This literature will be valuable to international and European researchers and academics and have a significant international impact. Although it is particularly suitable for postgraduate students and research workers, this book will also interest undergraduate students who require an overview of the traditional and more recent practical applications of fungal systems and biotechnology.

School of Biotechnology, Dublin City University, Dublin, Ireland

Vijai Kumar Gupta

Foreword by Lobna Hajji-Hedfi

The complex and frequently mysterious interactions between fungi and their plant hosts have consistently intrigued scientists from various fields. In *Advances in Lichens and Non-Flowering Plants Fungal Endophytes Research: Volume 1—Biodiversity and Ecological Perspective*, editors Prof. Ahmed M. Abdel-Azeem (Suez Canal University, Egypt) and Dr. Francesca Degola (Parma University, Italy) have meticulously assembled a pioneering anthology of research exploring the concealed realm of fungal endophytes linked to lichens and nonflowering plants. This volume, part of Springer's distinguished Fungal Biology Series, exemplifies the increasing acknowledgment of fungal endophytes as essential elements of biodiversity, ecological stability, and prospective biotechnological resources.

Fungal endophytes, frequently disregarded, are crucial for host physiology, stress resilience, and ecosystem stability. However, their diversity, distribution, and ecological roles remain inadequately investigated, especially in nonflowering plants—a deficiency this volume aims to rectify. The 20 chapters herein encompass a wide array of subjects, ranging from the enigmatic variety of endolichenic fungi to the intricate relationships between endophytes and primordial plant lineages, including ferns, lycophytes, and gymnosperms. The incorporation of research from varied ecosystems, such as the Brazilian Amazon and marine macroalgae, highlights the worldwide significance of these microbial symbionts.

The book's strength is its multidisciplinary approach, combining traditional mycological techniques with advanced high-throughput sequencing to elucidate the complexities of fungal endophyte communities. The chapters titled "The Fungal Endophytes: A Dilemma" and "Where Are the Undocumented and Unrecorded Endophytic Fungi?" contest prevailing paradigms and advocate for more extensive taxonomic and ecological research. Simultaneously, insights into the biotechnological potential of endophytes associated with medicinal ferns and the temporal dynamics of Ginkgo biloba underscore the practical importance of this research.

Given the threats of climate change and habitat destruction to global biodiversity, it is increasingly imperative to comprehend the resilience and adaptability of fungal endophytes. This volume synthesizes existing knowledge and identifies crucial research frontiers, including the effects of environmental changes on endophytic communities. The editors and contributors deserve commendation for their endeavors to connect fundamental and applied mycology, providing insights that will appeal to ecologists, microbiologists, conservationists, and biotechnologists.

I am honored to endorse this timely and thorough work. This book reveals the concealed fungal networks in lichens and non-flowering plants, facilitating future discoveries and underscoring the necessity of conserving these symbiotic relationships in a swiftly evolving environment.

Regional Centre of Agricultural
Research of Sidi Bouzid, Tunis, Tunisia

Lobna Hajji-Hedfi

Preface

The past decades have been characterized by a sensible increase in investigations on fungal endophytes, as their hidden but significant impact on host plants, ecological equilibrium, and biodiversity has consistently intrigued researchers in life sciences. These enigmatic microorganisms that flourish within plant tissues, in fact, are able to establish intricate—and frequently symbiotic—relationships that deeply influence ecosystems in ways we are just beginning to comprehend. By moving from water to land, the ancestors of modern plants faced multiple challenges, evolutionary forces that helped to shape their physiology and metabolism, making them talented colonizers that resourcefully succeeded in adapting to the most wide-ranging environmental variables: nowadays, we know that the association with microorganisms has been pivotal for the most profitable exploitation of nutrients and response to abiotic stressors. However, if plant-microbial interactions have been largely investigated in crops as well as other higher species, partnerships of nonflowering plants with endophytic populations still remain poorly studied.

Advances in Lichens and Non-Flowering Plants Fungal Endophytes Research: Volume 1—Biodiversity and Ecological Perspective investigates the complex relationships among fungi and nonflowering plant species, a domain that has traditionally garnered less attention than angiosperms. A particular focus on lichens, that are composite organisms formed by the collaboration between fungi (mycobionts) and photosynthetic partners (photobionts, algae or cyanobacteria) and thus exemplifying the mutualistic symbioses, is provided. Recent research has revealed a more intricate network of interactions involving endophytic fungi within lichen thalli, affecting their physiology, resilience, and ecological adaptability. Likewise, nonflowering plants—such as bryophytes, ferns, and gymnosperms—possess varied endophytic communities that enhance their survival in adverse environments. These associations are essential for nutrient cycling, stress resilience, and the synthesis of bioactive compounds with prospective pharmaceutical and agricultural uses.

This volume, the inaugural installment in a series concerning fungal endophytes in lichens and nonflowering plants, emphasizes their biodiversity and ecological functions; it consolidates advanced research from mycologists, ecologists, and plant biologists to integrate existing knowledge and underscore emerging trends. The chapters examine the diversity of endophytic fungi across various host systems, their evolutionary adaptations, and their roles in ecosystem functioning.

A primary theme of this book is the extraordinary diversity of fungal endophytes linked to lichens and nonflowering plants. In contrast to the extensively researched grass and angiosperm endophytes, these fungi display distinct taxonomic and functional characteristics, frequently associated with the ancient lineages of their hosts. Advancements in molecular methodologies, such as high-throughput sequencing and metagenomics, have transformed our capacity to identify and characterize these elusive fungi, uncovering previously unrecognized lineages and ecological niches.

From an ecological standpoint, fungal endophytes are crucial in facilitating plant-environment interactions. They strengthen host resilience to abiotic stresses including drought, extreme temperatures, and heavy metal toxicity, making them essential contributors in ecosystems confronting climate change and environmental degradation. Moreover, certain endophytes synthesize secondary metabolites that repel herbivores or pathogens, thereby indirectly influencing community dynamics and species distributions. In lichen systems, endophytic fungi can affect thallus morphology, metabolite synthesis, and reproductive strategies, thereby introducing additional complexity to these intricate symbioses.

This book examines the challenges and future trajectories in fungal endophyte research. Notwithstanding considerable advancements, numerous inquiries persist without resolution. In what ways do endophyte communities differ across various biogeographical regions? What mechanisms govern host specificity and fungal colonization? How can we utilize these fungi for biotechnological and conservation objectives? This volume seeks to inspire additional investigation and collaboration in this dynamic field by synthesizing recent findings and promoting interdisciplinary dialogue.

We express our profound appreciation to the contributing authors, whose knowledge and commitment have rendered this book feasible. Their research not only enhances scientific comprehension but also emphasizes the necessity of safeguarding these concealed microbial networks, which are essential to global biodiversity. We recognize the researchers who have advanced the field of fungal endophyte studies, along with the institutions and funding organizations that facilitate this essential research.

This volume aims to be a valuable resource for students, researchers, and enthusiasts as we explore the complexities of fungal endophytes in lichens and non-flowering plants. The hidden domain of these fungi possesses significant potential—for scientific advancement, ecological balance, and sustainable resolutions to urgent environmental issues.

Ismaliia, Egypt — Ahmed M. Abdel-Azeem
Parma, Italy — Francesca Degola

Acknowledgments

All the authors are sincerely acknowledged for contributing up-to-date information on the *Advances in Lichens and Non-Flowering Plants Fungal Endophytes Research: Volume 1—Biodiversity and Ecological Perspective*. The Editors are thankful to all the Authors for their valuable contributions.

All Editors would like to thank their families who were very patient and supportive during this journey. Our sincere thanks to the whole Springer team who were directly or indirectly involved in the compilation of this book. We are grateful to the many people who helped to bring this book to light. The Editors wish to thank Project Coordinator, Springer, for generous assistance, constant support, and patience in initializing the volume.

We are very sure that this book will interest scientists, graduates, undergraduates, and postdocs who are investigating on "Endophytic Fungi."

Contents

About the Editors

Ahmed M. Abdel-Azeem (Zemo) is a Full Professor of mycology at the Botany and Microbiology Department, Faculty of Science, Suez Canal University in Egypt. He is an affiliated professor at the Research Institute of the University of the Bucharest (ICUB) and an affiliated research fellow in the Centre for Mineral Biogeochemistry at the University of the Free State, South Africa. He is the founder and president of the Arab Society for Fungal Conservation and president of the African Mycological Association (AfriMA). He is an executive committee member of the International Mycological Association (IMA). He was a founding member of the International Society for Fungal Conservation, a member of the European Mycological Association, and a member of the Mycological Society of America. He founded Egypt's National Fungus Day, African Annual Fungus Day, International Microorganisms Day for Egypt, and many mycologists' networks, including African, Egyptian, Iraqi, Algerian, Libyan, and Pakistani networks. He founded Pan Arab Mycologists. He is particularly interested in ecology, taxonomy, biology, secondary metabolites, endophytes, climate change, and fungal biodiversity, as well as conservation, with a particular focus on members of the phylum Ascomycota. He studies the fungal deterioration of cultural heritage like Abydos Middle Cemetery, Cheops Ceremonial Unveiled Boat, Al-Aqsa Mosque in Occupied Palestine, and San El-Hagar stony monuments. He has the necessary vision combined with proven experience leading and successfully completing more than

15 complex and ambitious local and international projects. He is a guest professor at many international universities in the USA, Italy, Malta, Romania, Spain, Greece, and South Africa. He received the President's FUNGAL CONSERVATION AWARD in 2013. He is a member of the IUCN Species Survival Commission Specialist Group for Cup Fungi, Truffles, and Their Allies. He is also a member of the IUCN's Fungal Conservation Committee (FunCC). He has published more than 125 research papers in international journals, 60 book chapters with international publishers, and 15 books.

Francesca Degola is an emerging figure in the field of plant biology and physiology, whose research lines fall prominently into the metabolic responses to both biotic and abiotic interactions of plants with their environment. After completing her studies in Molecular Biology, she pursued a PhD degree in Plant Biology at the University of Parma, Italy, where she is currently a Tenure-Track Researcher and Professor of Botany. Her doctoral studies were focused on the strategies for the management of mycotoxins contamination in crops: from fungal biocompetition to the design of natural-based, newly synthesized compounds, her contributions in this field have led to several high-profile publications. Following her doctoral dissertation, she held various postdoctoral positions funded by projects aimed at the investigation of plants response to both biotic and abiotic environmental interactors (such as heavy metals, fungal pathogens, and beneficial endophytes), in terms of proteome and metabolome modulation. She further explored the evolutionary relevance that some of these responses have represented during terrestrialization, bringing new awareness about the relationship between plants and their ecosystems. Throughout her career, she has been a prolific researcher, publishing numerous articles (up to 60) in highly impacted, peer-reviewed journals, for some of which she also serves as Editor and Guest Editor.

Overall, her work often emphasizes the importance of interdisciplinary approaches to environmental challenges, integrating insights from genetics, evolution, physiology, and ecology.

The Fungal Endophytes: A Dilemma

Laith Khalil Tawfeeq Al-Ani, Lobna Hajji-Hedfi, Mavis Agyeiwaa Acheampong, and Dalia Sukmawati

Contents

L. K. T. Al-Ani (✉)
School of Biology Science, Universiti Sains Malaysia, Pulau Pinang, Malaysia

L. Hajji-Hedfi
Regional Centre of Agricultural Research of Sidi Bouzid, Sidi Bouzid, Tunisia

Research Laboratory of Agricultural Production Systems and Sustainable Development (LR03AGR02), Carthage University, Tunis, Tunisia

M. A. Acheampong
Department of Crop Science, University of Ghana, Accra, Ghana

D. Sukmawati
Faculty of Mathematics and Natural Sciences, Department of Biology, Universitas Negeri Jakarta, Rawamangun, Jakarta Timur, Indonesia

Universitas Negeri Jakarta Culture Collection (UNJCC), Faculty of Mathematics and Natural Sciences, Universitas Negeri Jakarta, Rawamangun, Indonesia

A. M. Abdel-Azeem, F. Degola (eds.), *Advances in Lichens and Non-Flowering Plants Fungal Endophytes Research*, Fungal Biology,
https://doi.org/10.1007/978-3-032-01506-8_1

1 Introduction

Endophytic fungi (EF) can colonise the tissue of plants throughout their life cycle or as part of it (Hajji-Hedfi et al. 2024) without causing any disease (Al-Ani et al. 2021a). EF belong to different phylum (Wijayawardene et al. 2020; Raza et al. 2023). Almost each plant on earth is associated with EF due to relevance between plant and fungi as interesting partners (Suryanarayanan 2013). EF can live in plants as facultative and obligate parasites. For facultative living, they colonise the living plants and complete another stage outside of plants as saprophyte or parasites. Abreu-Tarazi et al. (2010) mentioned the association of some EF in rhizosphere of plants in certain stages of their life cycle, living as endophytes in another plant part within the life cycle. Fungi can be beneficial (as biocontrol agents) (Al-Ani et al. 2020a) or harmful to plants (as pathogens) living as facultative endophytes in plants. Facultative EF can improve plant growth and act as biocontrol agents against plant disease and pests (Lorena et al. 2021). Other EF are obligate organisms living in plants throughout their life cycle. These EF can therefore be compatible with plant by growth, reproduction, and dispersal into plant generations (Gouda et al. 2016).

Thus, biological control is considered a suitable method in controlling plant diseases instead of synthetic chemical pesticides, particularly those that included plant growth-promoting rhizobacteria (PGPR), plant growth-promoting fungi (PGPF), plant extracts (Al-Ani et al. 2012; Mohammed et al. 2012; Al-Ani 2017a; Jatoi et al. 2020; Soumare et al. 2021), as well as plant activators (Mohammed et al. 2013) and edible mushroom (Castañeda-Ramírez et al. 2020; Comans-Pérez et al. 2021; Aguilar-Marcelino et al. 2022). PGPR have been used in controlling plant viral pathogens (Al-Ani 2006; Al-Ani and Al-Ani 2011) and plant fungal pathogens (Mohammed et al. 2011, 2012, 2014; Rahman et al. 2023). *Trichoderma harzianum* could control the grey mold disease of rose (*Rosa indica* L.) (Jatoi et al. 2022), whilst the several yeast groups are utilised as beneficial microbes (Sukmawati et al. 2024a, b).

However, EF can alter the chemistry of plant by producing secondary compounds. EF showed the ability to change the leaf chemistry and induced the biosynthesis of essential oil of host plant (Estrada et al. 2013; Du et al. 2022). EF can produce many kinds of secondary compounds with different activity such as enhancing the defence and growth of plant (Al-Ani et al. 2020b), remediating for abiotic stresses (As salt, heavy metals, and drought) (Lalarukh et al. 2022a, b; Meshram et al. 2023), and antibiotic production which could inhibit different kinds of plant pathogens including viruses, nematodes, fungi, and bacteria (Aguilar-Marcelino et al. 2020a; Al-Ani et al. 2024). Aly et al. (2013) emphasised that there is a beneficial relationship between EF and their host plants leading to the production of bioactive compounds that play a role in plant–microbe interactions. The activity of EF in plants involves stimulation and secretion of secondary compounds that can be utilised in sustainable agriculture (Al-Ani 2018a). Al-Ani (2019a) suggested three levels for endophytic fungal genera that are isolated from plants comprising low, middle, and high levels. Thirteen interesting endophytic fungal genera were found in medicinal plants, and these include *Trichoderma*, *Aspergillus*, *Alternaria*, *Cladosporium*, *Chaetomium*, *Curvularia*, *Diaporthe*, *Phoma*, *Penicillium*, *Colletotrichum*, *Pestalotiopsis*, *Phomopsis*, and *Fusarium* (Al-Ani 2019a, b). Four of these fungal genera are considered more promising as patent isolates comprising *Neotyphodium* sp., *Fusarium* sp., *Muscodor* sp., and *Curvularia* sp. (Al-Ani 2019c). This study aimed to provides a dilemma of existence of EF in plants.

2 Diversity of Endophytic Fungi

Fossil records from the Early Devonian Rhynie Chert suggest that EF have been associated with plants for over 400 million years (Taylor et al. 1995, 2005; Krings et al. 2007), although the first description of endophytes and subsequent isolation of EF (from darnel ryegrass *Lolium temulentum*) were documented in the nineteenth century (de Barry 1866; Guerin 1898; Oulhen et al. 2016). Currently, the number of EF is estimated at over 1.3–3 million, with majority housed within the phylum *Ascomycota* (Dreyfuss and Chapela 1994; Arnold 2007; Rodriguez et al. 2009; Rashmi et al. 2019; Bhunjun et al., 2024; Rungjindama and Jones 2024).

A global compilation list identified 2771 EF belonging to 877 genera with ascomycetes accounting for 87% of all the reported studies (Rashmi et al. 2019). The most predominant genera amongst the EF were "*Alternaria*, *Aspergillus*, *Colletotrichum*, *Fusarium*, *Penicillium*, and *Phoma*" (Rashmi et al. 2019). The aforementioned ascomycetous endophytic fungal genera and the others were isolated mainly from leaf samples in diverse environments, ranging from aquatic, xeric, arctic, temperate, and tropical ecosystems (Zhang and Yao 2015; Rashmi et al. 2019; Yadav 2020). basidiomycetes accounted for 10% of the global EF list and were mostly recovered as root endophytes in arctic regions, whilst *Chytridiomycetes* and *Mucoromycetes* accounted for the remaining 3%. This finding corroborates earlier compilations with selected EF studies, as well as recent isolations where ascomycetous EF exceeded 90% (Zhang and Yao 2015; Supaphon et al. 2017; An et al. 2020; Khalmuratova et al. 2020; De Silva et al. 2021; Ettinger and Eisen 2021). The lower occurrence of the basidiomycetes and the other fungal endophytes (mainly *Chytridiomycetes* and *Mucoromycetes*) has been attributed to the media used, isolation protocols, incubation period, and slow growth of basidiomycetes, amongst others (Rungjindama and Jones 2024). Whilst the discovery of the paclitaxel (taxol)-producing endophytic basidiomycetous fungus, *Ceriporiopsis andreanae* (formerly *Taxomyces andreanae*), from the Pacific yew tree (*Taxus brevifolia*) and its industrial production revolutionised EF research, the most studied endophytic fungal association falls within the ascomycetes—clavicipitaceous grass endophytes, particularly *Epichloë* spp. (Schardl et al. 2004; Tanaka et al. 2005; Márquez et al. 2012; Verma et al. 2019). Nonetheless, all plants generally harbour EF, but their diversity varies greatly depending on the type of host plant and habitat. Generally, tropical forests, with diverse plants, warmer temperatures, and abundant rainfall, which are conducive for fungal growth and development, harbour great diversity of EF compared with boreal forests (Suryanarayanan et al. 2002; Arnold et al. 2003; Rodriguez et al. 2009; U'Ren et al. 2012; Roy and Banerjee 2018; Zheng et al. 2016; Oita et al. 2021). The fewer EF identified from extreme environments—Arctic or Antarctic (Botnen et al. 2014; Zhang and Yao 2015; Pedraza et al. 2018; Yu et al. 2019), saline (Brotman et al. 2013; Manjunatha et al. 2022), and arid (Yu et al. 2019; Georgousaki et al. 2022) areas have adaptations to their respective regions.

3 Endophytic Fungi from Mutualists and Potential Latent Pathogens

The intracellular and intercellular spaces of plants are suitable for colorisation by EF between mutualists to latent pathogens. Fungi either as endophytes or as plant pathogens could inhabit the plant tissue in the same environmental niches (Nwe et al. 2024). EF can be changing in phase between the aggressive as high virulent to mutualism according to state of host. Some EF can attack the host plant when weakened as opportunistic fungi to cause disease (Sieber 2007). Three levels of activity of EF can be exhibited in the host plant. This can show as visible or invisible effect on host plant (Kogel et al. 2006). Davis and Shaw (2008) opined that fungal group that are endophytes interacts with plant as mutualists and latent pathogens (Commensals). The activity of endophytic fungi can change from mutualists (Beneficial) or latent pathogens (avirulent) to virulent and vice versa, depending on the physiological status of the host plant as well as prevailing environmental conditions (Alam et al. 2021). *Colletotrichum tofieldiae* could switch between pathogenicity to endophyte within the same host depending on the environmental conditions of host (Newfeld et al. 2025). Meanwhile some plant fungal pathogens could attack plants (non-host plant or host) without causing any effect on the growth and physiology of plant (As incompatibility interaction), as part of the endophyte-microbe system (Al-Ani and Furtado 2020). However, the mutualistic interaction of EF could cause mutual benefit for the fungi and the host as symbiosis (Jia et al. 2016). EF in mutualistic phase could divide to two group such as, first, constitutive mutualism (living as systemic), and, second, inducible mutualist (useful as biocontrol agent) (Carroll 1988). *Fusarium pseudograminearum* as isolated endophytic fungi showed mutualistic interaction with some trees which could improve the plant growth, exchange of nutrient, uptake of phosphorus and nitrogen, and upregulation of PtSweet1 gene (transporter gene expression) for root sugar (Yang et al. 2025). On the other hand, the latent phase is part of the life cycle for some plant fungal pathogens (Rai and Agarkar 2014). Many important fungal genera such as *Fusarium*, *Alternaria*, *Phomopsis*, *Cladosporium*, *Colletotrichum*, and *Diaporthe* and some species within the family *Botryosphaeriaceae* live as endophyte in a lot of plants, whilst remaining pathogenic to several others. Sessa et al. (2018) found two endophyte fungal species belong to *Diaporthe* genus (*Diaporthe infecunda* and *Diaporthe serafiniae*) isolated from apple which could be pathogenic by causing necrotic lesions on detached twigs of this tree crop. Indeed, EF belonging to plant fungal pathogen group which do not cause any disease are only being non-pathogenic or avirulent state in the host from which they were isolated from. Many non-pathogenic *Fusarium oxysporum* f. sp. *cubense* Tropical race 4 (FocTR4), other races of Foc, and other *Fusarium* species isolated from banana tree (Berangan cv. Intan) from different sites in Malaysia may be latent pathogen for other cultivar of banana (Al-Ani 2017b). *Colletotrichum acutatum* isolated from two plant weed genera *Vicia* and *Conyza* as non-pathogenic fungi is a pathogen of anemone and strawberry (Freeman et al. 2001).

The transformation in level of activity of fungi from latent pathogen to active pathogen depends on ecological state such as stress condition environmental and ecological time. *Botryosphaeriaceae* family including *Diplodia*, *Botryosphaeria* (anamorph *Fusicoccum*), *Guignardia*, *Neofusicoccum*, *Dothidotthia* (anamorph *Dothiorella*), *Pseudofusicoccum*, and *Lasiodiplodia* are common endophytes and latent pathogens but aggressive pathogens for some host plants during stress (Slippers and Wingfield 2007). Marsberg et al. (2017) reported that *Botryosphaeria dothidea* can grow in woody tissues of plants (asymptomatic) without causing any damage to tissues but is a latent pathogen under unfavourable environmental conditions Oren et al. (2003) reported *Fusarium verticillioides* as an endophyte in maize which later became pathogen with time.

The type of enzyme produced by the EF can also contribute to their transition from endophytes to latent pathogen. Sunitha et al. (2013) suggested that the activity of fungal species is dependent on the kind of enzyme it produces. EF can produce cuticle degrading enzymes such as cellulase, pectinase and xylanase, which assist in the colonisation of plant tissues (Ramos-Garza et al. 2016). The latent pathogens could produce proteinase (Reddy and Basappa 1997). However, some EF exist in weeds in two phases—switching between biocontrol agent to pathogen. Romero et al. (2001) isolated several fungal species from leaves of *Parthenium hysterophorus* (weed) including *Alternaria helianthi*, *Alternaria zinniae*, *Cylindrocarpon* sp., *Fusarium* sp., *Curvularia brachyspora*, *Periconia* sp., *Nigrospora oryzae*, and *Penicillium funiculosum*, with some being endophyte, biocontrol agents (against weeds) as well as latent pathogens.

4 Pathogenicity of the Plant Pathogens

Fungal pathogen can attack plant until the occurrence of disease according to some steps: adherence, penetration, invading, colonising, and spreading. Infection starts by attachment of propagules on susceptible host plant followed by their direct penetration after germination or via mechanical forces into plant (Schafer 1994). The chemical weapons produced aid infection for instance production of toxins, growth regulators (auxins, cytokinins, gibberellins, abscisic acid, ethylene), and enzymes (cutinases, cellulases, pectinases, etc.) (Sharma and Gautam 2019). Some fungi can live as endophytes and then shift to a pathogenic form in same host. However, some plant fungal pathogen cannot be considered as endophytic species though facultative living as examples the genera *Phytophthora* (e.g. *Phytophthora infestans*), and *Aspergillus* g (e.g. *Aspergillus flavus* and *A. terreus*) as the former cause huge economic losses to crops whilst the latter continue to pose risk to grains (aflatoxin production) (Attitalla et al. 2010a, b; Raza et al. 2022). Therefore, the integrative approaches of omics technologies can be useful to study comparative genomics by determining molecules such as effectors (different in numbers of effector genes), secondary metabolites (phytotoxic), and CAZymes (cellulases, hemicellulases, and metalloproteases; Newfeld et al. 2025). The production of fungal biofilms has also shown a role in the pathogenicity (Aguilar-Marcelino et al. 2021).

Therefore, three points are needed to complete occurrence the disease of plant by the fungal pathogens: (i) the mechanisms of fungal pathogen to cause the disease for plant, (ii) overcoming of plant defences and degradation of the antifungal compounds such as phytoalexins, and (iii) establishment of the suitable environmental conditions for the growth and sporulation of the fungal pathogen. The lack of one of these points during the infection time will not lead to the development of disease temporarily or permanently.

Some fungal pathogens are highly sensitive to UV radiation hence may not survive in the field and thus exhibit non-pathogenicity as the radiation inhibited growth of fungal pathogen (De Souza and Al-Ani 2023). However, in the greenhouse environment, the same fungi showed full pathogenic activity (Fry and Myers 1981).

Other factors could also have potential effect on fungal pathogenicity. Viruses-infected fungi called mycoviruses could cause defect in fungal pathogenicity (Kotta-Loizou 2021). The occurrence of hypovirulence phenomenon in various fungal species such as *Diaporthe ambigua*, *Fusarium graminearum*, *Botrytis cinerea*, *Sclerotinia homoeocarpa*, *Sclerotinia sclerotiorum*, and *Ophiostoma novo-ulmi* are well documented (Hong et al. 1999; Preisig et al. 2000; Chu et al. 2002; Castro et al. 2003; Deng et al. 2003; Zhang et al. 2020). The occurrence of mutations in some *Fusarium* isolates could cause defect in the capability of some isolates such of *Fusarium sacchari* and *Fusarium subglutinans* to cause pathogenicity (Mohammed and Al-Ani 2021). On the other hand, the defect in pathogenicity subsequently leading to non-pathogenicity of the fungi could result in their possible utilisation as beneficial agent by having some abilities such as (A) inducing of plant immunity, (B) competition on nutrient and space in niches, and (C) producing some antibiotics and hydrolytic enzymes (Al-Ani 2019d).

5 Natural Compounds of Endophytic Fungi

The discovery of the paclitaxel-producing endophytic fungus *Ceriporiopsis andreanae* (formerly *Taxomyces andreanae*), in Pacific yew (*T. brevifolia*) (Stierle et al. 1993) and its subsequent isolation from other *Taxus* spp. and other EF, followed by industrial production of this most potent anticancer metabolite (Isah, 2015; Gakuubi et al. 2021) intensified research on EF. Research over the past three decades has highlighted that EF are well endowed with useful bioactive compounds such as alkaloids, steroids, terpenoids, peptides, polyketones, flavonoids, quinols, and phenols, with antimicrobial, anticancer, and immunosuppressive activities, amongst other plant growth promoting roles (Yu et al. 2010; Mousa and Raizada 2013; Yadav et al. 2015; Liao et al. 2025). Thus, these compounds are highly sought after and explored immensely in the pharmaceutical and agricultural industries. The alkaloid camptothecin, which is also another potent antineoplastic drug, has been isolated from several EF including *Neocosmospora solani* and *Fusarium solani* in *Camptotheca acuminata* (Nyssaceae) (Promputtha et al. 2007; Kusari et al. 2009; Shweta et al. 2010) and *Alternaria alternata*, *Phomopsis* sp., and *Fomitopsis* sp. in *Miquelia dentata* (Shweta et al. 2013), whilst ergoflavin (dimeric xanthenes) a novel antineoplastic has also been isolated from an endophytic fungus in *Mimusops elengi* (Deshmukh et al. 2009). Cryptocandin (lipopeptide) obtained from the endophytic fungus *Cryptosporiopsis* cf. *quercina* is potent against some important human fungal pathogens, including *Candida albicans* and *Trichophyton* sp., as well as several plant pathogenic fungi, including *Sclerotinia sclerotiorum* and *Botrytis cinerea* (Strobel et al. 1999). Secondary compounds such as trichorzianines, harzianins, peptaibols, 6-pentyl-α-pyrone, viridiofungin A, harzianic acid, trichoharzianol, and harzianopyridone are also produced by many different species of *Trichoderma* (Al-Ani 2019e). *Xylaria* spp. isolated as endophytes in several plants possessed numerous bioactives including griseofulvin (spirobenzofuran) and sordaricin (tetracyclic diterpenoid) used as antimycotic agent against *C. albicans* and other human and animal diseases (Park et al. 2005; Pongcharoen et al. 2008; Liao et al. 2025).

However, with respect to environmental and industrial applications, laccase produced by EF such as *Chaetomium* (Sara et al. 2016) is used for humification during composting (Chefetz et al. 1998), delignification (Yadav et al. 2019), bioremediation of environmental pollutants (Yadav et al. 2019), paper pulp bleaching, and decolorisation and detoxification of industrial effluents (Kunamneni et al. 2008; Viswanath et al. 2014;). In agriculture, three endophytic *Trichoderma* spp. (*Trichoderma longibrachiatum*, *Trichoderma asperellum*, and *Trichoderma atroviride*) produced pectinase and cutinase solubilised phosphorus (Sallam et al. 2021). The three isolates produced siderophores and indole acetic acid which reduce the availability of iron to *Rhizoctonia solani*, improved growth of soybean plants, and reduced in vivo infection by 57–63%, relative to untreated plants (Sallam et al. 2021). *Trichoderma erinaceum* was isolated from ginger, which produced polyketide group 6-n-pentyl-2H-pyran-2-one (6PAP), β-1,3-glucanase and chitinase, inhibited *Sclerotium rolfsii* growth by 64%, and reduced infection by 58% in pot experiments (Suebrasri et al. 2020). These natural compounds could also be used in the biopesticides industry to control insects (Adetunji et al. 2019).

6 Utilising Endophytic Fungi as Biological Control Agents

Endophytic fungi include several groups of nematophagous fungi, entomopathogenic fungi (EPF), symbiotic fungi, non-pathogenic fungi, beneficial fungi (mycoparasitic fungi), amongst others. EF can be a great alternative to synthetic chemical pesticides as the chemicals have negative impacts on ecosystem (Aguilar-Marcelino et al. 2023; Hajji-Hedfi et al. 2025). Cumagun et al. (2025) isolated three EF from some plants in the rainforest of the Philippines comprising *Phomopsis* sp., *Tinctoporellus epimiltinus*, and *Fusarium proliferatum* which could control *Colletotrichum musae* causing anthracnose disease of banana fruit by producing some volatile compounds. The ability of fungi to produce siderophore as Fe-chelating compounds affecting growth of many plant pathogens has also been reported (Al-Ani et al. 2021b). Several authors have highlighted the need to apply molecular techniques in utilising many EF as a suitable method in sustainable agriculture (Aguilar-Marcelino et al. 2020b; Sharma et al. 2020; Singh et al. 2021).

Endophytic entomopathogenic fungi (EEPF) are now well sought after in Integrated Pest Management (IPM) (Al-Ani 2019f; Quesada Moraga 2020) as they are now documented to be capable of reducing both insect pests and diseases of crops (Vega 2018). *Beauveria bassiana* and *Metarhizium* sp. (EEF) showed the capability in changing of emissions of plant volatile compounds that enhanced the plant resistance against pests and pathogens of plants (González-Mas et al. 2021; Aravinthraju et al. 2024; Bhattacharyya et al. 2024), as well as, enhancing the plant growth (Panwar and Szczepaniec 2024). *Fusarium proliferatum* and *B. bassiana* showed efficacy in managing adults of *Tribolium confusum* (Al-Ani et al. 2018). Two isolates of EEPF as *Metarhizium brunneum* and *Beauveria bassiana* completely (100%) inhibited the mycelium growth of two important fungal pathogens of olive (*Phytophthora megasperma* and *Verticillium dahlia*) (Lozano-Tovar et al. 2017). *Metarhizium* spp. and *B. bassiana* exhibited efficacy in controlling *Rhizoctonia solani* that caused two diseases of cucumber (damping-off and root rot) (Rhouma et al. 2024) whilst the suppression of arthropod insect pests by EEPF are well documented (Leckie 2002; Cherry et al. 2004; Posada et al. 2007; Powell et al. 2009; Qayyum et al. 2015; Klieber & Reineke 2016; Pelizza et al. 2017; Sánchez-Rodríguez et al. 2018; Vega 2018).

Endophytic beneficial fungi (EBF): Some endophytic fungi have can be beneficial in plants by various mode of action comprising two ways, first, directly as via parasitism, antibiosis, competition, production of siderophore and lytic enzyme, and, second, indirectly through induce systemic resistance (ISR), systemic acquired resistance (SAR), and enhancing plant growth (Akram et al. 2023). There are many EBF available in plant tissues. Three EBF isolates including *Trichoderma*

(*Trichoderma crassum* and *Trichoderma afroharzianum*) and dark septate endophytes (*Gelasinospora brevispora*) showed great antagonism of three fungal pathogens (*Fusarium oxysporum* B1, *F. solani* B2, and *Fusarium acuminatum* B3) through competition and inhibition of the mycelial growth (Yao et al. 2025). Dark septate endophytic fungi reduced white root rot disease on *Hevea brasiliensis* (Dalimunthe et al. 2023). EBF such as *Trichoderma* and mycorrhizal were efficient in controlling plant parasitic nematodes (Poveda et al. 2020). *Trichoderma* spp. are more efficacious as biocontrol agents of plant diseases as they are more tolerant of different climate zones (Al-Ani 2018b; Mitrović et al. 2023) and therefore used widely in the control of different plant diseases (Al-Ani and Mohammed 2020). *Trichoderma* and non-pathogenic *Fusarium* spp. exhibited antagonism against *Fusarium oxysporum* f. sp. *cubense* tropical race 4 (FocTR4) (Al-Ani and Salleh, 2010; Al-Ani et al. 2013a, b; Al-Ani and Albaayit 2018a, b). The hexane extract of the culture filtrates for *Pochonia chlamydosporia* could reduce of mycelium growth of three plant fungal pathogens, *Fusarium solani*, *F. oxysporum*, and *Macrophomina phaseolina* significantly (Siddiqui and Shaukat 2003). Mycorrhizal are also more beneficial for plants through symbiosis (Bhattacharyya et al. 2023; Mitra et al. 2023; Yelikbayev et al. 2024).

Nonetheless, plant parasitic nematodes (PPNs) caused some important diseases for different plants (Al-Ani et al. 2025a, b). Endophytic nematophagous fungi (ENF) can control plant parasitic nematodes (PPNs) by utilising two modes of action directly and indirectly (Al-Ani et al. 2022; Villar-Luna et al. 2022; de Freitas Soares et al. 2023). *Acremonium sclerotigenum* (ENF) could suppress the root knot nematode *Meloidogyne incognita* of tomato in vitro and in vivo (Yao et al. 2023). Many secondary metabolites (> 50) with nematicidal properties obtained from nematophagous fungi are being explored as larvicidal agent of mosquito (Kudesia et al. 2025). *Arthrobotrys oligospora* could control plant fungal pathogen and nematodes indirectly by enhancing the plant defences, plant growth, and plant vigour (Singh et al. 2013). Sarker et al. (2020) mentioned the ability of Bio-nematicide + BAU-Biofungicide in enhancing the plant growth and biocontrol of *Meloidogyne incognita*.

7 Methodology in Study of Endophytic Fungi and Problems

Earlier works on the detection, systematics, and diversity of EF relied on morphocultural characteristics of strains obtained, following their isolation. Isolation is preceded by systematic sampling from plant tissues (usually leaves, stems and roots) (Fernandes et al. 2018; Wang et al. 2019; Yao et al. 2019; Dos Reis et al. 2022) followed by surface sterilisation (in ethanol 30–70%, for 2–5 min, followed by sodium hypochlorite for 0.5–8 min, and finally in sterile water) (Li et al. 2012; Greenfield et al. 2015; Dos Reis et al. 2022) and plating of tissue segments in appropriate media (mainly potato dextrose agar (PDA) or malt extract agar supplemented with antibiotics; Zhang and Yao 2015) and incubation at 25–28 °C for 3–6 weeks (Li et al. 2016; Yao et al. 2017; Zheng et al. 2021). Identification of isolates was based on morphology of colonies and sporulating structures.

This conventional culture-based isolation procedure, although results in successful recovery of EF, has limitations: (i) inadequate sample size may affect species abundance and diversity in different locations (Connolly 2005; Rashmi et al. 2019); (ii) endophytes in tissue segments may be killed if the chemical sterilant is too strong (Rungjindama and Jones 2024); (iii) ascomycetous EF if present may grow faster on the nutrient-rich media usually used (e.g. PDA), thereby outcompeting others which are slow-growing, for instance, basidiomycetes (Rungjindama and Jones 2024); (iv) it does not give accurate identification of isolates obtained (Lücking et al., 2020); (v) non-sporulating endophytes (sterile mycelia) which are frequently isolated may be difficult to identify (Guo et al. 2000; Wang and Guo 2007; Elgorban et al. 2019); and (vi) it excludes unculturable EF (Wu et al. 2019; Fan et al. 2020; Lücking et al. 2020).

The development of molecular tools such as DNA sequencing particularly high-throughput sequencing, molecular phylogeny, and genomics amongst others has led to significant advancements in detection, diversity, and secondary metabolite studies of EF (Chi et al. 2019; De Silva et al. 2021; Sagita et al. 2021). Next-generation sequencing and genomics studies have now enabled determination of diversity even from environmental samples (del Carmen Orozco-Mosqueda and Santoyo 2021). In addition, these culture-independent-based methods have overcome some of the limitations with conventional methods, as the identity of the exact species can be determined depending on the specific tool deployed (Promputtha et al. 2005; Jeewon et al. 2013; Doilom et al. 2017), non-culturable and non-sporulating EF can be identified (Promputtha et al. 2005), and higher detection of other fungal lineages besides the dominant ascomycetes have been observed (Vaz et al. 2020; Abdel-Wahab et al. 2021; Høyer and Hodkinson, 2021), hence giving a much more complete overview of endophytic communities and the number of previously unknown or unrecoverable species. The downside with use of the culture-independent method is the absence of the endophytic fungal cultures following detection; it is therefore recommended to use both methods particularly for culturable organisms to be able get isolates for follow-up studies.

8 Conclusions

The nature of the plant tissues is suitable for living by a lot of fungi as endophytes. Fungal endophytes produce toxins, growth regulators, and enzymes but generally do not cause damage of plant tissues. Occurrence of a plant disease is needed to confront plant defences depending on prevailing suitable environmental conditions. Nonetheless fungal endophytes can be used as biocontrol agents and can equally be latent pathogens. Finally, the activity of fungal isolate/s and their relationship with plant/s is dependent on the gene of fungal isolate, the environmental conditions, as well as the interaction between the gene and environment.

References

Abdel-Wahab MA, Bahkali AH, Elgorban AM, Jones EBG (2021) High-throughput amplicon sequencing of fungi and microbial eukaryotes associated with the seagrass *Halophila stipulacea* (Forssk.) Asch. from Al-Leith mangroves, Saudi Arabia. Mycol Prog 20:1365–1381. https://doi.org/10.1007/s11557-021-01744-2

Abreu-Tarazi MF, Navarrete AA, Andreote FD, Almeida CV, Tsai SM, Almeida M (2010) Endophytic bacteria in long-term in vitro cultivated "axenic" pineapple microplants revealed by PCR–DGGE. World J Microbiol Biotechnol 26:555–560. https://doi.org/10.1007/s11274-009-0191-3

Adetunji CO, Egbuna C, Tijjani H, Adom D, Al-Ani LKT, Patrick-Iwuanyanwu KC (2019) Homemade preparations of natural biopesticides and applications. In: Egbuna C, Sawicka B (eds) Natural remedies for pest, disease and weed control. Elsevier Science, pp 179–186

Aguilar-Marcelino L, Al-Ani LKT, Castañeda-Ramirez GS, Garcia-Rubio V, Ojeda-Carrasco QJJ (2020a) Microbial technologies to enhance crop production for future needs. In: Rastegari AA, Yadav AN, Awasthi AK, Yadav N (eds) New and future developments in microbial biotechnology and bioengineering. Elsevier, pp 29–47

Aguilar-Marcelino L, Mendoza-de-Gives P, Al-Ani LKT, Lopez-Arellano ME, Gomez-Rodrıguez O, Villar-Luna E et al (2020b) Using molecular techniques applied to beneficial microorganisms as biotechnological tools for controlling agricultural plant pathogens and pest. In: Sharma V, Salwan R, Al-Ani LKT (eds) Molecular aspects of plant beneficial microbes in agriculture. Elsevier Science, Cambridge, pp 333–349

Aguilar-Marcelino L, Al-Ani LKT, Soares FEF, Moreira ALE, Téllez-Téllez M, Castañeda-Ramírez GS et al (2021) Formation, resistance, and pathogenicity of fungal biofilms: current trends and future challenges. In: Yadav AN (ed) Recent trends in mycological research, Volume 1: agricultural and medical perspective, fungal biology. Springer Nature, Berlin, pp 411–438

Aguilar-Marcelino L, Al-Ani LKT, Soares FEF, Braga FR, Padilla AVV, Sharma A (2022) Wild mushrooms: characteristics, nutrition, and processing. In: Dhull SB, Bains A, Chawla P, Sadh PK (eds) Wild mushrooms: characteristics, nutrition, and processing. Taylor and Francis, CRC Press, pp 471–492

Aguilar-Marcelino L, Al-Ani LKT, Wong-Villarreal A, César Sotelo-Leyva C (2023) Persistence of pesticides residues with chemical food preservatives in fruits and vegetables. In: Singh J, Pandey A, Singh S, Garg VK, Ramamurthy P (eds) Current developments in biotechnology and bioengineering pesticides: human health, environmental impacts and management, 1st edn. Elsevier, pp 99–118

Akram S, Ahmed A, He P, He P, Liu Y, Wu Y, Munir S, He Y (2023) Uniting the role of endophytic fungi against plant pathogens and their interaction. J Fungi 9(1):72. https://doi.org/10.3390/jof9010072

Alam B, Lǐ J, Gě Q, Khan MA, Gōng J, Mehmood S, Yuán Y, G□ng W (2021) Endophytic fungi: from symbiosis to secondary metabolite communications or vice versa? Front Plant Sci 12: 791033. https://doi.org/10.3389/fpls.2021.791033

Al-Ani LKT (2006) Induce resistance against cucumber mosaic virus by *Pseudomonas fluorescens* Migula. M.Sc., Department of Plant Protection, College of Agriculture, University of Baghdad, Baghdad, Iraq, p 90

Al-Ani LKT (2017a) PGPR: a good step to control several of plant pathogens. In: Singh HB, Sarma BK, Keswani C (eds) Advances in PGPR research. CABI, pp 398–410

Al-Ani LKT (2017b) Potential of utilizing biological and chemical agents in the control of Fusarium wilt of banana. PhD School of Biology Science, Universiti Sains Malaysia, Pinang, Malaysia, p 259

Al-Ani LKT (2018a) *Trichoderma*: beneficial role in sustainable agriculture by plant disease management. In: Egamberdieva D, Ahmad P (eds) Plant Microbiome: Stress Response. Microorganisms for Sustainability, vol 5. Springer, Singapore, pp 105–126

Al-Ani LKT (2018b) *Trichoderma* from extreme environments: physiology, diversity, and antagonistic activity. In: Egamberdieva D, Birkeland N-K, Panosyan H, Li W-J (eds) Extremophiles in Eurasian Ecosystems: Ecology, Diversity, and Applications. Microorganisms for Sustainability, vol 8. Springer, Singapore, pp 388–403
Al-Ani LKT (2019a) The importance of endophytic fungi from the medicinal plant: diversity, natural bioactive compounds, and control of plant pathogens. In: Egamberdieva D, Tiezzi A (eds) Medically Important Plant Biomes Source of Secondary Metabolites. Springer, Singapore, pp 189–238
Al-Ani LKT (2019b) A patent survey on *Trichoderma* spp. (from 2007-2017). In: Singh HB, Keswani C, Singh SP (eds) Intellectual Property Issues in Microbiology. Springer, Singapore, pp 163–192
Al-Ani LKT (2019c) Recent patents on endophytic fungi and their international market. In: Singh HB, Keswani C, Singh SP (eds) Intellectual Property Issues in Microbiology. Springer, Singapore, pp 271–284
Al-Ani LKT (2019d) Secondary metabolites of nonpathogenic *Fusarium* spp.; scope in agriculture. In: Singh HB, Keswani C, Reddy MS, Royano ES, García-Estrada C (eds) Secondary metabolites of plant growth promoting rhizomicroorganisms – discovery and applications. Springer, Singapore, pp 59–76
Al-Ani LKT (2019e) Bioactive secondary metabolites of *Trichoderma* spp. for efficient management of phytopathogens. In: Singh HB, Keswani C, Reddy MS, Royano ES, García-Estrada C (eds) Secondary metabolites of plant growth promoting rhizomicroorganisms – discovery and applications. Springer, Singapore, pp 125–143
Al-Ani LKT (2019f) Entomopathogenic fungi in IP landscape. In: Singh HB, Keswani C, Singh SP (eds) Intellectual property issues in microbiology. Springer, Singapore, pp 223–238
Al-Ani RA, Al-Ani LKT (2011) Induced of systemic resistance in cucumber plants against cucumber mosaic virus (CMV) by *Pseudomonas fluorescens* Migula. Arab J Plant Protect 29: 36–42
Al-Ani LKT, Albaayit SFA (2018a) Antagonistic of some *Trichoderma* against *Fusarium oxysporum* f. sp. *cubense* tropical race 4 (FocTR4). International conference on Research in Education and Science, ICRES April 28—May 1, Marmaris, Turkey, pp 271 (Abstract)
Al-Ani LKT, Albaayit SFA (2018b) Antagonistic of some Trichoderma against Fusarium oxysporum f. sp. cubense tropical race 4 (FocTR4). Eurasia Proc Sci Technol Eng Math 2: 35–38
Al-Ani LKT, Furtado EL (2020) The effect of incompatible plant pathogen on the host plant. In: Sharma V, Salwan R, Al-Ani LKT (eds) Molecular aspects of plant beneficial microbes in agriculture. Elsevier Science, Cambridge, pp 47–57
Al-Ani LKT, Mohammed AM (2020) Versatility of *Trichoderma* in plant disease management. In: Sharma V, Salwan R, Al-Ani LKT (eds) Molecular Aspects of Plant Beneficial Microbes in Agriculture. Elsevier Science, Cambridge, pp 159–168
Al-Ani, LKT, Salleh B (2010) Control of Fusarium wilt of banana by non pathogenic F*usarium oxysporum*. PPSKH colloquium, Pust Pengajian Sains Kajihayat/School of Biological Sciences, USM, 2–4 June, p 10
Al-Ani LKT, Negim E.-S, Mohammed AM, Salleh B, Saleh MI (2012) Antifungal activity of novel binary grafting polymers. 1st USM – KAZNU international conference on: challenges of teaching and chemistry research in institutions of higher learning, 11–13 July, p 44
Al-Ani LKT, Salleh B, Mohammed AM, Ghazali AHA, Al-Shahwany AW, Azuddin NF (2013a) Biocontrol of Fusarium wilt of banana by non-pathogenic *Fusarium* spp. International symposium on tropical fungi, ISTF, IPB International Convention Center, Bogor, Indonesia; 09/2013, pp 50–51
Al-Ani LKT, Salleh B, Ghazali AHA (2013b Biocontrol of Fusarium wilt of banana by *Trichoderma* spp. 8th PPSKH colloquium, Pust Pengajian Sains Kajihayat/School of Biological Sciences, USM, 5–6 June, p 13

Al-Ani LKT, Yonus MI, Mahdii BA, Omer MA, Taher JK, Albaayit SFA, Al-Khoja SB (2018) First record of use *Fusarium proliferatum* fungi in direct treatment to control the adult of wheat flour *Tribolium confusum*, as well as, use the entomopathogenic fungi *Beauveria bassiana*. Ecol Environ Conserv 24(3):29–34

Al-Ani LKT, Aguilar-Marcelino L, Fiorotti J, Sharma V, Sarker MS, Raza W et al (2020a) Biological control agents and their importance for the plant health. In: Singh JS, Vimal SR (eds) Microbial services in ecological restoration. Elsevier, pp 13–36

Al-Ani LKT, Franzinoc T, Aguilar-Marcelino L et al (2020b) The role of microbial signals in plant growth and development: current status and future prospects. In: Rastegari AA, Yadav AN, Awasthi AK, Yadav N (eds) New and future developments in microbial biotechnology and bioengineering. Elsevier, pp 225–242

Al-Ani LKT, Surono Aguilar-Marcelino L, Salazar-Vidal VE, Becerra AG, Raza W (2021a) Role of useful fungi in agriculture sustainability. In: Yadav AN (ed) Recent trends in mycological research, fungal biology. Springer Nature, Berlin, pp 1–44

Al-Ani LKT, Aguilar-Marcelino L, Becerra AG, Salazar-Vidal VE (2021b) Fe-chelating compounds producing fungal communities and their applications. In: Yadav AN (ed) Recent trends in mycological research, Volume 1: agricultural and medical perspective, fungal biology. Springer Nature, Berlin, pp 135–157

Al-Ani LKT, Soares FEDF, Sharma A et al (2022) Strategy of nematophagous fungi in determining the activity of plant parasitic nematodes and their prospective role in sustainable agriculture. Front Fungal Biol 3:863198. https://doi.org/10.3389/ffunb.2022.863198

Al-Ani LKT, Mitrović I, Zimowska B, Sarker MS, Dautt-Castro M, Amjad SF et al (2024) Metabolites of fungi and applications in the agricultural. In: Siddiqui S, Meshram V, Patil RS, Gupta M (eds) Present status and prospects of plant associated fungi. Cambridge Scholars Publishing, pp 351–397

Al-Ani LKT, Aguilar-Marcelino L, Santos-Villalobos SDL, Montoya-Martínez AC, Pineda-Alegría JA, Mitrovic I (2025a) Identification and characterization of plant parasitic nematodes using molecular techniques. In: Akhtar MS, Hakeem KR (eds) Plant-nematode interactions and their management strategies. CRC Press

Al-Ani LKT, Aguilar-Marcelino L, Cruz-Arévalo J, Grifaldo-Alcántara PF, Palomares-Pérez M (2025b) Progress and development of plant nematology and future consequences. In: Akhtar MS, Hakeem KR (eds) Plant-nematode interactions and their management strategies. CRC Press

Aly AH, Debbab A, Proksch P (2013) Fungal endophytes – secret producers of bioactive plant metabolites. Int J Pharm Sci 68:499–505

An C, Ma S, Shi X, Xue W, Liu C, Ding H (2020) Diversity and antimicrobial activity of endophytic fungi isolated from *Chloranthus japonicus* Sieb in Qinling mountains. China Int J Mol Sci 21(17):5958. https://doi.org/10.3390/ijms21175958

Aravinthraju K, Shanthi M, Murugan M, Srinivasan R, Maxwell LA, Manikanda Boopathi N, Anandham R (2024) Endophytic entomopathogenic fungi: their role in enhancing plant resistance, managing insect pests, and synergy with management routines. J Fungi 10

Arnold AE (2007) Understanding the diversity of foliar endophytic fungi: progress, challenges, and frontiers. Fungal Biol Rev 21(2–3):51–66. https://doi.org/10.1016/j.fbr.2007.05.003

Arnold AE, Mejía LC, Kyllo D, Rojas EI, Maynard Z, Robbins N, Herre EA (2003) Fungal endophytes limit pathogen damage in a tropical tree. Proc Natl Acad Sci 100(26): 15649–15654. https://doi.org/10.1073/pnas.253348310

Attitalla IH, Mansour SE, Mohamed WS, Al-Ani LKT, Mohammed AM, Faturi MY et al (2010a) Influence of *Aspergillus flavus* and *Aspergillus terreus* on the protein value of the two varieties of peanut grains. In: International Mycotoxin conference, MycoRed, Penang, Malaysia, 1–4 December, p 177

Attitalla IH, Al-Ani LKT, Nasib MA, Balal IAA, Zakaria M, El-Maraghy SSM, Karim SMR (2010b) Screening of fungi associated with commercial grains and animal feeds in Al-Bayda Governorate, Libya. World Appl Sci J 9(7):746–756

Bhattacharyya PN, Islam NF, Sarma B, Nath BC, Al-Ani LKT, Lesueur D (2023) Frankia-actinorhizal symbiosis: a non-chemical biological assemblage for enhanced plant growth, nodulation and reclamation of degraded soils. Symbiosis 92. https://doi.org/10.1007/s13199-023-00956-2

Bhattacharyya PN, Nath BC, Sarma B, Al-Ani LKT, Borgohain DJ, Garganese F et al (2024) Interaction of *Metarhizium anisopliae* against emergent insect pest problems in the North-Eastern tea industry. In: Deshmukh SK, Sridhar KR (eds) entomopathogenic fungi: prospects and challenges. Springer Nature Singapore, pp 319–357. https://doi.org/10.1007/978-981-97-5991-0_13

Bhunjun CS, Phukhamsakda C, Hyde KD, McKenzie EH, Saxena RK, Li Q (2024) Do all fungi have ancestors with endophytic lifestyles? Fungal Divers 125:73–98. https://doi.org/10.1007/s13225-023-00516-5

Botnen S, Vik U, Carlsen T, Eidesen PB, Davey ML, Kauserud H (2014) Low host specificity of root-associated fungi at an Arctic site. Mol Ecol 23(4):975–985. https://doi.org/10.1111/mec.12646

Brotman Y, Landau U, Cuadros-Inostroza A, Tohge T, Fernie AR, Chet I et al (2013) *Trichoderma*-plant root colonization: escaping early plant defense responses and activation of the antioxidant machinery for saline stress tolerance. PLoS Pathog 9(4):e1003221. https://doi.org/10.1371/journal.ppat.1003221

Carroll G (1988) Fungal endophytes in stems and leaves: from latent pathogen to mutualistic symbiont. Ecology 69(1):2–9

Castañeda-Ramírez GS, Torres-Acosta JFJ, Sánchez JE, Mendoza-de-Gives P, González-Cortázar M, Zamilpa A et al (2020) The possible biotechnological use of edible mushroom bioproducts for controlling plant and animal parasitic nematodes. Biomed Res Int 2020:1–12. https://doi.org/10.1155/2020/6078917

Castro M, Kramer K, Valdivia L, Ortiz S, Castillo A (2003) A double-stranded RNA mycovirus confers hypovirulence associated traits to Botrytis cinerea. FEMS Microbiol Lett 228:87–91. https://doi.org/10.1016/S0378-1097(03)00755-9

Chefetz B, Chen Y, Hadar Y (1998) Purification and characterization of laccase from *Chaetomium thermophilium* and its role in humification. Appl Environ Microbiol 64(9):3175–3179. https://doi.org/10.1128/AEM.64.9.3175-3179.1998

Cherry AJ, Banito A, Djegui D, Lomer C (2004) Suppression of the stem-borer *Sesamia calamistis* (Lepidoptera; Noctuidae) in maize following seed dressing, topical application and stem injection with African isolates of *Beauveria bassiana*. Int J Pest Manag 50(1):67–73

Chi WC, Chen W, He CC, Guo SY, Cha HJ, Tsang LM et al (2019) A highly diverse fungal community associated with leaves of the mangrove plant *Acanthus ilicifolius* var *xiamenensis* revealed by isolation and metabarcoding analyses. PeerJ 7. https://doi.org/10.7717/peerj.7293

Chu YM, Jeon JJ, Yea SJ, Kim YH, Yun SH, Lee YW et al (2002) Double-stranded RNA mycovirus from *Fusarium graminearum*. Appl Environ Microbiol 68:2529–2534. https://doi.org/10.1128/AEM.68.5.2529-2534.2002

Comans-Pérez RJ, Sánchez JE, Al-Ani LKT, González-Cortázar M, Castañeda-Ramírez GS et al (2021) Biological control of sheep nematode *Haemonchus contortus* using edible mushrooms. BioControl 152:104420. https://doi.org/10.1016/j.biocontrol.2020.104420

Connolly SR (2005) Process-based models of species distributions and the mid-domain effect. Am Nat 166(1):1–11

Cumagun CJR, Adiova JM, Jakobs R, Rathgeb A, Karlovsky P, Müller C (2025) Metabolic profiling of endophytic fungi acting as antagonists of the banana pathogen *Colletotrichum musae*. PLoS One 20(1):e0310442

Dalimunthe CI, Soekarno BPW, Al-Ani LKT, Munif A, Sriherwanto C, Nurdebyandaru N (2023) First report of inhibitory abilities of dark septate endophytic fungi against white root rot disease on *Hevea brasiliensis* seedlings in nursery conditions. Egypt J Biol Pest Control 33(1):1–12

Davis EC, Shaw AJ (2008) Biogeographic and phylogenetic patterns in diversity of liverwortassociated endophytes. American Journal of Botany 95(8):914–924 de Barry A (1866) Morphologie und physiologie der pilze, flechten und myxomyceten, vol 1. Engelmann

de Freitas Soares FE, Ferreira JM, Genier HLA, Al-Ani LKT, Aguilar-Marcelino L (2023) Biological control 2.0: use of nematophagous fungi enzymes for nematode control. J Nat Pest Res 4:100025

De Silva NI, Maharachchikumbura SSN, Thambugala KM, Bhat DJ, Karunarathna SC, Tennakoon DS et al (2021) Morpho-molecular taxonomic studies reveal a high number of endophytic fungi from *Magnolia candolli* and *M. garrettii* in China and Thailand. Mycosphere 12(1):163–237. https://doi.org/10.5943/mycosphere/12/1/3

De Souza GF, Al-Ani LKT (2023) Effects of gamma radiation and electron beam on samples of the peanut candies contaminated with *Aspergillus flavus*. In: Conference: VIII. International scientific and vocational studies congress – (BILMES 2023), Turkey

del Carmen O-MM, Santoyo G (2021) Plant-microbial endophytes interactions: scrutinizing their beneficial mechanisms from genomic explorations. Curr Plant Biol 25:100189. https://doi.org/10.1016/j.cpb.2020.100189

Deng F, Xu R, Boland G (2003) Hypovirulence-associated double-stranded RNA from Sclerotinia homoeocarpais conspecific with *Ophiostoma novo-ulmi mitovirus* 3a-Ld. Phytopathology 93: 1407–1414. https://doi.org/10.1094/PHYTO.2003.93.11.1407

Deshmukh SK, Mishra PD, Kulkarni-Almeida A, Verekar S, Sahoo MR, Periyasamy G et al (2009) Anti-inflammatory and anticancer activity of ergoflavin isolated from an endophytic fungus. Chem Biodivers 6:784–789. https://doi.org/10.1002/cbdv.200800103

Doilom M, Manawasinghe IS, Jeewon R, Jayawardena RS, Tibpromma S, Hongsanan S et al (2017) Can ITS sequence data identify fungal endophytes from cultures? A case study from *Rhizophora apiculata*. Mycosphere 8:1869–1892

Dos Reis JBA, Lorenzi AS, do Vale HMM (2022) Methods used for the study of endophytic fungi: a review on methodologies and challenges, and associated tips. Arch Microbiol 204(11):675. https://doi.org/10.1007/s00203-022-03283-0

Dreyfuss MM, Chapela IH (1994) Potential of fungi in the discovery of novel, low–molecular weight pharmaceuticals. In: Gullo VP (ed) Discovery of novel natural products with therapeutic potential. Butterworth-Heinemann, London, pp 49–80. https://doi.org/10.1016/B978-0-7506-9003-4.50009-5

Du T-Y, Dao C-J, Mapook A, Stephenson SL, Elgorban AM, Al-Rejaie S, Suwannarach N, Karunarathna SC, Tibpromma S (2022) Diversity and biosynthetic activities of agarwood associated fungi. Diversity 14(3):211. https://doi.org/10.3390/d14030211

Elgorban AM, Bahkali AH, Al Farraj DA, Abdel-Wahab MA (2019) Natural products of *Alternaria* sp., an endophytic fungus isolated from *Salvadora persica* from Saudi Arabia. Saudi. J Biol Sci 26(5):1068–1077. https://doi.org/10.1016/j.sjbs.2018.04.010

Estrada C, Wcislo WT, Van Bael SA (2013) Symbiotic fungi alter plant chemistry that discourages leafcutting ants. New Phytologist 198(1):241–251

Ettinger CL, Eisen JA (2021) Correction: Fungi, bacteria and oomycota opportunistically isolated from the seagrass, *Zostera marina*. PLoS One 16(5):e0251536. https://doi.org/10.1371/journal.pone.0251536

Fan S, Miao L, Li H, Lin A, Song F, Zhang P (2020) Illumina-based analysis yields new insights into the diversity and composition of endophytic fungi in cultivated *Huperzia serrata*. PLoS One 15(11):e0242258. https://doi.org/10.1371/journal.pone.0242258

Fernandes GW, Oki Y, Belmiro MS, Resende FM, Corrêa Junior A, de Azevedo JL (2018) Multitrophic interactions among fungal endophytes, bees, and *Baccharis dracunculifolia*: resin tapering for propolis production leads to endophyte infection. Arthropod Plant Interact 12:32–337. https://doi.org/10.1007/s11829-018-9597-x

Freeman S, Horowitz S, Sharon A (2001) Pathogenic and nonpathogenic lifestyles in *Colletotrichum acutatum* from strawberry and other plants. Phytopathology 91(10):986–992

Fry WE, Myers DF (1981) Hydrogen cyanide metabolism by fungal pathogens of cyanogenic plants. In: Vennesland B, Conn EE, Knowles CJ, Westley J, Vissing F (eds) Cyanide in biology. Academic, London, p 32134

Gakuubi MM, Munusamy M, Liang ZX, Ng SB (2021) Fungal endophytes: a promising frontier for discovery of novel bioactive compounds. J Fungi 7(10):786. https://doi.org/10.3390/jof7100786

Georgousaki K, González–Menéndez V, Tormo JR, Tsafantakis N, Mackenzie TA, Martín J, Gumeni S, Trougakos IP, Reyes F, Fokialakis N, Genilloud O (2022) Comoclathrin, a novel potent skin-whitening agent produced by endophytic *Comoclathris* strains associated with Andalusia desert plants. Sci Rep 12:1649. https://doi.org/10.1038/s41598-022-05448-9

González-Mas N, Gutiérrez-Sánchez F, Sánchez-Ortiz A, Grandi L, Turlings TC, Manuel Muñoz-Redondo J et al (2021) Endophytic colonization by the entomopathogenic fungus *Beauveria bassiana* affects plant volatile emissions in the presence or absence of chewing and sap-sucking insects. Front Plant Sci 12

Gouda S, Das G, Sen SK, Shin H-S, Patra JK (2016) Endophytes: a treasure house of bioactive compounds of medicinal importance. Front Microbiol 7:1538. https://doi.org/10.3389/fmicb.2016.01538

Greenfield M, Pareja R, Ortiz V, Gómez-Jiménez MI, Vega FE, Parsa S (2015) A novel method to scale up fungal endophyte isolations. Biocontrol Sci Tech 25(10):1208–1212. https://doi.org/10.1080/09583157.2015.1033382

Guerin P (1898) Sur la présence d'un champignon dans l'ivraie. J Bot 12:230–238

Guo LD, Hyde KD, Liew ECY (2000) Identification of endophytic fungi from *Livistona chinensis* based on morphology and rDNA sequences. New Phytol 147(3):617–630. https://doi.org/10.1046/j.1469-8137.2000.00716.x

Hajji-Hedfi L, Rhouma A, Al-Ani LKT, El-Abeid SE, Ibrahim DSS, Acheampong MA, Bousselma A, Ortega-Villaizán AG, Pollmann S (2024) Exploring endophytic fungal diversity: review of traditional and molecular techniques. In: Abdel Azeem AM, Yadav AN, Yadav N (eds) Developments in applied microbiology and biotechnology, endophytic fungi. Academic Press, pp 1–24. https://doi.org/10.1016/B978-0-323-99314-2.00005-X

Hajji-Hedfi L, Dali S, Rhouma A, Al-Ani LKT, Chihani-Hammas N, Khlif A, Bargougui O (2025) Microbial inoculants for plant resilience performance: roles, prospects, and challenges. J Plant Dis Pro 132(2):88. https://doi.org/10.1007/s41348-025-01080-7

Hong Y, Dover SL, Cole TE, Brasier CM, Buck KW (1999) Multiple mitochondrial viruses in an isolate of the Dutch elm disease fungus Ophiostoma novo-ulmi. Virology 258:118–127. https://doi.org/10.1006/viro.1999.9691

Høyer AK, Hodkinson TR (2021) Hidden fungi: combining culture-dependent and-independent DNA barcoding reveals inter-plant variation in species richness of endophytic root fungi in *Elymus repens*. J Fungi 7(6):466. https://doi.org/10.3390/jof7060466

Isah T (2015) Natural sources of taxol. Br J Pharm Res 6(4):214–227

Jatoi GH, Muhammad S, Menlo WA, Al-Ani LKT, Haseenullah Abro MA, Gadhi MA, Awan NW, Reki MA (2020) Efficacy of different essential oils, fungicides and biocontrol agents against Aspergillus Niger the causal agent of fruit rot in Pomegranate. Int J Biosci 16(3):51–65

Jatoi G, Abro M, Ahmed S, Al-Ani LKT, Ali U, Jatoi M et al (2022) Preliminary selection and evaluation of fungicides and natural compounds to control grey mold disease of rose caused by botrytis cinerea. Int J Phytopathol 11(1):49–58

Jeewon R, Ittoo J, Mahadeb D, Jaufeerally-Fakim Y (2013) DNA based identification and phylogenetic characterisation of endophytic and saprobic fungi from *Antidesma madagascariense*, a medicinal plant in Mauritius. J Mycol 781914. https://doi.org/10.1155/2013/781914

Jia M, Chen L, Xin HL, Zheng CJ, Rahman K, Han T, Qin LP (2016) A friendly relationship between endophytic fungi and medicinal plants: a systematic review. Frontiers in microbiology 7:906

Khalmuratova I, Choi DH, Woo JR, Jeong MJ, Oh Y, Kim YG et al (2020) Diversity and plant growth-promoting effects of fungal endophytes isolated from salt-tolerant plants. J Microbiol Biotechnol 30(11):1680. https://doi.org/10.4014/jmb.2006.06050

Klieber J, Reineke A (2016) The entomopathogenic *Beauveria bassiana* has epiphytic and endophytic activity against the tomato leaf miner *Tuta absoluta*. J Appl Entomol 140(8):580–589

Kogel KH, Franken P, Heückelhoven R (2006) Endophyte or parasite – what decides? Curr Opin Plant Biol 9:358–363. https://doi.org/10.1016/j.pbi.2006.05.001

Kotta-Loizou I (2021) Mycoviruses and their role in fungal pathogenesis. Curr Opin Plant Biol 63: 10–18

Krings M, Taylor TN, Hass H, Kerp H, Dotzler N, Hermsen EJ (2007) Fungal endophytes in a 400-million-yr-old land plant: infection pathways, spatial distribution, and host responses. New Phytol 174(3):648–657. https://doi.org/10.1111/j.1469-8137.2007.02008.x

Kudesia N, Najitha Banu A, Raut AM, Assouguem A (2025) Role of entomopathogenic fungi and nematophagous fungi in mosquito control. CABI Reviews 20(1):0023. https://doi.org/10.1079/cabireviews.2025.0023

Kunamneni A, Camarero S, García-Burgos C, Plou FJ, Ballesteros A, Alcalde M (2008) Engineering and applications of fungal laccases for organic synthesis. Microb Cell Factories 7:1–17. https://doi.org/10.1186/1475-2859-7-32

Kusari S, Zühlke S, Spiteller M (2009) An endophytic fungus from *Camptotheca acuminata* that produces camptothecin and analogues. J Nat Prod 72(1):2–7

Lalarukh I, Al-Dhumri SA, Al-Ani LKT, Hussain R, Al Mutairi KA, Mansoora N et al (2022a) A combined use of rhizobacteria and Moringa leaf Extract mitigates the adverse effects of drought stress in wheat (*Triticum aestivum* L.). Front Microbiol 13:813415. https://doi.org/10.3389/fmicb.2022.813415

Lalarukh I, Al Dhumri SA, Al-Ani LKT, Shahbaz M, Amjad SF, Mansoora N, Al Shammari WB, Alhusayni FS, Almutari MM (2022b) Combined application of poultry litter biochar with NPK fertilizer mitigates the effects of stress in wheat (*Triticum aestivum* L.) grown on chromium contaminated soil. Plant Soil 480:201–212. https://doi.org/10.1007/s11104-022-05574-3

Leckie BM (2002) Effects of *Beauveria bassiana* mycelia and metabolites incorporated into synthetic diet and fed to larval *Helicoverpazea*, and detection of endophytic *Beauveria bassiana* in tomato plants using PCR and ITS primers. MSc thesis. University of Tennessee Knoxville, Tennessee, USA

Li HY, Shen M, Zhou ZP, Li T, Wei YL, Lin LB (2012) Diversity and cold adaptation of endophytic fungi from five dominant plant species collected from the Baima Snow Mountain, Southwest China. Fungal Divers 54:79–86. https://doi.org/10.1007/s13225-012-0153-1

Li JL, Sun X, Chen L, Guo LD (2016) Community structure of endophytic fungi of four mangrove species in southern China. Mycol 7(4):180–190

Liao C, Doilom M, Jeewon R, Hyde KD, Manawasinghe IS, Chethana KW et al (2025) Challenges and update on fungal endophytes: classification, definition, diversity, ecology, evolution and functions. Fungal Divers 131:1–67. https://doi.org/10.1007/s13225-025-00550-5

Lorena BB, Javiera O, Franco CJ (2021) Facultative fungal endophytes and their potential for the development of sustainable agriculture. In: Lorena B-B, Javiera O (eds) Microbial management of plant stresses: current trends, application and challenges, 1st edn. Woodhead Publ Elsevier, pp 1–12. https://doi.org/10.1016/B978-0-323-85193-0.00014-0

Lozano-Tovar MD, Garrido-Jurado I, Quesada-Moraga E, Raya-Ortega MC, Trapero-Casas A (2017) *Metarhizium brunneum* and *Beauveria bassiana* release secondary metabolites with antagonistic activity against *Verticillium dahliae* and *Phytophthora megasperma* olive pathogens. Crop Prot 100:186–195

Lücking R, Aime MC, Robbertse B, Miller AN, Ayapanas HA, Aoki T et al (2020) Unambiguous identification of fungi: where do we stand and how accurate and precise is fungal DNA barcoding? Int Mycol Assoc (IMA). Fungus 11(1):14. https://doi.org/10.1186/s43008-020-00033-z

Manjunatha N, Manjunatha N, Li H, Sivasithamparam K, Jones MG, Edwards I et al (2022) Fungal endophytes from salt-adapted plants confer salt tolerance and promote growth in wheat (*Triticum aestivum* L.) at early seedling stage. Microbiology 168(8):001225. https://doi.org/10.1099/mic.0.001225

Márquez SS, Bills GF, Herrero N, Zabalgogeazcoa I (2012) Non–systemic fungal endophytes of grasses. Fungal Ecol 53:289–297

Marsberg A, Kemler M, Jami F, Nagel JH, Postma-Smidt A, Naidoo S, Wingfield MJ, Crous PW, Spatafora JW, Hesse CN, Robbertse B (2017) *Botryosphaeria dothidea*: a latent pathogen of global importance to woody plant health. Mol Plant Pathol 18(4):477–488

Meshram V, Elazar M, Maymon M, Sharma G, Shawahna R, Charuvi D, Freeman S (2023) Endophytic Fusarium clavum confers growth and salt tolerance in Cucumis melo. Environ Exp Bot 206:105–153

Mitra D, Nayeri FD, Sansinenea E, Ortiz A, Bhatta BB, Adeyemi NO, Janeeshma E, Tawfeeq Al-Ani LK, Sharma SB, Boutaj H et al (2023) Unraveling arbuscular mycorrhizal fungi interaction in rice for plant growth development and enhancing phosphorus use efficiency through recent development of regulatory genes. J Plant Nutr 46:1–37

Mitrović I, Vučurović D, Al-Ani LKT, Mitrović B, Bajić B, Dodić S, Živanov ST (2023) Production of Trichoderma harzianum K179 bioagent for maize diseases control: complete laboratory stage bioprocess development. J Appl Microbiol 134(6):lxad115

Mohammed AM, Al-Ani LKT (2021) Identification and production of Beauvericin by Fusarium subglutinans and F. sacchari from sugarcane. Braz Arch Biol Technol 64:e21200088

Mohammed AM, Al-Ani LKT, Bekbayeva L, Salleh B (2011) Biological control of *Fusarium oxysporum* f. sp. *cubense* by *Pseudomonas fluorescens* and BABA in vitro. World Appl Sci J 15(2):189–1 91

Mohammed AM, Negim E-S, Al-Ani LKT, Salleh B, Saleh MI (2012) Utilization of amino–azines polymerase as antifungal activity for banana. In: 1st USM–KAZNU international conference on: challenges of teaching and chemistry research in institutions of higher learning 11–13 July, p 29

Mohammed AM, Al-Ani LKT, Salleh B (2013) Potential management of *Fusarium oxysporum* f. sp. *cubense*, the banana wilt pathogen by using *Pseudomonas* and beta-amino-butyric acid (BABA). In: International symposium on tropical fungi, ISTF, IPB International Convention Center, Bogor, Indonesia; 09/2013, p 37

Mohammed AM, Al-Ani LKT, Salleh B, Ghazali AMA (2014) Determining plant growth promoting and biocontrol factor of bacterial culture media. In: The 3rd conference on Pests management. Crop Protection Research Centre, Sudan, 3–4 February, p 103

Mousa WK, Raizada MN (2013) The diversity of anti–microbial secondary metabolites produced by fungal endophytes: an interdisciplinary perspective. Front Microbiol 4:1–2. https://doi.org/10.3389/fmicb.2013.00065

Newfeld J, Ujimatsu R, Hiruma K (2025) Uncovering the host range–lifestyle relationship in the endophytic and anthracnose pathogenic genus *Colletotrichum*. Microorganisms 13(2):428. https://doi.org/10.3390/microorganisms13020428

Nwe LL, Casonato S, Jones EE (2024) Endophytic fungal isolates from apple tissue: Latent pathogens lurking within? Fungal Biol 128(4):1836–1846

Oita S, Ibáñez A, Lutzoni F, Miadlikowska J, Geml J, Lewis LA et al (2021) Climate and seasonality drive the richness and composition of tropical fungal endophytes at a landscape scale. Commun Biol 4(1):313. https://doi.org/10.1038/s42003-021-01826-7

Oren L, Ezrati S, Cohen D, Sharon A (2003) Early events in the *Fusarium verticillioides*–maize interaction characterized by using a green fluorescent protein–expressing transgenic isolate. Appl Environ Microbiol 69(3):1695–1701

Oulhen N, Schulz BJ, Carrier TJ (2016) English translation of Heinrich Anton de Bary's 1878 speech, 'Die Erscheinung der Symbiose' ('De la symbiose'). Symbiosis 69:131–139. https://doi.org/10.1007/s13199-016-0409-8

Panwar N, Szczepaniec A (2024) Endophytic entomopathogenic fungi as biological control agents of insect pests. Pest Manag Sci 80(12):6033–6040

Park JH, Choi GJ, Lee HB, Kim KM, Jung HS, Lee S et al (2005) Griseofulvin from *Xylaria* sp. strain F0010, an endophytic fungus of *Abies holophylla* and its antifungal activity against plant pathogenic fungi. J Microbiol Biotechnol 15:112–117

Pedraza RO, Torres-Díaz C, Lavin P, Retamales-Molina P, Atala C, Acuña-Rodríguez I et al (2018) Root fungal endophytes improve the growth of antarctic plants through an enhanced nitrogen acquisition (No. e26774v1). PeerJ:26774. https://doi.org/10.7287/peerj.preprints.26774v1

Pelizza SA, Mariottini Y, Russo LM, Vianna FM, Scorsetti AC, Lange CE (2017) *Beauveria bassiana* (Ascomycota: Hypocreales) introduced as an endophyte in corn plants and its effects on consumption, reproductive capacity, and food preference of *Dichroplus maculipennis*(-Orthoptera: Acrididae: Melanoplinae). J Insect Sci 17(2):1–6

Pongcharoen W, Rukachaisirikul V, Phongpaichit S, Kühn T, Pelzing M, Sakayaroj J et al (2008) Metabolites from the endophytic fungus *Xylaria* sp. PSU–D14. Phytochemistry 69:1900–1902. https://doi.org/10.1016/j.phytochem.2008.04.003

Posada F, Aime MC, Peterson SW, Rehner SA, Vega FE (2007) Inoculation of coffee plants with the fungal entomopathogen *Beauveria bassiana* (Ascomycota: Hypocreales). Mycol Res 111(6):748–757

Poveda J, Abril-Urias P, Escobar C (2020) Biological control of plant–parasitic nematodes by filamentous fungi inducers of resistance: *Trichoderma*, mycorrhizal and endophytic fungi. Front Microbiol 11:992

Powell WA, Klingeman WE, Ownley BH, Gwinn KD (2009) Evidence of endophytic *Beauveria bassiana* in seed-treated tomato plants acting as a systemic entomopathogen to larval *Helicoverpa zea* (Lepidoptera: Noctuidae). J Entomol Sci 44(4):391–396

Preisig O, Moleleki N, Smit WA, Wingfield BD, Wingfield MJ (2000) A novel RNA mycovirus in a hypovirulent isolate of the plant pathogen Diaporthe ambigua. J Gen Virol 81:3107–3114. https://doi.org/10.1099/0022-1317-81-12-3107

Promputtha I, Jeewon R, Lumyong S, McKenzie EH, Hyde KD (2005) Ribosomal DNA fingerprinting in the identification of non-sporulating endophytes from *Magnolia liliifera* (Magnoliaceae). Fungal Divers 20:167–186

Promputtha I, Lumyong S, Dhanasekaran V, McKenzie EHC, Hyde KD, Jeewon R (2007) A phylogenetic evaluation of whether endophytes become saprotrophs at host senescence. Microb Ecol 53:579–590

Qayyum MA, Wakil W, Arif MJ, Sahi ST, Dunlap CA (2015) Infection of *Helicoverpa armigera* by endophytic *Beauveria bassiana* colonizing tomato plants. Biol Control 90:200–207

Quesada Moraga E (2020) Entomopathogenic fungi as endophytes: their broader contribution to IPM and crop production. Biocontrol Sci Tech 30:864–877

Rahman V, Meena KR, Al-Ani LKT, Singh A, Kumar A (2023) *Bacillus velezensis* strain improvement to control *Helminthosporium maydis* causing southern corn leaf blight disease in maize. Eur J Plant Pathol 167:1–13

Rai M, Agarkar G (2014) Plant–fungal interactions: What triggers the fungi to switch among lifestyles? 42(3):428–438. https://doi.org/10.3109/1040841X.2014.958052

Ramos-Garza J, Rodríguez-Tovar AV, Flores-Cotera LB, Rivera-Orduña FN, Vásquez-Murrieta MS, Ponce-Mendoza A et al (2016) Diversity of fungal endophytes from the medicinal plant *Dendropanax arboreus* in a protected area of Mexico. Ann Microbiol 66:991–1002

Rashmi M, Kushveer JS, Sarma VV (2019) A worldwide list of endophytic fungi with notes on ecology and diversity. Mycosphere 10(1):798–1079. https://doi.org/10.5943/mycosphere/10/1/19

Raza W, Ghazanfar MU, Asif M et al (2022) Morphological characterization of *Phytophthora infestans* and its growth on different growth media. Sarhad J Agric 38(4):1189–1202

Raza M, Razaghi P, Al-Ani LKT, Aguilar-Marcelino L (2023) Underpinning the phylogeny and taxonomy of Phytofungi through computational biology. In: Haq IU, Ijaz S, Khan IA (eds) Phytomycology and molecular biology of plant pathogen interactions. CRC Press, Taylor and Francis Group, pp 149–169

Reddy VSO, Basappa SC (1997) Preparation of sweet potato flour and its fermentation to ethanol. Journal of Food Science and Technology 34(2):108–112

Rhouma A, Hajji-Hedfi L, Bouqellah NA, Khaire PB, Dali S, Bargougui O et al (2024) Uniting the role of entomopathogenic fungi against *Rhizoctonia solani* JG Kühn, the causal agent of cucumber damping–off and root rot diseases. Phyton 93(11):2857–2881

Rodriguez RJ, White JF Jr, Arnold AE, Redman ARA (2009) Fungal endophytes: diversity and functional roles. New Phytol 182(2):314–330. https://doi.org/10.1111/j.1469-8137.2009.02773.x

Romero A, Carrion G, Rico-Gray V (2001) Fungal latent pathogens and endophytes from leaves of *Parthenium hysterophorus* (Asteraceae). Fungal Divers 7:81–87

Roy S, Banerjee D (2018) Diversity of endophytes in tropical forests. In: Pirttilä A, Frank A (eds) Endophytes of forest trees. Forestry sciences, vol 86. Springer, Cham. https://doi.org/10.1007/978-3-319-89833-9_3

Rungjindamaa N, Jones EBG (2024) Why are there so few basidiomycota and basal fungi as endophytes? Rev J Fungi 10(1):67. https://doi.org/10.3390/jof10010067

Sagita R, Quax WJ, Haslinger K (2021) Current state and future directions of genetics and genomics of endophytic fungi for bioprospecting efforts. Front Bioeng Biotechnol 9. https://doi.org/10.3389/fbioe.2021.649906

Sallam N, Ali EF, Seleim MA, Khalil Bagy HM (2021) Endophytic fungi associated with soybean plants and their antagonistic activity against *Rhizoctonia solani*. Egypt J Biol Pest Control 31:1–9. https://doi.org/10.1186/s41938-021-00402-9

Sánchez-Rodríguez AR, Raya-Díaz S, Zamarreño AM, García-Mina JM, Del Campillo MC, Quesada-Moraga E (2018) An endophytic *Beauveria bassiana* strain increases spike production in bread and durum wheat plants and effectively controls cotton leafworm (*Spodoptera littoralis*) larvae. Biol Control 116:90–102

Sara B, Noreddine KC, Jacqueline D (2016) Production of laccase without inducer by *Chaetomium* species isolated from Chettaba forest situated in the East of Algeria. Afr J Biotechnol 15(7):207–213. https://doi.org/10.5897/AJB2015.15001

Sarker MS, Mohiuddin KM et al (2020) Effect of Bionematicide and Bau–biofungicide against root–knot (*Meloidogyne* spp.) of Soybean. Malaysian J Sustain Agric 4(2):44–48

Schafer W (1994) Molecular mechanisms of fungal pathogenicity to plants. Annu Rev Phytopathol 32(1):461–477

Schardl CL, Leuchtmann A, Spiering MJ (2004) Symbioses of grasses with seedborne fungal endophytes. Annu Rev Plant Biol 55:315–340. https://doi.org/10.1146/annurev.arplant.55.031903.141735

Sessa L, Abreo E, Lupo S (2018) Diversity of fungal latent pathogens and true endophytes associated with fruit trees in Uruguay. J Phytopathol 166(9):633–647

Sharma N, Gautam AK (2019) Early pathogenicity events in plant pathogenic fungi: a comprehensive review. Biol Forum Int J 11(1):24–34

Sharma V, Salwan R, Al-Ani LKT (2020) Molecular aspects of plant beneficial microbes in agriculture. Elsevier science, Cambridge, p 454

Shweta S, Zuehlke S, Ramesha BT, Priti V, Kumar PM, Ravikanth G et al (2010) Endophytic fungal strains of *Fusarium solani*, from *Apodytes dimidiate* E. Mey. ex Arn (Icacinaceae) produce camptothecin, 10-hydroxycamptothecin and 9-methoxycamptothecin. Phytochemistry 71(1):117–122. https://doi.org/10.1016/j.phytochem.2009.09.030

Shweta S, Gurumurthy BR, Ravikanth G, Ramanan US, Shivanna MB (2013) Endophytic fungi from *Miquelia dentata* Bedd, produce the anti–cancer alkaloid, camptothecine. Phytomed 20(3–4):337–342. https://doi.org/10.1016/j.phymed.2012.11.015

Siddiqui IA, Shaukat SS (2003) Combination of *Pseudomonas aeruginosa* and *Pochonia chlamydosporia* for Control of Root-Infecting Fungi in Tomato. J Phytopathol 151(4):215–222

Sieber TN (2007) Endophytic fungi in forest trees: are they mutualists? Fungal Biol Rev 21(2–3):75–89

Singh UB, Sahu A, Sahu N, Singh RK, Renu S, Singh DP et al (2013) Arthrobotrys oligospora–mediated biological control of diseases of tomato (*Lycopersicon esculentum Mill.*) caused *by Meloidogyne incognita* and *Rhizoctonia solani*. J Appl Microbiol 114(1):196–208

Singh S, Kumar V, Dhanjal DS, Sonali DV, Thotapalli S, Singh J, Al-Ani LK, Aguilar-Marcelino L (2021) Rhizosphere biology: a key to agricultural sustainability. In: Yadav AN, Singh J,

Singh C, Yadav N (eds) Current trends in microbial biotechnology for sustainable agriculture. Springer Nature Singapore, pp 161–182. https://doi.org/10.1007/978-981-15-6949-4_7

Slippers B, Wingfield MJ (2007) Botryosphaeriaceae as endophytes and latent pathogens of woody plants: diversity, ecology and impact. Fungal Biol Rev 21(2–3):90–106

Soumare A, Diédhiou AG, Arora NK, Al-Ani LKT, Ngom M, Fall S et al (2021) Potential role and utilization of plant growth promoting microbes in plant tissue culture. Front Microbiol 12: 649878. https://doi.org/10.3389/fmicb.2021.649878

Stierle A, Strobel G, Stierle D (1993) Taxol and taxane production by *Taxomyces andreanae*, an endophytic fungus of Pacific yew. Science 260(5105):214–216. https://doi.org/10.1126/science.8097061

Strobel GA, Miller RV, Martinez-Miller C, Condron MM, Teplow DB, Hess WM (1999) Cryptocandin, a potent antimycotic from the endophytic fungus *Cryptosporiopsis* cf. *quercina*. Microbiol 145(8):1919–1926. https://doi.org/10.1099/13500872-145-8-1919

Suebrasri T, Somteds A, Harada H, Kanokmedhakul S, Jogloy S, Ekprasert J et al (2020) Novel endophytic fungi with fungicidal metabolites suppress Sclerotium disease. Rhizosphere 16. https://doi.org/10.1016/j.rhisph.2020.100250

Sukmawati D, Adisyahputra A, Al-Ani LKT, Al Husna SN, Afifah ZN, Sriherwanto C et al (2024a) Pichia kudriavzevii UNJCC Y–137 and *Candida tropicalis* UNJCC Y–140 isolated from Durio kutejensis as potential probiotic agents. Food Sci Biotechnol 33. https://doi.org/10.1007/s10068-024-01609-w

Sukmawati D, Nurmayanti F, Supiyani A, Kurniati TH, Adisyahputra A, Tawfeeq Al-Ani LK et al (2024b) Molecular identification and antifungal activity of *Candida chrysomelidarum* and *Rhodotorula toruloides* isolated from indonesian indigenous fruit *Flacourtia inermis roxb.* (lobi–lobi) against pathogenic mold *Aspergillus flavus*. Trends Sci 20(10):8321

Sunitha VH, Devi DN, Srinivas C (2013) Extracellular enzymatic activity of endophytic fungal strains isolated from medicinal plants. World J Agric Sci 9(1):01–09

Supaphon P, Phongpaichit S, Sakayaroj J, Rukachaisirikul V, Kobmoo N, Spatafora JW (2017) Phylogenetic community structure of fungal endophytes in seagrass species. Bot Mar 60(4): 489–501. https://doi.org/10.1515/bot-2016-0089

Suryanarayanan TS (2013) Endophyte research: going beyond isolation and metabolite documentation. Fungal Ecol 6(6):561–568

Suryanarayanan TS, Murali TS, Venkatesan G (2002) Occurrence and distribution of fungal endophytes in tropical forests across a rainfall gradient. Canadian J Bot 80(8):818–826. https://doi.org/10.1139/b02-069

Tanaka A, Tapper BA, Popay A, Parker EJ, Scott B (2005) A symbiosis expressed non-ribosomal peptide synthetase from a mutualistic fungal endophyte of perennial ryegrass confers protection to the symbiotum from insect herbivory. Mol Microbiol 57:1036–1050. https://doi.org/10.1111/j.1365-2958.2005.04747.x

Taylor TN, Remy W, Hass H, Kerp H (1995) Fossil arbuscular mycorrhizae from the early Devonian. Mycologia 87:560–573. https://doi.org/10.1080/00275514.1995.12026569

Taylor TN, Kerp H, Hass H (2005) Life history biology of early land plants: deciphering the gametophyte phase. Proc Nat Acad Sci 102(16):5892–5897. https://doi.org/10.1073/pnas.0501985102

U'Ren JM, Lutzoni F, Miadlikowska J, Laetsch AD, Arnold AE (2012) Host and geographic structure of endophytic and endolichenic fungi at a continental scale. Am J Bot 99(5):898–914. https://doi.org/10.3732/ajb.1100459

Vaz ABM, Fonseca PLC, Silva FF, Quintanilha-Peixoto G, Sampedro I, Siles JA et al (2020) Foliar mycoendophytome of an endemic plant of the Mediterranean biome (*Myrtus communis*) reveals the dominance of basidiomycete woody saprotrophs. PeerJ 8:e10487. https://doi.org/10.7717/peerj.10487

Vega FE (2018) The use of fungal entomopathogens as endophytes in biological control: a review. Mycol 110(1):4–30

Verma SK, Kharwar RN, Gond SK, Kingsley KL, White JF (2019) Exploring endophytic communities of plants: methods for assessing diversity, effects on host development and potential biotechnological applications. In: Verma SK, White JA (eds) Seed endophytes: biology and biotechnology. Springer, pp 55–82. https://doi.org/10.1007/978-3-030-10504-4_4

Villar-Luna E, Gómez-Rodríguez O, Villar-Luna H, Aguilar-Marcelino L, Al-Ani LKT, Fernández-Herrera E (2022) Eco-friendly management of false root-knot nematode *Nacobbus aberrans*: an overview. In: Chaudhary KK, Meghvansi MK (eds) Sustainable management of nematodes in agriculture, volume 1: Org MGNT. Springer Nature, Berlin, pp 327–342

Viswanath B, Rajesh B, Janardhan A, Kumar AP, Narasimha G (2014) Fungal laccases and their applications in bioremediation. Enzyme Res 2014. https://doi.org/10.1155/2014/163242

Wang Y, Guo LD (2007) A comparative study of endophytic fungi in needles, bark, and xylem of *Pinus tabulaeformis*. Botany 85(10):911–917. https://doi.org/10.1139/B07-084

Wang L, Xiang WL, Ren LiLi RL, Li ChunChun LC, Gao ChengLong GC, Liu XiaoBo LX et al (2019) Effects of endophytic fungi diversity in different coniferous species on the colonization of *Sirex noctilio* (Hymenoptera: Siricidae). Sci Rep 9(1):5077. https://doi.org/10.1038/s41598-019-41419-3

Wijayawardene NN, Hyde KD, Al-Ani LKT, Tedersoo L, Tedersoo L, Haelewaters D et al (2020) Outline of fungi and fungi-like taxa. Mycosphere 11(1):1060–1456. https://doi.org/10.5943/mycosphere/11/1/8

Wu B, Hussain M, Zhang W, Stadler M, Liu X, Xiang M (2019) Current insights into fungal species diversity and perspective on naming the environmental DNA sequences of fungi. Mycology 10(3):127–140. https://doi.org/10.1080/21501203.2019.1614106

Yadav AN (2020) Plant microbiomes for sustainable agriculture: current research and future challenges. In: Yadav AN, Singh J, Rastegari AA, Yadav N (eds) Plant microbiomes for sustainable agriculture. Sustainable development and biodiversity. Springer, Berlin, pp 475–482

Yadav M, Vivekanand V (2019) *Chaetomium globosporum*: a novel laccase producing fungus for improving the hydrolyzability of lignocellulosic biomass. Heliyon 5(3). https://doi.org/10.1016/j.heliyon.2019.e01353

Yadav R, Singh AV, Joshi S, Kumar M (2015) Antifungal and enzyme activity of endophytic fungi isolated from *Ocimum sanctum* and *Aloe vera*. Afr J Microbiol Res 9(29):1783–1788. https://doi.org/10.5897/AJMR2015.7451

Yadav M, Singh A, Balan V, Pareek N, Vivekanand V (2019) Biological treatment of lignocellulosic biomass by *Chaetomium globosporum*: process derivation and improved biogas production. Int J Biol Macromol 128:176–183. https://doi.org/10.1016/j.ijbiomac.2019.01.118

Yang N, Shan X, Wang K, Lu J, Zhu Y, Regina RS et al (2025) A fusarioid fungus forms mutualistic interactions with poplar trees that resemble ectomycorrhizal symbiosis. IMA Fungus 16:e143240

Yao YQ, Lan F, Qiao YM, Wei JG, Huang RS, Li LB (2017) Endophytic fungi harbored in the root of *Sophora tonkinensis* Gapnep: diversity and biocontrol potential against phytopathogens. Microbiol Open 6(3). https://doi.org/10.1002/mbo3.437

Yao H, Sun X, He C, Maitra P, Li XC, Guo LD (2019) Phyllosphere epiphytic and endophytic fungal community and network structures differ in a tropical mangrove ecosystem. Microbiome 7:1–15. https://doi.org/10.1186/s40168-019-0671-0

Yao Y, Huo J, Ben H, Gao W, Hao Y, Wang W et al (2023) Biocontrol efficacy of endophytic fungus, *Acremonium sclerotigenum*, against *Meloidogyne incognita* under in vitro and in vivo conditions. Biologia 78(11):3305–3313

Yao L, Ji C, Wang Z, Liu Y, Li W, Wang M et al (2025) Enhancing *Astragalus mongholicus* performance through endophytic fungi Improvement for Fusarium wilt resistance. Physiol Mol Plant Pathol 136:102532

Yelikbayev B, Correa EJA, Al-Ani LK, Duarte NF, Pagano MC (2024) Revitalizing degraded rural landscapes: approaches and strategies for ecological restoration. Environ Rep Int J 6. https://doi.org/10.51470/ER.2024.6.2.23

Yu H, Zhang L, Li L, Zheng C, Guo L, Li W et al (2010) Recent developments and future prospects of antimicrobial metabolites produced by endophytes. Microbiol Res 165(6):437–449. https://doi.org/10.1016/j.micres.2009.11.009

Yu X, Zhang W, Lang D, Zhang X, Cui G, Zhang X (2019) Interactions between endophytes and plants: beneficial effect of endophytes to ameliorate biotic and abiotic stresses in plants. J Plant Biol 62:1–3. https://doi.org/10.1007/s12374-018-0274-5

Zhang T, Yao YF (2015) Endophytic fungal communities associated with vascular plants in the high arctic zone are highly diverse and host–plant specific. PLoS One 10(6). https://doi.org/10.1371/journal.pone.0130051

Zhang H, Xie J, Fu Y, Cheng J, Qu Z, Zhao Z et al (2020) A 2-kb mycovirus converts a pathogenic fungus into a beneficial endophyte for brassica protection and yield enhancement. Mol Plant 13:1420–1433

Zheng YK, Qiao XG, Miao CP, Liu K, Chen YW, Xu LH et al (2016) Diversity, distribution and biotechnological potential of endophytic fungi. Ann Microbiol 66:529–542. https://doi.org/10.1007/s13213-015-1153-7

Zheng H, Qiao M, Xu J, Yu Z (2021) Culture-based and culture-independent assessments of endophytic fungal diversity in aquatic plants in Southwest China. Front Fungal Biol 2:692549. https://doi.org/10.3389/ffunb.2021.692549

Endophytic Fungi from Lichens and Non-flowering Plants: Biodiversity and Ecological Prospects

Salma S. Higazy, Amal Belaid, and Ahmed M. Abdel-Azeem

Contents

S. S. Higazy (✉)
Botany and Microbiology Department, Faculty of Science, Suez Canal University, Ismailia, Egypt

A. Belaid
Biology Department, Selcuk University, Konya, Turkey

A. M. Abdel-Azeem
Botany and Microbiology Department, Faculty of Science, Suez Canal University, Ismailia, Egypt

Research Institute of University of the Bucharest (ICUB), University of Bucharest, Bucharest, Romania

The Centre for Mineral Biogeochemistry, The University of the Free State, Bloemfontein, Free State, South Africa

A. M. Abdel-Azeem, F. Degola (eds.), *Advances in Lichens and Non-Flowering Plants Fungal Endophytes Research*, Fungal Biology,
https://doi.org/10.1007/978-3-032-01506-8_2

1 Introduction

Microbiota known as fungal endophytes live inside the tissues of living plants and do not immediately cause harm (Bacon and White 2000). However, as the host ages, they may develop into a pathogen (Rodriguez and Redman 2008). Following the colonization of host tissue, endophytes enter a dormant (latent) phase that may persist for the entire lifespan of the host plant or beyond, remaining inactive until the host attains a developmental stage conducive to the fungus or until environmental conditions become favourable for fungal activity (Sieber 2007). Endophytes were initially identified in 1904. Endophytes primarily disseminate to their host plants via airborne spores. Nevertheless, certain endophytes can be transmitted vertically by seeds to subsequent plant generations (Hartley and Gange 2009), but they garnered significant attention only recently due to their recognized ecological and therapeutic significance (Gunatilaka 2006). It is believed that they reside in every plant species on Earth (Strobel and Daisy 2003). Recent developments in screening technology (Zhang et al. 2006; Aly et al. 2010) have demonstrated the significant potential of endophytes as a primary source of biologically active chemicals for medicinal and agricultural applications, several of which will be examined in this study. Thus, there exists considerable potential to identify new natural products from endophytic bacteria, which have been inadequately researched and flourish in various habitats and ecosystems (Guo et al. 2008). Moreover, microbial natural products provide a substantial and predominantly underutilized reservoir of distinctive chemical structures that have been honed via (co-)evolution and presumably serve significant functions in communication or adaptation to environmental and habitat changes (Gunatilaka 2006). This review examines the diversity, host specificity, interactions with host plants, communication systems, and regulation of secondary metabolite production in endophytes. Instances of plant-derived secondary metabolites generated by fungal endophytes with therapeutic significance are also examined.

2 Endophytic Fungi Diversity

Because of the vast number of microbial species that are known or assumed to exist, assessing microbial diversity is a significant difficulty in modern microbiology (Gamboa et al. 2003). Bacteria have evolved to live in almost every biological niche on the planet (Gunatilaka 2006). Every one of the roughly 300,000 known plant species is known to have at least 1, and frequently hundreds, of endophytic strains (Strobel and Daisy 2003; Huang et al. 2018). When it comes to examining endophyte diversity, one of the main drawbacks of cultivation-dependent methods is their propensity to focus on dominant, quickly developing species, which might result in the removal of uncommon species with unique ecological requirements (Sun et al. 2011a, b); Unterseher and Schnittler 2009). Advancements in dependable DNA-based techniques have improved the categorization of taxonomic units, enabling more precise identification of sterile morphotypes and unculturable fungi. Utilizing these methodologies has enhanced our comprehension of the extensive diversity of fungal endophytes (Tejesvi et al. 2011; Arnold et al. 2007; Arnold and Lutzoni 2007). Endophytes have historically been classified according to their morphological characteristics, which may inadequately represent their diversity, especially at the subspecies level. Endophytes were isolated from mature, asymptomatic *Quercus* sp. leaves for a study investigating endophyte diversity and host specificity in significant ecological hosts within the biodiverse Madrean Archipelago. Family Fagaceae, Pinaceae, Cupressaceae, and Platycladus orientalis found in the Tucson, Arizona, USA, area. The nuclear ribosomal internal transcribed spacer region was utilized for genotyping due to its greater variability compared to other rDNA regions. Lineages among recognized endophyte species were discerned by examining sequence variations across four prevalent endophytic genera: *Botryosphaeria*, *Colletotrichum*, *Mycosphaerella*, and *Xylaria*. This study demonstrated a 1–2% compositional variation among conspecifics (Arnold et al. 2007). This emphasizes that the suitable method for investigating the variety of endophyte communities goes beyond simple species classification and delves into genotype-level differences (Petrini 1991). Small plant tissue samples exhibit notable amounts of intraspecific genetic heterogeneity as well as species variety. According to studies conducted on Norway spruce needles, each needle contained many genotypes of *Lophodermium piceae* (Müller et al. 2001). Denaturing gradient gel electrophoresis (DGGE) in conjunction with 18S rRNA sequence analysis was used in a follow-up study to examine the fungal diversity in *Magnolia liliifera* leaves. Total genomic DNA was extracted as part of the experiment, and fungal DNA was amplified using certain primers. 14 taxonomic units were discovered by the PCR-DGGE study, identifying fungal taxa that were not detectable by morphological or culture-based techniques. According to Duong et al. (2006), this suggests that certain endophytes might not thrive in artificial environments. Similarly, *Rhododendron tomentosum* leaf tissues included 11 unculturable endophytic fungus strains (Tejesvi et al. 2011).

Even in seemingly identical research conditions, endophyte composition diversity among individual host plants of the same species is frequently seen and is controlled by the physiological state of the host Sieber and Hugentobler (1987), Ghimire et al. (2011), and González and Tello (2011). Because of their broad range, host plants are exposed to certain environmental circumstances that have a substantial impact on endophyte diversity (Vega et al. 2010; Arnold 2007; Higgins et al. 2007). Research indicates that endophyte variety, prevalence, and host range typically increase with latitude (Arnold 2007; Arnold and Lutzoni 2007). Endophyte diversity decreased linearly from tropical to boreal environments, according to a study that included 1202 endophytic fungal isolates from various host species in six different geographic locales, including lowland tropical forests in Panama and boreal forests in Québec, Canada. Tropical endophyte assemblages, despite having fewer classes, had a wider variety of species, endophyte communities in higher latitudes showed less diversity among different *Ascomycota* classes (Arnold and Lutzoni 2007).

Interpretations of fungal diversity, species composition, and endophyte infection frequency are greatly influenced by isolation techniques (Arnold 2007; Sun et al. 2011a, b). Compared to larger leaf pieces, more fungal morphospecies were isolated from leaf tissue when smaller fragments were utilized. It is often advised to reduce fragment size and increase sample amount in order to effectively measure fungal endophyte diversity (Gamboa et al. 2003). Since their first isolation from *Lolium temulentum* seeds in 1898, endophytic fungi in plants have been known to exist for over a century. Endophytic fungi were not well documented between 1890 and 1980 (Hyde and Soytong 2008). These fungi have been identified in practically every tracheophyte that has been examined since the 1980s, including both woody and herbaceous species. Endophytic fungi are ubiquitous and crucial to microbial biodiversity, demonstrating the large range of plant species (Qin et al. 2011).

Because these plants are likely to harbour endophytes with unique bioactive properties, the selection of plant hosts for endophyte investigations is deliberate, focusing on species that occupy particular ecological niches or have ethnobotanical value (Strobel and Daisy 2003). Since it is the first step towards producing pure cultures, the methodology employed to isolate endophytic fungi holds significant importance. The isolation of fungi is influenced by various factors such as the species of the host, the techniques used for sampling, the interactions that occur between the host and endophytes as well as among different endophytes, the type and age of the tissue, the geographic location, environmental conditions, the selection of culture media, and the methods employed for surface sterilization (Zhang et al. 2006). Hallmann et al. (2006) investigated various isolation strategies, including plant sampling, surface sterilization, and media selection, as detailed by Strobel and Daisy (2003).

A vital initial measure in averting infection by external microbes is surface sterilization. Common sterilants comprise potassium permanganate (2%), hydrogen peroxide (3%), ethanol (70–90%), and sodium hypochlorite (2–10%). A frequently used approach involves the sequential application of ethanol and sodium hypochlorite (Rivera-Orduña et al. 2011; Li et al. 2012b; Xiong et al. 2013). Surfactants such as Triton X-100, Tween 20, and Tween 80 can also enhance the efficacy of sterilization (Zhang et al. 2006; Qin et al. 2011). Qin et al. (2009) introduced a five-step method that uses sodium thiosulfate to neutralize residual hypochlorite and increase endophyte recovery, while described a three-step surface sterilization procedure. Because different species and developmental stages exhibit varying sensitivities, sterilization techniques must be customized for specific plant tissues (Qin et al. 2011).

Plant tissues are meticulously cut into tiny pieces after sterilization, and they are subsequently put on culture media. Stems and roots range in size from 0.3 to 0.5 cm, while leaf segments are normally 0.5 × 0.5 cm. To increase the chances of effective isolation, it has been proposed to use a blender to break up tissues (Qin et al. 2009). Czapek agar (CZA), Sabouraud dextrose agar (SAB), potato dextrose agar (PDA), and malt extract agar (MEA) are common culture media (Rivera-Orduña et al. 2011; Sun et al. 2011a, b). Low-nutrient media, including synthetic low-nutrient agar (SNA) and carnation leaf agar (CLA), have been effective for certain endophytes

(Arnold and Herre 2003). The diversity of isolates can be improved by utilizing a variety of media. ITS and 18S rDNA sequencing are among the molecular and morphological techniques that are typically employed to achieve taxonomic identification (Zhang et al. 2006).

3 Endophytic Fungal Distribution

Endophytic fungal populations inhabit the tissues of various tree species, each exhibiting distinct structural characteristics. Endophytic fungal colonization occurs significantly more frequently in stems compared to leaves. There are significant differences in the populations of fungi found in foliar and root tissues. Employed DGGE and random cloning methodologies to investigate the endophytes in the leaf and root tissues of *Bletilla ochracea* (Orchidaceae). The diversity of leaves ($H' = 2.354$) was greater than that of roots ($H' = 1.560$). This trend was further corroborated by later studies (Fisher et al. 1994; Sun et al. 2011a, b; Li et al. 2012a; Zheng et al. 2013). A possible explanation is that leaves are deciduous, while stems are perennial (Li et al. 2012a). Kharwar et al. (2011) noted that the leaves of the medicinal plant *Mansoa alliacea* exhibited the greatest rate of endophytic fungal colonization at 72.22%, exceeding the 67.78% reported in the stems. Identical plant species may exhibit variations in endophytic communities across different geographic regions. Fisher et al. (1994) observed substantial changes in the endophytic fungus populations within the leaves of *Quercus ilex* from England, Majorca, and Switzerland. Vega et al. (2010) indicated that coffee plants from Colombia, Hawaii, Mexico, and Puerto Rico contained a rich assortment of endophytic fungus.

Studies on endophytic fungal communities in *Taxus chinensis* var. mairei yielded comparable findings, particularly in the Chinese provinces of Jiangxi, Zhejiang, and Chongqing (Wu et al. 2013). Variation in endophytic fungi associated with specific species is influenced by the age and tissue type of host plants (Arnold and Herre 2003; Mei et al. 2014). However, fungi from the same plant species within a defined region generally show similarities in abundance and distribution. The aging of the host plant is associated with an increase in the diversity and prevalence of endophytic fungi. This trend is attributable to two primary factors: first, mature trees exhibit increased vulnerability to recurrent infections as a result of the dispersal of fungi through air and rain; second, alterations in the physiological condition of plants and the structural integrity of bark in older trees facilitate the formation of new pathways in plant tissues, thereby enabling fungal colonization (Fisher et al. 1994).

The height of the crown and the extent of canopy cover may influence the dissemination of endophytic fungal populations (Arnold and Herre 2003). Petrini et al. (1982) assert that samples collected from densely wooded regions exhibit greater species diversity and abundance of endophytic fungus compared to those obtained from more open environments. Environmental factors affecting endophytic fungal communities comprise ambient humidity (Lau et al. 2013), seasonal variations (Mishra et al. 2012), altitude (Davey et al. 2013), precipitation (Zimmerman and Vitousek 2012), temperature (Li et al. 2012b), interactions with other plant communities (Victoria et al. 2007), and environmental pollution (Li et al. 2012a; Zheng et al. 2013).

Biogeography examines the spatial and temporal distribution of biodiversity, seeking to comprehend species' habitats, population densities, and the determinants of these patterns. Since the seventeenth century, biologists have examined the spatial distribution of biodiversity among flora and fauna. Recently, researchers have focused on the spatial dispersion of microorganisms. Ecologists frequently reference Baas Becking's century-old assertion, "everything is everywhere, the environment selects" (the EisE hypothesis), in discussions of microbial biogeography (Baas Becking 1934; Green et al. 2008).

This leads to the question: does the biogeographical distribution and spatial variability of endophytic fungi align with the idea that "the environment selects"? The EisE hypothesis seems to be supported by findings from a recent study by Zimmerman and Vitousek (2012), which examined endophytic fungal communities in the leaves of a single tree species (*Metrosideros polymorpha*) across short geographic distances in Hawaii, encompassing a wide range of environmental gradients (500–5500 mm rainfall/year; 10–22 °C mean annual temperature). However, additional research is required to confirm this claim. Exploring biogeography not only advances fungal biodiscovery but also extends beyond purely academic interest.

4 Using Fungal Endophytes to Manage Abiotic Stress

Abiotic stressors greatly hinder plant growth, making a robust system or mechanism for stress tolerance crucial. Many bacteria are known to derive nutrients from various plants, with some interactions benefiting the host and others causing harm. Ripa et al. (2019) define "endophytes" as microorganisms that live within plant tissues without causing significant harm or offering benefits beyond basic survival. Every plant hosts one or more endophytes that provide nutrients, improve environmental adaptability, and confer resistance to both biotic and abiotic stresses.

Fungi that reside within plants, referred to as endophytes, typically occupy the intercellular spaces of petioles, leaves, and stems. Highly resilient, fungal endophytes spend their entire lifecycle integrated into a plant's ecosystem and can assist in adapting to new environmental conditions (Vandenkoornhuyse et al. 2015). The resilience and sustainability of modern agriculture rely heavily on fungal endophytes. Fig. 1 (Redman et al. 2021) depicts their partnership with host plants to perform various roles within agricultural systems. Their capacity to solubilize phosphate and supply nitrogen to host plants is vital, as it progressively enhances vegetative growth and fortifies interconnected plants to build resistance (Yadav 2017).

Under stress, endophytic fungi like *Penicillium* sp. release less of the compounds that promote growth. Gibberellin is a phytohormone that suaeda japonica secretes.

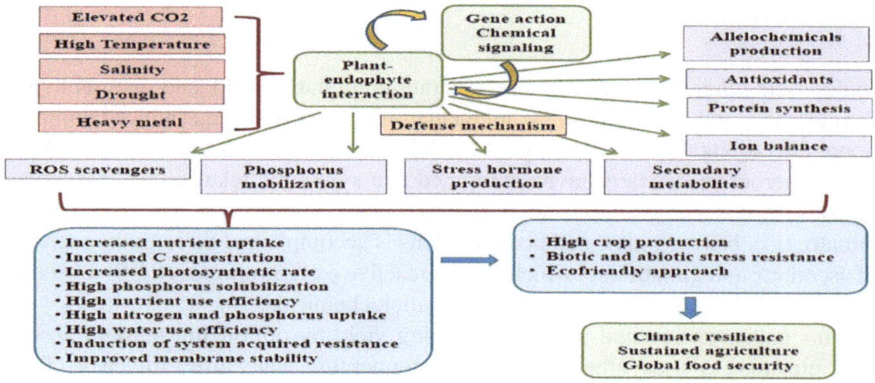

Fig. 1 According to Anamika Verma (2022), abiotic resistance is conferred via the symbiotic relationship between plants and endophytes

Triterpenoid saponins and ginsenosides are ingredients in *Panax ginseng*, which are made by a variety of fungal species including *Aspergillus*, *Penicillium*, *Fusarium*, and Ascomycota. found that these chemicals promote root development, which in turn increases plants' resistance to internal and external challenges. investigated how different fungi generate gibberellins. Results like this demonstrate that gibberellins are useful for increasing harvest yields.

Phoma sp. secretes growth hormones in *Tinospora cordifolia*. Likewise, the *Phaseolus vulgaris*-associated *Trichoderma atroviride*, *Trichoderma polysporum*, and *Trichoderma harzianum* generate bioactive volatile and non-volatile metabolites, phosphate-solubilizing compounds, and proteolytic enzymes. Moreover, *Mucor* sp., found in *Brassica campestris*, enhances the production of phosphate solubilization, indole-3-acetic acid (IAA), and ACC deaminase, all of which facilitate plant growth.

Fungal endophytes promote plant growth in adverse environments by facilitating siderophore formation, phosphate solubilization, and the synthesis of antibiotics, phytohormones, and nutrients. Salicylic acid, jasmonic acid, and abscisic acid serve as defence signalling molecules produced in response to abiotic stresses. Numerous soil types exhibit phosphorus deficiency, which is regarded as a vital component for plant development.

Iron hydroxides and aluminium hydroxides become immobilized in acidic soil due to their insolubility, rendering them inaccessible to plants, even when applied as inorganic fertilizers. Redox-active iron serves as a cofactor for numerous enzymes and is an essential micronutrient; however, it exists in soil as an insoluble compound (ferric hydroxide). Fungal endophytes enhance iron uptake and phosphate solubilization by synthesizing low-molecular-weight siderophores with a high affinity for iron (Fe^{3+}), hence facilitating plant growth. Endophytes enhance biomass synthesis, nitrogen uptake, and trash biodegradation.

4.1 Carbon Dioxide

Endophytic fungi enhance carbon sequestration, biomass, yield, and internal hyphal development and colonization by augmented photosynthesis and induced compositional alterations under increasing CO_2 (Galani et al. 2022). Yadav (2017) indicate that numerous researchers have determined that symbiotic relationships are advantageous in this context since they enhance plant development in crops such as tomato, rice, beans, grains, and potatoes. This is accomplished through the secretion of ascorbate and glutathione, which reduce reactive oxygen species (ROS) levels and promote the synthesis of antioxidants and allelochemicals.

This enhances survival while maintaining yield by optimizing nutrient absorption, elevating photosynthetic rates, and augmenting water use efficiency. These components also augment phytohormone production, elevate gene expression, promote protein functionality, create metabolites, produce signalling molecules, and activate stress-responsive pathways (Ma et al. 2016).

Nonetheless, other contradicting observations have indicated diminished root and shoot growth, illustrating that endophytes exhibit variable impacts on identical host plants throughout diverse habitats, particularly under specific growth conditions. Moreover, plant type influences colonization; specifically, C3 plants necessitate greater carbon for biomass production, diminishing their potential to benefit from colonization, while C4 plants utilize more carbohydrates, thereby enhancing their propensity to establish associations with endophytic fungi (Tang et al. 2009).

Because endophytes use molecular mechanisms to activate stress-responsive genes, synthesize antistress metabolites, and produce ROS scavenger molecules, current research suggests that the role of microbiota in biological and ecological processes presents a challenge, especially concerning potential agricultural applications.

4.2 Drought

A major stressor that diminishes agricultural output and directly threatens the food security objectives of emerging nations is drought. Under dry conditions, plants exhibit several physiological and biochemical responses, including stomatal closure, heightened respiration, growth suppression, and reduced photosynthesis. Researchers have identified two categories of drought stress-inducible genes that enhance plant tolerance. The initial group comprises proteins, whereas the subsequent group includes regulatory proteins engaged in signal transduction (Shinozaki and Yamaguchi-Shinozaki 2006). The fungal endophyte *Piriformospora indica* (family *Sebacinaceae*) colonizes the outer layer of host plant roots, enhancing plant growth, elevating chlorophyll levels, and improving yield under drought stress conditions. Ripa et al. (2019) examined the function of these fungal endophytes in conferring drought stress tolerance in wheat under different temperature conditions.

The fungus genera *Fusarium*, *Trichoderma*, *Penicillium*, *Aspergillus*, and *Alternaria* were examined for their impacts on plant growth parameters and stress resilience. The majority were identified as offering wheat drought tolerance. Compared to non-symbiotic plants, those connected with endophytic fungi—such as rice, tomatoes, and dune grass—utilized far less water and exhibited greater biomass. The phenomenon of drought tolerance in endophyte-infected plants, as opposed to non-infected plants, may be attributed to the development of a thicker cuticle, reduced leaf conductance, a deceleration of the transpiration stream, or an increased accumulation of solutes in the tissues (Ripa et al. 2019). The characterization of *Trichoderma* sp. demonstrated alterations in gene expression patterns associated with *Theobroma cacao*, suggesting that *Trichoderma* sp. may promote tolerance to abiotic stressors, including drought.

The interaction between plants and water is influenced by drought, heat, and salinity stress, leading to intricate plant responses, including heightened osmolyte synthesis. Marquez et al. (2007) assert that solute potential and matrix potential significantly impact osmotic potential. Symbiotic fungus likely enhances the matrix potential, which is crucial for plants to store moisture and improve drought resistance. Consequently, symbiotic plants utilize considerably less water than non-symbiotic plants (Marquez et al. 2007).

4.3 Salinity

Salinity is a significant abiotic element that constrains plant development and output globally. Antioxidant enzymes, specifically reactive oxygen species (ROS) scavengers such as glutathione, ascorbate, and tocopherol, along with superoxide dismutases (SOD), catalases (CAT), ascorbate- or thiol-dependent peroxidases (APX), glutathione reductases (GR), dehydroascorbate reductases (DHAR), and mono-dehydroascorbate reductases (MDHAR), are associated with plants' capacity to endure salt stress. Reactive oxygen species (ROS) can be directly neutralized by superoxide dismutase (SOD), catalase (CAT), and ascorbate peroxidase (APX), or indirectly by replenishing the cell with ascorbate and glutathione. The root fungal endophyte *Piriformospora indica* enhanced biomass production in barley under salt stress by augmenting antioxidative capacity and elevating the levels of lipid peroxidase, dehydroascorbate reductase, catalase, and glutathione reductase enzymes.

The fungus *Exophiala* sp. functions as an endophyte. LHL08, isolated from cucumber roots, enhances rice's tolerance to drought and salinity stress. Likewise, *Serendipita vermifera* and *Piriformospora indica* exhibited resilience to saline stress, resulting in enhanced growth and yield. The amalgamation of these biological agents exhibited significant impacts.

Under saline and drought circumstances, the endophytic fungus *Fusarium culmorum* (FcRed1) exhibited a symbiotic association with rice seedlings, leading to enhanced biomass, plant height, root length, and osmolyte accumulation relative to non-symbiotic rice seedlings. It is essential to elucidate how fungal endophytes enhance plant defence mechanisms against abiotic stress (Ma et al. 2016).

4.4 High Temperatures

Endophytic fungi may influence the impact of elevated temperatures on plants by altering, regulating, or modifying their physiological, biochemical, and metabolic processes, contingent upon the fungi's biology, developmental stage, and the cultivar of the host plant. Environmental factors can alter metabolite production by enhancing the quantity and diversity of endophytic populations. Numerous researchers (Lata et al. 2018) have examined various isolated fungal endophytes, their diverse interactions that confer advantages to the host under high-temperature stress, and their distinct functions in environmental acclimatization and tolerance enhancement. A study has demonstrated that the fungal endophyte *Curvularia protuberata* confers thermotolerance to panic grass and tomatoes. *Paraphaeosphaeri aquadriseptata* confers heat tolerance to *Arabidopsis* sp. by synthesizing an inhibitor of heat shock protein 90 (HSP90). A study on *Stenotrophomonas maltophilia* SBP-9 demonstrates its capacity to enhance wheat growth under saline conditions, suggesting that viral infection may induce heat tolerance in the plant (Marquez et al. 2007; Yadav 2017). A thorough assessment of the intriguing and enigmatic capabilities of various fungal endophytes to generate host metabolites is necessary for prospective application in many contexts (Galani et al. 2022).

5 Fungal Endophytes for the Management of Biologic Stress

Because of their ability to both aid in plant growth processes and offer resistance against a variety of biotic stresses, including nematode attacks and insect herbivory, endophytic fungi have gained importance in agriculture.

Agents of Biocontrol By protecting their host plants from diseases and pests, fungus endophytes act as biocontrol agents. Through the production of bioactive or antiherbivorous compounds, or through intricate metabolic interactions between the fungus and the plant after infection, endophytes deter herbivory. These protective substances may reduce insect performance (antibiosis) or prevent feeding (antixenosis). Certain physiologically active substances are secreted by endophytic fungus without inflicting any noticeable damage to the tissues they inhabit. Alkaloids, terpenoids, isocoumarin derivatives, quinones, flavonoids, chlorinated metabolites, phenols, and phenolic acids are some of the functional groups into which defensive chemicals can be divided.

Agents that Cause Nematodes Fungal endophytes produce compounds that are harmful to nematodes, which in population while enhancing banana yield with the use of *Fusarium oxysporum*, makes them nematicidal agents. The endophytic fungus *Fusarium oxysporum* showed reduced nematode efficiency against *Meloidogyne* *incognita*, a plant parasite, in tomato plants, according to Diedhiou et al. (2003) According to Schwarz et al., different endophytic fungi that were isolated from the organs of aerial plants produced 3-hydroxypropionic acid (HPA). A bioactive compound obtained through bioactivity-guided fractionation of fungal extracts demonstrated selective nematocidal efficacy against *M. incognita*, with LD50 values ranging from 12.5 to 15 μg/ml. Felde et al. also discovered that the endophytic fungal isolates *Trichoderma atroviride* and *Fusarium* were co-inoculated. The burrowing nematode *Radopholus similis* (Cobb), a significant parasitic nematode impacting bananas, can be diminished.

6 Endolichenic Fungi's Function Inside the Lichen Microbiome

The functional roles and biotechnological possibilities of bacterial lichen partners are more understood than those of endolichenic fungi. These may provide significant insights for tackling endolichenic fungus issues. Diverse bacterial communities are consistently linked with lichens, enhancing lichen symbiosis by (1) facilitating nutrient absorption, (2) augmenting lichen resilience to biotic and abiotic stressors, (3) detoxifying metabolites, and (4) degrading older thallus components. Regarding the function of endolichenic fungi in lichen growth and survival, little is known. The age and chemical makeup of the host tissue have an impact on the endophyte community in plants (Arnold and Herre 2003). However, it is still unknown if a lichen's age and chemical makeup have an impact on the species composition and colonization density of its endolichenic fungi. By increasing their resistance to abiotic stressors including drought, salt, and extreme temperatures, endophytes help their host plants become more ecologically resilient (Sun et al. 2011c). In some cases, endophyte-mediated activation of host genes responsive to stress results in increased host fitness. By altering the metabolism of growth hormones, endophytes may increase a plant's resistance to abiotic stress.

Lichens are characterized by their extreme habitats and remarkable resilience to abiotic stresses; certain species endure protracted desiccation, elevated temperatures (90 °C and −196 °C), and intense UV radiation. Several organisms, including *Xanthoria elegans*, are subjected to actual space conditions and Martian atmospheres. Determining if endolichenic fungus, akin to plant endophytes, assists their lichen hosts in enduring abiotic stress is crucial. In certain plants, endophytes enhance the host plant's tolerance to specific biotic stresses, thereby mitigating the costs associated with their maintenance (by decreasing photosynthate levels). Endophytes, for example, inhibit diseases detrimental to their host plants (Arnold et al. 2003) and insect pests. In the plant host, endophyte infection enhances the expression of many defence-related genes.

While it is established that numerous genes in both the mycobiont and photobiont are upregulated in lichens during the initial interaction phase, it remains unclear if endolichenic fungi similarly affect gene regulation in their lichen hosts as endophytes do. A variety of antibacterial compounds are synthesized by the endolichenic microbiota. The lichen *Lobaria pulmonaria* hosts a bacterial community that produces diverse microbial antagonistic compounds, as demonstrated by metagenomic and culture-based approaches. Of the 62 isolates of endolichenic fungi evaluated for antifungal and antialgal metabolite production, 45 exhibited antifungal activity, 37 displayed antifungal properties, and 30 showed combined antifungal and antialgal activities. It would be intriguing to explore the link between the lichen's photobiont and the endolichenic fungus that synthesizes antialgal compounds.

The quantity of invertebrates is augmented when secondary compounds are removed from the lichen thallus. These investigations indicate that the lichen thallus's capacity to endure biotic stress may be affected by the resident bacteria and fungi. A comprehensive analysis of the metabolite profiles of several endolichenic fungi is essential to validate their impact on the modification of a lichen's fitness. Over 1000 secondary metabolites have been isolated from lichens inhabiting regions from the Arctic to the tropics, rendering lichens a compelling source of distinctive secondary metabolites.

Secondary metabolites from the acetyl polymalonyl, mevalonic, and shikimate pathways make up between 0.1% and 5% of the dry weight of lichen thallus. Many of these metabolites have a variety of interesting bioactivities for technological applications, and most of them are derived from fungi. These metabolites also serve ecological functions like managing symbiosis. Along with lichens, endolichenic fungi are an important source of novel bioactive chemicals (Kellogg and Raja 2017). They are similar to plant endophytes. It would then be wise to assess endolichenic fungi for their ability to produce new chemicals with unique biological roles.

The interactions between endolichenic fungus and their host lichen, as well as with other species supported by lichens, require more elucidation. The endolichenic fungal community of a lichen must be either resistant to these metabolites or possess the ability to detoxify them, as numerous lichens synthesize antifungal compounds. It may be presumed that alone those fungi constitute the specific endolichenic fungal assemblage of a given lichen. The capacity of endolichenic fungi to decompose resistant compounds and biotransform metabolites into innovative pharmaceuticals can be evaluated (Wang and Dai 2011). The bacterial community of the Alpine lichen *Solorina crocea* alters during infection by a fungus. suggest that the in vivo generation and detoxification of antimicrobial chemicals by endolichenic fungi may result in interspecific competition, thus, changing the mix of species in the bacterial and fungal groups inside the lichen thallus. To comprehend the interactions among the diverse components of the lichen microbiome, it is essential to investigate competition and co-metabolism utilizing contemporary techniques such as isotopologue profiling, considering that endolichenic fungi generate antimicrobial metabolites.

We can learn more about the endolichenic fungi's function in lichen secondary metabolite profile formation and their possible significance within the lichen microbiome if we study them at the gene level. The unique ability of endolichenic fungus to create a wide variety of secondary metabolites, similar to that of plant endophytes, may have evolved in response to selection pressures caused by interactions within the lichen microbiome. The natural product production genes in fungi typically remain inactive when grown in the typical laboratory environment. Genomic mining, proteomic study of secondary metabolism, genetic modification of biosynthetic pathway genes, and the application of epigenetic modifiers are now being developed as ways to awaken these latent secondary metabolite gene clusters. The use of cutting-edge methods to study endolichenic fungi has the potential to yield new secondary metabolites with medicinal applications. The biotic weathering of rocks and biogeochemical processes like nutrient cycling and soil formation are both aided by lichens, which are pioneer colonizers of rocks. Heavy metals and radioactive substances can be accumulated by them as well. Endolichenic fungi have not yet had their exact function in lichen-related processes understood, but fungi are known to aid in rock erosion by producing carbonic and other acids, and they can also tolerate radionuclides and heavy metals (Cordero and Casadevall 2017).

According to, some bacteria that are associated with lichens help break down the components of lichen thalli, which in turn helps move biomass. The outer layers of dead lichen thalli may also be decomposed by endolichenic fungus. One outstanding example of evolutionary mutualism is the lichen community, which provides support for a wide variety of microbes. Lichen thallus microenvironments are the result of intricate interactions between these microorganisms and their environment. In such an environment, bacteria could be a source of biotechnologically relevant compounds.

7 Production of Essential Enzymes for Agriculture

Fungal endophytes play a vital role in the digestion of organic waste and the replenishment of soil nutrients. Fungal pathogens produce various extracellular hydrolases, including cellulase, laccase, pectinase, phosphatase, lipase, xylanase, and proteinase. These enzymes facilitate the evasion of pathogens, while the host organism utilizes its resources. Enzymes transport smaller molecules into cells for metabolic functions and to facilitate symbiotic relationships with the host. The molecules comprise proteins, polysaccharides, sugar-derived polymers, lignin, and organic phosphates. Sunitha et al. identified over 50 endophytic fungal strains capable of producing enzymes including amylase, laccase, cellulase, pectinase, lipase, and protease. He et al. found that endophytic fungus might help with nutrient cycling by breaking down organic substances such lignin, cellulose, and hemicelluloses. The endophytic fungus of the medicinal plant *Cardiospermum halicacabum* can stimulate plant development in adverse environments by secreting a variety of extracellular enzymes, as previously shown by Chathurdevi and Gowrie. Fungal chitinases play a crucial role in breaking down and reusing chitin, a linear homopolymer of β-1,4-N-acetylglucosamine present in the cell walls of fungi, exoskeletons of insects, and shells of crustaceans. It has been demonstrated that a number of fungal endophytes found in Southern Indian plant leaves are capable of producing chitinases. Additionally, it has been discovered that the maize endophyte *Acremonium zeae* produces the extracellular enzyme hemicellulase, which could help transform lignocellulosic biomass into fermentable sugars.

8 Uses of Endophytic Fungi in Biotechnology

Endophytic fungi isolated from diverse plant sources have garnered prominence in recent decades as significant natural product sources for industrial, biomedical, and agricultural applications. The extracellular hydrolase enzymes produced by these fungi—namely, pectinases, cellulases, lipases, amylases, laccases, xylanases, and proteases—serve as defensive measures against infections and facilitate nutritional absorption for the host. Medicinal plants include endophytic fungi that produce hydrolytic enzymes (Khan et al. 2017).

Extracellular enzymes decompose various macromolecules into simpler constituents, such as lignin, proteins, carbohydrates, and sugar-based polymers. The synthesis of these enzymes has been assessed qualitatively and quantitatively by techniques such as advanced spectrophotometric methods and agar plate assays (Yadav et al. 2017). Microbial endophytes have been identified in lower organisms such as mosses and algae, on the xylem, bark, and leaves of grasses and other higher plants.

Several microorganisms, including fungus, find an ecological haven in lichen thalli, which has received little attention. The endolichenic fungus lives in close proximity to the lichens' thalli, much like the endophytic fungi that colonize plants. It has been discovered that lichens host endolichenic fungus as parasymbionts. Lichen thalli are home to a plethora of fungus that produces fruiting bodies with similar fructifications. Lichenicolous fungi are located on the outside of internal structures, while endolichenic fungi are located within their tissues.

Some studies recovered endolichenic fungal strains from the lichen thalli of *Pseudocyphellaria* sp., *Usnea* sp., and *Parmotrema* species, whereas others employed the name "endolichenic" due to their isolation from interior tissues. In many regions, lichens hold economic importance as traditional sources of food and medicine. The lichen thalli of *Pseudocyphellaria* sp., *Parmotrema* sp., and *Usnea* sp. were utilized to identify endolichenic fungal strains that exhibited biological activity, including antifungal properties against prevalent plant diseases. The lack of endolichenic fungi in surface-sterilized lichen thalli was evidenced by the retrieval of these fungi from tissue segments of the lichen thalli.

Environmental contaminants were determined to be the catalyst for the invasion of these fungus. Many studies have focused on endophytic microorganisms linked to higher plants; however, a considerable study vacuum remains in the isolation of endolichenic fungus from lichen thalli. Fruticose lichens, acting as symbionts or pathogens of higher plants and lichen-associated fungi, were also discovered to harbour fungus associated with soil and litter symbiosis or pathogenicity. Vinayaka et al. report over 1000 species and 300 genera of obligatory lichenicolous fungi that engage with lichen species through various biological interactions, including parasitic, saprophytic, parasymbiotic, and obligatory symbiotic relationships. An inquiry was conducted in the forests of Bhadra Wildlife Sanctuary, Southern India, to extract endolichenic fungi from the internal tissues of macrolichens. Eleven species of foliose and fruticose lichens were studied, and 30 different endolichenic fungal species were identified.

It is significant that 24 out of the isolated fungal species developed fructifications. Among the 30 isolates analysed, 24 species were identified, with the majority belonging to the *Ascomycota* phylum. The fungal species associated with the thalli of the lichen include *Chaetomium globosum*, *Cladosporium cladosporioides*, *Colletotrichum dematium*, *Fusarium heterosporum*, *Phoma* sp., and *Xylaria* spp. In the Western Ghats of southern India, 10 lichen species yielded 389 fungal strains from 38 species. Endolichenic fungi were found in all examined lichens, irrespective of their habitat or growth form. The *Ascomycotina* group was used to classify them majority of these fungi. The most common taxa identified in five or more lichen species were Chaetomium sp., *Hypoxylon investiens*, *Nemania bipapillata*, *Nodulisporium* sp., *Paecilomyces* sp., and *Sporormiella intermedia*. Seven of the 28 sporulating fungi belonged to the *Xylariales* family, and they colonized all lichens with a notably higher overall frequency. The study, which was carried out between May 2013 and January 2014, looked at the relationships between endolichenic fungi and 11 lichen species from five families that are located in Bhadra Wildlife Sanctuary in South India. Out of 486 segments, 30 endolichenic taxa were identified. Eight of the 11 lichen species had *Chaetomium globosum*, making it the most prevalent endolichenic fungal species in the screened lichen thalli. The endolichenic fungi showed the highest levels of colonization in *Parmotrema tinctorum* and *Ramalina pacifica*.

Twenty-four of the 30 fungal species that were analysed were able to sporulate and were identified down to the genus and species level. *Fusarium* species, *Xylaria* species, and *Chaetomium globosum* were separated from lichen species' thalli. In Beijing's Baihua Mountain, Li et al. investigated the endophytic fungi linked to seven lichen species from five distinct families. The lichens were found to have a wide variety of endophytic fungi.

Fifteen ascomycetes, sixteen hyphomycetes, one coelomycete, and one yeast species were identified from 488 thallus segments. In six out of seven lichens that were tested, *Scopulariopsis* sp. emerged as the dominating species. The lichen *Dermatocarpon miniatum* was found to be primarily populated by *Geniculosporium serpens*. Only in *Xanthoria mandschurica* was the species *Sporormiella* discovered. Endophytic fungus varies greatly among lichen species and has a high degree of variability across lichen thalli.

9 Prospects for Future Research and Development in the Exploitation of Beneficial Endophytic Fungi

Using endophytes to make crops more resistant to climate change-related abiotic stress is a clear win-win, but there must be a major push to find novel endophyte strains that work with different crops and changing environments. Integrating methods from ecology, genetics, microbiology, metabolomics, and molecular biology with a focus on understanding the interaction mechanisms between endophytic fungi (EFs) and crops under abiotic stress should be the goal of future study in this field. Our aim is to create systems that can quickly identify and isolate strains of EFs that are resistant to or tolerant of stress, and we also want to find ways to understand the molecular mechanisms by which EFs work to help people.

The genetic and enzymatic processes that endophytic fungi (EFs) use to survive harsh environments and form mutually beneficial partnerships with plants must be defined. Isolating, growing, and storing EF strains calls for microbiological methods to accomplish this aim. It is critical to create EF collections that are available worldwide, particularly for disadvantaged groups, because they have the ability to mitigate some of the impacts of climate change on food production, food security, and the decrease of hunger.

To improve understanding of the cellular interactions between endophytic fungi (EFs) and their host plants, advanced microscopic techniques, such as Raman/IR microscopy and photon microscopy, should be utilized. The use of molecular tools, including DNA sequencing, metagenomics, transcriptomics, and proteomics, is crucial for a thorough assessment of the evolution, phylogeny, genetic and functional characteristics, as well as gene expression and protein profiles of beneficial EFs. These approaches can be applied to study the genetic, functional, physiological, and systemic responses of plants in symbiotic relationships with EFs, emphasizing improvements in plant fitness, growth, crop yield, stress tolerance, and resilience under diverse environmental conditions.

Moreover, standardized protocols for functional assays, including in vitro analyses, greenhouse experiments, and field studies, must be established to investigate specific interactions between endophytes and plants, such as the production of plant-growth-promoting compounds, pathogen biocontrol effectiveness, nutrient exchange, and resilience to biotic and abiotic stresses. These techniques will be implemented to create a global repository of effective strains with validated attributes to enhance plant growth and agricultural productivity, employing data on the crop, environmental factors, and the specific impacts of each strain.

Genetic engineering and bioinformatics are further methods to improve the interaction between EFs and crops under abiotic stress. By analysing omics data using bioinformatic techniques, one can identify the genes, metabolic pathways, and regulatory networks that promote beneficial interactions between EFs and crops, as well as the molecular mechanisms that allow EFs to withstand unfavourable environmental conditions. Furthermore, bioinformatics can enhance the prediction of potential benefits and ecological roles of diverse EF strains by the application of genomic, transcriptomic, proteomic, and metabolomic data.

In contrast, specific genes or metabolic pathways of the plant and fungus may be modified by genetic engineering methods such as RNA interference, gene knock-outs, or gene overexpression to examine their roles in the symbiosis. Moreover, the generation of fungal mutants may clarify the functions of certain genes related to colonization, nutrient exchange, and the synthesis of bioactive compounds.

Therefore, a multidisciplinary strategy is necessary to maximize the benefits of using EFs in alleviating abiotic stresses in crops. This method should address the issue from multiple perspectives in order to achieve short-term advantages with reachable and quantifiable targets.

References

Arnold AE, Henk DA, Eells RL, Lutzoni F, Vilgalys R (2007) Diversity and phylogenetic affinities of foliar fungal endophytes in loblolly pine inferred by culturing and environmental PCR. Mycologia 99:185–206. https://doi.org/10.1080/15572536.2007.11832578

Arnold AE, Mejía LC, Kyllo D, Rojas EI, Maynard Z, Robbins N, Herre EA (2003) Fungal endophytes limit pathogen damage in a tropical tree. Proc Natl Acad Sci U.S.A. 100:15649–15654.https://doi.org/10.1073/pnas.2533483100

Duong LM, Jeewon R, Lumyong S and Hyde KD (2006) DGGE coupled with ribosomal DNA gene phylogenies reveal uncharacterized fungal phylotypes. Fungal Divers

Ghimire SR, Charlton ND, Bell JD, Krishnamurthy YL, Craven KD (2011) Biodiversity of fungal endophyte communities inhabiting switchgrass (Panicum virgatum L.) growing in the native tallgrass prairie of northern Oklahoma. Fungal Divers 47:19–27. https://doi.org/10.1007/s13225-010-0085-6

González V, Tello ML (2011) The endophytic mycota associated with Vitis vinifera in central Spain. Fungal Divers 47:29–42. https://doi.org/10.1007/s13225-010-0073-x

Müller MM, Valjakka R, Suokko A, Hantula J (2001) Diversity of endophytic fungi of single Norway spruce needles and their role as pioneer decomposers. Molecular Ecology 10:1801–1810. https://doi.org/10.1046/j.1365-294X.2001.01304.x

Aly AH, Debbab A, Kjer J, Proksch P (2010) Fungal endophytes from higher plants: a prolific source of phytochemicals and other bioactive natural products. Fungal Divers 41:1–16. https://doi.org/10.1007/s13225-010-0034-4

Arnold AE, Herre EA (2003) Canopy cover and leaf age affect colonization by tropical fungal endophytes: ecological pattern and process in Theobroma cacao (Malvaceae). Mycologia 95: 388–398. https://doi.org/10.1080/15572536.2004.11833083

Arnold AE, Lutzoni F (2007) Diversity and host range of foliar fungal endophytes: are tropical leaves biodiversity hotspots? Ecology 88:541–549. https://doi.org/10.1890/05-1459

Bacon CW, White JF (2000) Microbial endophytes. Dekker, New York, pp 341–388

Cordero RJB, Casadevall A (2017) Functions of fungal melanin beyond virulence. Fungal Biol Rev 31:99–112. https://doi.org/10.1016/j.fbr.2016.12.003

Diedhiou PM, Hallmann J, Oerke E-C, Dehne H-W (2003) Effects of arbuscular mycorrhizal fungi and a non-pathogenic fusarium oxysporum on Meloidogyne incognita infestation of tomato. Mycorrhiza 13:199–204. https://doi.org/10.1007/s00572-002-0215-4

Davey ML, Heegaard E, Halvorsen R, Kauserud H, Ohlson M (2013) Amplicon-pyrosequencing-based detection of compositional shifts in bryophyte-associated fungal communities along an elevation gradient. Mol Ecol 22:368–383. https://doi.org/10.1111/mec.12122

Fisher PJ, Petrini O, Petrini LE, Sutton BC (1994) Fungal endophytes from the leaves and twigs of Quercus ilex L. from England, Majorca and Switzerland. New Phytol 127:133–137. https://doi.org/10.1111/j.1469-8137.1994.tb04267.x

Galani YJH, Hansen EMØ, Droutsas I, Holmes M, Challinor AJ, Mikkelsen TN, Orfila C (2022) Effects of combined abiotic stresses on nutrient content of European wheat and implications for nutritional security under climate change. Sci Rep 12:5700. https://doi.org/10.1038/s41598-022-09538-6

Gunatilaka AAL (2006) Natural Products from Plant-Associated Microorganisms: Distribution, Structural Diversity, Bioactivity, and Implications of Their Occurrence J Nat Prod 69:509–526. https://doi.org/10.1021/np058128n

Guo B, Wang Y, Sun X, Tang K (2008) Bioactive natural products from endophytes: A review Appl Biochem Microbiol 44:136–142. https://doi.org/10.1134/S0003683808020026

Gamboa MA, Laureano S, Bayman P (2003) Measuring diversity of endophytic fungi in leaf fragments: Does size matter? Mycopathol 156:41–45. https://doi.org/10.1023/A:1021362217723

Green JL, Bohannan BJM, Whitaker RJ (2008) Microbial Biogeography: From Taxonomy to Traits. Sci 320:1039–1043. https://doi.org/10.1126/science.1153475

Hallmann J, Berg G, Schulz B (2006) Isolation procedures for endophytic microorganisms. In: Schulz BJE, Boyle CJC, Sieber TN (eds) Microbial root endophytes, soil biology. Springer, Berlin Heidelberg, Berlin, Heidelberg, pp 299–319. https://doi.org/10.1007/3-540-33526-9_17

Huang Y-L, Zimmerman N, Arnold A (2018) Observations on the early establishment of foliar endophytic fungi in leaf discs and living leaves of a model Woody angiosperm, Populus trichocarpa (Salicaceae). JoF 4:58. https://doi.org/10.3390/jof4020058

Hartley SE, Gange AC (2009) Impacts of Plant Symbiotic Fungi on Insect Herbivores: Mutualism in a Multitrophic Context Annu Rev Entomol 54:323–342. https://doi.org/10.1146/annurev.ento.54.110807.090614

Higgins KL, Arnold AE, Miadlikowska J, Sarvate SD, Lutzoni F (2007) Phylogenetic relationships, host affinity, and geographic structure of boreal and arctic endophytes from three major plant lineages. Mol Phylogenetics Evol 42:543–555. https://doi.org/10.1016/j.ympev.2006.07.012

Hyde KD, Soytong K (2008) The fungal endophyte dilemma. Fungal Divers 33:163–173

Kellogg JJ, Raja HA (2017) Endolichenic fungi: a new source of rich bioactive secondary metabolites on the horizon. Phytochem Rev 16:271–293. https://doi.org/10.1007/s11101-016-9473-1

Khan AL, Shahzad R, Al-Harrasi A, Lee I-J (2017) Endophytic microbes: a resource for producing extracellular enzymes. In: Maheshwari DK, Annapurna K (eds) Endophytes: crop productivity and protection, vol 2. Springer International Publishing, Cham, pp 95–110. https://doi.org/10.1007/978-3-319-66544-3_5

Kharwar RN, Verma SK, Mishra A, Gond SK, Sharma VK, Afreen T, Kumar A (2011) Assessment of diversity, distribution and antibacterial activity of endophytic fungi isolated from a medicinal plant Adenocalymma alliaceum Miers. Symbiosis 55:39–46. https://doi.org/10.1007/s13199-011-0142-2

Lata R, Chowdhury S, Gond SK, White JF (2018) Induction of abiotic stress tolerance in plants by endophytic microbes. Lett Appl Microbiol 66:268–276. https://doi.org/10.1111/lam.12855

Li H-Y, Shen M, Zhou Z-P, Li T, Wei Y, Lin L (2012a) Diversity and cold adaptation of endophytic fungi from five dominant plant species collected from the Baima Snow Mountain, Southwest China. Fungal Divers 54:79–86. https://doi.org/10.1007/s13225-012-0153-1

Li H-Y, Wei D-Q, Shen M, Zhou Z-P (2012b) Endophytes and their role in phytoremediation. Fungal Divers 54:11–18. https://doi.org/10.1007/s13225-012-0165-x

Lau MK, Arnold AE, Johnson NC (2013) Factors influencing communities of foliar fungal endophytes in riparian woody plants. Fungal Ecol 6:365–378. https://doi.org/10.1016/j.funeco.2013.06.003

Ma Y, Rajkumar M, Zhang C, Freitas H (2016) Beneficial role of bacterial endophytes in heavy metal phytoremediation. J Environ Manag 174:14–25. https://doi.org/10.1016/j.jenvman.2016.02.047

Márquez LM, Redman RS, Rodriguez RJ, Roossinck MJ (2007) A virus in a fungus in a plant: three-way Symbiosis required for thermal tolerance. Science 315:513–515. https://doi.org/10.1126/science.1136237

Mei L, Zhu M, Zhang D-Z, Wang Y-Z, Guo J, Zhang H-B (2014) Geographical and temporal changes of foliar fungal endophytes associated with the invasive plant Ageratina adenophora. Microb Ecol 67:402–409. https://doi.org/10.1007/s00248-013-0319-8

Mishra A, Gond SK, Kumar A, Sharma VK, Verma SK, Kharwar RN, Sieber TN (2012) Season and tissue type affect fungal endophyte communities of the Indian medicinal plant Tinospora cordifolia more strongly than geographic location. Microb Ecol 64:388–398. https://doi.org/10.1007/s00248-012-0029-7

Pancher M, Ceol M, Corneo PE, Longa CMO, Yousaf S, Pertot I, Campisano A (2012) Fungal endophytic communities in grapevines (Vitis vinifera L.) respond to crop management. Appl Environ Microbiol 78:4308–4317. https://doi.org/10.1128/AEM.07655-11

Panjiar N, Mishra S, Yadav AN, Verma P (2017) Functional foods from cyanobacteria: an emerging source for functional food products of pharmaceutical importance. In: Gupta VK et al (eds) . Wiley, Microbial functional foods and nutraceuticals, pp 21–37. https://doi.org/10.1002/9781119048961.ch2

Petrini O (1991) Fungal endophytes of tree leaves. In: Andrews JH, Hirano SS (eds) Microbial ecology of leaves, brock/springer series in contemporary bioscience. Springer, New York, pp 179–197. https://doi.org/10.1007/978-1-4612-3168-4_9

Petrini O, Stone J, Carroll FE (1982) Endophytic fungi in evergreen shrubs in western Oregon: A preliminary study. Can J Bot 60:789–796. https://doi.org/10.1139/b82-102

Qin S, Xing K, Jiang J-H, Xu L-H, Li W-J (2011) Biodiversity, bioactive natural products and biotechnological potential of plant-associated endophytic actinobacteria. Appl Microbiol Biotechnol 89:457–473. https://doi.org/10.1007/s00253-010-2923-6

Qin J-C, Zhang Y-M, Gao J-M, Bai M-S, Yang S-X, Laatsch H, Zhang A-L (2009) Bioactive metabolites produced by Chaetomium globosum, an endophytic fungus isolated from Ginkgo biloba. Bioorganic & Med Chem Lett 19:1572–1574. https://doi.org/10.1016/j.bmcl.2009.02.025

Redman RS, Kim YO, Cho S, Mercer M, Rienstra M, Manglona R, et al (2021) A symbiotic approach to generating stress tolerant crops. Microorg 9:920. https://doi.org/10.3390/microorganisms9050920

Ripa FA, Cao W, Tong S, Sun J (2019) Assessment of plant growth promoting and abiotic stress tolerance properties of wheat endophytic fungi. Biomed Res Int 2019:1–12. https://doi.org/10.1155/2019/6105865

Rivera-Orduña FN, Suarez-Sanchez RA, Flores-Bustamante ZR, Gracida-Rodriguez JN, Flores-Cotera LB (2011) Diversity of endophytic fungi of Taxus globosa (Mexican yew). Fungal Divers 47:65–74. https://doi.org/10.1007/s13225-010-0045-1

Rodriguez R, Redman R (2008) More than 400 million years of evolution and some plants still can't make it on their own: plant stress tolerance via fungal symbiosis. J Exp Bot 59:1109–1114. https://doi.org/10.1093/jxb/erm342

Shinozaki K, Yamaguchi-Shinozaki K (2006) Gene networks involved in drought stress response and tolerance. J Exp Bot 58:221–227. https://doi.org/10.1093/jxb/erl164

Sieber TN (2007) Endophytic fungi in forest trees: are they mutualists? Fungal Biol Rev 21:75–89. https://doi.org/10.1016/j.fbr.2007.05.004

Sieber VT, Hugentobler C (1987) Endophytische Pilze in Blättern und Ästen gesunder und geschädigter Buchen (Fagus sylvatica L.). European J For Pathol 17:411–425. https://doi.org/10.1111/j.1439-0329.1987.tb01119.x

Strobel G, Daisy B (2003) Bioprospecting for microbial endophytes and their natural products. Microbiol Mol Biol Rev 67:491–502. https://doi.org/10.1128/MMBR.67.4.491-502.2003

Sun X, Guo L-D, Hyde KD (2011a) Community composition of endophytic fungi in Acer truncatum and their role in decomposition. Fungal Divers 47:85–95. https://doi.org/10.1007/s13225-010-0086-5

Sun Y, Luo H, Li Y, Sun C, Song J, Niu Y, Zhu Y, Dong L, Lv A, Tramontano E, Chen S (2011b) Pyrosequencing of the Camptotheca acuminata transcriptome reveals putative genes involved in camptothecin biosynthesis and transport. BMC Genomics 12:533. https://doi.org/10.1186/1471-2164-12-533

Sun X, Guo L-D, Hyde KD (2011c) Community composition of endophytic fungi in Acer truncatum and their role in decomposition. Fungal Divers 47:85–95. https://doi.org/10.1007/s13225-010-0086-5.

Tang J, Xu L, Chen X, Hu S (2009) Interaction between C4 barnyard grass and C3 upland rice under elevated CO2: impact of mycorrhizae. Acta Oecol 35:227–235. https://doi.org/10.1016/j.actao.2008.10.005

Tejesvi MV, Kajula M, Mattila S, Pirttilä AM (2011) Bioactivity and genetic diversity of endophytic fungi in Rhododendron tomentosum Harmaja. Fungal Divers 47:97–107. https://doi.org/10.1007/s13225-010-0087-4

Unterseher M, Schnittler M (2009) Dilution-to-extinction cultivation of leaf-inhabiting endophytic fungi in beech (Fagus sylvatica L.) – Different cultivation techniques influence fungal biodiversity assessment. Mycol Res 113:645–654. https://doi.org/10.1016/j.mycres.2009.02.002

Vandenkoornhuyse P, Quaiser A, Duhamel M, Le Van A, Dufresne A (2015) The importance of the microbiome of the plant holobiont. New Phytol 206:1196–1206. https://doi.org/10.1111/nph.13312

Verma A, Shameem N, Jatav HS, Sathyanarayana E, Parray JA, Poczai P, Sayyed RZ (2022) Fungal endophytes to combat biotic and abiotic stresses for climate-smart and sustainable agriculture. Front Plant Sci 13. https://doi.org/10.3389/fpls.2022.953836

Vega FE, Simpkins A, Aime MC, Posada F, Peterson SW, Rehner SA, Infante F, Castillo A, Arnold AE (2010) Fungal endophyte diversity in coffee plants from Colombia, Hawai'i, Mexico and Puerto Rico. Fungal Ecol 3:122–138. https://doi.org/10.1016/j.funeco.2009.07.002

Victoria NM, Collantes M, Cabral D (2007) Environmental effects on grass-endophyte associations in the harsh conditions of south Patagonia: Environment affects endophytes incidence in native grasses. FEMS Microbiol Ecol 61:164–173. https://doi.org/10.1111/j.1574-6941.2007.00319.x

Wu H, Yang HY, You XL, Li YH (2013) Diversity of endophytic fungi from roots of Panax ginseng and their saponin yield capacities. SpringerPlus 2:1–9

Wang Y, Dai C-C (2011) Endophytes: a potential resource for biosynthesis, biotransformation, and biodegradation. Ann Microbiol 61:207–215. https://doi.org/10.1007/s13213-010-0120-6

Xiong Z-Q, Yang Y-Y, Zhao N, Wang Y (2013) Diversity of endophytic fungi and screening of fungal paclitaxel producer from Anglojap yew, Taxus x media. BMC Microbiol 13:71. https://doi.org/10.1186/1471-2180-13-71

Yadav A (2017) Exploring the potential of endophytes in agriculture: a Minireview. APAR 6. https://doi.org/10.15406/apar.2017.06.00221

Zimmerman NB, Vitousek PM (2012) Fungal endophyte communities reflect environmental structuring across a Hawaiian landscape. Proc Natl Acad Sci 109:13022–13027. https://doi.org/10.1073/pnas.1209872109

Zhang HW, Song YC, Tan RX (2006) Biology and chemistry of endophytes Nat Prod Rep 23:753. https://doi.org/10.1039/b609472b

Zheng YK, Chen C, Ren D, Gu YF (2013) Genetic diversity of the plant endophytic fungi in a Pb-Zn mine area in Hanyuan, Sichuan Province, China. J Sichuan Agric Univ 31(3):308–313

Diversity and Distribution Patterns of Endolichenic Fungi

Sana, Palvisha Javeed, and Munazza Kiran

Contents

1 Introduction

Lichens, as the pioneer organisms, hold dominant position in the initial developmental stages of a variety of ecosystems experiencing primary or secondary succession. Lichens are not only important components of present-day ecosystems particularly in Antarctic and Arctic regions, rock, arid, and coastal regions but are common in the tropics. Lichens are regarded as an intricate miniature ecosystem

Sana · P. Javeed · M. Kiran (✉)
Department of Botany, Division of Science & Technology, University of Education, Lahore, Pakistan
e-mail: munazza.kiran@ue.edu.pk

A. M. Abdel-Azeem, F. Degola (eds.), *Advances in Lichens and Non-Flowering Plants Fungal Endophytes Research*, Fungal Biology,
https://doi.org/10.1007/978-3-032-01506-8_3

(Honegger 1991) due to their close interactions with various organisms such as fungi, algae, cyanobacteria, eubacteria, and micro-invertebrates. Primary association usually involves a fungus (ascomycetes or basidiomycetes) and a photoautotroph (*Chlorophyta* or cyanobacteria) to form lichen thallus, others make secondary, obligate, or opportunistic associations. The lichen thallus, which resembles plant tissue, serves as an interesting ecological habitat for numerous bacteria (Cardinale et al. 2008; Hodkinson and Lutzoni 2009; Bates et al. 2011; Grube et al. 2015) and eukaryotes (e.g., fungi, arthropods, and nematodes) (Bates et al. 2011; Park et al. 2014). In asymptomatic thalli of lichens, fungal symbionts resembling endophytes (i.e., endolichenic fungi) appear ubiquitous (Suryanarayanan et al. 2005; Arnold et al. 2009; Ding et al. 2009; U'Ren et al. 2010), wherein they associate primarily with photobionts (Arnold et al. 2009).

Recently, there has been a debate regarding yeast as an obligate, third partner in lichen formation (Spribille et al. 2016); however, some studies support this idea, while others do not agree to it (Mark et al. 2020). Due to the increasing uncertainty over the presence of yeast and other microorganisms, the lichens are redefined as "a self-sustaining ecosystem formed through the interaction of an exhabitant fungus and an extracellular arrangement containing one or more photosynthetic partners and an unspecified number of other microscopic organisms" (Hawksworth and Grube 2020). Among the secondary associates, the lichenicolous and endolichenic fungi (ELF) are interesting groups particularly, due to their taxonomic and evolutionary significance (Oh et al. 2020), as well as with a potential for bioprospecting (Maduranga et al. 2018). Lichens possess many extraordinary traits including longevity, high diversity, productivity of secondary metabolites, tolerance to extreme environments, and sensitivity toward pollution. Reports of lichen growth rates range from less than a millimeter to a few centimeters per year, and their longevity can span decades or centuries, up to millennia, depending on the species and the environmental conditions. Over 700 secondary metabolites have been isolated from lichens, with the majority (>90%) that are unique to this symbiotic relationship, and many of these possess valuable medicinal properties. Fungal partners can secrete these metabolites in such a large amount that they can account for more than 50% of the dry weight of some lichens (Miadlikowska et al. 2004).

2 Endolichenic Fungi

Endolichenic fungi are an emerging category of endosymbiotic microorganisms and comprise of fungal strains that inhabit asymptomatic lichen thalli, much like endophytic fungi live within healthy plant tissues. They are predominantly filamentous; they reside inside the lichen thallus without showing any noticeable symptoms of their presence without causing any harm to the host. The ELF establish a close association with the photobiont of the lichen for their nutrition (Arnold et al. 2009). They share similarities with endophytic fungi of vascular plants. However, it remains uncertain whether there is any role of ELF in increasing the fitness of lichens or not (Suryanarayanan et al. 2017). Endolichenic fungi are species that are non-mycobiont living in the lichen thallus. They are distinct from lichenicolous fungi in terms that they do not produce any visible symptoms. The relationship between ELF and lichens is unclear, and it is uncertain whether they have any kind of intimate association. The ELF share many similarities with endophytic fungi of vascular plants as they inhabit interior of the lichens, causing no visible disease symptoms and are horizontally transmitted (Arnold et al. 2009; Kannangara et al. 2009; U'Ren et al. 2012). Moreover, similar to endophytic fungi, they are capable of producing secondary metabolites including alkaloids, quinones, furanones, pyrones, benzopyranoids, xanthones, terpenes, steroids, peptides, and alicyclic compounds (Paranagama et al. 2007; He et al. 2012; Yang et al. 2012, Li et al. 2015; Samanthi et al. 2015; Kellogg and Raja 2016; Yuan et al. 2016). These metabolites display a range of unique bioactivities such as antibacterial, antifungal, cytotoxic, and antioxidant properties (Kellogg and Raja 2016; Suryanarayanan et al. 2017). Research on ELF has revealed that their presence in lichens is found across diverse environments, from the arctic to the tropics (Arnold et al. 2009; Suryanarayanan et al. 2005, 2017; Li et al. 2007; Kannangara et al. 2009; Tripathi and Joshi 2015). However, knowledge of these fungi remains limited because only a small fraction of the estimated 18,500 lichen species (Nash 2008) has been investigated for their ELF communities (Tripathi et al. 2014a). Most of the existing literature on ELF primarily focuses on their ability to produce unique secondary metabolites with some biological activity (Kellogg and Raja 2016). To gain deeper understanding of biology of ELF, it is essential to explore their role in the establishing and sustaining the lichen symbiosis and their interactions with lichen microbiome. Moreover, comprehensive studies of endolichenic fungi require employment of molecular methods.

3 Similarities and Differences Between Endolichenic Fungi and Endophytes

Endolichenic fungi live in close association with algal photobionts within asymptomatic lichen thalli and share many similarities in terms of taxonomy, diversity, transmission mode, and evolutionary history with fungal endophytes of plants. These similarities have raised questions regarding the distinctiveness of ELF compared with endophytes (U'Ren et al. 2010). Despite these similarities, endophytic and endolichenic communities differ significantly from one another. Previous studies have observed variations in incidence, diversity, and composition among biogeographic regions and taxonomically related host differences (Fisher et al. 1995; Arnold et al. 2003; Arnold 2007; Arnold and Lutzoni 2007; Hoffman and Arnold 2008). Although a latitudinal gradient of isolation frequency and diversity was reported for ELF, similar patterns had not been studied before (Arnold et al. 2009). Endolichenic diversity increases linearly with the length of the growing season, but it is not influenced by annual precipitation. In the temperate zone, higher abundance and diversity of ELF are likely due to the longer exposure of lichen thalli to spore rain relative to the shorter lifetimes of leaves (some thalli can live up to 100 years) (Arnold et al. 2009). The networks of endophytic fungi are less nested, less connected, and more modular than endolichenic networks. Relative to endolichenic networks, endophytic networks have consistently more asymmetric interactions with host exerting stronger effect on their symbiont and vice versa. Endophytic fungi exhibit particularly high diversity in earlier-occurring plant lineages such as mosses and lycophytes, especially when compared to angiosperms. In contrast, endolichenic diversity does not show significant variation based on mycobiont order. While the ELF have been extensively studied for their chemical properties, a number of times, only a limited number of ELF have thus far been investigated for their secondary metabolites. However, studies have demonstrated that ELF is also valuable source of structurally diverse and biologically active small-molecule metabolites (Arnold et al. 2009).

4 Origin and Diversity of ELF

The origin of ELF is not well-known as we know about the endophytic fungi. The phylogenetic determination of trophic transition of ascomycetes reported by Arnold et al. (2009) somehow revealed the origins of endophytism and evolution of ELF. They used the ancestral state reconstruction method and explored that the ELF played a vital role in the evolution of endophytism within *Ascomycota*. Their results also declare that ELF show a rapid evolutionary rate for the transition of fungi into endophytic associations in plants. They also reported that the ELF are localized inside the lichen thalli.

Fossilized lichen thalli, from as early as the lower Devonian period, had revealed the evidence of endolichenic fungal association in them (Honegger et al. 2013). This finding suggests that survival within lichen structures has been a successful one over a long evolutionary time span. It's noteworthy that ELF belong to the same major *Ascomycota* lineages as endophytic fungi of plants (Arnold et al. 2009). These ELF are distinct from the other fungal partners associated with lichen (Lutzoni and Miadlikowska 2009) and the lichenicolous fungi that inhabit lichens (Arnold et al. 2009). Molecular research conducted by U'Ren et al. (2010) confirmed that ELF represent a distinct ecological group and they are not merely accidental colonizers of the lichen structures. Endolichenic fungi have been reported in a wide range of lichen types encompassing various growth forms (e.g., crustose, foliose, fruticose), substrates (e.g., epiphytic, saxicolous, terricolous), and ecosystems (e.g., boreal, temperate, tropical, and Antarctic) (Bates et al. 2011; Zhang et al. 2016).

The ELF have global distribution and are known from every lichen species spanning from the arctic to tropics. It is estimated that there could be approximately 13,500 species of ELF to occur in the world (Kellogg and Raja 2016). The diversity of ELF is evident in various studies. For instance, Petrini et al. (1990) isolated 506 fungal taxa from 17 lichen species; of these, 306 were isolated only once. An intensive study of two lichens *Peltigera praetextata* (Florke ex Sommerf.) Zopf and *Xanthoparmelia taractica* (Kremp.) hale from a coniferous forest (Aosta valley, Italy) yielded different fungi belonging to 117 taxa (Girlanda et al. 1997). Suryanarayanan et al. (2005) isolated 242 ELF from 500 thalli segments of five corticolous lichens. Li et al. (2007) isolated a total of 32 taxa from 488 thalli segments, with 31 ascomycetes in 1 year. In another study, Zhang et al. (2016) recovered 247 fungal isolates from 2712 thalli segments of 17 lichen species. Govinda Rajulu et al. (2020) isolated a total of 389 species from 10 lichen species of Western Ghats, Southern India. These reports underscore the immense diversity of ELF. Rashmi et al. (2019) provided a comprehensive list of endophytic fungi reported from the world; however, it included very few ELF (e.g., *Nodulisporium hyalosporum* S.C. Agarwal & J.K. Misra). The compilation of all the ELF studies carried out so far indicated that about 500 ELF have been isolated till date. Among them only 135 were identified up to species level with the majority remaining identified up to the genus or class level or treated as unidentified. It is difficult to identify them under routine microscopic studies as the ELF do not produce sufficient morphological characters such as coloration, conidia, or spores in culture (Tripathi and Joshi 2019). While DNA-based molecular techniques offer some assistance, their effectiveness is hampered by the lack of sequences in the GeneBank or library related to ELF. Usually, researchers prioritize the isolation of ELF for bioactive molecules often giving less attention toward their taxonomy. Our understanding of the species diversity of ELF remains quite limited, although a pyro-sequencing study showed that numerous fungi are present in lichen thalli (Bates et al. 2011). A comprehensive analysis of fungal species inhabiting alpine rock lichens using multiple genetic markers uncovered the presence of previously undiscovered fungal strains within the Chaethothyriomycetes and Dothideomycetes lineages (Muggia et al. 2016). More recently, the application of molecular techniques has led to the discovery of 11 new species of fungi that form lichen (Leavitt et al. 2016). However, some recent studies have indicated that ELF are not different from endophytes and the DNA-based techniques are able to unravel their identity (Govinda Rajulu et al. 2020).

5 Taxonomic Distribution

Endolichenic species may belong to *Ascomycota*, *Basidiomycota*, and *Mucoromycota*, but most commonly they belong to *Ascomycota* (Arnold et al. 2009; U'Ren et al. 2010; Zhang et al. 2016; Suryanarayanan et al. 2017; Oh et al. 2020). At the class level, some studies show that ELF are abundant in *Sordariomycetes* (in France, North America, China, Japan, Sri Lanka, Norway, and South Korea) (Li et al. 2007; Arnold et al. 2009; Zhang et al. 2016; Lagarde et al. 2018; Maduranga et al. 2018; Masumoto and Degawa 2019; Oh et al. 2020). Some studies reveal the dominance of *Dothideomycetes*, *Leotiomycetes*, or *Pezizomycetes* (in Germany, the USA, Antarctica, and Italy) (Petrini et al. 1990;

U'Ren et al. 2010; Girlanda et al. 1997, 2019). However, there is no specific pattern of ELF dominance reported depending on the continent, country, or climate zone (Oh et al. 2020). The ELF discovered till now belongs to the total three phyla, 32 orders, 70 families, and 112 genera (Wijayawardene et al. 2020). Among 112 genera the most common are *Alternaria* Nees, *Aspergillus* P. Micheli, *Chaetomium* Kunze, *Cladosporium* Link, *Curvularia* Boedijn, *Fusarium* Link, *Penicillium* Link, *Trichoderma* Pers., and *Xylaria* Hill ex Schrank while *Alternaria alternata* (Fr.) Keissl., *Aspergillus flavus* Link, *Aspergillus niger* Tiegh., *Neocosmospora solani* (Mart.) L. Lombard and Crous, and *Trichoderma harzianum* Rifai are the most frequently occurring ELF species.

The taxonomic status of some of the important genera is as follows:

Phylum: *Ascomycota*, **Class:** *Dothideomycetes*, **Order:** *Pleosporales*, **Family:** *Pleosporaceae*, **Genus:** *Alternaria*, *Curvularia*.
Phylum: *Ascomycota*, **Class:** *Dothideomycetes*, **Order:** *Capnodiales*, **Family:** *Cladosporiaceae*, **Genus:** *Cladosporium*.
Phylum: *Ascomycota*, **Class:** *Dothideomycetes*, **Order:** *Eurotiales*, **Family:** *Aspergillaceae*, **Genus:** *Aspergillus*, *Penicillium*.
Phylum: *Ascomycota*, **Class:** *Sordariomycetes*, **Order:** *Hypocreales*, **Family:** *Nectriaceae*, **Genus:** *Fusarium*.
Phylum: *Ascomycota*, **Class:** *Sordariomycetes*, **Order:** *Hypocreales*, **Family:** *Hypocreaceae*, **Genus:** *Trichoderma*.
Phylum: *Ascomycota*, **Class:** *Sordariomycetes*, **Order:** *Xylariales*, **Family:** Xylariaceae, **Genus:** *Xylaria*.
Phylum: *Ascomycota*, **Class:** *Sordariomycetes*, **Order:** *Sordariales*, **Family:** Chaetomiaceae, **Genus:** *Chaetomium*.

6 Factors Affecting the Distribution of ELF

Ecoregions, photobiont type, and host taxonomy impart great influence on the distribution and community structures of ELF. For example, the genera belonging to *Peltigerales* (e.g., *Collema* P. Browne, *Leptogium* (Ach.) Gray, and *Peltigera* Wild.) are found abundantly in ecoregions with high elevation and precipitation, while most genera in *Lecanorales* (e.g., *Cladonia* P. Browne, *Myelochroa* (Asahina) Elix and Hale, *Parmotrema* A. Massal., and *Stereocaulon* (Schreb.) Schrad. are generally not common in such ecoregions. *Cladonia* and *Stereocaulon* are abundant in ecoregions with moderate environmental conditions except for high isothermality and annual range of temperature, while *Parmotrema* and *Physcia* (Schreb.) Michx. are mostly found in the ecoregions that had low elevation with high mean temperature. The photobiont type also significantly influence the ELF richness and communities' composition (Oh et al. 2020) as ELF have to extract nutrients from photobiont (Suryanarayanan and Thirunavukkarasu 2017). The lichens having two types of photobionts show low ELF richness. *Peltigerales* and *Stereocaulon* are distinct from other lichens regarding the type of photobiont. The cyanobacteria act as the primary photobiont of most *Peltigerales* species. On the other hand, *Stereocaulon* have two types of photobionts: green algae as primary photobiont while cyanobacteria as secondary photobiont (Oh et al. 2020). Moreover, the ELF community structure also changes by changing the type of photobiont. Mostly, ELF species have a saprotrophic mode, but it is not active within the thallus of lichens which might be due to the taxonomic resemblance between ELF and endophytic fungi as endophytic fungus also requires suitable environment to start the process of decomposition (Oh et al. 2020). Some ELF also behave as coprophilus fungi and resides in the feces like *Preussia* Fuckel and *Sordaria* Ces. and De Not. (Richardson 2004; Newcombe et al. 2016). Feces provide nitrogen enriched environment to the lichens that have cyanobacteria as photobiont (Rai and Rai 2018). Some ELF species, i.e., *Sordaria*, *Daldinia* Ces. and De Not. and *Xylaria*, produce melanin (Engh et al. 2007; Stadler et al. 2014; Lagashetti et al. 2019) and are found in harsh environments as melanin has protective role against harsh environmental conditions and protect them against UV-rays, high temperature, high pH, and extreme drought conditions (Gostin^car et al. 2012). Therefore, the melanin-producing potential of ELF can serve as adaptation for survival of lichens in extreme environmental conditions.

7 Effect of Thallus Size on ELF Diversity

The diversity of ELF community is largely affected by the size of thallus fragment. The smaller fragments of thalli have more ELF diversity followed by medium and larger fragments. Even different parts of a thallus, having different age, also show significant differences in ELF diversity. The smaller-sized thallus has more ELF diversity in the middle and central part as compared to the young marginal part because the oldest lichen thallus contains not only early lichenized fungal species but also the newly colonized fungi. Some species taxa are even also confined to the specific thallus part, i.e., *Phlebia acerina* Peck and *Cercophora caudata* (Sacc.) N. Lundq. can be isolated from the middle part of the lichen thallus. *Hypoxylon howeanum* Peck and *Rosellinia limonispora* Ellis and Everh. can be detected from the central part while *Virgaria nigra* (Link) Nees from the marginal part of the thallus. Lichens do not possess any cuticle layer; therefore lichens possess open systems to take up water from their surfaces. This could be one of the reasons that most lichen inhabiting fungi are not restricted to any specific part of the lichen thallus. Moreover, ELF diversity is also strongly influenced by individual lichens collected at different sites (Arnold et al. 2009).

8 Host Specificity

Endolichenic species generally do not show host specificity. *Peltigera praetextata* (Flörke ex Sommerf.) Zopf, *Xanthoparmelia taractica* (Kremp.) Hale, *Bulbothrix meizospora* (Nyl.) Hale., *Heterodermia flabellata* (Fée) D.D. Awasthi, *Heterodermia hypochraea* (Vain.) Swinscow and Krog, *Parmotrema crinitum* (Ach.) M. Choisy, *Parmotrema praesorediosum* (Nyl.) Hale, *Parmotrema nilgherrense* (Nyl.) Hale, *Parmotrema reticulatum* (Taylor) M. Choisy, *Parmotrema grayanum* (Hue) Hale, *Parmotrema thomsonii* (Stirt.) A. Crespo, Divakar and Elix, *Parmotrema tinctorum* (Dèspr. ex Nyl.) Hale, *Physcia dilatata* Nyl., *Usnea aciculifera* Vain, *Flavoparmelia caperata* (L.) Hale, *Leptogium burnetiae* C.W. Dodge, *Cladonia fruticulosa* Kremp. are the host lichens of *Alternaria alternata* (Fr.) Keissl.

Bulbothrix meizospora, Heterodermia flabellata, H. hypochraea, Parmotrema crinitum, P. nilgherrense, P. praesorediosum, P. reticulatum, P. thomsonii, P. grayanum, Physcia dilatata, Flavoparmelia caperata, and *Leptogium burnetiae* are the host lichens for *Aspergillus flavus* Link. Similarly, *Aspergillus niger* Tiegh resides inside the *Heterodermia flabellata, Parmotrema thomsonii, P. cristiferum* (Taylor) Hale, *P. ravum* (Krog & Swinscow) Sérus, *Physcia dilatata, Roccella montagnei, Ramalina arabum* (Dill. ex Ach.) Meyen & Flot, *Teloschistes flavicans* (Sw.) Norman, *Usnea galbinifera* Asahina (Girlanda et al. 1997; Suryanarayanan et al. 2005; Li et al. 2007; Tripathi et al. 2014b; Vinayaka et al. 2016; Padhi et al. 2019). Studies also suggest that the host generality in ELF is higher than endophytic fungi that inhabit the plants (Chagnon et al. 2016). Generally, the lichens are referred as poikilohydric organisms which means that they frequently face the alternating dry and wet conditions (Kranner et al. 2008) and got affected by reactive oxygen species (Minibayeva and Beckett 2001). These conditions can be harsh for ELF inhabiting the lichen thallus. On the other hand, the fungal communities show very less specificity in harsh conditions, i.e., the host specificity of fungal symbionts in plants (ectomycorrhizal, endophytes, or ericoid fungi) got decreased in arctic and alpine conditions (Ryberg et al. 2011; Walker et al. 2011; Timling et al. 2012). So, this trend of ELF toward generality could be the result of harsh environmental conditions of the lichen thallus.

8.1 Some Lichen Hosts of ELF

Arthoniaceae.
Arthonia sp.
Arthonia antillarum (Fée) Nyl.
Arthonia parantillarum Aptroot.
Acarosporaceae.
Acarospora molybdina (Ach.) Trevis.

Caliciaceae.
Dirinaria consimilis (Stirt.) D.D. Awasthi.
Collemataceae.
Leptogium askotense D.D. Awasthi.
L. burnetiae Dodge.
Leptogium saturninum (Dicks.) Nyl.
Graphidaceae.
Diorygma hieroglyphicum (Pers.) Staiger and Kalb.
Cladoniaceae.
Cladonia sp.
Cladonia arbuscula (Wallr.) Flot.
Cladonia borealis S. Stenroos.
Cladonia grisea (Ahti) Trass.
Cladonia coniocraea (Flörke) Spreng.
Cladonia evansii Abbayes.
Cladonia fruticulosa Kremp.
Cladonia ochrochlora Flörke.
Cladonia pocillum (Ach.) O.J. Rich.
Stereocaulon sp.
Stereocaulon alpinum Laurer.
Stereocaulon botryosum Ach.
Stereocaulon vesuvianum Pers.
Hydnaceae.
Multiclavula sp.
Lecanoraceae.
Lecanora sp.
Ochrolechiaceae.
Ochrolechia frigida (Sw.) Lynge.
Lobariaceae.
Lobaria discolor (Bory) Hue.
Lobaria japonica (Zahlbr.) Asahina.
Peltigeraceae.
Lobaria kurokawae Yoshim.
Lobaria retigera (Bory) Trevis.
Peltigera canina (L.) Willd.
Peltigera elisabethae var. *mauritzii* (Gyeln.) J.C. Wei.
Peltigera praetextata (Fl örke ex Sommerf.) Zopf Pseudocyphellaria sp.
Ricasolia quercizans (Michx.) Stizenb.
Pertusariaceae.
Pertusaria laeviganda Nyl.
Physciaceae.
Amandinea sp.
Dirinaria picta (Sw.) Clem. & Shear.
Heterodermia diademata (Taylor) D.D. Awasthi.
Heterodermia flabellata (Fée) D.D. Awasthi.

Heterodermia hypochraea (Vain.) Swinsc & Krog.
Heterodermia incana (Stirt.) D.D. Awasthi.
Heterodermia obscurata (Nyl.) Trevis.
Heterodermia podocarpa (Bél) D.D. Awasthi.
Phaeophyscia hispidula (Ach.) Moberg.
Physcia sp.
Physcia dilatata Nyl.
Polyblastidium dendriticum (Pers.) Kalb.
Pyxine cocoes (Sw.) Nyl.
Placynthiaceae.
Placynthium asperellum (Ach.) Trevis.
Pyrenulaceae.
Pyrenula sp.
Prenula ochraceoflava (Nyl.) R.C. Harris.
Prenula parvinuclea (Meyen & Flot.) Aptroot.
Porinaceae.
Porina tetracerae (Ach.) Müll. Arg.
Roccellaceae.
Opegrapha arabica (Müll. Arg.) Vain.
Opegrapha medusulina Nyl.
Roccella montagnei Bél.
Ramalinaceae.
Bacidina sp.
Ramalina sp.
Ramalina arabum (Dill. ex Ach.) Meyen & Flot.
Ramalina calicaris (L.) Röhl.
Ramalina conduplicans Vain.
R. pacifica Asahina.
R. sinensis Jatta.
Teloschistaceae.
Teloschistes flavicans (Sw.) Norman.
Zeroviella mandschurica (Zahlbr.) S.Y. Kondr. & Hur.
Parmeliaceae.
Allantoparmelia alpicola (Th. Fr.) Essl.
Bulbothrix meizospora (Nyl.) Hale.
Canoparmelia texana (Tuck.) Elix & Hale.
Cetraria islandica (L.) Ach. Elix.
P. tinctorum (Despr. ex Nyl.) Hale.
Pseudevernia intensa (Nyl.) Hale and W.L. Culb. Pseudephebe pubescens (L.) M. Choisy.
Punctelia borreri (Turner) Krog.
Punctelia rudecta (Ach.) Krog.
Punctelia subrudecta (Nyl.) Krog.
Usnea sp.
U. aciculifera Vain.

Usnea cavernosa Tuck.
U. galbinifera Asahina.
Usnea longissima Ach.
Usnea stigmatoides G. Awasthi.
Xanthoparmelia angustiphylla (Gyelnik) Hale.
Umbilicariaceae.
Umbilicaria sp.
Umbilicaria aprina Nyl.
Umbilicaria arctica (Ach.) Nyl.
Umbilicaria torrefacta (Lightf.) Schrad.
Verrucariaceae.
Dermatocarpon miniatum (L.) W. Mann.

9 Role of Endolichenic Fungi in Lichen Microbiome

In comparison to ELF, there is more sustainable information available regarding the functional roles and biotechnological potential of bacterial associates found in lichens (Grube and Berg 2009; Suzuki et al. 2016). This information could provide valuable insights and serve as starting point for addressing the roles of ELF. Lichens associated with bacterial colonies contribute in lichen symbiosis by facilitating nutrient supply, enhancing lichen's resistance to both abiotic and biotic stress, detoxifying metabolites, and breaking down older parts of the thallus (Grube et al. 2015). Specific functions of ELF in the development and survival of lichens are not known. The endolichenic microbiome is known to produce a variety of antimicrobial compounds. Lichens grow in harsh environments and display wide tolerance to abiotic stressors. Some lichens can survive extended exposure to desiccation, extreme temperatures (90 °C) and low (−196 °C) temperatures, and high UV radiation (De Los et al. 2005; Grube and Berg 2009). A few of them such as *Xanthoria elegans* (Link) Th. Fr. can even withstand exposure to simulated Martian atmosphere and real space conditions (De Vera et al. 2010). It would be of interest to investigate whether endolichenic fungi, like endophytes of plants, contribute to lichen hosts ability to tolerate these abiotic stressors.

Lichens are an attractive source of unique secondary metabolites (Kumar et al. 2014), with over 1000 such compounds identified in lichens spanning from the arctic to the tropical regions (Stocker-Wörgötter 2008). Interestingly, apart from the lichens themselves, the ELF like the plant endophytes (Suryanarayanan et al. 2005; Kaushik et al. 2014; Chen et al. 2016), also hold promise as a valuable reservoir of unique bioactive molecules (Chang et al. 2015; Kellogg and Raja 2016). As a result, it would be highly worthwhile to systematically screen endolichenic fungi for their ability to produce innovative metabolites with specific biological activities. Another aspect to explore involves the interactions between ELF and the host lichen as well as between ELF and other groups of organisms supported by lichens. Many lichens produce antifungal metabolites (Basile et al.

2015); it is reasonable to expect that the endolichenic fungal community should either possess insensitivity to these metabolites or possess detoxification mechanism. It could be assumed that only such fungi would comprise the endolichenic fungal community within given lichen.

Bacteria associated with lichens aid in breaking down parts of lichen thalli to facilitate biomass mobilization (Grube and Berg 2009); endolichenic fungi may also play a role in decomposing dead parts of lichen thalli. The evolution of endolichenic fungal association may have occurred in parallel along with the lichens as has been construed for bacterial association with lichens (Grube and Berg 2009). Given the limited number of lichen species studied for their endolichenic fungi, it is reasonable to assume that there are many undescribed fungal species within this ecological group of fungi (Muggia et al. 2016). It is also possible that the ELF harbor many novel genes whose products could have technological applications. To gain a deeper appreciation of the role of lichens as influencers of ecological processes, it is essential to understand the contributions of ELF—an aspect that often goes unnoticed (Asplund and Wardle 2017).

10 Role of ELF in Lichen Symbiosis

The role of ELF in lichen symbiosis has not been fully described; however, some hypotheses have been put forward. For example, the existence of ELF is thought to make a vital ecological contribution toward lichens by aiding their formation and growth as well as by protecting the host against insects (Yu et al. 2018). Endolichenic fungi are rich source of bioactive metabolites that have a wide range of biological activities. Depending on the distinct habitats of host lichens, the residing ELF might lead to essential metabolites. Metabolomics revealed that the contrast between bioactivities of host lichens and ELF is assisted by the exploration of distinct metabolites produced by these organisms. Endolichenic fungi are thought to contribute to lichen symbiosis by providing such metabolites that could stop the growth of harmful invaders and competitor microbes. In some cases, natural products or extracts obtained from ELF show high antimicrobial potential even though these bioactivities are not naturally seen within their ecological niche. Many such compounds have been extracted from ELF that show antimicrobial potential. For example, funiculosone, mangrovamide, and ravenilin are three different compounds that have been extracted from *Talaromyces funiculosus* (Thom) Samson, N. Yilmaz, Frisvad, and Seifert that resides inside the *Diorygma hieroglyphicum*. All these compounds are effective against *Escherichia coli* and *Staphylococcus aureus* (Padhi et al. 2019). Similarly, *Nigrospora sphaerica* (Sacc.) E.W. Mason reside inside the lichen *Parmelinella wallichiana* (Taylor) Elix and Hale. It is a source of two important compounds, namely, alternariol and alternariol-9-methyl. Alternariol shows highest antimicrobial potential against *E. coli*, *S. aureus*, and *Bacillus subtilis*, while alternariol-9-methyl is effective against *S. aureus*, *B. subtilis*, and *Pseudomonas fluorescens* (Gu 2009). ELF yield several important alkaloids. Lichen

Everniastrum nepalense (Taylor) Hale ex Sipman is the host for ELF *Chaetomium globosum* Kunze. The solid cultures of this ELF isolated from the same host lichen yields the vital cytochalasan alkaloid chaetoglobosin Y. It has a macrocyclic ring with isoindolone moiety (Zheng et al. 2014). Similarly, *Tolypocladium cylindrosporum* W. Gams is also an ELF that resides inside the lichen *Lethariella zahlbruckneri* (Du Rietz) Krog. The rice cultures of this ELF yield a new alkaloid tolypyridone A that is a pyridine-type alkaloid. It also yields three vital derivatives of tetramic acid [tolypocladenols A_1, A_2, and B] (Li et al. 2015). Anthraquinones are also integral components produced by ELF. *Aspergillus versicolor* (Vuill.) Tirab is an ELF strain. It yields many vital derivatives of anthracene. These are versicolorin B, 8-O-methylversicolorin A and 8-O-methylversicolorin B (Dou et al. 2014). There is another ELF *Chaetomium elatum* Kunze that contains five important xanthoquinodins. All these compounds possess cytotoxic effects against five different cancer cell lines (HL-660, SMMC-7721, A-549, MCF-7, and SW480) (Chen et al. 2013). ELF yield several important quinones. *Pleosporales* sp. is the well-known endolichenic fungus. Six novel metabolites were isolated from its culture. These compounds involve some cyclohexenones, derivatives of triphenyl and cucurbitacins (Jiao et al. 2015). Among cyclohexenones, ambuic acid is highly functionalized. It was firstly isolated from *Pestalotiopsis* sp. Steyaert and *Monochaetia* sp. (Sacc.) Allesch; both of them is an ELF (Li et al. 2001). Ambuic acid has strong antifungal activity. It also possesses moderate quorum-sensing inhibitory activity and antibacterial activity (Nakayama et al. 2009). The crude extract of *Pestalotiopsis* sp. that inhabits the *Clavaroid* sp. of lichen yields six novel ambuic acid derivatives (Ding et al. 2009). Another dimeric quinone that resembles to torreyanic acid was also gained from the same fungal strain. Few perylene quinonoid pigments were also identified. They mostly originate from ascomycete fungi (Zhou and Liu 2010). These pigments get converted into excited triplet states after absorbing light energy. These pigments are cytotoxic metabolites (Morgan et al. 2009). *Corynespora* sp. Güssow is an endolichenic fungal strain which is found in the cavern beard lichen *Usnea cavernosa* Tuck. It yields three novel herbarin-derived adducts which are 8-hydroxyherbarin, 7-desmethylherbarin, and 1-hydroxydehydroherbarin (Paranagama et al. 2007; Wijeratne et al. 2010). Corynesporol (Paranagama et al. 2007) and 9-O-methylscylatol A (Wijeratne et al. 2010) are biosynthetically related compounds. Furanones and pyrones are also secreted by ELF. Endolichenic fungus *Peziza* sp. Ray resides inside the lichen *Xanthoparmelia* sp. (Vain.) Hale. Its crude extract yields seven novel derivatives of furanone metabolites (Zhang et al. 2014). These compounds are pezizolides A–G. Fungal strain *Nodulisporium* sp. Preuss is symbiotic with the lichen *Everniastrum* sp. Hale. Rice culture of this fungal strain yields novel derivatives of a-pyrone (Zhao et al. 2015). Similarly, *Nectria* sp. (Fr.) Fr., resides inside the lichen *Pamelia* sp. Ach, and its rice culture produces novel derivatives necopyrone A, B, D, and E. The a-pyrones have week activity against *A. niger* and *Candida albicans* (C.P. Robin) Berkhout (Zhao et al. 2015). ELF yield a great deal of xanthones. Lichen *Xanthoria mandschurica* (Zahlbr.) Asahina is the host lichen of *Coniochaeta* sp. (Sacc.) Cooke. It yields a great deal of xanthones and xanthone derivatives. The isolated cultures of *Coniochaeta* sp. produce xanthone mandschurica A (Wang et al. 2010). A similar role was observed for endophytic fungi in plant, even though the similar secondary metabolites are produced by host plant and their associated endophytes (Ludwig-Müller 2015). Actually, the bioactivity and the chemical composition of most of the ELF have not been studied properly, leaving behind a huge number of compounds that demand investigation (Gao et al. 2016). Their potential as producers of natural products needs exploration. Actually, there are many factors that adversely affect the isolation of these compounds from ELF. One of the major problems is to distinguish the endolichenic microbiota and other strains like mycobiont of the lichens, airborne microbes, lichenicolous fungi, and epiphytic microorganisms. Apparent harm to endolichenic population could be restricted by carefully collecting the samples and implication of proper surface sterilization (Strobel and Daisy 2003). Another major drawback is the selectively isolating the ELF that grow faster than others. Its evidence can be found in most of the processes where different species of microbes are isolated and subcultures (Sarasan et al. 2017). Due to their slowly growing cultures, many important ELF as well as their metabolites may have been left out from analysis. This limitation can be overcome by growing the ELF in diluted, separated-out cultures and by supplying the handsome amount of time and differing conditions to slow developers to grow (Wethalawe et al. 2012). Fungi produce many compounds in minute amount. Due to this reason, the separation, determination, characterization, and even detection of their bioactivities becomes difficult (Guo et al. 2008). For example, sometimes the solvent extract of ELF shows great antimicrobial properties, and none of the isolated metabolites shows similar potential (Wethalawe et al. 2012). If new technologies will develop out, a new era could begin for the use of these natural products for the well-being of human health.

11 Importance

Recent research has unveiled a fascinating connection between certain groups of endophytic fungi within endophyte-like (endolichenic) fungi that can be isolated from the interior of apparently healthy lichens. Similarly, to endophytes, these ELF display remarkable diversity and have been isolated from every lichen thallus investigated so far. Interestingly, these tend to form associations with green algal photobionts in lichen thalli, which might have facilitated host switching as new land plants became available substrates through time. Endolichenism thus seems to be an incubator for the evolution of endophytism. In turn, endophytism is evolutionarily dynamic, with endophytic lineages frequently transitioning between pathogenic and nonpathogenic states. Endophytic fungi residing within plant tissues play various functional roles on plants including affecting disease resistance and susceptibility (Campanile et al. 2007), augmenting growth, improving the ability to exploit extreme environments, and aiding on the decomposition of plant litter (Botnen et al. 2014). Therefore, it is interesting to study the functional roles of ELF and determine whether they share similar roles with endophytic fungi.

12 Conclusions

Lichens are the pioneer organisms that play a dominant role in the early stages of development in a variety of ecosystems undergoing primary or secondary succession. Fossilized lichen thalli from as early as the Lower Devonian period had revealed the evidence of endolichenic fungal associations in them. Lichens are a rich source of unique secondary metabolites with over 1000 such compounds identified in lichens ranging from the arctic to the tropical regions. Similar to endophytic fungi, endolichenic fungi also have capability to produce wide range of secondary metabolites including alkaloids, quinones, furanones, pyrones, benzopyranoids, xanthones, terpenes, steroids, peptides, and alicyclic compounds to perform a range of bioactivities. The ELF are found worldwide and have been identified in every lichen species from the arctic to tropics. It is estimated that there could be approximately 13,500 species of ELF to occur in the world. However, the estimated diversity of ELF remains poorly documented, and many sequences deposited in GenBank lack sufficient taxonomic information. Our knowledge of the species diversity of endolichenic fungi is limited, although a pyrosequencing study showed the presence of numerous fungi that are present in lichen thalli. To better understand the taxonomic placement, ecological roles and evolutionary details of these fungi, culture-independent techniques and high-throughput sequencing methods can serve as valuable tools.

References

Arnold AE (2007) Understanding the diversity of foliar fungal endophytes: progress, challenges, and frontiers. Fungal Biol Rev 21:51–66. https://doi.org/10.1016/j.fbr.2007.05.003

Arnold AE, Lutzoni F (2007) Diversity and host range of foliar fungal endophytes: are tropical leaves biodiversity hotspots? Ecology 88:541–549. https://doi.org/10.1890/05-1459

Arnold AE, Mejia L, Kyllo D, Rojas E, Maynard Z, Herre EA (2003) Fungal endophytes limit pathogen damage in a tropical tree. Proc Natl Acad Sci USA 100:15649–15654. https://doi.org/10.1073/pnas.2533483100

Arnold AE, Miadlikowska J, Higgins KL, Sarvate SD, Gugger P, Way A et al (2009) A phylogenetic estimation of trophic transition networks for ascomycetous fungi: are lichens cradles of symbiotrophic fungal diversification? Syst Biol 58(3):283–297. https://doi.org/10.1093/sysbio/syp001

Asplund J, Wardle DA (2017) How lichens impact on terrestrial community and ecosystem properties. Biol Rev 92(3):1720–1738. https://doi.org/10.1111/brv.12305

Basile A, Rigano D, Loppi S, Di Santi A, Nebbioso A, Sorbo S et al (2015) Antiproliferative, antibacterial and antifungal activity of the lichen *Xanthoria parietina* and its secondary metabolite parietin. Int J Mol Sci 16:7861–7875

Bates ST, Cropsey GWG, Caporaso JG, Knight R, Fierer N (2011) Bacterial communities associated with the lichen symbiosis. Appl Environ Microb 77:1309–1314

Botnen S, Vik U, Carlsen T, Eidesen PB, Davey ML, Kauserud H (2014) Low host specificity of root-associated fungi at an Arctic site. Mol Ecol 23:975–985

Campanile G, Ruscelli A, Luisi N (2007) Antagonistic activity of endophytic fungi towards *Diplodia corticola* assessed by in vitro and in planta tests. Eur J Plant Pathol 117:237–246

Cardinale M, de Castro JV, Müller H, Berg G, Grube M (2008) In situ analysis of the bacterial community associated with the reindeer lichen *Cladonia arbuscula* reveals predominance of Alphaproteobacteria. FEMS Microb Ecol 66:63–71

Chagnon PL, U'Ren JM, Miadlikowska J, Lutzoni F, Arnold EA (2016) Interaction type influences ecological network structure more than local abiotic conditions: evidence from endophytic and endolichenic fungi at a continental scale. Oecologia 180:181–191

Chang W, Zhang M, Li Y, Li X, Gao Y, Xie Z et al (2015) Lichen endophyte derived pyridoxatin inactivates Candida growth by interfering with ergosterol biosynthesis. Biochim Biophys Acta 1850:1762–1771

Chen GD, Chen Y, Gao H, Shen LQ, Wu Y, Li XX et al (2013) Xanthoquinodins from the endolichenic fungal strain *Chaetomium elatum*. J Nat Prod 76(4):702–709. https://doi.org/10.1021/np400041y

Chen C, Hu SY, Luo DQ, Zhu SY, Zhou CQ (2016) Potential antitumor agent from the endophytic fungus *Pestalotiopsis photiniae* induces apoptosis via the mitochondrial pathway in HeLa cells. Oncol Rep 30:1773–1781

De Los RA, Wierzchos J, Sancho LG, Green TGA, Ascaso C (2005) Ecology of endolithic lichens colonizing granite in continental Antarctica. Lichenologist 37:383–395. https://doi.org/10.1017/S0024282905014969

De Vera JP, Möhlmann D, Butina F, Lorek A, Wernecke R, Ott S (2010) Survival potential and photosynthetic activity of lichens under Mars-like conditions: a laboratory study. Astrobiology 10:215–227. https://doi.org/10.1089/ast.2009.0362

Ding G, Li S, Fu S, Liu J, Wei Che Y (2009) Ambuic acid and torreyanic acid derivatives from the endolichenic fungus *Pestalotiopsis* sp. J Nat Prod 72:182–186. https://doi.org/10.1021/np800733y

Dou Y, Wang X, Jiang D, Wang H, Jiao Y, Lou H et al (2014) Metabolites from *Aspergillus versicolor*, an endolichenic fungus from the lichen *Lobaria retigera*. Drug Discov Ther 8(2):84–88. https://doi.org/10.5582/ddt.8.84

Engh I, Nowrousian M, Kück U (2007) Regulation of melanin biosynthesis via the dihydroxynaphthalene pathway is dependent on sexual development in the ascomycete *Sordaria macrospora*. FEMS Microbiol Lett 275:62–70. https://doi.org/10.1111/j.1574-6968.2007.00867.x

Fisher PJ, Graf F, Petrini LE, Sutton BC, Wookey PA (1995) Fungal endophytes of Dryas octopetala from high arctic semidesert and from the Swiss alps. Mycol 87:319–323. https://doi.org/10.1080/00275514.1995.12026536

Gao H, Zou J, Li J, Zhao H (2016) Endolichenic fungi: a potential treasure trove for discovery of special structures and bioactive compounds. In: Atta-Ur-Rahman (ed) Studies in natural products chemistry. Elsevier, Amsterdam, pp 347–397. https://doi.org/10.1016/B978-0-444-63602-7.00011-4

Girlanda M, Isocrono D, Bianco C, Luppi-Mosca AM (1997) Two foliose lichens as microfungal ecological niches. Mycology 89:531–536. https://doi.org/10.1080/00275514.1997.12026814

Girlanda CR, Santiago IF, da Costa CL, Câmara PEAS, Silva MC, Stech M, Rosa CA, Rosa LH (2019) Fungi associated with plants and lichens of Antarctica. In: Rosa LH (ed) Fungi of Antarctica. Springer International Publishing, Cham, pp 165–199. https://doi.org/10.1007/978-3-030-18367-7_8

Gostin^car C, Muggia L, Grube M (2012) Polyextremotolerant black fungi: Oligotrophism, adaptive potential, and a link to lichen symbioses. Front Microbiol 3:390. https://doi.org/10.3389/fmicb.2012.00390

Govinda Rajulu MB, Thirunavukkarasu N, Kumar SS, Kaur T, Reddy MS, Suryanarayanan TS (2020) Endolichenic fungal diversity associated with some lichens of Western Ghats. Planta Med 86:960–966. https://doi.org/10.1055/a-1045-1989

Grube M, Berg G (2009) Microbial consortia of bacteria and fungi with focus on the lichen symbiosis. Fungal Biol Rev 23:72–85. https://doi.org/10.1016/j.fbr.2009.10.001

Grube M, Cernava T, Soh J, Fuchs S, Aschenbrenner I, Lassek C et al (2015) Exploring functional contexts of symbiotic sustain within lichen-associated bacteria by comparative omics. ISME J 9: 412–424. https://doi.org/10.1038/ismej.2014.138

Gu W (2009) Bioactive metabolites from *Alternaria brassicicola* ML-P08, an endophytic fungus residing in *Malus halliana*. World J Microbiol Biotechnol 25:1677–1683. https://doi.org/10.1007/s11274-009-0062-y

Guo B, Wang Y, Sun X, Tang K (2008) Bioactive natural products from endophytes: a review. Appl Biochem Microbiol 44:136–142. https://doi.org/10.1134/S0003683808020026

Hawksworth DL, Grube M (2020) Lichens redefined as complex ecosystems. New Phytol 227(5): 128–1283. https://doi.org/10.1111/nph.16630

He JW, Chen GD, Gao H, Yang F, Li XX, Peng T et al (2012) Heptaketides with antiviral activity from three endolichenic fungal strains *Nigrospora* sp., *Alternaria* sp. and *Phialophora* sp. Fitoterapia 83:1087–1091. https://doi.org/10.1016/j.fitote.2012.05.002

Hodkinson BP, Lutzoni F (2009) A microbiotic survey of lichen-associated bacteria reveals a new lineage from the *Rhizobiales*. Symbiosis 49:163–180. https://doi.org/10.1007/s13199-009-0049-3

Hoffman M, Arnold AE (2008) Geography and host identity interact to shape communities of endophytic fungi in cupressaceous trees. Mycol Res 112:331–344. https://doi.org/10.1016/j.mycres.2007.10.014

Honegger R (1991) Fungal evolution: symbiosis and morphogenesis. In: Margulis L, Fester R (eds) Symbiosis as a source of evolutionary innovation. The MIT Press, Cambridge, pp 319–340

Honegger R, Axe L, Edwards D (2013) Bacterial epibionts and endolichenic actinobacteria and fungi in the lower devonian lichen *Chlorolichenomycites salopensis*. Fungal Biol 117:512–518. https://doi.org/10.1016/j.funbio.2013.05.003

Jiao Y, Li G et al (2015) New metabolites from endolichenic fungus *Pleosporales* sp. Chem Biodivers 12:1095–1104. https://doi.org/10.1002/cbdv.201400279

Kannangara BTSDP, Rajapaksha RSCG, Paranagama PA (2009) Nature and bioactivities of endolichenic fungi in *Pseudocyphellaria* sp., *Parmotrema* sp. and *Usnea* sp. at Hakgala montane forest in Sri Lanka. Lett Appl Microbiol 48:203–209. https://doi.org/10.1111/j.1472-765X.2008.02512.x

Kaushik NK, Murali TS, Sahal D, Suryanarayanan TS (2014) A search for antiplasmodial metabolites among fungal endophytes of terrestrial and marine plants of southern India. Acta Parasitol 59:745–757. https://doi.org/10.2478/s11686-014-0307-2

Kellogg JJ, Raja HA (2016) Endolichenic fungi: a new source of rich bioactive secondary metabolites on the horizon. Phytochem Rev 16:271–293. https://doi.org/10.1007/s11101-016-9473-1

Kranner I, Beckett R, Nash TH (2008) Desiccation-tolerance in lichens: a review. Bryologist 111: 576–593. https://doi.org/10.1639/0007-2745-111.4.576

Kumar J, Dhar P, Tayade AB, Gupta D, Chaurasia OP, Upreti DK et al (2014) Antioxidant capacities, phenolic profile and cytotoxic effects of saxicolous lichens from trans-Himalayan cold desert of Ladakh. PLoS One 9:e98696. https://doi.org/10.1371/journal.pone.0098696

Lagarde A, Jargeat P, Roy M, Girardot M, Imbert C, Millot M et al (2018) Fungal communities associated with *Evernia prunastri*, *Ramalina fastigiata* and *Pleurosticta acetabulum*: three epiphytic lichens potentially active against *Candida* biofilms. Microbiol Res 211:1–12. https://doi.org/10.1016/j.micres.2018.03.006

Lagashetti AC, Dufossé L, Singh SK, Singh PN (2019) Fungal pigments and their prospects in different industries. Microorganisms 7:604. https://doi.org/10.3390/microorganisms7120604

Leavitt SD, Esslinger TL, Divakar PK, Crespo A, Lumbsch HT (2016) Hidden diversity before our eyes: delimiting and describing cryptic lichen-forming fungal species in camouflage lichens (Parmeliaceae, Ascomycota). Fungal Biol 120:1374–1391. https://doi.org/10.1016/j.funbio.2016.06.001

Li JY, Harper JK, Grant DM, Tombe BO, Bashyal B, Hess WM et al (2001) Ambuic acid, a highly functionalized cyclohexenone with antifungal activity from *Pestalotiopsis* spp. and *Monochaetia* sp. Phytochemistry 56(5):463–468. https://doi.org/10.1016/S0031-9422(00)00408-8

Li WC, Zhou J, Guo SY, Guo LD (2007) Endophytic fungi associated with lichens in Baihua Mountain of Beijing, China. Fungal Divers 25:69–80

Li XB, Li L, Zhu RX, Li W, Chang WQ, Zhang LL et al (2015) Tetramic acids and pyridone alkaloids from the endolichenic fungus *Tolypocladium cylindrosporum*. J Nat Prod 78:2155–2160. https://doi.org/10.1021/np501018w

Ludwig-Müller J (2015) Plants and endophytes: equal partners in secondary metabolite production? Biotechnol Lett 37:1325–1334. https://doi.org/10.1007/s10529-015-1814-4

Lutzoni F, Miadlikowska J (2009) Lichens. Quick guide. Curr Biol 19:R502–R503

Maduranga K, Attanayake RN, Santhirasegaram S, Weerakoon G, Paranagama PA (2018) Molecular phylogeny and bioprospecting of Endolichenic fungi (ELF) inhabiting in the lichens collected from a mangrove ecosystem in Sri Lanka. PLoS One 13(8):e0200711. https://doi.org/10.1371/journal.pone.0200711

Mark K, Laanisto L, Bueno CG, Niinemets Ü, Keller C, Scheidegger C (2020) Contrasting co-occurrence patterns of photobiont and cystobasidiomycete yeast associated with common epiphytic lichen species. New Phytol 227(5):1362–1375. https://doi.org/10.1111/nph.16475

Masumoto H, Degawa Y (2019) The effect of surface sterilization and the type of sterilizer on the genus composition of lichen-inhabiting fungi with notes on some frequently isolated genera. Mycoscience 60:331–342. https://doi.org/10.1016/j.myc.2019.07.004

Miadlikowska J, Arnold AE, Lutzoni F (2004) Diversity of cryptic fungi inhabiting healthy lichen thalli in a temperate and tropical forest. Ecol Soc 89:349–350

Minibayeva F, Beckett RP (2001) High rates of extracellular superoxide production in bryophytes and lichens, and an oxidative burst in response to rehydration following desiccation. New Phytol 152:333–341. https://doi.org/10.1046/j.0028-646X.2001.00256.x

Morgan BJ, Dey S, Johnson SW, Kozlowski MC (2009) Design, synthesis, and investigation of protein kinase C inhibitors: total syntheses of (+)-calphostin D, (+)-phleichrome, Cercosporin, and new photoactive perylenequinones. J Am Chem Soc 131(26):9413–9425. https://doi.org/10.1021/ja902324j

Muggia L, Fleischhacker A, Kopun T, Grube M (2016) Extremotolerant fungi from alpine rock lichens and their phylogenetic relationships. Fungal Divers 76:119–142. https://doi.org/10.1007/s13225-015-0343-8

Nakayama J, Uemura Y, Nishiguchi K, Yoshimura N, Igarashi Y, Sonomoto K (2009) Ambuic acid inbhitis the biosynthesis of cyclic pentide quormones in gram-positive bacteria. Antimicrob Agents Ch 53(2):580–586. https://doi.org/10.1128/aac.00995-08

Nash TH (2008) Lichen biology, 2nd edn. Cambridge University Press, Cambridge

Newcombe G, Campbell J, Griffth D, Baynes M, Launchbaugh K, Pendleton R (2016) Revisiting the life cycle of dung fungi, including *Sordaria fimicola*. PLoS One 11:e0147425. https://doi.org/10.1371/journal.pone.0147425

Oh SY, Yang JH, Woo JJ, Oh SO, Hur JS (2020) Diversity and distribution patterns of Endolichenic fungi in Jeju Island, South Korea. Sustainability 12:3769. https://doi.org/10.3390/su12093769

Padhi S, Masi M, Cimmino A, Tuzi A, Jena S, Tayung K, Evidente A (2019) Funiculosone, a substituted dihydroxanthene-1,9- dione with two of its analogues produced by an endolichenic fungus *Talaromyces funiculosus* and their antimicrobial activity. Phytochemistry 157:175–183. https://doi.org/10.1016/j.phytochem.2018.10.031

Paranagama PA, Kithsiri Wijeratne EM, Burns AM, Marron MT, Gunatilaka MK, Arnold AE et al (2007) Heptaketides from *Corynespora* sp. inhabiting the cavern beard lichen, *Usnea cavernosa*: first report of metabolites of an endolichenic fungus. J Nat Prod 70:1700–1705. https://doi.org/10.1021/np070466w

Park CH, Kim KM, Elvebakk A, Kim OS, Jeong G, Hong SG (2014) Algal and fungal diversity in Antarctic lichens. J Eukaryot Microbiol 62:196–205. https://doi.org/10.1111/jeu.12159

Petrini O, Hake U, Dreyfuss MM (1990) An analysis of fungal communities isolated from fruticose lichens. Mycologia 82:444–451. https://doi.org/10.1080/00275514.1990.12025907

Rai AN, Rai AN (2018) Cyanobacterial-fungal symbioses: the Cyanolichens. In: Rai AN (ed) CRC handbook of symbiotic cyanobacteria. CRC Press, Boca Raton, pp 9–41

Rashmi M, Kushveer JS, Sarma VV (2019) A worldwide list of endophytic fungi with notes on ecology and diversity. Mycosphere 10:798–1079

Richardson MJ (2004) Coprophilous fungi from Iceland. Acta Bot Isl 14:77–102

Ryberg M, Andreasen M, Björk RG (2011) Weak habitat specificity in ectomycorrhizal communities associated with *Salix herbacea* and *Salix polaris* in alpine tundra. Mycorrhiza 21:289–296. https://doi.org/10.1007/s00572-010-0335-1

Samanthi KA, Wickramarachchi S, Wijeratne EM, Paranagama PA (2015) Two new bioactive polyketides from *Curvularia trifolii*, an endolichenic fungus isolated from Usnea sp., in Sri Lanka. J Natl Sci Found 43:201–224. https://doi.org/10.4038/jnsfr.v43i3.7950

Sarasan M, Puthumana J, Job N, Han J, Lee JS, Philip R (2017) Marine algicolous endophytic fungi-a promising drug resource of the era. J Microbiol Biotechnol 27:1039–1052

Spribille T, Tuovinen V, Resl P, Vanderpool D, Wolinski H, Aime MC et al (2016) Basidiomycete yeasts in the cortex of ascomycete macrolichens. Science 353:488–492. https://doi.org/10.1126/science.aaf8287

Stadler M, Læssøe T, Fournier J, Decock C, Schmieschek B, Tichy HV et al (2014) Polyphasic taxonomy of *Daldinia* (*Xylariaceae*). Stud Mycol 77:1–143. https://doi.org/10.3114/sim0016

Stocker-Wörgötter E (2008) Metabolic diversity of lichen-forming ascomycetous fungi: culturing, polyketide and shikimate metabolite production, and PKS genes. Nat Prod Rep 25:188–200. https://doi.org/10.1039/B606983P

Strobel G, Daisy B (2003) Bioprospecting for microbial endophytes and their natural products. Microbiol Mol Biol Rev 67:491–502. https://doi.org/10.1128/mmbr.67.4.491-502.2003

Suryanarayanan TS, Thirunavukkarasu N (2017) Endolichenic fungi: the lesser known fungal associates of lichens. Mycology 8:189–196. https://doi.org/10.1080/21501203.2017.1352048

Suryanarayanan TS, Thirunavukkarasu N, Hariharan GN, Balaji P (2005) Occurrence of nonobligate inside lichen thalli. Sydowia 57:120–130

Suryanarayanan TS, Govindarajulu MB, Rajamani T, Tripathi M, Joshi Y (2017) Endolichenic fungi in lichens of Champawat district, Uttarakhand, northern India. Mycol Prog 16:205–211. https://doi.org/10.1007/s11557-016-1268-7

Suzuki MT, Parrot D, Berg G, Grub M, Tomasi S (2016) Lichens as natural sources of biotechnologically relevant bacteria. Appl Microbiol Biotechnol 100:583–595. https://doi.org/10.1007/s00253-015-7114-z

Timling I, Dahlberg A, Walker DA, Gardes M, Charcosset JY, Welker JM et al (2012) Distribution and drivers of ectomycorrhizal fungal communities across the North American Arctic. Ecosphere 3:1–25. https://doi.org/10.1890/ES12-00217.1

Tripathi M, Joshi Y (2015) Endolichenic fungi in Kumaun Himalaya: a case study. In: Upreti DK, Divakar PK, Shukla V, Bajpai R (eds) Recent advances in lichenology, vol 2. Springer, New Delhi, pp 111–120. https://doi.org/10.1007/978-81-322-2235-4_6

Tripathi M, Joshi Y (2019) Taxonomic descriptions of Endolichenic fungi. In: Endolichenic Fungi: present and future trends. Springer, Singapore. https://doi.org/10.1007/978-981-13-7268-1_7

Tripathi M, Gupta RC, Joshi Y (2014a) Assessment of endolichenic fungal diversity in some forests of Kumaun Himalaya. Curr Sci 107:745–748

Tripathi M, Gupta RC, Joshi Y (2014b) *Spegazzinia tessarthra* isolated as a true endophyte from lichen *Heterodermia flabellata*. Indian Phytopathol 67:109–110

U'Ren JM, Lutzoni F, Miadlikowska J, Arnold AE (2010) Community analysis reveals close affinities between endophytic and endolichenic fungi in mosses and lichens. Microb Ecol 60: 340–353. https://doi.org/10.1007/s00248-010-9698-2

U'Ren JM, Lutzoni F, Miadlikowska J, Laetsch AD, Arnold AE (2012) Host and geographic structure of endophytic and endolichenic fungi at a continental scale. Am J Bot 99:898–914

Vinayaka K, Krishnamurthy YL, Shivakumar B, Prashith KTR (2016) Association and variation of Endophytic fungi among some macrolichens in Central Western Ghats, Southerns India. IJCAMS 5:115–124

Walker JF, Aldrich-Wolfe L, Riffel A, Barbare H, Simpson NB, Trowbridge J et al (2011) Diverse helotiales associated with the roots of three species of arctic ericaceae Timling provide no evidence for host specificity. New Phytol 191:515–527

Wang Y, Zheng Z, Liu S, Zhang H, Li E, Guo L (2010) Oxepinochromenones, furochromenone, and their putative precursors from the endolichenic fungus *Coniochaeta* sp. J Nat Prod 73:920–924. https://doi.org/10.1021/np100071z

Wethalawe AN, Alwis YV, Udukala DN, Paranagama PA (2012) Antimicrobial compounds isolated from Endolichenic fungi: a review. Molecules 26:3901. https://doi.org/10.3390/molecules26133901

Wijayawardene NN, Hyde KD, Al-Ani LKT, Tedersoo L, Haelewaters D, Rajeshkumar KC et al (2020) Outline of fungi and fungus like-taxa. Mycosphere 11(1):1060–1456. https://doi.org/10.5943/mycosphere/11/1/8

Wijeratne EMK, Bashyal BP, Gunatilaka MK, Arnold AE, Gunatilaka A (2010) Maximizing chemical diversity of fungal metabolites: biogenetically related heptaketides of the endolichenic fungus *Corynespora* sp. J Nat Prod 73(6):1156–1159. https://doi.org/10.1021/np900684v

Yang F, Chen GD, Gao H, Li XX, Wu Y, Guo LD, Yao XS (2012) Two new naphthalene derivatives from an endolichenic fungal strain *Scopulariopsis* sp. J Asian Nat Prod Res 14: 1059–1063

Yu NH, Park SY, Kim JA, Park CH, Jeong MH, Oh SO et al (2018) Endophytic and endolichenic fungal diversity in maritime Antarctica based on cultured material and their evolutionary position among Dikarya. Fungal Syst Evol 3:171–184

Yuan C, Guo YH, Wang HY et al (2016) Allelopathic polyketides from an endolichenic fungus *Myxotrichum* sp. by using OSMAC strategy. Sci Rep 6:19350

Zhang K, Ren J, Ge M, Li L, Guo L, Chen D, Che Y (2014) Mono- and bis-furanone derivatives from the endolichenic fungus *Peziza* sp. Fitoterapia 92:79–84. https://doi.org/10.1016/j.fitote.2013.10.011

Zhang T, Wei XL, Wei YZ, Liu HY, Yu LY (2016) Diversity and distribution of cultured endolichenic fungi in the Ny-Ålesund region, Svalbard (high Arctic). Sci Rep 20:461–470

Zhao Q, Wang CX, Yu Y, Wang GQ, Zheng QC, Chen GD (2015) Nodulisporipyrones A-D, new bioactive a-pyrone derivatives from *Nodulisporium* sp. J Asian Nat Prod Res 17(5):567–575. https://doi.org/10.1080/10286020.2015.1040776

Zheng QC, Kong MZ, Zhao Q, Chen GD, Tian HY, Li XX (2014) Chaetoglobosin Y, a new cytochalasan from *Chaetomium globosum*. Fitoterapia 93:126–131. https://doi.org/10.1016/j.fitote.2013.12.022

Zhou ZY, Liu JK (2010) Pigments of fungi (macromycetes). Nat Prod Rep 27:1531–1570

Diversity of Endophytic Fungi of Saxicolous Lichens

Saba Abdulameer Kadhim AL-Ziadi and Huda Raheem Hashim

Contents

S. A. K. AL-Ziadi (✉)
Faculty of Science, Ecology Department, AL-Qadisiyah University, Al Diwaniyah, Iraq
e-mail: saba.abdulameer@qu.edu.iq

H. R. Hashim
Faculty of Basic Education, Sciences Department, University of Al-Muthanna, Samawah, Iraq

A. M. Abdel-Azeem, F. Degola (eds.), *Advances in Lichens and Non-Flowering Plants Fungal Endophytes Research*, Fungal Biology,
https://doi.org/10.1007/978-3-032-01506-8_4

1 Introduction

Lichens are organisms with symbiotic relationships that consist of a photosynthesis partnership, which could be cyanobacteria or green algae, and a fungal companion, known as mycobiont. Lichens' poikilohydric and poikilothermic characteristics indicate that they are unable to actively control their internal water state and warmth (Lutzoni & Miadlikowska 2009).

According to earlier studies on lichens as markers of changes in the environment, their variety and richness vary depending on the type of climate, from subtropics to drylands (Vallese et al. 2024; Koch et al. 2016), but they can also be found in colder climates (Green et al. 2011). Aridity is a significant climate characteristic that is crucial for determining the taxonomy and diversity of functions across broad gradients in the environment. This lends credence to the idea that climatic differences significantly influence the distribution of lichens (Nascimbene et al. 2016).

Lichens can withstand a variety of extreme environmental factors, such as strong light exposure, limited water supply, and high temperatures; however, certain species are more sensitive to changes in their immediate environment than others. As a result, lichen diversity and richness are thought to be reliable markers of climate change (Verrall & Pickering 2020).

Endophytes are microorganisms, primarily fungus, that live their whole life cycle inside the tissues of plants without showing any symptoms of illness (Wilson 1995). An essential part of fungal biodiversity is endophytes. To present, about 100,000 fungal species have been identified, making up about 7% of the 1.5 million species thought to exist on Earth (Hawksworth 2004). Endophytes have developed particular connections with their host plants. One of the most varied groups of eukaryotes in terms of taxonomy, ecology, and functions is the fungal kingdom. The wide range of secondary metabolites that fungi make, such as organic acids and enzymes, allows them to colonize different substrates and break down a variety of materials. These substrates include lichen thalli, which support particular endophytic fungi populations. Due to their horizontal transfer and lack of obvious disease signs in lichen thalli, endophytic fungi of lichens are somewhat comparable to endophytic fungi of vascular plants (Arnold et al. 2009 ; U'Ren et al. 2010). (U'Ren et al. 2012; Tripathi and Joshi 2019).

The communities of endophytic fungi in several geographic zones and habitats have been the subject of extensive research in recent decades. The studies used both culture-based and culture-independent molecular approaches and examined a range of lichen species and morphologies on various substrates (Suryanarayanan and Thirunavukkarasu 2017; Oh et al. 2020; Santiago et al. 2021; Muggia et al. 2021; Michalet et al. 2024).

Lichen thalli, however, are still regarded as mycologically understudied habitats that may harbor new fungal taxa (Muggia et al. 2021). Furthermore, endophytic fungi are attractive for research in pharmacology and biotechnology because they generate a range of beneficial secondary metabolites (Tripathi and Joshi 2019; Gao et al. 2016; Singh et al. 2017; Santiago et al. 2021).

2 Saxicolous Lichens

Saxicolous lichens are naturally able to colonize rocky surfaces under harsh conditions and are also reported to be involved in the weathering process (Jung and Büdel et al. 2021) According to reports, they can live on a range of architectural materials, including limestone, marble, and sandstones (Chesnokov et al. 2018), While saxicolous lichens, such as *Parmelia sulcata*, *Lobothallia alphoplaca*, and *Melanelia disjuncta*, show noticeable cytotoxic action, *Xanthoria elegans* and *Xanthoparmelia stenophylla* have been reported to be antioxidants (Mukemre et al. 2021; Simonyan et al. 2020). Saxicolous lichens, such as *Umbilicaria phaea*, *Rhizoplaca melanophthalma* (Kovasi et al. 2022), *Acarospora fuscata* (Altarawneh and Salamon 2021), *Aspicilia calcarea* (Topal et al. 2023), *Caloplaca aff. brouardii*, *Xanthoparmelia mexicana*, etc. (Puy-Alquiza et al. 2017), have demonstrated their potential as biomonitoring agents in addition to their medicinal uses. The potential of lichens to aid in weathering, investigating lichenometry, and terraform has also been investigated (Table 1).

Table 1 Examples of saxicolous lichens (Arya et al. 2024)

Saxicolous lichen	Rock	Type references
Aspicilia intermutans	Granite	Ascaso et al. (1998), Halda and Hur (2016)
Lecidea auriculata		
Thelenella haradae		
Xanthoria elegans	Volcanic andesite	Ascaso et al. (1990)
Lecidea lapicida		
Rhizocarpon geographicum	Volcanogenic sediment	Ascaso et al. (1990)
Bacidia stipata		
Trapelia obtegens	Ultramafic, non-ultramafic	Rajakaruna et al. (2012)
Thelocarpon laureri		
Rhizocarpon saurinum		
Caloplaca oblongula		
Buellia ocellata		
Buellia aethalea		
Parmelia conspersa	Slate rock surface	Armstrong (1981)
Aspicilia calcarea	Chalk	Grishkan and Temina (2019)
Caloplaca oasis		
Lecanora albescens		
Caloplaca crenulatella	Basaltic	Grishkan and Temina (2019)
Diploschistes actinostomus		
Lecanora muralis		
Aspicilia caesiocinerea		
Tephromela grumosa		

(continued)

Table 1 (continued)

Saxicolous lichen	Rock	Type references
Lecidea silacea	Hornfels	Kossowska (2022)
Porpidia melinodes		
Lambiella furvella		
Stereocaulon		
Rhizocarpon oederi		
Xanthoria parietina	Coastal	Kahng et al. (2004)
Caloplaca marina		
Caloplaca aractina	Siliceous	Kinalioglu (2010)
Xanthoria calcicola		
Lecanora carlottiana	Calcareous cliff	Lewis and Sliwa (2012)

3 Applications of Saxicolous Lichens

3.1 Cytotoxicity Activity

Saxicolous lichens have their unique chemical composition, certain species shown antioxidant, antibacterial, and anticancer effects. (Arya et al. 2024; Tiwari et al. 2021). When tested in various cancer cell lines, extracts of *Parmelia sulcata* shown promising anticancer activity and were reported to have significant anticancer activity (Studzińska-Sroka et al. 2021). Using three cancer cell lines—L2OB, RD, and Hep2c—the cytotoxic potential of *Parmelia* lichens, specifically *P. conspersa* (saxicolous) and *Parmelia perlata*, was assessed using the MTT assay, with an IC50 value ranging from 76.33 to 163.39 μg/mL. Certain lichens were found to include salazinic acid (Lewi and Sliwa 2012), atranorin (Zonca et al. 2023), chloroatranorin (Zhao et al. 2021), norstictic acid (Galun 1988), usnic acid (Joly et al. 2009), and stictic acid (Garvie et al. 2008), which may be the source of their medicinal properties, according to phytochemical studies (Manojlovic et al. 2021). 14 saxicolous lichens from the Ladakh region were assessed for their potential to be cytotoxic (using RKO and HepG2 cancer cell lines) and antioxidant (using FRAP, DPPH, ABTS, and NO assays). Among the lichen species that were chosen, methanolic extracts from *Lobothallia alphoplaca* and *Xanthoparmelia stenophylla* had significant outcomes in antioxidant activity using FRAP and ABTS. These extracts also exhibited positive results in a cytotoxic experiment employing cancer cell lines. The n-hexane fraction of *Rhizoplaca chrysoleuca* indicated the highest NO radical scavenging capacity, while the aqueous extract of the saxicolous lichen *Melanelia disjuncta* showed the DPPH method's best antioxidant activity (Kovasi et al. 2022).

3.2 Antimicrobial Activity

Lecanora muralis lichen extracts were evaluated against *Bacillus subtilis*, *S. typhimurium*, *Escherichia coli*, and *S. aureus*. When tested against gram-positive bacteria, acetone, methanol, and ethanolic extracts all had inhibitory activity; however, acetone extract demonstrated the highest activity, with a minimum inhibitory concentration (MIC) of 57 μg/μL against *B. subtilis* (Sargsyan et al. 2023). Parmo tremareticulatum (Taylor) M. Choisy was tested in vitro for antifungal activity against Aspergillus niger, Aspergillus flavus, Fusarium oxysporum, and Fusarium solani. The antifungal effects of lichen were tested in edinacetone, methanol, and chloroform solvents using the disc diffusion method. Percentage inhibition (PI%). Acetone extracts demonstrated the highest efficacy against all infections, followed by methanol. Chloroform extracts showed the least action. Thus, the current analysis suggested that the tested lichen extract has potential antifungal activity and may be explored further in the production of effective antifungal biomolecules. (Babiah et al. 2014). The methanol, ethanol, chloroform, and aqueous extracts of the lichen Parmotrema nilgherrense that were gathered from Nainital, Kumaun Himalaya, have been tested for their antibacterial properties. Using the agar-well method, the extracts were evaluated against five pathogenic bacteria: Xanthomonas phaseoli, Escherichia coli, Agrobacterium tumefaciens, Erwinia chrysanthemi, and Bacillus subtilis. With the exception of the aqueous extract, all P. nilgherrense preparations had a mark zone of inhibition (ZOI), indicating their effectiveness. The results obtained demonstrated that P. nilgherrense extracts have broad-spectrum efficacy against a panel of bacteria that cause the most prevalent illnesses in plants and animals. (Sati and Savita et al. 2011). *Aspicilia cinerea*, a saxicolous lichen, was among the several lichens (*Ochrolechia androgyna*, *Collema cristatum*, *Physcia caesia*, and *Physcia aipolia*) whose antifungal and antibacterial properties were identified. Methanolic and acetone extracts of *A. cinerea* demonstrated antibacterial and antifungal action, with MIC values ranging from 1.56 to 6.25 mg/ml for antibacterial activity and 0.78 to 25 mg/ml for antifungal activity. *Bacillus subtilis*, *B. cereus*, *Staphylococcus aureus*, *Staphylococcus cohnii*, *Gordonia rubripertincta*, *Morganella morganii*, *Salmonella enterica*, *Enterobacter aerogenes*, *Proteus vulgaris*, *Yersinia pseudotuberculosis*, *Saccharomyces cerevisiae*, *Candida tropicalis*, *Candida albicans*, and *Candida parapsilosis* were used to test the antimicrobial and antifungal properties of various extracts (ethanol, chloroform, and water) of the lichen *Protoparmeliopsis muralis* in vitro. The ethanol and chloroform extracts were shown to have greater antibacterial potential than the aqueous extract (Aydin and Kinalioglu 2018).

3.3 Antioxidant Activity

The pharmacological ability and bioactive composition of Xanthoria lichens (X. candelaria, X. elegans, and X. parietina), which are widely farmed and

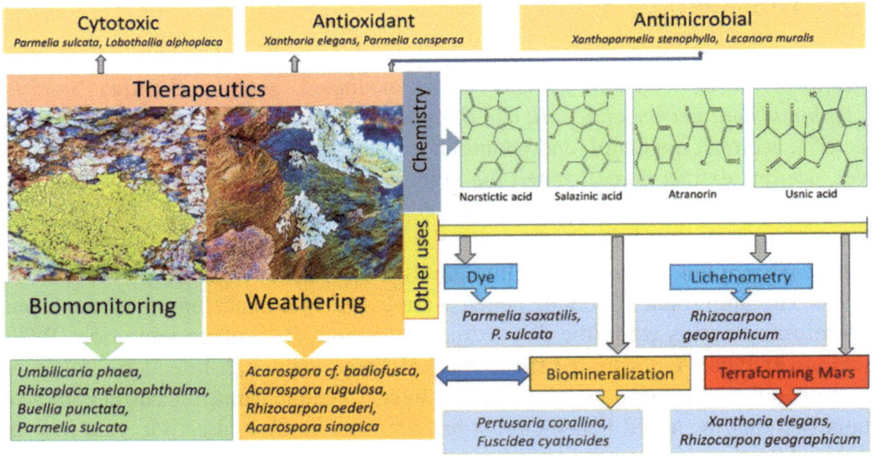

Fig. 1 Applications of saxicolous lichens. (Arya et al. 2024)

traditionally used as medicine in North Eastern Anatolia, Turkey, have distinct microclimatic andecological zones. The extracts contained high quantities of parietin (35-49 mg/g), modest levels of other fatty acids, and a volatile molecule called α-terpinene. The extracts demonstrated significant antioxidant activity via reducing and scavenging mechanisms (Mükemre et al. 2021). In vitro antioxidant and enzyme inhibitory models were used to determine the pharmacological potential of numerous *Xanthoria* species, including *X. elegans* (saxicolous), *X. parietina*, and *Xanthoria candelaria*. Among the principal chemicals proven from the species are parietin (Sonina and Androsova 2020), palmitic acid (Nokes et al. 2022), stearic acid (Ladyman 1996), oleic acid (Sonina et al. 2017), linoleic acid, and α-linolenic acid. The chosen lichens performed well in particular scenarios, making them viable candidates for use as antioxidant and enzyme inhibitory agents (Mukemre et al. 2021). Using the 1,1-diphenyl-2-picrylhydrazyl technique, a methanolic extract of saxicolous lichen thalli of *Xanthoparmelia stenophylla* demonstrated 68% free radical scavenging activity (Simonyan et al. 2020) (Fig. 1).

4 Environmental Influences on Lichen Diversity

4.1 The Effect of Drought on Lichen Diversity

Prior studies on lichens as environmental change indicators have demonstrated that, both in dry lands and subtropics, but particularly in cold climates, their diversity and abundance are influenced by the type and variability of the climate (Koch et al. 2016; Green et al. 2011). Aridity is a significant climatic variable that is crucial for lichen taxonomy and a variety of functions across broad environmental gradients. This lends credence to the theory that climatic gradients significantly influence the distribution of lichens (Green et al. 2011; Nascimbene et al. 2016). Saxicolous lichens are an essential species of this planet due to their remarkable capacity to adapt to the most harsh temperatures on Earth. They can survive in a variety of rock types, including sandstone, granite, volcanic andesite, siliceous, ultramafic, non-ultramafic, basaltic, hornfels, coastal, calcareous cliffs, and so on. These extremophiles have not only demonstrated their endurance from Antarctica to the deserts, but can also resist the extreme conditions of outer space (Arya et al. 2024).

Climate variations, including variations in temperature and precipitation, affect both vascular plants and the range and diversity of lichens in the boreal region (Fitzpatrick et al. 2008; Boch et al. 2016). But because of their poikilohydric nature, which makes them extremely sensitive to temperature and water availability fluctuations, lichens make particularly fascinating model organisms. But because of their poikilohydric nature, which makes them extremely sensitive to temperature and water availability fluctuations, lichens make particularly fascinating model organisms. Both precipitation and nonprecipitation water intakes are significant hydration sources (Raggio et al. 2017). Climate variations, including temperature and precipitation, have an effect on vascular plants and lichen availability and variety in the polar region (IPCC 2013, 2014). But because of their poikilohydric nature, which makes them extremely sensitive to temperature and water availability fluctuations, lichens make particularly fascinating model organisms. Both precipitation and nonprecipitation freshwater intakes are crucial sources for hydration (Gutiérrez et al. 2017). According to the Intergovernmental Panel on Climate Changes, there will be more frequent and severe droughts in addition to exceptionally high temperatures during the next few decades (IPCC 2014; Gutiérrez et al. 2017). These predictions are expected to have an impact on the distribution, quantity, and variety of lichens (Gutiérrez et al. 2017). While lichens are a reliable indication of climatic change, as evidenced by research (Nascimbene et al. 2016; Pinho et al. 2010; Colesie et al. 2014), very few have examined how lichen diversity reacts to aridity (Branquinho and Matos 2015; Concostrina-Zubiri et al. 2018). For the drylands of Mexico, which make up almost 65% of the nation's surface area, this is a reality (Semarnat et al. 2016). According to Maestre et al. (2011), two traits of drylands that reveal how arid an ecosystem are low water availability and high temperatures. Due to a global trend toward warmer and drier air, this will affect lichen-dominated dryland biocrust communities and their role in regulating ecosystem functioning globally (Reed et al. 2019; Baldauf et al. 2021). The absence of available water is known as aridity, and it can be measured using a variety of indices. The distribution, variety, and development of plant species have all been assessed using this index (Sáenz-Romero et al. 2010; Soto-Correa et al. 2014). Similar to plants, the physiology of lichens is directly impacted by water scarcity through lichen water relations (Raggio et al. 2017), such as thallus hydration and desiccation (Gauslaa et al. 2012), which determines the lichens' ability to establish and thrive. But lichen diversity is expected to show remarkable resistance to water availability, which is a unique trait. It is expected that lichen diversity will show discernible changes throughout aridity gradients because lichen tolerance to water availability is a distinctive characteristic (Giordani 2019).

4.2 The Effect of Fires on the Growth of Lichens

In addition to being a natural occurrence in many ecosystems worldwide, particularly in grassy formations (Kalfas 2024) fire is also a valuable tool for management, impacting and preserving a wide variety of vegetation types (Dowdy et al. 2022). Fire is most likely the primary factor restricting species composition and physiognomy in some grassland areas of southern Brazil (Pillar and Quadros 1997), particularly on grassland-forest ecotone zones (Müller et al. 2012). However, because there are few experimental and observational researches on this topic, the consequences of fire on species and functional diversity are still unknown (Overbeck et al. 2007, Müller et al. 2007). It is well recognized that different grassland ecosystem types, structures, and levels of flammability exhibit different fire behavior (Miller et al. 2018). Similar types of grasslands may also experience significant variations in fire intensity (Fidelis et al. 2010). As a result, it's critical to understand the causes and effects of these variations in community patterns following fire occurrences (Johansson and Reich 2005). The presence of rocky outcrops and other spatial variation within the grasslands affects fire intensity and may facilitate the emergence of distinct assemblages, above or next to rocks (Clarke 2002; Carlucci et al. 2011). In the southern Brazilian grasslands, rocky outcrops are somewhat prevalent (Carlucci et al. 2011), and lichens are frequently found covering them. Because they use certain lichen acids to start the slow erosion of the rocks, these lichens are pioneer species that are crucial to ecosystems (Valencia and Ceballos 2002). Other species, such mosses and eventually vascular plants may then be able to establish themselves as a result of this alteration in the rock's surface (Cozzolino et al. 2022). Lichens are found in a large geographic range (Valencia and Ceballos 2002), and they are primarily affected by light and humidity (Marcelli 1992), in addition to the physical and chemical properties of their substrate (Brodo 1973; Marcelli 1996). Lichens are greatly affected by fire disturbances, and most species are generally significantly reduced in burned areas (Holt and Severns 2005). The slow rate of colonization by most species and the low rate of thallus growth may account for their susceptibility to fire (Schulten 1985). Furthermore, lichens are likely dry and metabolically inactive after being partially burned because they need moisture to be photosynthetically active, inactive, which could have a significant impact on their capacity to regenerate from tiny pieces (Romagni and Gries 2000). Despite the fact that lichens are crucial components of ecosystems and aid in the cycling of nutrients, lichen recolonization following fire disruption has received little attention (Knops et al. 1996). The effects of fire on lichen communities have only been the subject of a small number of studies to date (Johansson and Reich 2005; Dunford et al. 2006), and none of them used a functional approach. The patterns of lichen communities may vary and depend on context, given that such effects may also depend on fire intensity and frequency, which are connected to the lichen substrate and the surrounding vegetation features.

5 Use Saxicolous Lichens as Bioindicators of Nitrogen

Saxicolous lichens have been utilized for terraforming, biomonitoring, weathering, lichenometry, and dyeing in addition to their therapeutic uses and textiles since ancient times. In parts of Western Ireland and the Highlands, it was rather popular to color wool from *Parmelia saxatilis* for using boiling water, dimethyl sulfoxide, and ammonia fermentation to dye silk cloth. According to reports, *Parmelia sulcata* gives textiles a reddish hue and lives in corticolous and saxicolous environments. It will be feasible to extend the collection of deposition-related biomonitoring data to high elevation wilderness regions if saxicolous lichen species are found to be trustworthy bioindicators of N deposition. Similar to epiphytic lichens in certain ways. In the western United States, lichen biomonitoring procedures frequently use Letharia vulpina as a bioindicator of N deposition (Jovan and Carlberg, 2007, Root et al. 2013). However, Letharia vulpina and other epiphytic (tree-dwelling) species are scarce or nonexistent in a large portion of the high elevation wilderness in the southern half of California. In remote locations like wilderness areas, deposition monitoring equipment is both expensive and unfeasible, which leaves gaps in our knowledge of deposition there. Some saxicolous (rock-dwelling) lichens have been suggested by previous studies as potential target species for biomonitoring in this area (Jovan et al. 2021). Because of the influence of their substrate rock and soils, saxicolous lichens typically have higher concentrations of N, S, and crustal elements like calcium, magnesium, iron , silicon, potassium, and aluminum than epiphytic lichens. This could confuse the elemental signal of air pollution (Fenn et al. 2007). In contrast, some elemental concentrations (antimony, palladium, copper, cadmium, tin, and zinc were linked to anthropogenic sources, while others were related to rock substrate (Zschau et al. 2003). In the western United States, however, saxicolous lichen N and S concentrations indicated N and S deposition (Fenn et al. 2007).

6 Role of Lichens in Weathering

It has been reported that a variety of saxicolous crustose lichen species, including A. carospora cf. badiofusca, A. sinopica, A. smaragdula, and A. rugulosa, can cause unusual rocks to weather (Purvis 2014). Lecidea, Bacidia stipata, and Xanthoria elegans. Rhizocarpon geographicum and R. lapicida, which live on volcanic sites, contribute in the weathering process (Ascaso et al.1990). Lichens have the ability to produce lichen acids, which interact with the mineral makeup of rocks to promote the weathering process. Lichens that live on rocks have the ability to accelerate the deterioration of their substrate. The physical and weathering of rocks is discovered to be influenced by these epilithic and endolithic lichens, which can pierce the exterior or interior surface of rocky substrates (Jie and Blume 2002). Lichens grow entirely or partially within a mineral substrate, while epilithic lichens use hyphae or rhizines to adhere extensively on a rock surface (Chen et al. 2000). When rhizines or individual fungal hyphae infiltrate mineral cleavage planes and crevices, the fungal tissue expands and contracts during wetting and drying, freezing and hawing, or swelling is caused by the deposition of organic and inorganic compounds. This process is known as biogeophysical weathering of rocks (Adamo and Violante 2000). It has been reported that a variety of saxicolous crustose lichen species, including *Acarospora* cf. *badiofusca*, *Acarospora sinopica*, *Acarospora smaragdula*, and *Acarospora rugulosa*, can cause unusual rocks to weather (Purvis 2014). *Lecidea*, *Bacidia stipata*, and *Xanthoria elegans*, *Rhizocarpon geographicum* and *R. lapicida*, which live on volcanic sites, contribute in the weathering process (Ascaso et al. 1998).

7 Diversity of Endophytic Fungi of Lichens

Lichens are linked to endolichenic fungi that live inside their thalli in addition to the fungal partner in this mutualism. Numerous endolichenic fungi are a source of novel compounds with distinctive bioactivities, and they seem to have evolved alongside lichens. The biology of endolichenic fungus and how they interact with the other elements of a lichen microbiome are little understood. Endophytic fungal diversity was found in the thalli of fossil lichens dating back to the Lower Devonian. (Suryanarayanan and Thirunavukkarasu 2017), fungal diversity, indicating that lichen thallus growth is an effective fungal tactic. The primary lineages of *Ascomycota* include endophytic fungi, to which plant endophytic fungi also belong They differ from the mycobiont, which is one of the other fungal companions of lichen (Arnold et al. 2010; Lutzoni and Miadlikowska 2009). Lichens are symbiotic organisms that contain a variety of microbes. Endolichenic fungi are lichen thalli dwellers with potential ecological and economic benefits due to their diverse secondary metabolites. Endolichenic fungi, like endophytic fungi, can have an impact on lichen diversity and ecosystem sustainability. This emphasizes the significance of knowing the diversity and structure of communities of endolichenic fungi, to evaluate the diversity and factors impacting the endolichenic fungi community structure. A total of 112 Endolichenic fungi species were found, with Xylariales in Sordariomycetes being the most frequent. The host taxonomy, as well as photobiont types and environmental conditions, had a considerable impact on the richness and organization of endolichenic fungi communities. Our findings indicate that more lichen species in a wider range of habitats should be studied in order to improve our understanding of endolichenic fungi diversity and ecology (Oh et al. 2020). There was minimal overlap between the endophytes of the plant that these lichens were developed on and the endophytic fungus of some lichens at the genus level (Suryanarayanan et al. 2005). To verify whether a method of selection is at work in the lichens' recruitment of endophytic fungus, more information is needed. The endophytic fungi rely on the photobiont for sustenance, just as other heterotrophic lichen companions. It is known that different lichen species have different photobiont species (Ruprecht et al. 2014). On Jeju Island in Korea, a varied endophytic fungus biota of (112) species was recovered from (79) lichen samples. These endophytic fungus biota can be considered among the most numerous at the regional level. Most investigations identified roughly 20-60 species on a regional scale. (Muggia et al. 2017; Yu et al. 2018) and because host lineage, environment, and geographical location all affect endophytic fungal diversity (U'Ren et al. 2012), this becomes even more significant. In contrast to endophytes, endophytic fungi exhibit generalist behavior and are not absolutely host specific, according to a recent investigation by Chagnon et al. (2016). *Xylaria*, *Colletotrichum*, *Pestalotiopsis*, and *Phomopsis* are a few endophytic fungus. *Chaetothyriomycetes* and *Dothideomycetes* contain fungal strains of new lineages, according to a multilocus genetic investigation of alpine rock lichen fungal. Approximately 200 endolichenic fungi species were isolated, the most ever discovered in a single investigation. The majority of endolichenic fungi species isolated in our study belonged to Ascomycota, which is consistent with prior research revealing Ascomycetes as the dominant endolichenic fungi species (Zhang et al. 2016; Suryanarayanan et al. 2017; U'Ren, et al. 2010). Sordariomycetes has the mostendolichenic fungi species, followed by Dothideomycetes. Previous research has also shown that Sordariomycetes are dominant in endolichenic fungi species (in China, France, Japan, North America, Norway, and Sri Lanka) (Arnold et al. 2009; Lagarde et al. 2018; Masumoto & Degawa 2019) In contrast, other research found that Dothideomycetes, Leotiomycetes, or Pezizomycetes were dominant (in Antarctica, Germany, Italy, and the United States) (De Carvalho et al. 2019; U'Ren, et al. 2010) Basidiomycete genomes are a valuable novel tool for studying fungal-fungal interactions in lichen symbioses (Tagirdzhanovaet al. 2021). Molecular data provide unparalleled insight into the variety of lichenized fungi, although morphologically cryptic species-level lineages delineated by sequence data frequently go undescribed, even in well-studied groups. These species can be recognized by differences in nucleotide position characteristics in the fungal barcoding marker (ITS), as well as high speciation probabilities. (Leavitt et al. 2016) Furthermore, finicky endophytic fungi that do not grow on regular growth medium may be isolated using lichen-enriched growth media (Biosca et al. 2016; Hussein et al. 2024). Hylogenetic ideas give a framework for investigating the evolution of heterotrophic lifestyles. Lichens, which are fungi that live in symbiotic relationships with algae, can be found in a variety of Dicaryomycotina species, including Ascomycetes and Basidiomycetes. A highly resolved parsimony study of DNA (SSU rDNA) sequences indicates that the lichen habit originated independently in at least five different genera of Ascomycetes and Basidiomycetes. Because lichen associations evolved from parasitic, mycorrhizal, or free-living saprobic fungi, mutualism and parasitism should not be interpreted as ends in symbiont evolution. (Gargas et al. 1995). It is uncertain how endophytic fungal assemblages fare in these interactions. Likewise, it is necessary to address the endophytic fungal state in maritime and "borderline" lichens. To comprehend the pattern of distribution and host specificity of endophytic fungi of various lichens flourishing in various settings, as well as the impact of biotic and abiotic factors on their diversity, more information is required.

8 Role of Endophytic Fungi in the Saxicolous Lichens

The functional activities and biotechnological possibilities of bacterial companions of lichens are more understood than those of endophytic fungi (Suzuki et al. 2016; 2009). These might be used as a starting point for dealing with endophytic fungal ones. According to Grube et al. (2015), a number of bacterial communities are consistently linked to lichens and support lichen symbiosis by assisting with nutrition delivery, enhancing lichen resistance to biotic and abiotic stress, detoxifying metabolites, and breaking down older thallus portions. The part endophytic fungi play in the growth and survival of lichens is unknown. There is no information on how endolichenic fungi contribute to lichen development and survival. While the age and chemistry of the host tissue influence endophyte assemblage in plants, it is unclear whether the age and chemical composition of a lichen influence the density of colonization and species composition of its endolichenic fungi (Suryanarayanan and Thennarasan 2004). By increasing their resistance to abiotic stressors including drought, salinity, and high temperatures, endophytes improve the environmental health of their host plants (Sun et al. 2010. Such enhanced host fitness is at least occasionally the consequence of endophyte-mediated activation of stress-related host genes (Sherameti et al. 2008). By changing the metabolism of growth hormones, endophytes may also improve a plant's resistance to abiotic stress (Kamran et al. 2022). Some lichens can withstand extended exposure to desiccation, high temperatures (90 °C and 196 °C), and intense UV radiation. Lichens thrive in hard settings and show a broad tolerance to abiotic stresses (Grube and Berg 2009; De Los Ríos et al. 2005). Some of them, such as *Xanthoria elegans*, can withstand exposure to both real space conditions and a Martian atmosphere simulation (Lorenz et al. 2023). If endophytic fungus, like plant endophytes, helps their lichen hosts tolerate abiotic stress, that would be interesting to know. In particular plants, the endophytes increase the host plant's resistance to specific biotic stresses, making up for the cost of harboring them (through the loss of photosynthates). Endophytes, for example, protect their host plants from infections and insect pests (Waqas et al. 2015; Estrada et al. 2015). Many defense-related genes are upregulated in the plant host as a result of endophyte infection (Mengistu 2020). Whether endolichenic fungus affect the regulation of genes in their lichen hosts, similar to endophytes, has not yet been determined. Numerous antibacterial compounds are produced by the endophytic microbiota. According to Cernava et al. (2015), a culture-based and metagenomic method revealed that the lichen *Lobaria pulmonaria* contains a bacterial population that produces a variety of microbial antagonistic chemicals. Regarding endophytic fungi, 45 and 37 isolates, respectively, demonstrated antialgal and antifungal activity, while 30 isolates shown both antialgal and antifungal activity out of the

62 isolates that were screened for the formation of antialgal and antifungal metabolites (Suryanarayanan et al. 2017; Waill et al. 2023). Lichens have a high medicinal potential, and various businesses are currently seeking to commercialise these unique compounds. Lichen produces secondary metabolites with a variety of biological functions, including antibacterial, antiviral, anticancer, antioxidant, antiherpes, allelochemical, insecticidal, and allergenic propertie (Waill et al. 2023). These findings imply that the lichen thallus's capacity to withstand biotic stress may be influenced by the bacteria and fungi that inhabit it. To verify their impact on changing a lichen's fitness, a thorough examination of the metabolite spectrum of the different endophytic fungi is required. According to Kumar et al. (2014), lichens are a desirable source of distinctive secondary metabolites. Lichens that grow from the Arctic to the tropics have yielded more than 1000 secondary metabolites thus far (Stocker-Wörgötter 2008). Secondary metabolites from the acetyl polymalonyl, mevalonic, and shikimate pathways make up between 0.1% and 5% of the dry weight of lichen thallus (Stocker-Wörgötter 2008). Many of these metabolites have a variety of technologically appealing bioactivities (Bačkorová et al. 2012), and the majority are of fungal origin (Abdel-Hameed et al. 2016; Stocker-Wörgötter 2008). They also serve ecological purposes, such as regulating symbiosis. Like plant endophytes (Kaushik et al. 2014; Suryanarayanan et al. 2009; Chen et al. 2016), endophytic fungi themselves are a good source of novel bioactive compounds in addition to lichens (Kellogg and Raja 2016; Chang et al. 2015). After that, it would be beneficial to check endophytic fungi for the production of novel compounds with particular biological activities. Interactions between endolichenic fungus and the host lichen as well as between endophytic fungi and other kinds of species that lichens support are another area that needs to be explored. The endophytic fungal population of a lichen should be insensitive to or capable of detoxifying the antifungal compounds that are produced by many lichens, Endophytic fungi produce industrially valuable enzymes and bioactive secondary metabolites, as well as the ability to breakdown or sequester inorganic and organic small molecules and macromolecules (for example, poisons, pollutants, and heavy metals) (Basile et al. 2015; Liu et al. 2021). One could presume that the endophytic fungal assemblage of a particular lichen would consist solely of these fungi. Such endophytic fungi could be evaluated for their capacity to break down resistant compounds and biotransform metabolites to create new medications (Strobel & Daisy 2003; Wang and Dai 2011). The bacterial community of the Alpine lichen *Solorina crocea* is known to change when it becomes infected by fungi (Grube et al. 2012). Similar to this, interspecific competition may ultimately determine the species composition of bacterial and fungal assemblages within the lichen thallus as a result of endophytic fungi's synthesis and detoxification of antimicrobial chemicals in vivo. Since endophytic fungi also generate antimicrobial metabolites (Suryanarayanan et al. 2017), it is necessary to investigate competition and cometabolism within the lichen microbiome using contemporary methods like isotopologue profiling in order to comprehend the interactions among the different components of the lichen microbiome (Götz et al. 2010; Bogale et al. 2024). Understanding endolichenic fungi's role in the secondary metabolite profile of lichens and their potential role

in the lichen microbiome would be made easier with gene-level analysis. Similar to plant endophytes, endolichenic fungi may be able to produce a variety of secondary metabolites because they possess novel genes. These genes may have developed as a result of selection pressure from interactions between various lichen microbiome entities. Under typical growing circumstances, the vast majority of fungal natural product genes are silent. To activate these cryptic secondary metabolite genes, a variety of techniques are used, such as proteomic analysis of secondary metabolism and genomic mining, genetic modification of genes involved in biosynthesis pathways, and the application of epigenetic modifiers (Lim et al. 2012; Singh 2023). Using these contemporary techniques to mine endolichenic fungi for novel secondary metabolites may result in the production of pharmaceutically significant compounds. Lichens are early colonizers of rocks that contribute to biological weathering of rocks as well as other biochemical processes including nutrient cycling and soil formation. Radionuclides and heavy metals are also accumulated by lichens (Gadd 2007; Zhang et al. 2024). While fungi are known to scavenge heavy metals, withstand radionuclides, and weather rocks by generating carbonic and other acids (Cordero and Casadevall 2017; Hoffland et al. 2004), it is unknown how endolichenic fungus contribute to these lichen-exhibited behaviors. Plant endophytic fungi provide a new source of a variety of enzymes, such as medicinal and biomass-degrading enzymes. Numerous of these fungi generate new forms of cellulase, laccase, lipase, pectinase, chitinases, and chitosanses as well as proteases (Govinda Rajulu et al. 2011; Thirunavukkarasu et al. 2017). Salt-tolerant xylanases and xylosidases are produced by endophytic fungi of marine plants (Thirunavukkarasu et al. 2015). Such enzymes have not been tested for in endophytic fungus. Lichen-associated bacteria break down portions of lichen thalli to facilitate the mobilization of biomass (Grube and Berg 2009). Degradation of dead lichen thalli (epinecral layers of lichen thalli) may also be facilitated by endophytic fungi. Lichens maintain a microbiome made up of many microorganisms and are an example of an evolutionary effective mutualism. A distinct microenvironment is created within the lichen thallus as a result of the intricate interactions between these bacteria and the environment. In such an environment, microbes may be a valuable source of metabolites of biotechnological significance. The lichen microbiome's endophytic fungus is the least researched in this regard. As has been assumed for bacterial and lichen connection, endophytic fungal interaction may have developed alongside the lichens (Morillas et al. 2022). Given the small number of lichen species that have been investigated in relation to endophytic fungi, it is logical to assume that this ecological category of fungi contains a large number of as-yet-undescribed fungal species (Muggia et al. 2016). The endophytic fungi may also include a large number of unique genes whose byproducts could be used in technological applications. This becomes even more significant when we take into account that the diversity of lichens is still poorly understood (Ebadi & Ebadi 2024). Understanding the function of endolichenic fungus is essential to appreciating lichens' frequently disregarded function as ecological process determinants (Asplund and Wardle 2017).

9 Conclusion

Lichens maintain a microbiome made up of many microorganisms and are an example of an evolutionary effective mutualism. A distinct microenvironment is created within the lichen thallus as a result of the intricate interactions their surroundings. In such an environment, microbes may provide valuable compounds of biotechnological significance.

The lichen microbiome's endophytic fungus is the least researched in this regard. Endophytic fungal relationship may have developed alongside lichens (Grube and Berg 2009). Given the small number of lichen species that have been investigated in relation to endolichenic fungi, it is logical to assume that this ecological category of fungi contains a large number of as-yet-undescribed fungal species (Muggia et al. 2016). Additionally, it's probable that the endophytic fungus includes a large number of unique genes whose products could be used in technological applications. This becomes even more significant when we take into account that the diversity of lichens is still poorly understood (Ebadi &Ebadi 2024). To grasp lichens' often-overlooked role as drivers of ecological processes, one must comprehend the function of endolichenic fungus.

References

Abdel-Hameed M, Bertrand RL, Piercey-Normore MD, Sorensen JL (2016) Putative identification of the usnic acid biosynthetic gene cluster by de novo whole-genome sequencing of a lichen-forming fungus. Fungal Biol 120:306–316

Altarawneh R, Salamon I (2021) Assessment of lichens as biomonitors of heavy metal pollution in selected mining area, Slovakia. Pak J Anal Environ Chem 22(1):53–59

Armstrong RA (1981) Field experiments on the dispersal, establishment and colonization of lichens on a slate rock surface. Environ Exp Bot 21(1):115–120

Arnold EM, Ward SR, Lieber RL, Delp SL (2010) A model of the lower limb for analysis of human movement. Ann Biomed Eng 38(2):269–279. https://doi.org/10.1007/s10439-009-9852-5. Epub 2009 Dec 3. PMID: 19957039; PMCID: PMC2903973

Arya V, Gill AK, Jamwal A (2024) Saxicolous lichens: from therapeutics to terraforming. In: Sustainable trends in engineering, sciences and social sciences. Research Foundation of India Publication

Ascaso C, Wierzchos J, De los Rios A (1998) In situ cellular and enzymatic investigations of saxicolous lichens using correlative microscopical and microanalytical techniques. Symbiosis 24(1998):221–234

Asplund J, Bokhorst S, Kardol P, Wardle DA (2015) Removal of secondary compounds increases invertebrate abundance in lichens. Fungal Ecol 18:18–25

Aydin S, Kinalioglu K (2018) Comparative antimicrobial activity of crude extracts of protoparmeliopsis muralis and parmotrema perlatum lichens. Gazi Univ J Sci 31(3)

Ascaso C, SanchoG & Rodriguez-Pascual C (1990) The weathering action of saxicolous lichens in maritime Antarctica. Polar Biol 11:33–39

Arnold AE, Miadlikowska J, Higgins KL, Sarvate SD, Gugger P, Way A, Hofstetter V, Kauff F, Lutzoni FA (2009) phylogenetic estimation of trophic transition networks for ascomycetous Fungi: Are lichens cradles of symbiotrophic Fungal diversification? Syst Biol 58:283–297

Adamo P, & Violante P (2000) Weathering of rocks and neogenesis of minerals associated with lichen activity, Applied Clay Science 16:(5–6)229–256

Asplund J, and Wardle DA (2017) How lichens impact on terrestrial community and ecosystem properties. Biol. Rev. 92:1720–1738. https://doi.org/10.1111/brv.12305

Bačkorová M, Jendželovskýa R, Kelloa M, Bačkorb M, Mikeša J, Fedoročko P (2012) Lichen secondary metabolites are responsible for induction of apoptosis in HT-29 and A2780 human cancer cell lines. Toxicol In Vitro 26:462–468

Baldauf S, Porada P, Raggio J, Maestre FT, Tietjen B (2021) Relative humidity predominantly determines long-term biocrust-forming lichen cover in drylands under climate change. J Ecol 109:1370–1385. https://doi.org/10.1111/1365-2745.13563

Basile A, Rigano D, Loppi S, Di Santi A, Nebbioso A, Sorbo S, Conte B, Paoli L, De Ruberto F, Molinari AM et al (2015) Antiproliferative, antibacterial and antifungal activity of the lichen Xanthoria parietina and its secondary metabolite parietin. Int J Mol Sci 16:7861–7875

Biosca EG, Flores R, Santander RD, Díez-Gil JL, Barreno E (2016) Innovative approaches using lichen enriched media to improve isolation and culturability of lichen associated bacteria. PLoS One 11(e0160328):pone.0160328. https://doi.org/10.1371/journal

Boch SD, Schöning I, Fischer M (2016) Acer pseudoplatanus, fagus sylvatica, fraxinus excelsior and quercus petraea. Biodivers Conserv 25(2):225–238. https://doi.org/10.1007/s10531-015-1037-y

Branquinho C, Matos P (2015) Lichens as ecological indicators to tack atmospheric changes: future challenges. In: Lindenmayer D, Barton P, Pierson J (eds) Indicators and surrogates of biodiversity and environmental change. Csiro, Melburne/London, p 216

Brodo IM (1973) Substrate ecology. In: Ahmadjian V (ed) The lichens. Academic Press, London, pp 401–441

Babiah PS, Upreti DK, & John (2014) An invitro analysis of antifungal potential of lichen species Parmotrema reticulatum against phytopathogenic fungi. Int J Curr Microbiol App Sci 3(12): 511–518

Bogale AT, Braun M, Bernhardt J, Zühlke D, Schiefelbein U, Bog M, Scheidegger C, Zengerer V, Becher D, Grube M, Riedel K, Bengtsson MM (2024) The microbiome of the lichen Lobaria pulmonaria varies according to climate on a subcontinental scale. Environ Microbiol Rep. 16(3): e13289. https://doi.org/10.1111/1758-2229.13289

Carlucci MB, Duarte LDS, Pillar VD (2011) Nurse rocks influence forest expansion over native grassland in southern Brazil. J Veg Sci 22:111–119. https://doi.org/10.1111/j.1654-1103.2010.01229.x

Cernava T, Müller H, Aschenbrenner IA, Grube M, Berg G (2015) Analyzing the antagonistic potential of the lichen microbiome against pathogens by bridging metagenomic with culture studies. Front Microbiol 6:620

Chagnon P-L, U'Ren JM, Miadlikowska J, Lutzoni F, Arnold AE (2016) Interaction type influences ecological network structure more than local abiotic conditions: evidence from endophytic and endolichenic fungi at a continental scale. Oecologia 180:181–191

Chang W, Zhang M, Li Y, Li X, Gao Y, Xie Z, Lou H (2015) Lichen endophyte derived pyridoxatin inactivates Candida growth by interfering with ergosterol biosynthesis. Biochim Biophys Acta 1850(9):1762–1771

Chen C, Hu SY, Luo DQ, Zhu SY, Zhou CQ (2016) Potential antitumor agent from the endophytic fungus Pestalotiopsis photiniae induces apoptosis via the mitochondrial pathway in HeLa cells. Oncol Rep 30:1773–1781

Chesnokov S, Konoreva L, Paukov A (2018) New species and records of saxicolous lichens from the Kodar range. Trans-Baikal Territory, Russia

Clarke PJ (2002) Habitat insularity and fire response traits: evidence from a sclerophyll archipelago. Oecologia 132:582–591. https://doi.org/10.1007/s00442-002-0962-0

Colesie C, Green A, Haferkamp I, Büdel B (2014) Habitat stress initiates changes in composition, CO2 gas exchange and C-allocation as life traits in biological soil crusts. ISME J 8:2115–2115. https://doi.org/10.1038/ismej.2014.47

Concostrina-Zubiri L, Matos P, Giodani P, Branquinho C (2018) Biocrust tissue traits as potential indicators of global change in the Mediterranean. Plant Soil 429:159–174. https://doi.org/10.1007/s11104-017-3483-7

Cordero RJB, Casadevall A (2017) Functions of fungal melanin beyond virulence. Fungal Biol Rev. https://doi.org/10.1016/j.fbr.2016.12.003

Cozzolino A, Adamo P, Bonanomi G, & Motti R (2022) The Role of Lichens, Mosses, and Vascular Plants in the Biodeterioration of Historic Buildings: A Review. Plants 11(24):3429. https://doi.org/10.3390/plants11243429

De Los Ríos A, Wierzchos J, Sancho LG, Green TGA, Ascaso C (2005) Ecology of endolithic lichens colonizing granite in continental Antarctica. Lichenologist 37:383–395

De Vera JP, Möhlmann D, Butina F, Lorek A, Wernecke R, Ott S (2010) Survival potential and photosynthetic activity of lichens under Mars-like conditions: a laboratory study. Astrobiology 10:215–227

Dunford JS, McLoughlin PD, Dalerum F, Boutin S (2006) Lichen abundance in the peatlands of northern Alberta: implications for boreal caribou. Ecoscience 13:469–474. https://doi.org/10.2980/1195-6860(2006)13%5B469:LAITPO%5D2.0.CO;2

Dowdy A, Purcell L, Boulter S, Moura LC (2022) Wildfires under Climate Change: A Burning Issue. In Frontiers 2022 Report: Emerging Issues of Environmental Concern; United Nations Environment Programme: New York, NY, USA, pp.23–49

De Carvalho CR, Santiago IF, da Costa Coelho L, Câmara PEAS, Silva MC, Stech M, Rosa CA, Rosa LH (2019) Fungi associated with plants and lichens of Antarctica. In Fungi of Antarctica; Rosa, L.H., Ed.; Springer International Publishing: Cham, Switzerland pp. 165–199

Estrada C, Degner EC, Rojas EI, Wcislo WT, Van Bael SA (2015) The role of endophyte diversity in protecting plants from defoliation by leaf-cutting ants. Curr Sci 109:55–61

Elkhateeb WA, & Dina E El-Ghwas, Ghoson M Daba (2023) Lichen Secondary Metabolites as Anti-Fungal Agent. J Pharm Drug Dev 2(1). https://doi.org/10.58489/2836-2322/01

Ebadi M, & Ebadi A (2024) Importance and Applications of Endophytic Fungi. IntechOpen. https://doi.org/10.5772/intechopen.1006383

Fidelis A, Delgado-Cartay MD, Blanco CC, Müller SC, Pillar VD, Pfadenhauer J (2010) Fire intensity and severity in Brazilian Campos grasslands. Interciencia 35:739–745

Fitzpatrick CM, Gove DA, Sanders JN, Dunn RR (2008) Climate change, plant migration, and range collapse in a global biodiversity hotspot: the banksia (Proteaceae) of Western Australia. Glob Chang Biol 14:1337–1352. https://doi.org/10.1111/j.1365-2486.2008.01559.x

Fenn ME, Geiser L, Bachman R, Blubaugh TJ, A Bytnerowicz A (2007) Atmospheric deposition inputs and effects on lichen chemistry and indicator species in the Columbia River Gorge, USA Environ Pollut 146:(1)77–91

Gadd GM (2007) Geomycology: biogeochemical transformations of rocks, minerals, metals and radionuclides by fungi, bioweathering and bioremediation. Mycol Res 111:3–49

Galun M (ed) (1988) CRC handbook of lichenology, vol 2. CRC Press, Boca Raton, pp 153–169

Gao LD et al (2016) Change of urinary cadmium and renal tubular protein in female works after cessation of cadmium exposure : response to the letter to the editor by Kawada (2016). Int Arch Occup Environ Health 90(3):307. https://doi.org/10.1007/s00420-017-1194-2. Epub 2017 Jan 12. PMID: 28083706

Garvie LA, Knauth LP, Bungartz F, Klonowski S, Nash TH (2008) Life in extreme environments: survival strategy of the endolithic desert lichen Verrucaria rubrocincta. Naturwissenschaften 95: 705–712

Gauslaa Y, Coxson DS, Solhaug KA (2012) The paradox of higher light tolerance during desiccation in rare old forest cyanolichens than in more widespread co-occurring chloro- and cephalolichens. New Phytol 195(4):812–822. https://doi.org/10.1111/j.1469-8137.2012.04221.x

Götz A, Eylert E, Eisenreich W, Goebel W (2010) Carbon metabolism of enterobacterial human pathogens growing in epithelial colorectal adeno carcinoma (Caco-2) cells. PLoS One 5:e10586

Govinda Rajulu MB, Thirunavukkarasu N, Suryanarayanan TS, Ravishankar JP, El Gueddari NE, Moerschbacher BM (2011) Chitinolytic enzymes from endophytic fungi. Fungal Divers 47:43–53

Green TA, Sancho LG, Pintado A, Schroeter B (2011) Functional and spatial pressures on terrestrial vegetation in Antarctica forced by global warming. Polar Biol 34:1643–1656. https://doi.org/10.1007/s00300-011-1058-2

Grishkan I, Temina M (2019) Interior of saxicolous lichens on diferent types of rocks as a habitat for microfungal communities in upper Galilee, Israel. Acta Mycol 54(1)

Grube M, Berg G (2009) Microbial consortia of bacteria and fungi with focus on the lichen symbiosis. Fungal Biol Rev 23:72–85

Grube M, Köberl M, Lackner S, Berg C, Berg G (2012) Host– parasite interaction and microbiome response: effects of fungal infections on the bacterial community of the alpine lichen Solorina crocea. FEMS Microbiol Ecol 82:472–481

Grube M, Cernava T, Soh J, Fuchs S, Aschenbrenner I, Lassek C, Wegner U, Becher D, Riede K, Sensen CW et al (2015) Exploring functional contexts of symbiotic sustain within lichen-associated bacteria by comparative omics. ISME J 9:412–424

Gutiérrez Gutiérrez M, Pando-Moreno M, González Rodríguez H, Mendoza Aguilar D (2017) Efecto del micrositio en la composición de costras biológicas del suelo en un área de matorral micrófilo del desierto chihuahuense, México. Interciencia 42(4):212–214

Giordani P (2019) Lichen Diversity and Biomonitoring: A Special Issue. Divers 11(9):171. https://doi.org/10.3390/d1109017153

Gargas A, DePriest PT, Grube M, Tehler A (1995) Multiple origins of lichen symbioses in fungi suggested by SSU rDNA Phylogeny Sci 268(5216):1492–5. https://doi.org/10.1126/science.7770775

Halda JP, Hur JS (2016) Thelenella haradae sp. nov., a saxicolous lichen from South Korea. Mycotax 131(4):805–809

Hawksworth DL (2004) Fungal diversity and its implications for genetic resource collections. Stud Mycol 50:9–18

Hoffland E, Thomas W. Kuyper, Håkan Wallander, Claude Plassard, Anna A. Gorbushina, Kurt Haselwandter, Sara Holmström, Renske Landeweert, Ulla S. Lundström, Anna Rosling, Robin Sen, Mark M. Smits, Patrick AW van Hees, Nico van Breemen (2004) The role of fungi in weathering. Front Ecol Environ 2:(5)258–264. https://doi.org/10.1890/1540-9295(2004)002[0258:TROFIW]2.0.CO;2

Holt EA, Severns PM (2005) The effects of prescribed burning on wet prairie lichen communities. Nat Areas J 25:130–136

Hussein JM, Myovela H, Tibuhwa DD (2024) Diversity of endophytic fungi from medicinal plant Oxalis latifolia and their antimicrobial potential against selected human pathogens. Saudi J Biol Sci 31(4):103958. https://doi.org/10.1016/j.sjbs.2024.103958

IPCC (2013) Climate change 2013: the physical science basis. In: Stocker TF, Qin D, Plattner GK, Tignor M, Allen SK et al (eds) Contribution of working group I to the fifth assessment report of the intergovernmental panel on climate change. Cambridge University Press, Cambridge/New York, p 1535

IPCC (2014) Summary for policymakers. In: Field CB, Barros VR, Dokken DJ, Mach KJ, Mastrandrea MD et al (eds) Climate change 2014: impacts, adaptation, and vulnerability. Part A: global and sectoral aspects contribution of working group II to the fifth assessment report of the intergovernmental panel on climate change. Cambridge University Press, Cambridge/New York, p 1132

Jie C, Blume HP (2002) Rock-weathering by lichens in Antarctic: patterns and mechanisms. J Geogr Sci 12:387–396

Johansson P, Reich PB (2005) Population size and fire intensity determine post-fire abundance in grassland lichens. Appl Veg Sci 8:193–198. https://doi.org/10.1658/1402-2001(2005)008%5B0193:PSAFID%5D2.0.CO;2

Joly K, Jandt RR, Klein DR (2009) Decrease of lichens in Arctic ecosystems: the role of wildfire, caribou, reindeer, competition and climate in North-Western Alaska. Polar Res 28(3):433–442

Jung P, & Büdel B (2021) 2.3 Lichens as pioneers on rock surfaces. Life at Rock Surfaces: Challenged by Extreme Light, Temperature and Hydration Fluctuations, edited by Burkhard Büdel and Thomas Friedl, Berlin, Boston: De Gruyter pp. 141–160. https://doi.org/10.1515/9783110646467-005

Jovan S, Carlberg T (2007) Nitrogen Content of Letharia vulpina tissue from Forests of the Sierra Nevada, California: Geographic Patterns and Relationships to Ammonia Estimates and Climate. Environ Monit Assess 129:243–251

Jovan S, Fenn ME, Buhler M, Bytnerowicz A, Kovasi A, Hutten M, DiMeglio E, Schweizer D (2021) Challenges characterizing N deposition to high elevation protected areas: A case study integrating instrument, simulated, and lichen inventory datasets for the Devils Postpile National Monument and surrounding region, USA Ecol Ind 122:107311

Kahng HY, Yoon BJ, Kim SH, Shin DJ, Hur JS, Kim HW, Koh YJ (2004) Introduction of saxicolous lichens distributed in coastal rocks of U-do islet in Jeju, Korea. J Microbiol 42(4): 292–298

Kaushik NK, Murali TS, Sahal D, Suryanarayanan TS (2014) A search for antiplasmodial metabolites among fungal endophytes of terrestrial and marine plants of southern India. Acta Parasitol 59:745–757

Kellogg JJ, Raja HA (2016) Endolichenic fungi: a new source of rich bioactive secondary metabolites on the horizon. Phytochem Rev 16(2):271–293. https://doi.org/10.1007/s11101-016-9473-1

Knops JMH, Nash TH III, Schlesinger WH (1996) The influence of epiphytic lichens on the nutrient cycling of an oak woodland. Ecol Monogr 66:159–176. https://doi.org/10.2307/2963473

Koch MN, Branquinho C, Matos P, Pinho P, Lucheta F et al (2016) The application of lichens as ecological surrogates of air pollution in the sub tropics: a case study in South Brazil. Environ Sci Pollut Res 23:20819–20834. https://doi.org/10.1007/s11356-016-7256-2

Kossowska M (2022) Hornfels rocks as a habitat for saxicolous lichen biota. A case study from the Sudety Mountains (SW Poland). Folia Cryptogamica Estonica 59:35–41

Kovasi A, McCune B, Jovan S (2022) The potential use of elemental content of saxicolous lichens as bioindicators of nitrogen deposition in the central and southern California mountains. Ecol Indic 144:109541

Kinalioglu K (2010) Epiphytic and saxicolous lichens of the District Centre of Arakli and in vicinity (Trabzon, Turkey). Pakistan J Biol Sci 13(12):588

Kovasi A, McCune B, & Jovan S (2022) The potential use of elemental content of saxicolous lichens as bioindicators of nitrogen deposition in the central and southern California mountains. Ecol Indic 144:109541

Kumar J, Dhar P, Tayade AB, Gupta D, Chaurasia OP, Upreti DK, Srivastava RB (2014) Antioxidant capacities, phenolic profile and cytotoxic effects of saxicolous lichens from trans-Himalayan cold desert of Ladakh. PLoS One 9(6):e98696

Kalfas D, Kalogiannidis S, Chatzitheodoridis F, & Margaritis N (2024) The Other Side of Fire in a Changing Environment: Evidence from a Mediterranean Country. Fire 7(2):36. https://doi.org/10.3390/fire7020036

Kamran M, Imran QM, Ahmed MB, Falak N, Khatoon A, & Yun BW (2022) Endophyte-Mediated Stress Tolerance in Plants: A Sustainable Strategy to Enhance Resilience and Assist Crop Improvement. Cells, 11(20): 3292

Ladyman JA (1996) Terrestrial cryptogams of pinyon-juniper woodlands in the southwestern United States: a review

Lagarde A, Jargeat P, Roy M, Girardot M, Imbert C, Millot M, Mambu L (2018) Fungal communities associated with Evernia prunastri, Ramalina fastigiata and Pleurosticta acetabulum: Three epiphytic lichens potentially active against Candida biofilms. Microbiol Res

Leavitt SD, Esslinger TL, Divakar PK, Crespo A, Lumbsch HT (2016) Hidden diversity before our eyes: delimiting and describing cryptic lichen-forming fungal species in camouflage lichens (Parmeliaceae, Ascomycota). Fungal Biol 120:1374–1391

Lewis CJ, Sliwa L (2012) Lecanora carlottiana, a new saxicolous lichen species from the Great Lakes region of North America. Bryologist 115(3):375–381

Lim FY, Sanchez JF, Wang CCC, Keller NP (2012) Toward awakening cryptic secondary metabolite gene clusters in filamentous fungi. Methods Enzymol 517:303–324

Liu X, Zhou ZY, Cui JL, Wang ML, Wang JH (2021) Biotransformation ability of endophytic fungi: from species evolution to industrial applications. Appl Microbiol Biotechnol 105(19): 7095–7113. https://doi.org/10.1007/s00253-021-11554-x

Lorenz C, Bianchi E, Poggiali G et al. (2023) Survivability of the lichen Xanthoria parietina in simulated Martian environmental conditions. Sci Rep 13:4893. https://doi.org/10.1038/s41598-023-32008-6

Lutzoni F, Miadlikowska J (2009) Lichens. Quick guide. Curr Biol 19:R502–R503

Maestre FT, Bowker MAY, Cantón Y, Castillo-Monroy AP, Cortina J et al (2011) Ecology and functional roles of biological soil crusts in semi-arid ecosystems of Spain. J Arid Environ 75: 1282–1291. https://doi.org/10.1016/j.jaridenv.2010.12.008

Manojlovic NT, Rancic AB, Décor R, Vasiljevic P, Tomovic J (2021) Determination of chemical composition and antimicrobial, antioxidant and cytotoxic activities of lichens Parmelia conspersa and Parmelia perlata. J Food Meas Charact 15:686–696

Marcelli MP (1992) Ecologia liquênica nos manguezais do sulsudeste brasileiro. Biblioteca Lichenologica 47:1–310

Marcelli MP (1996) Biodiversity assessment in lichenized fungi: the necessary naïve roll makers. In: Bicudo CEM, Menezes NA (eds) Biodiversity in Brazil: a first approach. CNPq, São Paulo, pp 93–107

Matos P, Pinho P, Aragón G, Martínez I, Nunes A et al (2015) Lichen traits responding to aridity. J Ecol 103(2):451–458. https://doi.org/10.1111/1365-2745.12364

Mejía LC, Herre EA, Sparks JP, Winter K, Garcia MN, Van Bael SA, Stitt J, Shi Z, Zhang Y, Guiltinan MJ et al (2014) Pervasive effects of a dominant foliar endophytic fungus on host genetic and phenotypic expression in a tropical tree. Front Microbiol 5:479

Muggia L, Fleischhacker A, Kopun T, Grube M (2016) Extremptolerent fungi from alpine rock lichens and their phylogenetic relationships. Fungal Divers 76:119–142

Muggia L, Quan Y, Gueidan C et al (2021) Sequence data from isolated lichen-associated melanized fungi enhance delimitation of two new lineages within Chaetothyriomycetidae. Mycol Prog 20:911–927. https://doi.org/10.1007/s11557-021-01706-8

Mukemre M, Zengin G, Turker RS, Aslan A, Dalar A (2021) Biological activities and chemical composition of Xanthoria lichens from Turkey. Int J Second Metab 8(4):376–388

Müller SC, Overbeck GE, Pfadenhauer J, Pillar VD (2007) Plant functional types of woody species related to fire disturbance in forest–grassland ecotones. Plant Ecol 189:1–14. https://doi.org/10.1007/s11258-006-9162-z

Müller SC, Overbeck G, Blanco CC, de Oliveira JM, Pillar VD (2012) South Brazilian forest-grassland ecotones: dynamics affected by climate, disturbance and woody species traits. In: Myster RW (ed) Ecotones between forest and grassland. Springer, New York, pp 167–187. https://doi.org/10.1007/978-1-4614-3797-0_7

Michalet R, Touzard B, Billard G, Choler P, Loucougaray G (2024) Changes in Taxonomic and Functional Composition of Subalpine Plant Communities in Response to Climate Change under Contrasting Conditions of Bedrock and Snow Cover Duration. J Veg Sci 35:e13253

Miller JED, Root HT, Safford HD (2018) Altered fire regimes cause long-term lichen diversity losses. Glob Chang Biol 24(10):4909–4918. https://doi.org/10.1111/gcb.14393

Muggia L, Kopun T, Grube M (2017) Effects of growth media on the diversity of culturable fungi from lichens. Molecules 22:824

Masumoto H, Degawa Y (2019) The effect of surface sterilization and the type of sterilizer on the genus composition of licheninhabiting fungi with notes on some frequently isolated genera. Mycoscience 60, 331–342

Mengistu AA (2020) Endophytes: Colonization, Behaviour, and Their Role in Defense Mechanism. Int J Microbiol 2020:6927219. https://doi.org/10.1155/2020/6927219

Morillas L, Roales J, Cruz C, & Munzi S (2022) Lichen as Multipartner Symbiotic Relationships. Encyclopedia, 2(3), 1421–1431. https://doi.org/10.3390/encyclopedia2030096

Nascimbene B, Casazza G, Benesperi R, Catalano I, Cataldo D et al (2016) Climate change forest the decline of epiphytic lobaria species in Italy. Biol Conserv 201:377–384. https://doi.org/10.1016/j.biocon.2016.08.003

Nokes LF, Haelewaters D, Pfister DH (2022) Exploration of marine lichenized fungi as bioindicators of Coastal Ocean Pollution in the Boston Harbor islands National Recreation Area. Rhodora 122(992):251–273

Oh DY, Kwek SS, Raju SS, Li T, McCarthy E, Chow E, Aran D, Ilano A, Pai CS, Rancan C, Allaire K, Burra A, Sun Y, Spitzer MH, Mangul S, Porten S, Meng MV, Friedlander TW, Ye CJ, Fong L (2020) Intratumoral $CD4^{+}$ T cells mediate anti-tumor cytotoxicity in human bladder cancer. Cell 181(7):1612–1625.e13. https://doi.org/10.1016/j.cell.2020.05.017. Epub 2020 Jun 3. PMID: 32497499; PMCID: PMC7321885

Overbeck GE, Müller SC, Fidelis A, Pfadenhauer J, Pillar VD, Blanco CC, Boldrini II, Both R, Forneck ED (2007) Brazil's neglected biome: the South Brazilian Campos. Perspect Plant Ecol Evol Syst 9:101–116. https://doi.org/10.1016/j.ppees.2007.07.005

Oh SY, Yang JH, Woo JJ, Oh SO, & Hur JS (2020) Diversity and Distribution Patterns of Endolichenic Fungi in Jeju Island, South Korea. Sustain 12:(9)3769. https://doi.org/10.3390/su12093769

Pillar VD, Quadros FLF (1997) Grassland-forest boundaries in southern Brazil. Coenoses 12:119–126

Pinho P, Máguas C, Branquinho C (2010) Modeling ecology of lichens communities based on photobiont type in relation to potential solar radiation and neighborhood land-use. In: Nash T III, Geiser L, McCune B, Triebel D, Tomescu AM, Sanders W (eds) Bibliotheca lichenologica. Biology of lichens-symbiosis, ecology, environmental monitoring cystematics and cyber applications, vol 105. J. Cramer, California, pp 149–160

Purvis OW (2014) Adaptation and interaction of saxicolous crustose lichens with. Bot Stud 55(1): 1–14

Puy-Alquiza MJ, Miranda-Aviles R, Zanor GA, Salazar-Hernandez MM, Ordaz- Zubia VY (2017) Study of the distribution of heavy metals in the atmosphere of the Guanajuato city: use of saxicolous lichen species as bioindicators. Ingeniería, Investigaci6n Tecnología 18(1):111–126

Purvis OW (2014) Adaptation and interaction of saxicolous crustose lichens with metals. Bot Stud 55:23. https://doi.org/10.1186/1999-3110-55-23

Raggio J, Green T, Sancho L, Pintado A, Colesie C et al (2017) Metabolic activity duration can be effectively predicted from macroclimatic data for biological soil crust habitats across Europe. Geoderma 306:10–17. https://doi.org/10.1016/j.geoderma.2017.07.001

Rajakaruna N, Knudsen K, O'Dell RE, Nathaniel P, Olday FC, Woolhouse S (2012) Investigation of the importance of rock chemistry for saxicolous lichen communities of the New Idria serpentinite mass, San Benito County, California, USA. Lichenologist 44(5):695–714

Reed SC, Delgado-Baquerizo M, Ferrenberg S (2019) Biocrust science and global change. New Phytol 223(3):1047–1051. https://doi.org/10.1111/nph.15992

Romagni JG, Gries C (2000) Post-fire recolonization of dominant epiphytic lichen species on Quercus hypoleucoides (Fagaceae). Am J Bot 87:1815–1820. https://doi.org/10.2307/2656834

Ruprecht U, Brunauer G, Türk R (2014) High photobiont diversity in the common European soil crust lichen Psora decipiens. Biodivers Conserv 23:1771–1785

Root HT, Geiser LH, Fenn ME, Jovan S, Hutten MA, Ahuja S, Dillman K, Schirokauer D, S. Berryman S, McMurray JA (2013) A simple tool for estimating throughfall nitrogen deposition in forests of western North America using lichens For Ecol Manage 306:(3)1–8

Sáenz-Romero C, Rehfeldt GE, Crookston NL, Duval P, St-Amant R et al (2010) Spline models of contemporary, 2030, 2060 and 2090 climates for Mexico and their use in understanding climate-change impacts on the vegetation. Clim Chang 102:595–623. https://doi.org/10.1007/s10584-009-9753-5

Santiago CD, Bustos Y, Jolie SA, Flores Toussaint R, Sosa SS, Raviv T, Cicchetti C (2021) The impact of COVID-19 on immigrant and refugee families: qualitative perspectives from newcomer students and parents. Sch Psychol 36(5):348–357. https://doi.org/10.1037/spq0000448. Epub 2021 Aug 26. PMID: 34435837

Sargsyan R, Khechikyan M, Panosyan H (2023) Study on antimicrobial activity of various extracts of saxicolous lichen Lecanora muralis. Ajastan Kensabanakan Handes 75(1)

Sati SC, & Savita Joshi SJ (2011) Antibacterial Activity of the Himalayan Lichen Parmotrema nilgherrense Extracts British Microbiol Res J 1(2):26–32

Schulten JA (1985) The effects of burning on the soil lichen community of a sand prairie. Bryologist 88:110–114. https://doi.org/10.2307/3242589

Semarnat. Secretaría de Medio Ambiente y Recursos Naturales (2016) Executive summary. Mexico state of the environment report compendium of environment statistics, key environmental indicators, environmental performance indicators and green growth indicators. Version 2015. SEMARNAT

Sherameti I, Venus Y, Drzewiecki C, Tripathi S, Dan VM, Nitz I, Varma A, Grundler FM, Oelmüller R (2008) PYK10, a β-glucosidase located in the endoplasmatic reticulum, is crucial for the beneficial interaction between Arabidopsis thaliana and the endophytic fungus Piriformosp aora indica. Plant J 54:428–439

Simonyan A, Sargsyan R, Panosyan H, Trchounian A (2020) Study of antibacterial and antioxidant activities of saxicolous lichen Xanthoparmelia stenophylla. Proc YSU B Chem Biol Sci 54:132–137

Singh RK, Chang HW, Yan D, Lee KM, Ucmak D, Wong K, Abrouk M, Farahnik B, Nakamura M, Zhu TH, Bhutani T, Liao W (2017) Influence of diet on the gut microbiome and implications for human health. J Transl Med 15(1):73. https://doi.org/10.1186/s12967-017-1175-y. PMID: 28388917; PMCID: PMC5385025

Singh G (2023) Linking Lichen Metabolites to Genes: Emerging Concepts and Lessons from Molecular Biology and Metagenomics. J Fungi (Basel) 9(2):160. https://doi.org/10.3390/jof9020160

Sonina AV, Androsova VI (2020) Coastal lichens: distribution patterns and adaptive strategies. In: Handbook of halophytes: from molecules to ecosystems towards biosaline agriculture. Springer, pp 1–22

Sonina AV, Rumjantseva AD, Tsunskaya AA, Androsova VI (2017) Adaptations of epilithic lichens to the microclimate conditions of the White Sea coast. Czech Polar Rep 7(2):133–143

Soto-Correa JC, Sáenz-Romero C, Lindig-Cisneros R, de-la-Barrera E (2014) Lupinus elegans kunth assisted migration on common garden field tests. Rev Fitotec Mex 37(2):107–116. https://doi.org/10.35196/rfm.2014.2.107

Stocker-Wörgötter E (2008) Metabolic diversity of lichen-forming ascomycetous fungi: culturing, polyketide and shikimate metabolite production, and PKS genes. Nat Prod Rep 25:188–200

Strobel G, Daisy B (2003) Bioprospecting for microbial endophytes and their natural products. Microbiol Mol Biol Rev 67(4):491–502

Studzińska-Sroka E, Majchrzak-Celińska A, Zalewski P, Szwajgier D, Baranowska-Wójcik E, Kaproń B, Plech T, Żarowski M, & Cielecka-Piontek J (2021) Lichen-Derived Compounds and Extracts as Biologically Active Substances with Anticancer and Neuroprotective Properties. Pharm 14(12):1293. https://doi.org/10.3390/ph14121293

Sun C, Johnson JM, Cai D, Sherameti I, Oelmüller R, Lou B (2010) Piriformospora indica confers drought tolerance in Chinese cabbage leaves by stimulating antioxidant enzymes, the expression of drought-related genes and the plastid-localized CAS protein. J Plant Physiol 167:1009–1017

Suryanarayanan TS, Thennarasan S (2004) Temporal variation in endophyte assemblages of Plumeria rubra leaves. Fungal Divers 15:197–204

Suryanarayanan TS, Thirunavukkarasu N (2017) Endolichenic fungi: the lesser known fungal associates of lichens. Mycology 8(3):189–196. https://doi.org/10.1080/21501203.2017.1352048

Suryanarayanan TS, Thirunavukkarasu N, Hariharan GN, Balaji P (2005) Occurrence of non-obligate microfungi inside lichen thalli. Sydowia 57:120–130

Suryanarayanan TS, Thirunavukkarasu N, Govindarajulu MB, Sasse F, Jansen R, Murali TS (2009) Fungal endophytes and bioprospecting. Fungal Biol Rev 23:9–19

Suryanarayanan TS, Govindarajulu MB, Rajamani T, Tripathi M, Joshi Y (2017) Endolichenic fungi in lichens of Champawat district, Uttarakhand, northern India. Mycol Prog 16:205–211

Suzuki MT, Parrot D, Berg G, Grub M, Tomasi S (2016) Lichens as natural sources of biotechnologically relevant bacteria. Appl Microbiol Biotechnol 100:583–595

Suryanarayanan TS, Govindarajulu MB, Rajamani T, Tripathi M, Joshi Y (2017) Endolichenic fungi in lichens of Champawat district, Uttarakhand, northern India. Mycol Prog 16:205–211.
Tagirdzhanova G, Saary P, Tingley JP, Díaz-Escandón D, Abbott DW, Finn RD, Spribille T (2021) Predicted Input of Uncultured Fungal Symbionts to a Lichen Symbiosis from Metagenome-Assembled Genomes. Genome Biol Evol 13(4):evab047. https://doi.org/10.1093/gbe/evab047
Thirunavukkarasu N, Jahnes B, Broadstock A, Rajulu MG, Murali TS, Gopalan V, Suryanarayanan TS (2015) Screening marine-derived endophytic fungi for xylan-degrading enzymes. Curr Sci 109:112–120
Thirunavukkarasu N, Suryanarayanan TS, Rajamani T, ParanetharanMS. (2017) A rapid and simple method for screening fungi for extracellular protease. Mycosphere 8:131–136
Tiwari P, Srivastava Y, Bae H (2021) Endophytes: Trend of pharmaceutical design of Endophytes as anti-infective. Curr Top Med Chem 21:1572–1586. https://doi.org/10.2174/1568026621666210524093234
Topal M, Topal EIA, Erdal OE, Aslan A (2023) Determination of some trace elements in various lichens as biomonitors of pollution and assessment of pollution status. Nigde Omer Halisdemir Üniversitesi Mühendislik Bilimleri Dergisi 12(3)
Tripathi M, Joshi Y (2019) Endolichenic fungi: present and future trends. Springer, Singapore
U'Ren JM, Lutzoni F, Miadlikowska J, Arnold AE (2010) Community analysis reveals close affinities between endophytic and endolichenic fungi in mosses and lichens. Microb Ecol 60: 340–353
U'Ren JM, Lutzoni F, Miadlikowska J, Laetsch AD, Arnold AE (2012) Host and geographic structure of endophytic and endolichenic fungi at a continental scale. Am J Bot 99(5):898–914. https://doi.org/10.3732/ajb.1100459. Epub 2012 Apr 26. PMID: 22539507
Valencia MC, Ceballos JA (2002) Hongos liquenizados. Universidad Nacional de Colombia, Bogotá
Verrall B, Pickering CM (2020) Alpine Vegetation in the Context of Climate Change: A Global Review of Past Research and Future Directions. Sci Total Environ 748:141344
Wang Y, Dai CC (2011) Endophytes: a potential resource for biosynthesis, biotransformation, and biodegradation. Ann Microbiol 61:207–215
Waqas M, Khan AL, Muhammad H, Shahzad R, Kang SM, Kim JG, Lee IJ (2015) Endophytic fungi promote plant growth and mitigate the adverse effects of stem rot: an example of Penicillium citrinum and aspergillus terreus. J Plant Interact 10:280–287
Wilson D (1995) Endophyte: the evolution of a term, and clarification of its use and definition. Oikos 73:274–276
Waill A Elkhateeb, Dina E El-Ghwas, Ghoson M. Daba (2023) Lichen Secondary Metabolites as Anti-Fungal Agent. J Pharm Drug Dev 2(1). https://doi.org/10.58489/2836-2322/01
Yu NH, Park SY, Kim JA, Park CH, Jeong MH, Oh SO, Hong SG, Talavera M, Divakar PK, Hur JS (2018) Endophytic and endolichenic fungal diversity in maritime Antarctica based on cultured material and their evolutionary position among Dikarya. Fungal Syst Evol 2:263–272
Zhao Y, Wang M, Xu B (2021) A comprehensive review on secondary metabolites and health-promoting effects of edible lichen. J Funct Foods 80:104283
Zonca V, Gladding J, Coccia E (2023) Lichens: toward a minimal resistance. Polity Press, Cambridge
Zschau T, Getty S, Gries C, Ameron Y, Zambrano A, Nash TH (2003) Historical and current atmospheric deposition to the epilithic lichen Xanthoparmelia in Maricopa County, Ariz Environ Pollut 125:(1)21–30
Zhang T, Wei XL, Wei YZ, Liu HY, Yu LY (2016) Diversity and distribution of cultured endolichenic fungi in the Ny-Ålesund Region, Svalbard (high Arctic). Extrem 20:461–470
Zhang W, Ran Q, Li H, & Lou H (2024) Endolichenic Fungi: A Promising Medicinal Microbial Resource to Discover Bioactive Natural Molecules—An Update. J Fungi 10(2):99. https://doi.org/10.3390/jof10020099

Endolichenic Fungi: The Lesser-Known Fungal Associates of Lichens

Qudsia Firdous, Samina Sarwar, and Hajra Khalil

Contents

Q. Firdous (✉)
Institute of Botany, University of the Punjab, Lahore, Pakistan

Department of Medicinal Chemistry, College of Pharmacy, University of Minnesota, Minneapolis, MN, USA

S. Sarwar
Botany Department, Lahore College for Women University of Lahore, Lahore, Pakistan

H. Khalil
Institute of Botany, University of the Punjab, Lahore, Pakistan

A. M. Abdel-Azeem, F. Degola (eds.), *Advances in Lichens and Non-Flowering Plants Fungal Endophytes Research*, Fungal Biology,
https://doi.org/10.1007/978-3-032-01506-8_5

1 Introduction

Lichens, enigmatic organisms that cover boulders, trees, and other surfaces, have long fascinated both scientists and nature enthusiasts (Krienitz 2018). Composed of a unique symbiotic relationship between fungi and algae or cyanobacteria, lichens thrive in some of the harshest environments on Earth, from the frozen tundra to the scorching deserts (Després et al. 2012). Recent studies have confirmed the presence of highly structured bacterial communities within lichens and provide evidence that these bacteria may play distinct functional functions within lichen symbioses (Bates et al. 2011; Wang et al. 2022). While the mutualistic bond between fungi and photobionts has been widely studied, there is a hidden world within lichens that remains relatively unexplored—the realm of endolichenic fungi (Tripathi et al. 2019).

Endolichenic fungi are a diverse group of microorganisms that inhabit the internal tissues of lichens without causing any apparent harm or displaying any discernible parasitic behavior (Naranjo-Ortiz and Gabaldón 2019). Phylogenetically they are different from lichenicolous fungi and more closely related to members of ascomycetes found as endophytes in vascular plants (Tripathi and Joshi 2019a, b). These lesser-known fungal associates are often overshadowed by the dominant fungal partner in the lichen symbiosis, making their presence easily overlooked (Raudabaugh et al. 2020). However, recent research has shed light on the important roles they play within these intricate organisms.

Within the lichen thallus, endolichenic fungi inhabit the intercellular spaces or colonize the cells of the fungal partner (Petrzik et al. 2019). Unlike the photobionts, which provide energy through photosynthesis, endolichenic fungi primarily depend on the resources supplied by the lichen's primary fungus (Muggia and Grube 2018). Despite this dependence, their presence does not seem to compromise the overall health or functionality of the lichen. Endolichenic fungi might have phenotypic effects on the overall thallus. This is certainly the case for the *Lecanoromycete Bryoria* sp. (Spribille et al. 2016), and drastic changes in lichen taxonomy may be likely.

The discovery and characterization of endolichenic fungi have presented scientists with new avenues of exploration in understanding the complex interactions within lichen communities. These fungi have been found to have a remarkable diversity, exhibiting unique adaptations and capabilities that contribute to the overall resilience and ecological success of the lichens they inhabit (Xu et al. 2022).

On the basis of the intra-ecology of the thallus, lichens are incredibly intricate, i.e., a body made as a result of the symbiotic association of two primary algal and fungal partners and participation of free-living bacteria and nonsymbiotic fungi in this association all of which led us to consider lichens as a miniature ecosystem (Farrar 1976; Seaward 1988; Bjelland and Ekman 2005).

In this chapter, we will delve into the fascinating world of endolichenic fungi, exploring their diversity, ecological roles, and potential interactions with their lichen hosts. We will also examine the methods used to study these elusive organisms and highlight some of the recent discoveries that have expanded our understanding of their ecological significance. By shedding light on these lesser-known fungal associates, we hope to enhance our appreciation of the intricate relationships that shape the remarkable world of lichens and their hidden fungal companions.

2 How Endolichenic Fungi Discovered?

The study of endolichenic fungi began in the late twentieth century when researchers began to investigate lichen-associated microbial communities in greater depth. Hofstetter et al. (2007) were the first to suggest that non-lichen fungal sequences commonly amplified from lichens may represent a previously unrecognized source of lichen-specific fungal diversity; these forms were later termed "endolichenic" by Miadlikowska et al. (2004) and Arnold et al. (2009a, b). Arnold et al. (2009a, b) considered them to be phylogenetically distinct from both lichen-forming mycobionts and lichenicolous fungi and incidental fungi on thallus surfaces, as well as an important evolutionary link to plant-associated endophytes (Lutzoni et al. 2001; Lawrey and Diederich 2003; Arnold et al. 2009a, b; Fleischhacker et al. 2015; Chagnon et al. 2016). The endolichenic fungi have been discovered within living, apparently healthy lichen thalli by various workers (Petrini et al. 1990; Girlanda et al. 1997; Suryanarayanan et al. 2005; Arnold et al. 2009a, b) and are known to coexist with the mycobiont (which forms the bulk of the lichen thallus) and live in very close association with the photobiont (the algal or cyanobacterial partner in the lichen symbiosis) (Arnold et al. 2009a, b).

3 Diversity and Classification of Endolichenic Fungi

The presence of endolichenic fungi has the potential to impact lichen diversity and ecosystem dynamics in a manner comparable to the influence of endophytic fungi on plant ecology and the long-term viability of ecosystems (Oh et al. 2020). This underscores the importance of comprehending the diversity and community structure of these fungi. There remains a significant knowledge gap regarding endolichenic biodiversity and the organization of local communities within phylogenetically diverse host assemblages in natural ecosystems, despite the growing attention towards their diversity and potential applications.

Endolichenic fungi are known to inhabit the lichen thallus in a concealed manner, necessitating specific methods for their detection and isolation (Petrini et al. 1990; Girlanda et al. 1997). In their study, Petrini et al. (1990) successfully isolated a total of 506 fungal species from a sample of 17 lichens that possessed fruiting structures. The study conducted by Girlanda et al. (1997) involved a thorough investigation of the lichen species *Xanthoparmelia taractica* and *Peltigera praetextata*. The findings of this study indicated that these two species exhibited distinct fungal assemblages, while their levels of biodiversity were found to be comparable. The extensive fungal diversity observed in lichen thalli can potentially be attributed to their permeability and diversity. In order to examine the fungal communities, Girlanda et al. (1997) obtained a total of 117 fungal isolates from two specific types of foliose lichens, namely, *Parmelia taractica* and *Peltigera praetexta*. Petrini et al. (1990) identified a significant proportion of individuals belonging to the *Pezizomycotina*, a major division within the *Ascomycota* fungal phylum. In their study, Li et al. (2007) identified a total of 32 endolichenic fungi from a sample of 488 lichen thalli segments collected from 5 distinct lichen groups found on Baihua mountain in Beijing. An endolichenic *Ascomycota* fungus was isolated. In a recent study, Arnold et al. (2009a, b) put forth compelling hypotheses regarding the phenomenon of endophytism and endolichenic fungi. These hypotheses were explored within the context of a phylogenetic analysis of trophic transitions in ascomycetes. The phylum *Ascomycota*, which boasts the highest species diversity, has developed endophytism as a result of the presence of endolichenic fungi. The findings of their study indicate that endolichenic fungi undergo rapid evolutionary changes, transitioning into plant endophytes. The initial discovery of the coexistence of endolichenic fungi within the thallus of lichens in close association with the photobiont was made by Arnold et al. (2009a; b). The study conducted by U'Ren et al. (2010) focused on investigating the ecosystems of endophytic fungi in moss and lichen through the utilization of ITS rDNA sequences. The researchers conducted an investigation on a total of ten lichen species in order to investigate the presence of endolichenic fungi. Their findings revealed that the *Lecanorales*, *Peltigerales*, and *Teloschistales* orders exhibited the highest frequencies of isolation. In their study, Cheon et al. (2013) conducted the isolation of endolichenic fungi from four different lichen species, namely *Stereocaulon* sp. 1 (with 1429 isolates), *Stereocaulon* 2 (with 1430 isolates), *Pezicula* (with 0,156 isolates), and *Graphis* (with 1245 isolates). In their study, Maduranga et al. (2018) conducted the isolation of 171 genotypes of endolichenic fungi from the lichens of mangrove plants and mangrove-associated plants in the Puttalam lagoon of Sri Lanka.

Several regional investigations have reported the presence of 20–60 species (Petrini et al. 1990; Yu et al. 2018; Vinayaka et al. 2016; Suryanarayanan et al. 2017; Maduranga et al. 2018; Masumoto and Degawa 2019; Tripathi and Joshi 2019a, b). In their study, Arnold et al. (2009a, b) identified a total of 200 distinct species of endolichenic fungi through their research conducted on the continent. The predominant group of fungal species found within lichens is *Ascomycota*, as reported by several studies (Arnold et al. 2009a, b; U'Ren et al. 2010; Zhang et al. 2016; Suryanarayanan et al. 2017). The taxonomic group *Sordariomycetes*, which constitutes 71.8% of the *Ascomycota* phylum, exhibited the highest abundance of endolichenic fungi. Subsequently, the group *Dothideomycetes* accounted for 13.6% of the total. Previous studies (Lagarde et al. 2018; Li et al. 2007) have indicated that *Sordariomycetes* constitute the predominant group of endolichenic fungi in various regions, including China, France, Japan, North America, Norway, and Sri Lanka. According to supplementary investigations conducted by Girlanda et al. (1997) and De Carvalho et al. (2019), the fungal taxa *Dothideomycetes*,

Leotiomycetes, and *Pezizomycetes* were found to be prevalent in the regions of Antarctica, Germany, Italy, and the United States. Endolichenic fungi do not dominate any country, continent, or climate zone.

A total of 61 isolates of endolichenic fungi were obtained from 45 Antarctic lichens, as reported by Yu et al. (2018). These isolates were identified as belonging to 21 different lichen species, which were classified into 10 families: *Candelariaceae*, *Cladoniaceae*, *Lecanoraceae*, *Parmeliaceae*, *Physciaceae*, *Pilocarpaceae*, *Ramalinaceae*, *Sphaerophoraceae*, *Stereocaulaceae*, and *Teloschistaceae*. In the Antarctic, a high percentage of endophytic (endolichenic) fungal species remain undescribed, as they identified as Leotiomycetes sp., fungal sp., *Capronia* sp., *Ascomycota* sp., *Chaetomium globosum*, *Phoma herbarum*, fungal endophyte, uncultured endophytic fungus clone, uncultured endophytic fungus clone, *Thelebolus* sp., *Phoma violacea*, uncultured fungus clone, fungal endophyte isolate, *Anthostomella leucospermi*, *Peziza varia*, *Barrmaelia macrospora*, *Coniochaetales* sp., *Sordariomycetes* sp., *Leotiomycetes* sp., *Helotiales* sp., *Helotiales* sp., *Xylariales* cf. JP60-5, *Coniochaetales* sp., uncultured fungus clone 3203 N2, *Hyalodendriella betulae*, *Cephalosporium* sp., *Xylaria* sp., *Biscogniauxia* sp., uncultured fungus (Yu et al. 2018).

A total of 500 endolichenic fungi were obtained from a sample of 114 officially recognized lichen species, which were classified into 22 distinct families. However, the identification process only yielded positive results for 135 of these species. This suggests that endolichenic fungi do not exhibit a strong preference for specific host species (Chakarwarti et al. 2020).

The findings from a metagenomic analysis of lichen thalli indicate that the conventional method employed for the isolation of endolichenic fungi is limited in its coverage, as it only targets a specific subset of these fungi. There is a need for improvement in the area of isolation. The *Sordariomycetes* taxonomic group is the predominant presence within the endolichenic fungus of *Parmotrema tinctorum*. In order to differentiate between distinct endolichenic fungi, a comprehensive foliose lichen thallus is necessary. In order to isolate a variety of endolichenic fungi, it is necessary to employ a moderate sterilization technique. Fragments of lichen thallus measuring 1 mm^2 were utilized for the purpose of isolating the predominant endolichenic fungal species. The size of the thallus fragment exhibited a more significant influence on the isolation of endolichenic fungi compared to factors such as age, the severity of surface sterilization, or nutritional demand. The populations of endolichenic fungi within lichen thalli exhibit significant diversity. However, the majority of endophytic and endolichenic fungi belong to the phyla *Ascomycota* (specifically *Arthoniomycetes*, *Sordariomycetes*, *Dothideomycetes*, *Lecanoromycetes*, *Leotiomycetes*, *Pezizomycetes*, and *Eurotiomycota*), *Basidiomycota*, and *Mucoromycota* (Arnold et al. 2009a, b; Park et al. 2015; Si et al. 2023).

The diversity and community structure of endolichenic fungi is influenced by host taxonomy, types of photobionts, and environmental conditions. In order to enhance our understanding of the diversity and ecological characteristics of endolichenic fungi, it is imperative to conduct further investigations on a broader range of lichen species across various habitats (Oh et al. 2020).

4 What Methods Can Be Employed to Ascertain the Presence of Endolichenic Fungi?

The objective exploration of species diversity in endolichenic fungi requires the use of DNA sequence-based sample identification due to the absence of visible reproductive structures and other distinctive phenotypic traits for classification. In order to achieve this objective, the researchers utilized DNA sequences from various genetic regions, including the nuclear internal transcribed region (ITS), nuclear ribosomal short subunit (nuSSU) and large subunit (nuLSU), and mitochondrial ribosomal short subunit (mtSSU) rDNA, as well as the two largest subunits of RNA polymerase II (RPB1 and RPB2), translation elongation factor-1α (TEF1), ACT, and CAL (Yu et al. 2018; Si et al. 2023).

5 Association Between Endolichenic Fungi and Their Host Lichen

Endolichenic fungi, known as endophytes in the context of plant biology, have the ability to inhabit lichen thalli (Girlanda et al. 1997; Li et al. 2007; Arnold et al. 2009a, b; Kannangara et al. 2009; U'Ren et al. 2010). The fungi in question exhibit a mutualistic relationship with algal photobionts within seemingly unaffected lichen thalli, resulting in extended periods of asymptomatic infection.

The island's moss mats are predominantly inhabited by *Usnea antarctica*, *Cladonia borealis*, and *Psilolechia lucida*, which serve as the dominant host lichens. According to Furbino et al. (2014), certain endophytes have the ability to select between lichens and bryophytes as their host organisms. Endophytic organisms derive nourishment from photobionts, which are symbiotic algae residing within lichen thalli. In contrast, it is plausible that lichens could potentially serve as a protective shield for endophytes against adverse environmental conditions.

Due to the limited body of research pertaining to endolichenic fungi, it is premature to draw definitive conclusions regarding their interactions with their host lichen. Nevertheless, according to Tripathi and Joshi's (2015) study, there is evidence suggesting that the facultative nature of *Parmotrema reticulatum*, a lichen species, may vary. This conclusion was reached after analyzing the endolichenic fungal diversity of specimens collected from five distinct locations, which revealed the presence of different fungal species that did not align with one another. In the event of an obligate association between endolichenic fungus and lichens, it would be expected that the identical fungal population would be consistently observed in each screening of the same lichen species. However, this anticipated outcome did not occur. The facultative mutualistic association in higher plant endophytes is contingent upon the overall health status of the host plant. In the event that the plant exhibits robust health, the endophytes establish a mutually beneficial symbiotic relationship. However, should the plant encounter stressors of either abiotic or biotic nature, the endophyte has the potential to transition into a parasitic or saprophytic state. Lichens exhibit a symbiotic relationship with various fungal species, some of which are endolichenic and engage in mutualistic interactions, while others are lichenicolous and exhibit parasitic behaviour. Examples of such fungal species include *Acremonium lichenicola* and *Fusarium* sp. In their study, Fernández-Mendoza et al. (2017) discovered a multitude of lichenicolous species within the mycobiomes of lichens, both those exhibiting signs of infection and those without such indications. Therefore, it is possible that contemporary lichenicolous fungi represent endolichenic fungi residing within a lichen thallus that is experiencing stress. According to Grabka et al. (2022), similar to the beneficial effects of plant endophytes on host lichen fitness and disease prevention, endolichenic fungi may also exhibit similar properties.

The release of secondary compounds by endolichenic fungi has been observed to contribute to the survival of lichens in challenging environments (Moreno et al. 2018; Elkhateeb and Daba 2021).

6 Do Lichenicolous Fungi and Endolichenic Fungi Exhibit Similarities?

Currently, the identification of endolichenics as lichenicolous species is constrained due to the limited number of sequenced lichenicolous species. Nevertheless, with the increasing publication of sequences pertaining to described species, it will become feasible to differentiate between those that were previously classified as endolichenic fungi but are in fact classified as described lichenicolous species (Diederich et al. 2018). However, the scarcity of endolichenic fungi within the order *Chaetothyriales* aligns with the findings of Arnold et al. (2009a, b), which suggest that lichenicolous fungi and endolichenic fungi frequently exhibit significant phylogenetic divergence. Moreover, the notion that lichenicolous fungi are abundantly present in both lichen-forming and non-lichenized clades of fungi is substantiated by research conducted by Lawrey and Diederich (2003). The utilization of phylogenetic placement is instrumental in elucidating the origins of these fungi. It becomes evident that lichenicolous fungi have multiple origins, stemming from both lichen-forming and non-lichenized ancestors. On the other hand, endolichenic fungi are more likely to have descended from non-lichenized ancestors (Arnold et al. 2009a, b). The terms "endolichenics" and "lichenicolous" pertain to different aspects of the same ecological communities. However, as further research is conducted on lichen mycobiomes and their inhabitants are identified, the differentiation between these two categories is expected to become less definitive (Diederich et al. 2018).

7 Survival of Endolichenic Fungi

Endolichenic fungi serve as a catalyst for the evolution of endophytic associations in plants. According to Arnold et al. (2009a, b), the frequency of transitions from an endolichenic lifestyle to an endophytic lifestyle was twice as high as the frequency of transitions from endophytism to endolichenism. Moreover, it is consistent with the findings that endolichenic fungi exhibit atypical colonization patterns in higher plants (Suryanarayanan et al. 2005). Considering the extensive range of lichenicolous fungi and the phylogenetic correlation between endolichenic fungi and endophytic fungi, it is arguable that the inclusion of lichens in studies assessing fungal diversity is imperative. The generation time of endophytic and endolichenic fungi is directly correlated with the lifespan of the host plant due to the absence of reproductive structures in the host plant. The presence of persistent leaves and lichens may impose limitations on the fungi that exist as endophytic and endolichenic symbionts, resulting in longer generation times when compared to saprophytic fungi. This could potentially exert significant selective pressure, discouraging transitions or reversions to symbiotic relationships. The transition from endolichenism to endophytism would be more advantageous due to the comparatively shorter generation times of leaves, which range from less than 1 to 15 years, in contrast to the long lifespan of lichen thalli, which can exceed 100 years. The viability of endophytism as a strategy may be sustained if endophytes exhibit characteristics of ecological opportunism, wherein they establish pathogenic infections in susceptible hosts while maintaining a nonvirulent symbiotic relationship with other hosts (Arnold et al. 2009a, b).

8 The Ecological Role of Endolichenic Fungi

The significance of these endolichenic fungi remains undetermined. However, there exists a substantial population of endolichenic fungi within the structures of lichen thalli, and it is widely believed that their existence serves a significant ecological function. This function includes aiding in the formation and growth of lichens, as well as providing protection against insect herbivores through the production of bioactive compounds (Li et al. 2007; Paranagama et al. 2007). As the ecological role of endolichenic fungi is not yet fully understood, researchers have identified several potential functions they may serve within the lichen ecosystem. Here are some possible ecological roles of endolichenic fungi:

1. Nutrient cycling: Endolichenic fungi may play a role in nutrient cycling within lichens. They can break down organic matter, releasing nutrients that can be utilized by both the fungal and algal/cyanobacterial partners. This nutrient recycling helps maintain the overall health and functioning of the lichen symbiosis.
2. Lichen disease dynamics: Some endolichenic fungi are known to be pathogens or parasites of lichens. They can cause diseases and affect the health and survival of lichens. By influencing the population dynamics of lichens, these fungi may play a role in regulating the abundance and distribution of lichen species.
3. Mutualistic interactions: While some endolichenic fungi are pathogens, others have been found to have mutualistic associations with lichens. They can contribute to the growth and survival of the lichen by providing additional nutrients or enhancing the lichen's tolerance to environmental stresses such as drought or UV radiation.
4. Genetic diversity and evolution: Endolichenic fungi can influence the genetic diversity and evolution of lichens. They can introduce genetic variation through gene transfer or recombination, potentially leading to the emergence of new lichen traits or adaptations. This genetic diversity contributes to the resilience and adaptability of lichens in changing environments.

It's important to note that our understanding of endolichenic fungi and their ecological roles is still developing. Further research is needed to fully unravel their contributions to lichen ecosystems and the broader implications for ecological processes.

9 Are Endolichenic Fungi a New Source of Rich Bioactive Secondary Metabolites?

The investigation of endolichenic fungi has garnered significant interest due to their potential as a reservoir of bioactive secondary metabolites. Secondary metabolites refer to chemical compounds that are synthesized by organisms and are not necessary for their primary metabolic functions. However, these compounds often serve significant ecological purposes, including defense against predators, competition for resources, and communication with other organisms (Karlovsky 2008). According to Bhagarathi et al. (2023), numerous secondary metabolites have demonstrated potential applications in the pharmaceutical and therapeutic fields. Lichens are organisms characterized by their slow growth rate and possess significant economic and ethnomedicinal importance. These organisms have been subjected to screening for their bioactive potential for an extensive period of time, as they possess a vast array of over 1000 secondary metabolites. However, due to their relatively slow growth rate, these organisms are not suitable for large-scale industrial production of pharmaceuticals. As previously stated by Strobel et al. (1996), it has been observed that endophytes have the ability to produce metabolites that are similar to those found in the host plant. If this phenomenon also applies to lichens, it would eliminate the necessity of harvesting lichen diversity. Instead, it would suffice to cultivate and ferment endolichenic fungi. The presence of endolichenic fungi has been observed to result in the production of a wide range of secondary metabolites that exhibit diverse biological activities. The metabolites identified in the aforementioned studies (Singh et al. 2017; Suryanarayanana and Thirunavukkarasu 2017; Shanmugam et al. 2018; Wethalawe et al. 2021; Padhi et al. 2023) exhibit a range of bioactive properties, including antimicrobial, antiviral, anticancer, antioxidant, anti-inflammatory, and potentially other effects. The distinctive ecological niche occupied by endolichenic fungi, which reside within the lichen symbiosis, has the potential to play a role in the synthesis of innovative and biologically active substances (Molehin et al. 2023).

The practice of investigating the recovery of endolichenic fungi has been in existence for over a decade, as evidenced by studies conducted by Petrini et al. (1990) and Girlanda et al. (1997). However, more recent studies have specifically focused on isolating secondary metabolites from these endolichenic fungi. Several other studies (Kannangara et al. 2009; Hwang et al. 2011; Kim et al. 2012; Padhi and Tayung 2015) have endeavored to elucidate the bioactivity of endolichenic fungi, albeit limited to the analysis of crude extracts. Consequently, comprehensive identification of the metabolites remains elusive. In their study, provided a description of the newly identified fungal order known as *Phaeomoniellales*, which belongs to the class *Eurotiomycetes*. Additionally, they classified certain endolichenic fungi as members of this order. U'Ren et al. (2016) employed a multigene phylogenetic analysis to examine the genetic relationships of various isolates of endolichenic fungi and fungal endophytes collected from the continental United States. The authors identified that some of these isolates potentially belong to previously unidentified species within the *Xylariaceae* family, indicating the necessity for additional investigation and research. U'Ren et al. (2016) also reached the conclusion that the *Xylariaceae*, a prominent and heterogeneous family within the *Pezizomycotina*, *Ascomycota*, harbors both symbiotrophic and saprotrophic fungi. In their study, Maduranga et al. (2018) documented the antioxidant, antilipase, and radical scavenging properties of several lichen species, namely, *Daldinia eschscholtzii*, *Diaporthe musigena*, and *Sordaria* sp. As of 2018, a total of 35 endolichenic microorganisms have been successfully cultured and extensively studied, resulting in the chemical identification of 196 previously unknown natural product structures among a larger pool of 351 secondary metabolites. These newly discovered compounds have demonstrated diverse biological activities. Preliminary studies have indicated that endolichenic fungi exhibit a comparable level of species diversity to the endophytic fungi found in the limited number of sites that have been examined thus far (Arnold et al. 2009a, b; U'Ren et al. 2010). Furthermore, these fungi have demonstrated equal potential for application in the fields of pharmaceuticals, agriculture, and industry (e.g., Paranagama et al. 2007; Ding et al. 2009).

Scientists have been exploring the bioactive potential of endolichenic fungi by isolating and characterizing their secondary metabolites. These studies have led to the discovery of novel compounds and compounds with structural similarities to known bioactive substances. Some endolichenic fungal metabolites have shown promising activity against human pathogens, cancer cells, or other disease targets.

Harnessing the bioactive potential of endolichenic fungi and their secondary metabolites may have applications in drug discovery, pharmaceutical development, and other biotechnological fields. However, it's important to note that further research is needed to fully explore the bioactivity, mechanisms of action, and potential therapeutic applications of these compounds.

10 Conclusion

A lichen is formed by the mutualistic association of a mycobiont (primary fungus) with one or more photobionts (green algae, cyanobacteria, or bacteria). Inside this structure, endolichenic fungi (secondary fungus) and other microorganisms reside without causing any apparent harm or benefit to the lichen itself. Endolichenic fungi have been largely overlooked but play vital roles in the biology, ecology, and evolution of lichen symbiosis. Endolichenic fungi exhibit a remarkable diversity, and their taxonomic composition can vary depending on the lichen species, geographical location, and environmental conditions. The majority of endolichenic fungi belong to the *Ascomycota* phylum, particularly in the classes *Dothideomycetes*, *Eurotiomycetes*, and *Leotiomycetes*. However, *Basidiomycota* and *Zygomycota* fungi have also been reported as endolichenic associates, although they are less common. Some endolichenic fungi may only be found within specific lichen species, while others have a broader range of hosts. Different regions or ecosystems may harbor distinct endolichenic fungal communities due to variations in climate, habitat types, and available lichen species. Endolichenic fungi may play a role in the ecology of lichens by interacting with their host and potentially influencing lichen physiology or metabolism. However, the specific ecological roles and interactions of most endolichenic fungi are still not well understood. The taxonomic diversity of endolichenic fungi is still not fully characterized. Further exploration and identification of novel species through advanced molecular techniques, such as DNA sequencing and metagenomics, would contribute to a more comprehensive understanding of their taxonomy and phylogenetic relationships. The economic potential of endolichenic fungi remains largely unexplored, but there are several areas where these fungi could have potential applications. While further research is needed to fully realize their economic value, here are a few potential avenues worth considering pharmaceutical and medicinal applications, bioremediation and environmental applications, agriculture and plant growth promotion, biotechnological applications, and novel natural products. Future research directions in the field of endolichenic fungi could concentrate on a variety of topics, such as taxonomic exploration, functional roles, host specificity and specificity factors, biotechnological potential, ecological interactions, distribution patterns, and dispersal mechanisms.

References

Arnold AE, Miadlikowska J, Higgins K et al (2009a) A phylogenetic estimation of trophic transition networks for ascomycetous fungi: are lichens cradles of symbiotrophic fungal diversification? Syst Biol 58:283–297

Arnold AE, Miadlikowska J, Higgins KL, Sarvate SD, Gugger P, Way A, Hofstetter V, Kauff F, Lutzoni F (2009b) A phylogenetic estimation of trophic transition networks for ascomycetous fungi: are lichens cradles of symbiotrophic fungal diversification? Syst Biol 58:283–297

Bates ST, Cropsey GW, Caporaso JG, Knight R, Fierer N (2011) Bacterial communities associated with the lichen symbiosis. Appl Environ Microbiol 77(4):1309–1314. https://doi.org/10.1128/AEM.02257-10

Bhagarathi LK, DaSilva PN, Subramanian G, Maharaj G, Kalika-Singh S, Pestano F et al (2023) An integrative review of the biology and chemistry of lichens and their ecological, ethnopharmacological, pharmaceutical and therapeutic potential. GSC Biol Pharm Sci 23(3): 092–119

Bjelland T, Ekman S (2005) Fungal diversity in rock beneath a crustose lichen as revealed by molecular markers. Microbial Ecology 49(4):598–603

Chagnon PL, U'Ren JM, Miadlikowska J et al (2016) Interaction type influences ecological network structure more than local abiotic conditions: evidence from endophytic and endolichenic fungi at a continental scale. Oecologia 180:181–191

Chakarwarti J, Nayaka S, Srivastava S (2020) Diversity of endolichenic fungi–a review. J Mycol 3(1):488–509

Cheon DM, Jang DS, Kim HY et al (2013) Detection of antifungal endolichenic fungi and antifungal compound. Korean J Microbiol 49(2):165–171

De Carvalho CR, Santiago IF, da Costa Coelho L, Câmara PEAS, Silva MC, Stech M, Rosa CA, Rosa LH (2019) Fungi associated with plants and lichens of Antarctica. In: Rosa LH (ed) Fungi of Antarctica. Springer International Publishing, Cham, pp 165–199

Després V, Huffman JA, Burrows SM, Hoose C, Safatov A, Buryak G et al (2012) Primary biological aerosol particles in the atmosphere: a review. Tellus B: Chem Phys Meteorol 64(1): 15598

Diederich P, Lawrey JD, Ertz D (2018) The 2018 classification and checklist of lichenicolous fungi, with 2000 non-lichenized, obligately lichenicolous taxa. Bryologist 121(3):340–425

Ding G, Li Y, Fu S et al (2009) Ambuic acid and torreyanic acid derivatives from the endolichenic fungus Pestalotiopsis sp. J Nat Prod 72(1):182–186

Elkhateeb W, Daba G (2021) Fungi over fungi, endophytic fungi associated with mushroom fruiting bodies and lichens. J Pharm Pharm Res 4:1–4

Farrar JF (1976) The uptake and metabolism of phosphate by the lichen Hypogymnia physodes. New Phytologist 77(1):127–134

Fernández-Mendoza A, Kopun FT, Grube M et al (2017) ITS1 metabarcoding highlights low specificity of lichen mycobiomes at a local scale. Mol Ecol 26:4811

Fleischhacker A, Grube M, Kopun T et al (2015) Community analyses uncover high diversity of lichenicolous fungi in alpine habitat. Micro Ecol 70:348–360

Furbino LE, Godinho VM, Santiago IF et al (2014) Diversity patterns, ecology and biological activities of fungal communities associated with the endemic macroalgae across the Antarctic peninsula. Microb Ecol 67:775–787

Girlanda M, Isocrono D, Bianco C, Luppi-Mosca AM (1997) Two foliose lichens as microfungal ecological niches. Mycologia 89:531–536

Grabka R, d'Entremont TW, Adams SJ, Walker AK, Tanney JB, Abbasi PA, Ali S (2022) Fungal endophytes and their role in agricultural plant protection against pests and pathogens. Plants 11(3):384

Hofstetter V, Miadlikowska J, Kauff F et al (2007) Phylogenetic comparison of protein-coding versus ribosomal RNA-coding sequence data: a case study of the Lecanoromycetes (Ascomycota). Mol Phylogenet Evol 44:412–426

Hwang HG, Kim YN, Baik KS et al (2011) Isolation and characteristics of endolichenic fungi producing antifungal compound. Korean J Microbiol 47(1):97–101

Kannangara BT, Rajapaksha RS, Paranagama PA (2009) Nature and bioactivities of endolichenic fungi in Pseudocyphellaria sp., Parmotrema sp. and Usnea sp. at Hakgala montane forest in Sri Lanka. Lett Appl Microbiol 48:203–209

Karlovsky P (2008) Secondary metabolites in soil ecology. Springer, Berlin Heidelberg, pp 1–19

Kim ES, Choi KS, Choi SK (2012) Detection of endolichenic fungi producing antifungal compound. Korean J Microbiol Biotechnol 40(1):23–29

Krienitz L (2018) Lesser flamingos: descendants of Phoenix. Springer

Lagarde A, Jargeat P, Roy M, Girardot M, Imbert C, Millot M, Mambu L (2018) Fungal communities associated with Evernia prunastri, Ramalina fastigiata and Pleurosticta acetabulum: three epiphytic lichens potentially active against Candida biofilms. Microbiol Res 211:1–12

Lawrey J, Diederich P (2003) Lichenicolous fungi: interactions, evolution, and biodiversity. Bryologist 106:80–120

Li WC, Zhou J, Guo SY et al (2007) Endophytic fungi associated with lichens in Baihua mountain of Beijing, China. Fungal Divers 25:69–80

Lutzoni F, Pagel M, Reeb V (2001) Major fungal lineages are derived from lichen symbiotic ancestors. Nature 411:937–940

Maduranga K, Attanayake RN, Santhirasegaram S et al (2018) Molecular phylogeny and bioprospecting of Endolichenic fungi (ELF) inhabiting in the lichens collected from a mangrove ecosystem in Sri Lanka. PLoS One 13(8):e0200711

Masumoto H, Degawa Y (2019) The effect of surface sterilization and the type of sterilizer on the genus composition of lichen-inhabiting fungi with notes on some frequently isolated genera. Mycoscience 60:331–342

Miadlikowska A, Arnold A, Lutzoni F (2004) Diversity of cryptic fungi inhabiting healthy lichen thalli in a temperate and tropical forest. Ecol Soc Am Annu Meet 89:349–350

Molehin OR, Adeniyi OS, Olowomofe TO, Ajayi EI (2023) Nanochitosan derived from marine endolichenic fungi. In: Next generation Nanochitosan. Academic Press, pp 125–136

Moreno LF, Vicente VA, de Hoog S (2018) Black yeasts in the omics era: achievements and challenges. Med Mycol 56:32–41

Muggia L, Grube M (2018) Fungal diversity in lichens: from extremotolerance to interactions with algae. Life 8(2):15

Naranjo-Ortiz MA, Gabaldón T (2019) Fungal evolution: major ecological adaptations and evolutionary transitions. Biol Rev 94(4):1443–1476

Oh S-Y, Yang JH, Woo J-J, Oh S-O, Hur J-S (2020) Diversity and distribution patterns of Endolichenic fungi in Jeju Island, South Korea. Sustainability 12(9):3769. https://doi.org/10.3390/su12093769

Padhi S, Tayung K (2015) In-vitro antimicrobial potentials of endolichenic fungi isolated from thalli of Parmelia lichen against some human pathogens. Beni Suef Univ J Basic Appl Sci 4(4): 299–306

Padhi S, Masi M, Mohanta YK, Saravanan M, Sharma S, Cimmino A et al (2023) In silico pharmacokinetics, molecular docking and dynamic simulation studies of endolichenic fungi secondary metabolites: an implication in identifying novel kinase inhibitors as potential anticancer agents. J Mol Struct 1273:134390

Paranagama PA, Wijeratne EMK, Burns AM et al (2007) Heptaketides from Corynespora sp. inhabiting the cavern beard lichen, Usnea cavernosa: first report of metabolites of an endolichenic fungus. J Nat Prod 70(11):1700–1705

Park CH, Kim KM, Elvebakk A et al (2015) Algal and fungal diversity in Antarctic lichens. J Eukaryot Microbiol 62:196–205

Petrini O, Hake U, Dreyfuss M (1990) An analysis of fungal communities isolated from fruticose lichens. Mycologia 82:444–451

Petrzik K, Koloniuk I, Sehadová H, Sarkisova T (2019) Chrysoviruses inhabited symbiotic fungi of lichens. Viruses 11(12):1120

Raudabaugh DB, Matheny PB, Hughes KW, Iturriaga T, Sargent M, Miller AN (2020) Where are they hiding? Testing the body snatchers hypothesis in pyrophilous fungi. Fungal Ecol 43: 100870

Seaward MR (1988) Contribution of lichens to ecosystems. CRC handbook of lichenology 2:107–129

Shanmugam P, Ponnusamy P, Balasubramanian G et al (2018) Screening of biologically potent endolichenic fungi isolated from selected lichens habitat on silver oak tree. Vegetos An Int J Pl Res 31(3):89–94

Si H, Wang Y, Liu Y, Li S, Bose T, Chang R (2023) Fungal diversity associated with thirty-eight lichen species revealed a new genus of Endolichenic fungi, *Intumescentia* gen. nov. (Teratosphaeriaceae). Journal of Fungi 9(4):423

Singh BN, Upreti DK, Gupta VK et al (2017) Endolichenic fungi: a hidden reservoir of next generation biopharmaceuticals. Trends Biotechnol 35(9):808–881

Spribille T, Tuovinen V, Resl P, Vanderpool D, Wolinski H, Aime MC, Schneider K, Stabentheiner E, Toome-Heller M, Thor G, Mayrhofer H, Johannesson H, McCutcheon JP (2016) Basidiomycete yeasts in the cortex of ascomycete macrolichens. Science 353:488–492

Strobel GA, Hess WM, Ford E et al (1996) Taxol from fungal endophytes and the issue of biodiversity. J Ind Microbiol 17:417–442

Suryanarayanan TS, Thirunavukkarasu N, Hariharan GN et al (2005) Occurrence of non-obligate microfungi inside lichen thalli. Sydowia 57(1):120–130

Suryanarayanan TS, Govindarajulu MB, Rajamani T, Tripathi M, Joshi Y (2017) Endolichenic fungi in lichens of Champawat district, Uttarakhand, northern India. Mycol Prog 16:205–211

Suryanarayanana TS, Thirunavukkarasu N (2017) Endolichenic fungi: the lesser known fungal associates of lichens. Mycology 8(3):189–196

Tripathi M, Joshi Y (2015) Endolichenic fungi in Kumaun Himalaya: a case study. In: Upreti DK, Divakar PK, Shukla V et al (eds) Recent advances in lichenology. Springer, New Delhi, pp 111–120

Tripathi M, Joshi Y (2019a) Endolichenic fungi: present and future trends. Springer

Tripathi M, Joshi Y (2019b) Endolichenic fungi: a case study from Uttarakhand. In: Tripathi M, Joshi Y (eds) Endolichenic Fungi: present and future trends. Springer Singapore Pte Ltd., Singapore, pp 119–145

Tripathi M, Gupta RC, Joshi Y (2014a) Spegazzinia tessarthra isolated as a true endophyte from lichen Heterodermia flabellata. Indian Phytopathol 67(1):109–110

Tripathi M, Gupta RC, Joshi Y (2014b) Physcia dilatata Nyl. (lichenized fungi, Physciaceae); a new host of Bipolaris australiensis (M.B. Ellis) Tsuda and Ueyama from Kumaun Himalaya, India. Proc Nat Acad Sci Lett 37(5):477–479

Tripathi M, Joshi Y, Gupta RC (2014c) Assessment of endolichenic fungal diversity in some forests of Kumaun Himalaya. Curr Sci 107(5):745–748

Tripathi M, Gupta RC, Joshi Y (2014d) Macrolichens as an excellent host of endophytic fungi: a case study in Banri Devi forest of Kumaun Himalaya, Uttarakhand, India. In: Jha P (ed) Biodiversity, conservation and sustainable development. New Academic Publishers, New Delhi, pp 141–146

Tripathi M, Joshi Y, Tripathi M, Joshi Y (2019) Introduction to endophytic fungi associated with lichens ie endolichenic fungi. In: Endolichenic Fungi: present and future trends, pp 27–47

U'Ren JM, Lutzoni F, Miadlikowska J et al (2010) Community analysis reveals close affinities between endophytic and endolichenic fungi in mosses and lichens. Microb Ecol 60:340–353

U'Ren JM, Lutzoni F, Miadlikowska J et al (2012) Host and geographic structure of endophytic and endolichenic fungi at a continental scale. Am J Bot 99:898–914

U'Ren JM, Riddle JM, Monacell JT et al (2014) Tissue storage and primer selection influence pyrosequencing-based inferences of diversity and community composition of endolichenic and endophytic fungi. Mol Ecol Resour 14:1032–1048

U'Ren JM, Miadlikowska J, Zimmerman NB et al (2016) Contributions of north American endophytes to the phylogeny, ecology, and taxonomy of Xylariaceae (Sordariomycetes, Ascomycota). Mol Phylogenet Evol 98:210–232
Vinayaka KS, Krishnamurthy YL, Banakar S et al (2016) Association and variation of endophytic fungi among some macrolichens in Central Western Ghats, southern India. Int J Curr Microbiol App Sci 5(6):115–124
Wang Q, Li J, Yang J, Zou Y, Zhao XQ (2022) Diversity of endophytic bacterial and fungal microbiota associated with the medicinal lichen Usnea longissima at high altitudes. Front Microbiol 13:958917
Wethalawe AN, Alwis YV, Udukala DN, Paranagama PA (2021) Antimicrobial compounds isolated from endolichenic fungi: a review. Molecules 26(13):3901
Xu H, Wang L, Feng X, Gong X (2022) Core taxa and photobiont-microbial interaction within the lichen Heterodermia obscurata (Physcsiaceae, Heterodermia). Symbiosis 86(2):187–204
Yu NH, Park S-Y, Kim JA, Park C-H, Jeong M-H, Oh S-O, Hong SG, Talavera M, Divakar PK, Hur J-S (2018) Endophytic and endolichenic fungal diversity in maritime Antarctica based on cultured material and their evolutionary position among Dikarya. Fungal Syst Evol 2:263–272
Zhang T, Wei XL, Wei YZ, Liu HY, Yu LY (2016) Diversity and distribution of cultured endolichenic fungi in the Ny-Ålesund region, Svalbard (high Arctic). Extremophiles 20:461–470

Diversity of Culturable Endophytic Fungi Associated with Bryophytes

Fatma A. Altamany, Mennatullah M. Allam, Mohamed A. Abdel-Azeem, Lobna Hajji-Hedfi, Alsayed E. Mekky, Salma S. Higazy, Youssef M. Salah, Norhan A. Soliman, Hager M. Fathy, Roaa S. Abdel-Monem, Sara A. Gezaf, Fatma A. Abo Nouh, Teroj A. Mohamed, Kholoud A. Elsayed, Alshaimaa S. Abdel-Salam, Emad Ahmed, Bola N. Aziz, Amany R. Ali, Ola F. Thabet, Mardian A. Adel, Amal M. Elshazly, and Ahmed M. Abdel-Azeem

Contents

F. A. Altamany (✉) · M. M. Allam · S. S. Higazy · Y. M. Salah · N. A. Soliman · H. M. Fathy · R. S. Abdel-Monem · K. A. Elsayed · A. S. Abdel-Salam · E. Ahmed · B. N. Aziz · A. R. Ali · O. F. Thabet · M. A. Adel · A. M. Elshazly
Faculty of Science, Botany and Microbiology Department, Suez Canal University, Ismailia, Egypt

M. A. Abdel-Azeem
Faculty of Pharmacy, Pharmacognosy Department, Sinai University – Arish Branch, Arish, Egypt

L. Hajji-Hedfi
Regional Centre of Agricultural Research of Sidi Bouzid, CRRA, Sidi Bouzid, Tunisia

Research Laboratory of Agricultural Production Systems and Sustainable Development (LR03AGR02), Carthage University, Tunis, Tunisia

A. E. Mekky
Faculty of Science Boys, Department of Botany and Microbiology, Al-Azhar University, Cairo, Egypt

S. A. Gezaf
Faculty of Science, Botany and Microbiology Department, Arish University, Arish, Egypt

F. A. Abo Nouh
Faculty of Science, Botany and Microbiology Department, Suez Canal University, Ismailia, Egypt

Natural and Medical Sciences Research Center, University of Nizwa, Nizwa, Oman

A. M. Abdel-Azeem, F. Degola (eds.), *Advances in Lichens and Non-Flowering Plants Fungal Endophytes Research*, Fungal Biology,
https://doi.org/10.1007/978-3-032-01506-8_6

1 Introduction

Endophytic fungi (EF), which began evolving over 400 million years ago (Kusari and Spiteller 2012), were first described nearly 200 years ago with *Sphaeria typhena* (now *Epichloe typhina*) (Khairalla et al. 2016). The concept of symbiosis, introduced by De Bary in 1879, has since revealed that most multicellular life on Earth is in symbiosis with microorganisms. Endophytes, which are ubiquitous among plants, establish mutualistic relationships with their hosts without causing harm (Anitha et al. 2013; Kusari and Spiteller 2012). They include fungi, bacteria, and viruses, with fungi being the most common and diverse (Hilszczanska 2016; Strobel and Daisy 2003).

Endophytic fungi belong to various phylogenetic groups within the *Dikarya*, including both *Clavicipitales*, predominantly in monocots, and non-clavicipitaceous species in both monocots and eudicots. Tropical regions are particularly rich in fungal diversity. While most research focuses on vascular plants, endophytes are also present in nonvascular plants like bryophytes, which are critical in ecosystems such as boreal forests, Antarctica and tundras (Davis et al. 2003a, b; U'Ren et al. 2010).

Bryophytes are nonvascular plants that include mosses, liverworts, and hornworts. They are characterized by their lack of vascular tissues (Hawksworth and Lücking 2017) which distinguishes them from vascular species. Bryophytes

T. A. Mohamed
Dental Basic Sciences Department, College of Dentistry, University of Duhok, Duhok, Iraq

A. M. Abdel-Azeem
Botany and Microbiology Department, Faculty of Science, Suez Canal University, Ismailia, Egypt

Research Institute of University of the Bucharest (ICUB), University of Bucharest, Bucharest, Romania

The Centre for Mineral Biogeochemistry, The University of the Free State, Bloemfontein, Free State, South Africa typically grow in moist environments and play crucial roles in ecological processes such as soil formation and water retention. Raven, (Frego 2007; Sun et al. 2011.

The diversity of bryophytes in forest ecosystems is influenced by climate, historical factors, forest health, and management practices (Frego 2007; Sun et al. 2011). Bryophytes are categorized into guilds based on their substrates, such as soil, deadwood, and rocks. Forest management impacts these substrates and, consequently, bryophyte diversity (Bengtsson et al. 2000; Paillet et al. 2010). Old-growth and sustainably managed forests support more bryophyte species, particularly those associated with deadwood.

In Europe, forest reserves exhibit high bryophyte diversity, particularly in ancient stands with varied deadwood stages (Sabovljevic et al. 2010). Canopy tree species also affect bryophyte richness (Rola et al. 2021). Bryophytes, especially those dependent on specific substrates, are sensitive to management practices and serve as indicators of forest integrity (Müller and Lenz 2006; Norden et al. 2007).

The diversity of endophytes in bryophytes is significant, with many unique species adapted to the specific conditions of these plants. This diversity is crucial for understanding the ecological roles and interactions within bryophyte communities (Porras-Alfaro and Bayman 2011).

And interactions between bryophytes and their endophytes involve various mutualistic, commensal, and sometimes pathogenic relationships (Porras-Alfaro and Bayman 2011). Endophytes in bryophytes can enhance their host's resistance to environmental stress and pathogens, contribute to nutrient acquisition, and influence the overall health and growth of the bryophyte (Degola et al. 2022). The symbiosis can also affect the bryophyte's ecological role, such as soil formation and water retention (Shaffer et al. 2018).

2 Endophytic Fungi

2.1 Definition

The term "endophytes" was first used by (de Bary 1866) De Bary in 1886, and he described it as "any organism that grows within plant tissues" (de Bary 1866). Endophytes are microorganisms that live in a symbiotic relationship with plants and reside within their healthy tissues. These microorganisms exhibit many symbiotic lives with the host plant, ranging from latent viruses or pathogens to prokaryotic bacteria, actinomycetes, and eukaryotic fungi (Wani et al. 2015; Kumar and Kaushik 2013). However, there have been occasional revisions and changes made to the definition of endophytes by various authors. Endophytes do not induce any symptoms in the host plants, in contrast to a variety of phytopathogens and mycorrhizas that do.

2.2 Endophytic Fungi and Their Secondary Metabolites

Long-term interactions exist between endophytic fungi and many endophytes that produce bioactive chemicals from their host plants (Kumar and Kaushik 2013). These interactions suggest that the evolutionary paths of endophytic fungi and host plants are closely related, particularly in their strategies for producing secondary metabolites. This phenomenon can be attributed to horizontal gene transfer (Soliman et al. 2013; Wang and Dai 2011), indicating that endophytic fungi may evolve into significant sources of new bioactive secondary metabolites.

Numerous bioactive substances with potential medicinal benefits have been identified, including gentiopicrin, vincristine, taxol, spiroquinazoline alkaloids, and cryptocin (Li et al. 1993). Recent studies have reported new bioactive compounds derived and other fungal endophytes. This growing body of research highlights the increasing interest in bioactive secondary metabolites from endophytic fungi, as evidenced by numerous recent reviews on the topic.

The documented compounds from these fungi are categorized into several structural types, including alkaloids, flavonoids, dihydroisocoumarins, cyclohexanones, benzofurans, phenylpropanoids, lipids, organic acids, peptides, pyridines, terpenoids, steroids, and quinones. Many of these new compounds exhibit antibacterial, insecticidal, antioxidant, anticancer, and cytotoxic properties. In order to find and create new medications, a lot of research is now being done on the screening of novel bioactive secondary metabolites from endophytic fungi (Xiao et al. 2014; Zhou et al. 2014a, b).

Medicinal plant endophytic fungi, such as those from *Dendrobium devonianum* and *Dendrobium thyrsiflorum*, exhibit cytotoxic (antitumor) and broad-spectrum antimicrobial bioactivities and have been shown to create compounds that prevent the growth of bacteria such *Staphylococcus aureus*, *Bacillus subtilis*, *Candida albicans*, *Aspergillus fumigatus*, *Cryptococcus neoformans*, and *Escherichia coli*. Additionally, human cancer cell lines such as HepG2, MCF7, SKVO3, HL-60, and 293-T are susceptible to the anticancer activity of fungal endophytes derived from agarwood (*Aquilaria sinensis*) (Cui et al. 2011).

Certain isolates were explored for their antimicrobial activity against *Candida albicans* and *Cladosporium sphaerospermum*, as well as growth inhibition on cancer cells MCF-7 and TK-10, as well as the biodiversity and bioactivities of endophytic fungi linked to the medicinal plant *Stryphnodendron adstringens*, as reviewed by, fungi that grow endophytically on medicinal plants possess both cytotoxic (antitumor) and broad-spectrum, antibacterial properties.

2.3 Classification of Endophytic Fungi

For a better understanding, the classification of endophytes has been a difficult task that has been ongoing for a number of years. Endophytes were originally classified into two groups: the first group consisted of endophytic fungi that show no symptoms during their whole life cycle within the host, whereas the second group showed symptoms on the host plant, such as root nodules. Another categorization of endophytes simply divided them into facultative endophytes and obligate endophytes, two distinct subgroups. According to previous understanding, the former were thought to be entirely dependent on their host for growth, survival, and transmission, while the latter were believed to be administered to the host from the external environment and also capable of independent existence.

According to Rodriguez et al. (2009), endophytic fungi can also be classified as clavicipitaceous or non-clavicipitaceous endophytes depending on the range of hosts, mode of transmission, plant tissue colonized, frequency of colonization, biodiversity (high or low), and habitat and non-habitat specific functions. These two groupings are part of four distinct classes, which are classes 1, 2, 3, and 4. According to Rodriguez et al. (2009), class 1 endophytes (Abdel-Azeem et al. 2019; Abo Nahas et al. 2021) are clavicipitaceous endophytes, which are primarily found in grasses. The remaining three classes are classified as non-clavicipitaceous endophytes, which are primarily found in non-grasses and higher vascular plants.

Endophytes are categorized into two broad categories, systemic or true endophytes and nonsystemic or transient endophytes, according to a recent classification. These categories are based on a variety of characteristics, including biology, functional diversity, taxonomy, evolution, and their mode of transmission (Wani et al. 2015). It is known that systemic endophytes live on the same plant during different seasons and that they spread more vertically (by seeds) than horizontally. Conversely, nonsystemic endophytes have a horizontal mode of transmission and fluctuate in both abundance and diversity inside the plant in response to changes in the external climate (Wani et al. 2015).

2.4 Endophytic Fungi Diversity and Abundance

Endophytic fungi can be found in various plant tissues, including roots, stems, leaves, branches, twigs, petioles, flowers, bark, fruits, seeds, scales, resin canals, and even meristems (Abdel-Azeem et al. 2019; Abo Nahas et al. 2021). The types of fungi present within a plant can vary depending on the location within the plant, the time of year, and the function of the fungi. For example, different fungi colonize the leaves, stems, and roots of alfalfa plants, producing different types of secondary metabolites (Lugtenberg et al. 2016).

While leaves, twigs, branches, and roots are the most commonly studied tissues for endophytic fungi, fewer studies have investigated the presence of endophytes in the trunks or fruits of woody plants. It's important to note that the composition of endophytic fungi can vary significantly between different organs or tissues of the same plant. (Fisher and Petrini 1990).

Endophytic fungi found in many healthy plant species range from crops, invasive plants, woody tree species—especially medicinal plants, mosses, ferns, and lichens. (Cui et al. 2011; Fisher et al. 1992; U'Ren et al. 2010). The diversity of endophytic fungi varies significantly depending on location. Regions with higher latitudes, like temperate zones, tend to have fewer fungal species compared to tropical regions. While tropical regions have a smaller number of fungal classes, they boast a much wider variety of endophytic species.

Leptosphaerulina, *Nigrospora*, *Phoma*, *Fusarium*, *Gibberella*, *Glomerella*, *Giardinia*, *Alternaria*, *Colletotrichum*, *Xylaria*, and *Phomopsis* are the genera that are most frequently isolated. For instance, from Taxus x, 81 fungi were isolated in media, which are categorized into the following 8 distinct genera: *Alternaria*, *Glomerella*, *Gibberella*, *Colletotrichum*, *Guignardia*, *Phomopsis*, *Phoma*, and *Nigrospora* (Sun et al. 2012; Xiong et al. 2013) discovered 1955 strains of three woody plants from China: *Ulmus macrocarpa*, *Quercus liaotungensis*, and *Betula platyphylla* that belonged to 61 taxa, which were identified based on morphological characteristics and DNA sequence data (*Alternaria*, *Xylaria*, *Fusarium*, *Phoma*, and so on).

Melanconis spp. dominated the endophytic fungal community. *Fusicoccum* spp. dominated in *B. platyphylla* and *Disculina* spp. in *Fusarium* species, *Alternaria* species, and *Q. liaotungensis* in *U. macrocarpa*, in that order. From the leaves of the invasive plant, (Arnold et al. 2000). The fungal endophyte—Google Scholar n.d.; Zhang et al. 2006; Zimmerman and Vitousek 2012). Isolated 463 endophytic fungi that were categorized into 38 genera and 112 operational taxonomic units (OTUs). *Ageratina adenophora*, along with *Xylaria* species, the most common isolates were *Colletotrichum* species, *Nemania* species, and *Phomopsis* species.

Researchers think that tropical and temperate region have the highest endophyte richness (Zimmerman and Vitousek 2012; Zhang et al. 2006; Arnold et al. 2000; Arnold and Lutzoni 2007). The breadth of endophytic fungus diversity in the southern boreal forest, Fisher's $\alpha = 2.6$, and in the tropical woodland. According to the review by Banerjee (2011), more than 80 different endophytic fungal genera were identified from tropical and subtropical plants, demonstrating the amazing endophyte. The endophytic fungal communities found in endophytes over a wide gradient from the lowland tropical to the Canadian arctic The center forest of Panama. Six plant species in total—21 in total—delineated away from the tropics in these localities.

Endophytic fungal diversity and their ecological functions in cold-climate plants have received less attention than in the case of tropical trees. Li et al. (2012b) identified 604 endophytic Fungi gathered in Southwest China's Baima Snow Mountain, from 5 different plant species, 4000–4300 m above sea level. Features of morphology and internal transcription spacer (ITS) sequencing investigation showed that 43 distinct taxa were obtained, wherein *Penicillium, Sirococcus*, and *Cephalosporium* and *Aspergillus* were the prevalent, widely dispersed genera.in each of the five types of plants. Zhang and colleagues (Zhang et al. 2013) looked at the diversity of endophytic fungus linked to bryophytes (*Chorisodontium aciphyllum*, *Barbilophozia hatcheri*, *Sanionia uncinata*) on King George Island's

Fildes Region Antarctica by sea. The total number of endophytic fungi discovered was 128, recognized by 21 different taxa, 15 of which are *Ascomycota*.

Another popular area of study for endophytic fungus is mangroves. A total of 343 endophytic fungi were isolated by (De Souza Sebastianes et al. 2013) from 3 mangrove plants (*Avicennia schaueriana*, *Rhizophora mangle*, *Laguncularia racemosa*) that are a part of at least at least 34 distinct genera, with the most prevalent being *Fusarium* species, *Diaporthe* species, *Colletotrichum* species, *Xylaria* species, and *Trichoderma* species. This suggests that the mangrove fungus community is remarkably diverse and abundance of endophytic fungus. Over 200 fungi that are endophytes, mainly genera of *Aspergillus* species, *Alternaria* species, *Fusarium* species, *Colletotrichum* species, *Cladosporium* species, *Penicillium* species, *Pestalotiopsis* species, *Paecilomyces* species, *Trichoderma*, *Phyllosticta* spp., *Phoma* spp., and *Phomopsis* spp. were found in mangroves and isolated (Cheng et al. 2009).

Xylariaceous ascomycetes are particularly common in tropical regions. Studies have shown that tropical plants harbor a greater diversity of endophytes compared to temperate plants. Interestingly, endophytes in temperate regions often show host specificity, meaning they are restricted to a particular plant species. In contrast, tropical endophytes are less host specific, indicating a broader range of plant hosts they can colonize (Chhipa and Deshmukh 2019).

3 Bryophytes

3.1 Definition

Bryophytes, which are typically small and lack true roots and a vascular system, are a unique group of land plants (embryophytes) exhibiting exceptional species diversity, surpassed only by flowering plants (angiosperms) (Christenhusz and Byng 2016). In their life cycle, the gametophyte phase is dominant and independent, while the sporophyte depends on the gametophyte for hydration and sustenance. The bryophyte gametophyte may have a thalloid morphology or develop structures analogous to roots, stems, and leaves, referred to as rhizoids, cauloids, and phyllodes.

This structural complexity places bryophytes between green algae and vascular plants in the evolutionary timeline, suggesting they emerged shortly after the first land plants. As a result, the evolutionary status of mosses, liverworts, and hornworts within terrestrial plants has garnered considerable attention, as discussed in Donoghue et al. (2021).

In the first evolution of terrestrial flora, bryophytes were among the earliest to acclimate to arid, nutrient-deficient settings and exposed to high levels of UV radiation. Land plants developed several adaptations, such as stomata, cuticles, water-conducting tissues, and root systems, to survive on land. These traits are expressed in varying degrees across the three bryophyte lineages, both morphologically and genetically. Since their evolution around 500 million years ago, bryophytes have successfully colonized diverse habitats, spanning deserts to aquatic ecosystems, and from tropical to cold climates.

Their species composition and distribution are intricately associated with microhabitats (e.g., forest floors, tree trunks, boulders) and microclimates (e.g., humidity, temperature, light intensity). As a result, bryophytes have evolved diverse adaptations in shape, physiology, and reproductive tactics, including desiccation tolerance and dormancy, which allow them to swiftly react to environmental changes. There can be no doubt that early land plant evolution transformed the planet but, until recently, how and when this was achieved was unclear. Coincidence in the first appearance of land plant fossils and formative shifts in atmospheric oxygen and CO_2 are a consequence of the scarcity of older terrestrial rocks. The elucidation of the chronology of land plant body plan assembly and its influence on global biogeochemical cycles has been hindered by ambiguities regarding the phylogenetic relationships among bryophytes and between bryophytes and tracheophytes, as well as the temporal framework of these occurrences. Recent genome and transcriptome sequencing initiatives, along with the utilization of advanced phylogenomic modeling techniques, have provided growing evidence for the Setaphyta clade of liverworts and mosses among monophyletic bryophytes. This study examines the evolution of anatomy, genes, genomes, and development within a phylogenetic framework, concluding that numerous innovations in vascular plants (tracheophytes) were already present in a rather complex last common ancestor of extant land plants (embryophytes). Molecular clock investigations suggest that embryophytes originated during the mid-Cambrian to early Ordovician period, aligning with theories of their function as geoengineers, which contributed to the onset of early Paleozoic glaciations.

Bryophytes also play a major role in energy flow and nutrient cycling within woodland environments, with mosses alone contributing 49% of the annual phytomass production and 75% of the forest's phosphorus supply (Frahm 2007). Additionally, bryophytes provide habitats for numerous invertebrates, fungi, and other microorganisms. Mosses that form mat-like colonies help prevent soil erosion and create seed beds that influence tree growth by either enhancing or inhibiting seedling survival.

The species composition and richness of bryophytes in forest ecosystems are primarily influenced by macroclimatic factors such as air temperature and precipitation (Sun et al. 2013a, b), along with longitudinal and altitudinal gradients microrelief diversity and forest integrity. Forest management and silviculture practices also play a significant role.

On a finer scale, microclimatic conditions such as shade, humidity, substrate availability, and light conditions determined by forest structure greatly affect bryophyte diversity and coverage (Márialigeti: The effects of stand structure on groundfl...—Google Scholar n.d.; Müller et al. 2019). Soil properties, particularly soil reaction, also influence the diversity of bryophyte communities on the forest floor (Vitt: Patterns of bryophyte diversity in peatlands...—Google Scholar n.d.). Lastly, human activities, such as habitat disturbances from logging, can significantly impact bryophyte diversity in forests (Caners et al. 2013; Nelson: Short-term effects of timber harvest and...—Google Scholar n.d.).

3.2 Diversity

In many temperate environments, bryophytes greatly increase diversity, making them the second most diverse group of terrestrial plants after flowering plants, contributing to overall ecosystem functioning (Kolari et al. 2006). These plants are poikilohydric and typically lack roots and, in many cases, a waxy cuticle, allowing them to absorb water and nutrients directly through their surfaces (Gignac 2001). In contrast to tracheophytes, bryophytes are particularly sensitive to microclimatic conditions, especially moisture, due to their inability to regulate water loss. They employ various physiological strategies and colony architectures to improve desiccation tolerance (Proctor: Poikilohydry and homoihydry: antithesis...—Google Scholar n.d.; Spitale: Switch between competition and facilitation...—Google Scholar n.d.).

Bryophytes are particularly vulnerable to environmental changes due to their simplistic structure, lacking roots and possessing an efficient vascular system, which allows them to absorb water and nutrients through their entire surface.

Consequently, they are significantly affected by global warming and climate change, leading to shifts in their distribution patterns, often moving to higher elevations or latitudes in search of more suitable habitats. Given their high sensitivity to climate conditions and their widespread presence—from tropical regions to polar areas and from sea level to high altitudes—they have been proposed as valuable organisms for conducting studies on latitudinal and altitudinal variations (Andrew et al. 2003).

Bryophyte species richness often surpasses that of vascular plants in humid artificial forests located in temperate regions (Gignac 2001). Thus, regional studies are crucial for advancing our understanding of bryophyte taxonomy and diversity.

Despite their significance, bryophytes are frequently overlooked in biodiversity assessments compared to vascular plants (Sun et al. 2013a, b). They are widely distributed and play essential roles in forest ecosystems, particularly in eastern China, a region known for its high concentration of endemic bryophytes in East Asia (Andrew et al. 2003).

However, these abundant bryophyte populations face serious threats from climate change, habitat loss, disturbances, and land-use changes (Callaghan 2021; He et al. 2016; Riffo-Donoso et al. 2021). Additionally, tourism development poses significant risks to their habitats (Hernández-Hernández et al. 2019), as activities like deforestation and road construction disrupt ecosystem balance and bryophyte survival in popular areas. Consequently, sensitive bryophyte species are particularly vulnerable to changing habitat conditions (赵德先 et al. 2020).

A study by Zhang et al. (2013) investigated the fungal diversity associated with three Antarctic bryophyte species: the liverwort *Barbilophozia hatcheri* and the mosses *Chorisodontium aciphyllum* and *Sanionia uncinata*. The researchers identified a total of 78 operational taxonomic units (OTUs) belonging to *Ascomycota*, 13 OTUs from *Basidiomycota*, 1 OTU from *Zygomycota*, and 1 OTU from an unidentified phylum. The dominant fungal orders observed included *Helotiales*, *Chaetothyriales*, *Eurotiales*, *Sebacinales*, and *Platygloeales*.

At the phylum level, the fungal communities varied significantly among the three bryophyte species. *Sanionia uncinata* exclusively hosted *Ascomycota*, while *Barbilophozia hatcheri* was predominantly colonized by *Basidiomycota*. In contrast, *Chorisodontium aciphyllum* exhibited a mix of both *Ascomycota* and *Basidiomycota*. Notably, no OTUs were shared across the three bryophytes, highlighting distinct fungal assemblages in each species. Additionally, *Sanionia uncinata* displayed a higher number of OTUs compared to the other two species, suggesting greater endophytic fungal diversity within this moss.

3.3 The Adaption of Bryophytes to Early Land Environment

As pioneer organisms, bryophytes face various environmental challenges, including desiccation, UV radiation, and temperature fluctuations (Degola et al. 2022), being also predominantly located on soil and rock surfaces that are inhabited by bacteria, fungi, and lichens (Beraldi-Campesi 2013; Graham et al. 2004; Heckman et al. 2001; Weber et al. 2016). Despite their simple morphology, bryophytes exhibit crucial adaptations for survival in terrestrial environments, such as a root system, water-conducting cells (WCCs), stomata, cuticle, and a multicellular embryo (Jill Harrison 2017; Pires and Dolan 2012; Szövényi et al. 2019). The morphological diversity among mosses, liverworts, and hornworts varies during early adaptation. The gametophytes of hornworts are very simple thalli, whereas those of mosses and most liverworts have stem-like and leaf-like structures. In gametophytes of bryophytes, the rhizoid-based rooting system primarily serves to anchor the plant, although some evidence suggests it may also play a role in water and nutrient uptakes (Jones and Dolan 2012).

In mosses, rhizoids are multicellular, while in liverworts and hornworts, they are unicellular. Water-conducting cells (WCCs), known as hydroids in mosses, are present in both mosses and liverworts but are lacking in hornworts (Ligrone et al. 2000, 2002). Most mosses and hornworts possess stomata on their sporophytes, but liverworts lack stomata. The principal role of stomata in mosses and hornworts appears to aid in sporophyte desiccation and spore dissemination, differing from the roles in vascular plants, where they assist in gas exchange and water retention (Chater et al. 2016; Duckett and Pressel 2017; McAdam et al. 2021; Renzaglia et al. 2017; Sussmilch et al. 2019). The spore-bearing capsules of their sporophytes are supported by setae in both liverworts and mosses, whereas hornwort sporophytes grow continuously from a basal meristem and lack setae.

3.4 Fungal Symbioses in Bryophytes

Liverworts Associations

The surprising finding that fungi intensively populate the branched, septate rhizoids of the Schistochilaceae family was a major advance in our knowledge of the relationships between liverworts and ascomycetes. Comprehensive cytological, cross-infection, and molecular investigations were motivated by this discovery (Pressel et al. 2008a, b). Rhizoscyphus was recognized as the endophytic fungus like the fungus *Cephaloziella* from Antarctica (Upson et al. 2007). A mutually balanced interaction was revealed by the rhizoids' healthy hyphae and host cytoplasm. According to cross-infection tests, fungi isolated from the host *Pachyschistochila* species could cause the distinctive septation and branching of rhizoids observed in the Schistochilaceae, as well as from other liverworts with enlarged rhizoids. Furthermore, rhizoid swelling was brought on by this fungus in *Cephalozia*, although rhizoids in the *Lepidoziaceae* frequently swollen before fungal infection (Duckett et al. 1991). Interestingly, experimental studies on the ascomycete "lichen" *Mniaecia jungermanniae*, initially believed to be the liverwort rhizoid fungus's ideal stage, turned out to have little effect on *Cephalozia*'s rhizoid development. Instead, it acted as a parasite, inducing the formation of giant perichaetia and apogamous sporophytes (Pressel and Duckett 2011).

Cytological and cross-infection studies in the *Jungermanniopsida* have demonstrated that the swollen rhizoids found in several leafy liverwort families, such as Lepidoziaceae, Cephaloziaceae, Calypogeiaceae, and Cephaloziellaceae, are densely populated with a specific ascomycete fungus, identifiable by its simple septa and Woronin bodies. For example, the rhizoid tips of *Calypogeia muelleriana* were found to be filled with hyphae, strongly suggesting the presence of *Rhizoscyphus ericae*, a fungus recognized for establishing ericoid mycorrhizas (Duckett et al. 1991; Duckett and Read 1995). The identification of this fungus was subsequently validated through comprehensive DNA sequencing analyses of Antarctic *Cephaloziella* populations (Chambers et al. 1999; Upson et al. 2007).

Basidiomycetes Associations: *Aneuraceae*

The only thalloid liverworts that have been discovered to date to have basidiomycetous endophytes that can be easily recognized as such by the presence of dolipore septa are the *Aneuraceae* (Bidartondo et al. 2003; Pressel et al. 2008a, b; Preussing et al. 2010). The interactions between endophytes and their liverwort hosts resemble those observed in orchids, where colonized cells develop hyphal coils. After the host digests these coils, there may be one or more reinfection cycles. Similar to *Marchantiophyta*, the extent of fungal colonization varies among taxa: *Aneura* species from Chile and New Zealand exhibit colonization that extends over two to five cell layers, while in the northern hemisphere species *Aneura pinguis*, it affects about half of the thallus. In *Cryptothallus*, nearly every cell is colonized.

This type of association is also found in *Verdoornia* (Pressel et al. 2008a) and in almost all examined specimens of *Aneura*, except for populations of *A. pinguis* in California and *Aneura pellioides* (Bidartondo and Duckett 2010). The sibling genus

Riccardia, on the other hand, has very few fungal relationships. However, in three endemic New Zealand *Riccardia* species where endophytes are present, the relationships differ significantly from those in *Aneura*. Two of these associations are described here for the first time. Conversely, a third species, *Riccardia intercellula*, is recognized solely from a singular collection and has yet to be rediscovered in its natural habitat (Brown and Braggins 1989).

In *Riccardia pennata*, notable basidiomycetous hyphal coils occupy the ventral epidermal cells without exhibiting any indications of host digestion. Even more striking is *Riccardia cochleata*, which exhibits unique and remarkable features in its fungal associations.

3.5 Ecological Role of Bryophytes

Bryophytes are ecologically important, enhancing forest biodiversity, water retention, soil enhancement, and nutrient cycling and serving as ecological indicators (Bryophyte Ecology by Janice M. Glime n.d.; Decomposability of lichens and bryophytes from across an elevational gradient under standardized conditions—van Zuijlen—2020—Oikos—Wiley Online Library n.d.; Gao et al. 2017; Moore et al. 2019; Rousk et al. 2017; Xiao and Bowker 2020).

Bryophytes play a crucial role in forest ecosystems, which, due to their complex structure, offer a variety of substrates and niches that support significant species richness and functional diversity (Steel et al. 2004). They perform several vital functions, including colonizing bare soil and efficiently absorbing, retaining, and gradually releasing moisture, thus acting as a water reservoir (Mosses & Liverworts by Porley, R., and N. Hodgetts | Hardcover | 2005 | HarperCollins | 9,780,002,202,121 | BIBLIO n.d.). In temperate forests, their ability to store water helps prevent rainwater runoff, ensuring that nutrients from rainwater are absorbed into the ecosystem (Frahm 2001).

Their responses to management practices and disturbances in temperate forests highlight their importance for forest management and conservation strategies (Horvat et al. 2017). Additionally, climatic factors such as air temperature and precipitation significantly influence the composition of bryophyte communities in these ecosystems (Sun et al. 2013a).

A significant factor in this evolution has been the growing recognition of the critical role bryophytes play in many ecosystem processes. While numerous papers have been published, summarizing textbooks and comprehensive reviews remain scarce, as is typical for relatively young disciplines. Although textbooks on bryophyte biology have always included some aspects of bryophyte ecology (e.g.,rahm 2001; Goffinet and Shaw 2009), dedicated books on bryophyte ecology were lacking for a long time.

A landmark in the field is the bryophyte ecology book edited by which is still a valuable resource. More recently, the continuously updated e-book by (Glime: Bryophyte ecology—Google Scholar n.d.) and the volume on bryophyte ecology and climate change (Tuba et al. 2011) have become available. This chapter does not aim to cover the entire discipline but will focus on autecology and selected topics in population, community, and systems ecology.

4 Endophytic Fungi Associated with Bryophytes

Studies examining the biodiversity of bryophytes, which include mosses (Musci), liverworts (Hepaticae), and hornworts (Anthocerotae), have recently attracted more attention, phytochemistry, ecological and evolutionary significance, as well as their applications in biotechnology and biomonitoring. The study of bryophyte-associated endophytes and epiphytes is especially fascinating since these plants flourish in unique biological niches that also harbor a wide variety of microorganisms (Koua et al. 2015).

It is clear that bryophytes and endophytes are closely related because early Devonian land plants resembled bryophytes (around 400 million years ago) formed endophytic symbioses similar to vesicular–arbuscular mycorrhizas (VAM) (Guo et al. 2008), even before the development of roots.

Brundrett (2002), Turnau et al. (1999). Endophytic hyphae of VAM fungus are commonly found in mosses, while hornworts have a symbiotic relationship similar to that of VAM with glomalean fungi that develop arbuscules inside their tissues.

Koua et al. (2015) focused on the divergent sections of the 16S rRNA gene using a PCR-DGGE technique that is culture-independent. However, it is important to note that a limitation of this study was the lack of surface sterilization of the plant material, which led to the analysis of both endophytes and epiphytes.

According to the study, the microbial community associated with *Haplocladium microphyllum*, which grew on densely populated soil, contained *γ-Proteobacteria* like *Citrobacter murliniae* and *Klebsiella terrigena*. Conversely, *Brachythecium buchananii* from the same environment contained various *γ-Proteobacteria*, including *Hafnia* sp., *Pseudomonas fluorescens*, *Klebsiella intermedia*, and *Pseudomonas fluorescens*, in addition to some *Firmicutes*, such as *Clostridium butyricum* and *Clostridium puniceum*.

However, only *γ-Proteobacteria*, such as *Klebsiella oxytoca*, *Dickeya dieffenbachiae*, *Serratia proteamaculans*, *Serratia grimesii*, *Pectobacterium wasabiae*, and *Pectobacterium betavasculorum*, were found in *Trachycystis microphylla*. The microbial communities of bryophytes from virgin rocks varied, according to the study. *Salmonella enterica* subsp. *enterica* and *Firmicutes* (*Anaerobacter polyendosporus*) were found in *Brachythecium plumosum*, whereas *Pseudomonas antarctica*, *Pseudomonas cedrina*, and *Firmicutes* (*Anaerobacter polyendosporus*, *Clostridium disporicum*, and *Clostridium saccharoperbutylacetonicum*) were found in *Hypnum plumaeforme*. Likewise, *γ-Proteobacteria* (*Erwinia rhapontici*, *C. murliniae*, and *Pantoea ananatis*) were present in *Reboulia hemisphaerica* subsp. *orientalis*. *Racomitrium japonicum*, a plant that thrived in controlled soil, was discovered to include exclusively *Firmicutes*, such as *C. puniceum* and *C. saccharoperbutylacetonicum.*

Certain bacterial genera were found to be similarly distributed throughout many ecosystems; however, the analysis of these microorganisms revealed distinct species-level variances irrespective of the ecology, indicating a host-dependent dynamic in microbial communities.

Notably, bryophytes from densely populated soil and virgin rocks exhibited higher bacterial biodiversity compared to those from managed soils (Deshmukh et al. 2022). Additionally, Xie et al. (2015) identified 93 fungal isolates from lichens and bryophytes while examining the variety of endophytic and endolichenic fungi in maritime Antarctica. The six *Ascomycota* classes that comprised the majority of these fungi were *Dothideomycetes*, *Eurotiomycetes*, *Lecanoromycetes*, *Leotiomycetes*, *Pezizomycetes*, and *Sordariomycetes* (Yu et al. 2018).

5 Diversity of Endophytic Fungi Associated with Bryophytes

Antarctica is regarded as an extreme environment, marked by intense stress and disturbances (Pugh, 1980) which are heightened by its low temperatures, high dryness, elevated levels of ultraviolet radiation, irregular nutrient availability, and drastic temperature changes, especially in microhabitats (The microbiology of Antarctic soils—NLM Catalog—NCBI n.d.). As the study show, King George Island, the largest of the South Shetland Islands within the maritime Antarctic zone, benefits from the moderating influence of the ocean, resulting in a milder climate (Lee et al. 2008). The island's flora is primarily composed of diverse bryophyte and lichen species, with just two vascular plant species present (Øvstedal and Lewis Smith n. d.). Microbial diversity and activity are influenced by extreme conditions, and it is widely recognized that species richness among microbes is typically lower in such environments compared to temperate regions (Wynn-Williams 1996).

Previous research on Antarctic fungi has predominantly focused on soil (Bailey and Wynn-Williams 1982; Boyd et al. 1966; Heal et al. 1967; Tubaki et al. 1965) and freshwater environments (Baublis et al. 1991; Del Frate and Caretta 1990; Fletcher et al. 1985; Ellis-Evans 1985). Studies have generally involved isolating microfungi from organic materials, such as plants and mosses (Baublis et al. 1991; Bradner et al. 2000; Del Frate and Caretta 1990; Fletcher et al. 1985). Research into Antarctic fungi is closely tied to the distribution of hosts, including birds, invertebrates, and vegetation, with particular attention to bryophyte and lichen communities (Tosi et al. 2002), Many studies focus on fungi associated with nonvascular plants like bryophytes, which include mosses, liverworts, and hornworts, recognized as crucial producers in Antarctic ecosystems (Hoshino et al. 2009; Upson et al. 2007; U'Ren et al. 2010; Zhang et al. 2013).

Microfungi were found to be most prevalent in mosses and the underlying soil in samples collected from mid-Victoria Land between 1988 and 1998 (Caretta et al. 1994; Del Frate and Caretta 1990). Mosses provide a favorable environment for certain nematode-trapping fungi belonging to the genus *Arthrobotrys* (Caretta et al. 1994). In the Antarctic region, mosses represent one of the most diverse microhabitats for microfungi, especially those that are indigenous and psychrophilic (Tosi et al. 2002).

According to study, 16 endophytic fungi were isolated from 13 bryophytes belonging to 7 species collected at the Korea Antarctic Research Station on Barton Peninsula, King George Island.

Investigated were the variety and social organization of culturable endophytic fungi (EF) related to bryophytes that were gathered from China's Dawei Mountain. The largest area in China with both broad-leaved and evergreen forests is Dawei Mountain Nature Reserve, which is situated in Yunnan Province in the country's southwest. Furthermore, the region experiences a humid subtropical climate with more than 1500 mm of precipitation annually. Over 2000 meters above sea level is home to a whole mountain forest ecosystem, ranging from tropical humid rainforest to seasonal rainforest, monsoonal evergreen broad-leaved forest, moss evergreen broad-leaved forest, and so on. The "Quaternary Glacier" period had little effect on it (Zhai D.C., Yang L.Q., and Zhu R.L.). As a result, there is a great diversity of plant species, and many ancient plant species have survived.

Three bryophyte species (*Hylocomiaceae*: *Hylocomium splendens*; *Polytrichum commune*; *Marchantia polymorpha*, *Marchantiaceae*; and *Marchantia polymorpha*, *Marchantiaceae*) were recovered from Dawei Mountain. Three bryophytes from tropical and temperate habitats have CRs ranging from 60.9% to 96.6%, according to Davis and Shaw (2008). A total of 678 endophytic fungi (EF) have been isolated from 30 bryophyte species. Of these, the CS of EF for three of the bryophytes varied from 0.44 to 0.58, above the values of three Antarctic-region bryophytes (0.18–0.40) (Zhang et al. 2013). A diverse collection of fungi known as fungal endophytes reside inside plants for a portion of their lives without producing obvious signs of illness (Wilson 1995). Most of these fungi are from the *Ascomycota* phylum, with others belonging to *Basidiomycota* and *Zygomycota* (Gazis and Chaverri 2010). They can be found in a wide range of plants and ecosystems, from tropical to polar regions (Arnold and Lutzoni n.d.; Rosa et al. 2009). These fungi have a major impact on plants in a number of ways, including changing their susceptibility to and resistance to disease (Rodriguez et al. 2009), affecting plant fitness (Rodriguez et al. 2009), and promoting the decomposition of plant litter (Sun et al. 2011). In spite of their significance, much about their diversity and ecological roles remains poorly understood.

Although endophytes in vascular plants are frequently the subject of research, they are also found in nonvascular plants such as bryophytes (mosses, liverworts, and hornworts), which are important contributors to the biomass of boreal forests and tundra ecosystems. A broad range of endophytes has been identified in mosses and liverworts across various forest types (Davis et al. 2003a, b; Davis and Shaw 2008; High diversity of fungi associated with living parts of boreal forest bryophytes n.d.; Kauserud et al. 2008; U'Ren et al. 2010). Additional research is necessary to examine the variety, distribution, and ecological functions of endophytic fungus linked to bryophytes throughout different areas, including Antarctica.

A specific study conducted in the Fildes Region of King George Island, maritime Antarctica, examined endophytic fungi in three bryophyte species: the liverwort *Barbilophozia hatcheri* and the mosses *Chorisodontium aciphyllum* and *Sanionia uncinata*.

By analyzing 1329 tissue segments from 14 samples, researchers isolated 128 endophytic fungi. *B. hatcheri, C. aciphyllum*, and *S. uncinata* all had colonization rates of 12.3, 12.1, and 8.7%, respectively. Morphological and genetic analyses led to the classification of the fungi into 21 taxa: 15 Ascomycota, 5 Basidiomycota, and 1 undetermined. The research revealed diverse indices (0.91–1.99) and minimal commonality in fungal components among the bryophytes. Growth temperature tests revealed that the fungi include psychrophiles, psychrotrophs, and one mesophile, indicating their adaptation to cold Antarctic conditions .

Antarctica's plant life, being geographically isolated and the coldest continent, is primarily composed of bryophytes (about 111 moss species and 27 liverwort species) and includes only 2 native vascular plants. Bryophytes are key primary producers in Antarctic ecosystems, significantly affecting thermal and hydrological cycles, thus playing a crucial role in ecological processes. Antarctica is therefore a perfect place to investigate the taxonomy, ecology, evolution, and adaption of endophytic fungi that are related to bryophytes. However, research on these fungi in Antarctica has been limited to a few bryophyte species and sites within the continental zone. When moving from the maritime to the continental Antarctic zone, the diversity and abundance of life decline. No investigations on endophytic fungi in the maritime zone are currently known. This study aims to examine the cold adaptability, variety, and distribution of endophytic fungi within three predominant bryophyte species in the Fildes Region, one of the most extensive ice-free regions in the marine Antarctic zone. These species include the mosses *Chorisodontium aciphyllum* and *Sanionia uncinata*, as well as the liverwort *Barbilophozia hatcheri*.

Moving from the maritime Antarctic to the continental zone tends to reduce the diversity and richness of species (Ruisi et al. 2007). There are currently no published investigations on the diversity of endophytic fungi found in bryophytes within the Antarctic maritime region. This study aims to examine the diversity, distribution, and cold-adaptive traits of endophytic fungi associated with three prevalent bryophyte species in the Fildes Region (62°12′–62°13′S and 58°56′–58°57′W), one of the largest ice-free areas in this region: the liverwort *Barbilophozia hatcheri* and the mosses *Chorisodontium aciphyllum* and *Sanionia uncinata* (Extracted from Ruisi et al. 2007).

Despite the harsh conditions in Antarctica, studies have documented the presence of non-lichenized fungi in soil (Arenz et al. 2014; Bridge and Newsham 2009) and in substrates from exotic plants (Hirose et al. 2013). This supports the idea that many Antarctic fungi are cosmopolitan with significant dispersal capabilities (Arenz et al. 2014), aligning with Baas Becking's hypothesis that "Everything is everywhere, but the environment selects" (De Wit and Bouvier 2006). A plausible hypothesis emerging from these concepts is that niche-based mechanisms—like trait-mediated environmental filtering linked to water availability, nitrogen content, and salinity—are more influential in shaping Antarctic fungal communities than non-niche spatial processes. However, research specifically focused on the Antarctic Peninsula is limited, and there have been no studies in continental Antarctica that explore community composition across various locations along latitudinal and environmental gradients. This gap highlights the need to examine how environmental factors and latitude affect fungal communities (Dennis et al. 2012; Newsham et al. 2016; Yergeau et al. 2006). It remains unclear whether the geographic distribution of fungal communities in continental Antarctica is more closely tied to environmental conditions or geographic location.

Similarly, in boreal forests, bryophytes support a wide range of fungal species. Molecular identification revealed 158 operational taxonomic units (OTUs) from living parts of bryophytes like *Hylocomium splendens*, with *Ascomycota* and *Basidiomycota* being predominant. This study highlighted the immense and largely unexplored fungal diversity in these environments, with minimal overlap among different bryophyte species (Botany 86 (11), 1326–1333, 2008).

The study by Kauserud et al. (2008) examines the diversity of fungi associated with living parts of three common bryophytes in boreal forests: *Hylocomium splendens*, *Pleurozium schreberi*, and *Polytrichum commune*. Despite bryophytes being key components of forest floor vegetation, their associated fungal diversity is poorly understood. Through molecular identification, the researchers analyzed 376 cloned ITS sequences, identifying 158 operational taxonomic units (OTUs) that correspond to various fungal species. Most OTUs (62.8%) were from *Ascomycota*, followed by *Basidiomycota* (32.0%) and smaller proportions from *Chytridiomycota* and *Glomeromycota*. Common fungal orders included *Helotiales* and *Agaricales*, with many OTUs detected only once and minimal overlap between bryophyte species, indicating a vast fungal diversity that warrants further exploration.

In a related study by Rosa et al. (2009), the endophytic fungi associated with the Antarctic grass *Deschampsia antarctica* were characterized. From 91 individual plants collected in Admiralty Bay, 26 fungal isolates were identified from 273 leaf fragments using ITS sequence analysis. The most common genera were *Alternaria* and *Phaeosphaeria*, along with *Entrophospora* and several unidentified *Ascomycete* species. Notably, some isolates showed low identity to known sequences, suggesting they might represent new species. This research is the first to document fungal endophytes in the leaves of *D. antarctica*.

References

"Bryophyte Ecology" by Janice M. Glime (n.d.) WWW Document. https://digitalcommons.mtu.edu/oabooks/4/. Accessed 2.1.25

Abdel-Azeem A, Abdel Azeem M, Khalil W, (2019) Endophytic fungi as a new source of antirheumatoid metabolites. pp 335–384. https://doi.org/10.1016/B978-0-12-813820-5.00021-0

Abo Nahas H, Darwish A, Nahas Y, Elsayed M, Abdel Azeem M, Abdel-Azeem A (2021) Fungi as a gold mine of antioxidants. pp 73–113. https://doi.org/10.1007/978-3-030-85603-8_2

Andrew NR, Rodgerson L, Dunlop M (2003) Variation in invertebrate–bryophyte community structure at different spatial scales along altitudinal gradients. J Biogeogr 30:731–746. https://doi.org/10.1046/j.1365-2699.2003.00849.x

Anitha D, Vijaya T, Chandramouli K, Pragathi V, Venkateswarlu D, Desaraju SB (2013) Isolation and characterization of endophytic fungi from endemic medicinal plants of Tirumala Hills. Int J Lifesci Biotechnol, Pharma Res

Arenz BE, Blanchette RA, Farrell RL (2014) Fungal diversity in antarctic soils. In: Cowan DA (ed) Antarctic terrestrial microbiology. Springer, Berlin/Heidelberg, pp 35–53. https://doi.org/10.1007/978-3-642-45213-0_3

Arnold AE, Lutzoni F (2007) Diversity and host range of foliar fungal endophytes: are tropical leaves biodiversity hotspots? Ecology 88:541–549. https://doi.org/10.1890/05-1459

Arnold AE, Lutzoni F (n.d.) Diversity and host range of foliar endophytes: Are tropical leaves biodiversity hotspots? Ecology 88:541–549 [WWW Document]. ResearchGate. https://www.researchgate.net/publication/6327947_Arnold_A_E_Lutzoni_F_Diversity_and_host_range_of_foliar_endophytes_Are_tropical_leaves_biodiversity_hotspots_Ecology_88_541-549. Accessed 2.1.25

Arnold AE, Maynard Z, Gilbert GS, Coley PD, Kursar TA (2000) Are tropical fungal endophytes hyperdiverse? Ecol Lett 3:267–274. https://doi.org/10.1046/j.1461-0248.2000.00159.x

Bailey AD, Wynn-Williams DD (1982) Soil microbiological studies at Signy Island, South Orkney Islands. Br Antarct Surv Bull 51:167–191

Banerjee D (2011) Endophytic fungal diversity in tropical and subtropical plants. Res J Microbiol 6: 54–62. https://doi.org/10.3923/jm.2011.54.62

Baublis JA, Wharton RA, Volz PA (1991) Diversity of micro-fungi in an Antarctic dry valley. J Basic Microbiol 31:3–12. https://doi.org/10.1002/jobm.3620310102

Bengtsson J, Nilsson S, Franc A, Menozzi P (2000) Biodiversity, disturbances, ecosystem function and management of European forests. For Ecol Manag 132:39–50. https://doi.org/10.1016/S0378-1127(00)00378-9

Beraldi-Campesi H (2013) Early life on land and the first terrestrial ecosystems. Ecol Process 2:1. https://doi.org/10.1186/2192-1709-2-1

Bidartondo MI, Duckett JG (2010) Conservative ecological and evolutionary patterns in liverwort–fungal symbioses. Proc R Soc B Biol Sci 277:485–492. https://doi.org/10.1098/rspb.2009.1458

Bidartondo MI, Bruns TD, Weiss M, Sérgio C, Read DJ (2003) Specialized cheating of the ectomycorrhizal symbiosis by an epiparasitic liverwort. Proc R Soc B Biol Sci 270:835–842. https://doi.org/10.1098/rspb.2002.2299

Boyd WL, Staley JT, Boyd JW (1966) Ecology of soil microorganisms of Antarctica. In: Antarctic soils and soil forming processes. American Geophysical Union (AGU), pp 125–159. https://doi.org/10.1029/AR008p0125

Bradner JR, Sidhu RK, Yee B, Skotnicki ML, Selkirk PM, Nevalainen KMH (2000) A new microfungal isolate, Embellisia sp., associated with the Antarctic moss Bryum argenteum. Polar Biol 23:730–732. https://doi.org/10.1007/s003000000161

Bridge P, Newsham K (2009) Soil fungal community composition at Mars oasis, a southern maritime Antarctic site, assessed by PCR amplification and cloning. Fungal Ecol 2:66–74. https://doi.org/10.1016/j.funeco.2008.10.008

Brown EA, Braggins JE (1989) A revision of the genus Riccardia S. F. Gray in New Zealand with notes on the genus Aneura Dum. J Hattori Bot Lab 66:1–132. https://doi.org/10.18968/jhbl.66.0_1

Brundrett MC (2002) Coevolution of roots and mycorrhizas of land plants. New Phytol 154:275–304. https://doi.org/10.1046/j.1469-8137.2002.00397.x

Callaghan DA (2021) Survival of the critically endangered Ditrichum cornubicum and dependence on conservation management intervention. J Bryol 43:175–180. https://doi.org/10.1080/03736687.2020.1834958

Caners RT, Ellen Macdonald S, Belland RJ (2013) Linking the biological traits of boreal bryophytes to forest habitat change after partial harvesting. For Ecol Manag 303:184–194. https://doi.org/10.1016/j.foreco.2013.04.019

Caretta G, Frate GD, Mangiarotti A (1994) A record of Arthrobotrys tortor Jarowaja and Engyodontium album (Limber) de Hoog from Antarctica. Bol Micológico 9:9–13. https://doi.org/10.22370/bolmicol.1994.9.0.1089

Chambers SM, Williams PG, Seppelt RD, Cairney JWG (1999) Molecular identification of *Hymenoscyphus* sp. from rhizoids of the leafy liverwort *Cephaloziella exiliflora* in Australia and Antarctica. Mycol Res 103:286–288. https://doi.org/10.1017/S0953756298007217

Chater CC, Caine RS, Tomek M, Wallace S, Kamisugi Y, Cuming AC, Lang D, MacAlister CA, Casson S, Bergmann DC, Decker EL, Frank W, Gray JE, Fleming A, Reski R, Beerling DJ (2016) Origin and function of stomata in the moss Physcomitrella patens. Nat Plants 2:16179. https://doi.org/10.1038/nplants.2016.179

Cheng Z, Pan J-H, Tang W, Chen Q, Lin Y (2009) Biodiversity and biotechnological potential of mangrove-associated fungi. J For Res 20:63–72. https://doi.org/10.1007/s11676-009-0012-4

Chhipa H, Deshmukh SK (2019) Diversity of Endophytic fungi and their role in artificial agarwood production in Aquilaria tree. In: Satyanarayana T, Deshmukh SK, Deshpande MV (eds) Advancing frontiers in mycology & mycotechnology: basic and applied aspects of fungi. Springer, Singapore, pp 479–494. https://doi.org/10.1007/978-981-13-9349-5_19

Christenhusz MJM, Byng JW (2016) The number of known plants species in the world and its annual increase. Phytotaxa 261:201. https://doi.org/10.11646/phytotaxa.261.3.1

Cui J, Guo S, Xiao P (2011) Antitumor and antimicrobial activities of endophytic fungi from medicinal parts of Aquilaria sinensis

Davis EC, Shaw AJ (2008) Biogeographic and phylogenetic patterns in diversity of liverwort-associated endophytes. Am J Bot 95:914–924. https://doi.org/10.3732/ajb.2006463

Davis EC, Franklin JB, Shaw AJ, Vilgalys R (2003a) Endophytic Xylaria (Xylariaceae) among liverworts and angiosperms: phylogenetics, distribution, and symbiosis. Am J Bot 90:1661–1667. https://doi.org/10.3732/ajb.90.11.1661

Davis et al (2003b) J Appl Poult Res 12(2):137–144. Feedipedia [WWW Document], n.d. https://www.feedipedia.org/node/2626. Accessed 2.1.25

de Bary A (1866) Morphologie und Physiologie der Pilze, Flechten und Myxomyceten, Engelmann

De Souza Sebastianes FL, Romão-Dumaresq AS, Lacava PT, Harakava R, Azevedo JL, De Melo IS, Pizzirani-Kleiner AA (2013) Species diversity of culturable endophytic fungi from Brazilian mangrove forests. Curr Genet 59:153–166. https://doi.org/10.1007/s00294-013-0396-8

De Wit R, Bouvier T (2006) 'Everything is everywhere, but, the environment selects'; what did baas Becking and Beijerinck really say? Environ Microbiol 8:755–758. https://doi.org/10.1111/j.1462-2920.2006.01017.x

Decomposability of lichens and bryophytes from across an elevational gradient under standardized conditions – van Zuijlen – 2020 – Oikos – Wiley Online Library (n.d.) WWW Document. https://nsojournals.onlinelibrary.wiley.com/doi/10.1111/oik.07257. Accessed 2.1.25

Degola F, Sanità di Toppi L, Petraglia A (2022) Bryophytes: how to conquer an alien planet and live happily (ever after). J Exp Bot 73(13):4267–4272. https://doi.org/10.1093/jxb/erac252

Del Frate G, Caretta G (1990) Fungi isolated from Antarctic material. Polar Biol 11:1–7. https://doi.org/10.1007/BF00236515

Dennis PG, Rushton SP, Newsham KK, Lauducina VA, Ord VJ, Daniell TJ, O'Donnell AG, Hopkins DW (2012) Soil fungal community composition does not alter along a latitudinal gradient through the maritime and sub-Antarctic. Fungal Ecol 5:403–408. https://doi.org/10.1016/j.funeco.2011.12.002

Deshmukh SK, Dufossé L, Chhipa H, Saxena S, Mahajan GB, Gupta MK (2022) Fungal endophytes: a potential source of antibacterial compounds. J Fungi 8:164. https://doi.org/10.3390/jof8020164

Donoghue PCJ, Harrison CJ, Paps J, Schneider H (2021) The evolutionary emergence of land plants. Curr Biol 31:R1281–R1298. https://doi.org/10.1016/j.cub.2021.07.038

Duckett JG, Pressel S (2017) The evolution of the stomatal apparatus: intercellular spaces and sporophyte water relations in bryophytes – two ignored dimensions. Philos Trans R Soc B Biol Sci 373:20160498. https://doi.org/10.1098/rstb.2016.0498

Duckett JG, Read DJ (1995) Ericoid mycorrhizas and rhizoid-ascomycete associations in liverworts share the same mycobiont: isolation of the partners and resynthesis of the associations in vitro. New Phytol 129:439–447. https://doi.org/10.1111/j.1469-8137.1995.tb04315.x

Duckett JG, Renzaglia KS, Pell K (1991) A light and electron microscope study of rhizoid-ascomycete associations and flagelliform axes in British hepatics with observations on the effects of the fungi on host morphology. New Phytol 118:233–257. https://doi.org/10.1111/j.1469-8137.1991.tb00975.x

Ellis-Evans JC (1985) Fungi from maritime Antarctic freshwater environments. Br Antarct Surv Bull 68:37–45

Fisher PJ, Petrini O (1990) A comparative study of fungal endophytes in xylem and bark of *Alnus* species in England and Switzerland. Mycol Res 94:313–319. https://doi.org/10.1016/S0953-7562(09)80356-0

Fisher W, Piazza CC, Bowman LG, Hagopian LP, Owens JC, Slevin I (1992) A comparison of two approaches for identifying reinforcers for persons with severe and profound disabilities. J Appl Behav Anal 25:491–498. https://doi.org/10.1901/jaba.1992.25-491

Fletcher LD, Kerry EJ, Weste GM (1985) Microfungi of Mac.Robertson and Enderby lands. Antarctica Polar Biol 4:81–88. https://doi.org/10.1007/BF00442904

Frahm JP (2001) Biologie der Moose. https://doi.org/10.1007/978-3-662-57607-6

Frahm J-P (2007) Diversity, dispersal and biogeography of bryophytes (mosses). In: Foissner W, Hawksworth DL (eds) Protist diversity and geographical distribution, topics in biodiversity and conservation. Springer, Dordrecht, pp 43–50. https://doi.org/10.1007/978-90-481-2801-3_4

Frego K (2007) Bryophytes as potential indicators of forest integrity. For Ecol Manag 242:65–75. https://doi.org/10.1016/j.foreco.2007.01.030

Gao B, Li X, Zhang D, Liang Y, Yang H, Chen M, Zhang Y, Zhang J, Wood AJ (2017) Desiccation tolerance in bryophytes: the dehydration and rehydration transcriptomes in the desiccation-tolerant bryophyte Bryum argenteum. Sci Rep 7:7571. https://doi.org/10.1038/s41598-017-07297-3

Gazis R, Chaverri P (2010) Diversity of fungal endophytes in leaves and stems of wild rubber trees (*Hevea brasiliensis*) in Peru. Fungal Ecol 3:240–254. https://doi.org/10.1016/j.funeco.2009.12.001

Gignac LD (2001) Bryophytes as indicators of climate change. Bryologist 104:410–420

Glime: Bryophyte ecology – Google Scholar (n.d.) WWW Document. https://scholar.google.com/scholar_lookup?title=Bryophyte%20ecology&author=JM%20Glime&publication_year=2006&book=Bryophyte%20ecology. Accessed 2.1.25

Goffinet B, Shaw A (2009) Bryophyte biology. https://doi.org/10.1017/CBO9781139171304

Graham L, Kodner R, Fisher M, Graham J, Wilcox L, Hackney J, Obst J, Bilkey P, Hanson D, Cook M (2004) Early land plant adaptations to terrestrial stress: a focus on phenolics. Evol Plant Physiol:155–169. https://doi.org/10.1016/B978-012339552-8/50010-X

Guo L, Wu J, Han T, Cao T, Rahman K, Qin L (2008) Chemical composition, antifungal and antitumor properties of ether extracts of Scapania verrucosa Heeg. And its Endophytic fungus Chaetomium fusiforme. Molecules 13:2114–2125. https://doi.org/10.3390/molecules13092114

Hawksworth D, Lücking R (2017) Fungal diversity revisited: 2.2 to 3.8 million species. Microbiol Spectr 5. https://doi.org/10.1128/microbiolspec.FUNK-0052-2016

He X, He KS, Hyvönen J (2016) Will bryophytes survive in a warming world? Perspect Plant Ecol Evol Syst 19:49–60. https://doi.org/10.1016/j.ppees.2016.02.005

Heal OW, Bailey AD, Latter PM (1967) Bacteria, fungi and protozoa in Signy Island soils compared with those from a temperate moorland. Philos Trans R Soc Lond Ser B Biol Sci 252:191–197. https://doi.org/10.1098/rstb.1967.0011

Heckman DS, Geiser DM, Eidell BR, Stauffer RL, Kardos NL, Hedges SB (2001) Molecular evidence for the early colonization of land by fungi and plants. Science 293:1129–1133. https://doi.org/10.1126/science.1061457

Hernández-Hernández R, Kluge J, Ah-Peng C, González-Mancebo JM (2019) Natural and human-impacted diversity of bryophytes along an elevational gradient on an oceanic Island (La Palma, Canarias). PLoS One 14:e0213823. https://doi.org/10.1371/journal.pone.0213823

High diversity of fungi associated with living parts of boreal forest bryophytes (n.d.) WWW Document. https://www.researchgate.net/publication/237153098_High_diversity_of_fungi_associated_with_living_parts_of_boreal_forest_bryophytes. Accessed 2.1.25

Hilszczanska D (2016) Endofity—charakterystyka i możliwości wykorzystania w leśnictwie. For Res Pap Śne Pr Badaw 77:276–282. https://doi.org/10.1515/frp-2016-0029

Hirose D, Matsuoka S, Osono T (2013) Assessment of the fungal diversity and succession of ligninolytic endophytes in *Camellia japonica* leaves using clone library analysis. Mycologia 105:837–843. https://doi.org/10.3852/12-385

Horvat V, Heras P, García-Mijangos I, Biurrun I (2017) Intensive forest management affects bryophyte diversity in the western Pyrenean silver fir-beech forests. Biol Conserv 215:81–91. https://doi.org/10.1016/j.biocon.2017.09.007

Hoshino T, Xiao N, Xiao N, Hoshino T, Tkachenko OB (2009) Cold adaptation in the phytopathogenic fungi causing snow molds. Mycoscience 50:26–38. https://doi.org/10.1007/S10267-008-0452-2

Jill Harrison C (2017) Development and genetics in the evolution of land plant body plans. Philos Trans R Soc B Biol Sci 372:20150490. https://doi.org/10.1098/rstb.2015.0490

Jones VAS, Dolan L (2012) The evolution of root hairs and rhizoids. Ann Bot 110:205–212. https://doi.org/10.1093/aob/mcs136

Kauserud H, Mathiesen C, Ohlson M (2008) High diversity of fungi associated with living parts of boreal forest bryophytes. Botany 86:1326–1333. https://doi.org/10.1139/B08-102

Khairalla A, Rosella S, Yagi S, Mohamed I, Laurain-Mattar D (2016) Endophytic fungi: occurrence, classification, function and natural products. In: Endophytic fungi: diversity, characterization and biocontrol

Kolari P, Pumpanen J, Kulmala L, Ilvesniemi H, Nikinmaa E, Grönholm T, Hari P (2006) Forest floor vegetation plays an important role in photosynthetic production of boreal forests. For Ecol Manag 221:241–248. https://doi.org/10.1016/j.foreco.2005.10.021

Koua FHM, Kimbara K, Tani A (2015) Bacterial-biota dynamics of eight bryophyte species from different ecosystems. Saudi J Biol Sci 22:204–210. https://doi.org/10.1016/j.sjbs.2014.07.009

Kumar S, Kaushik N (2013) Metabolites of endophytic fungi as novel source of biofungicide: a review. Phytochem Rev 11:507–522. https://doi.org/10.1007/s11101-013-9271-y

Kusari S, Spiteller M (2012) Metabolomics of endophytic fungi producing associated plant secondary metabolites: progress, challenges and opportunities. In: Roessner U (ed) Metabolomics. INTECH. https://doi.org/10.5772/31596

Lee JS, Lee HK, Hur J-S, Andreev M, Hong SG (2008) Diversity of the lichenized fungi in King George Island, Antarctica, revealed by phylogenetic analysis of partial large subunit rDNA sequences. J Microbiol Biotechnol 18:1016–1023

Li J, Saunders JC, Gilmour RS, Silver M, Fowden AL (1993) Insulin-like growth factor-II messenger ribonucleic acid expression in fetal tissues of the sheep during late gestation: effects of cortisol. Endocrinology 132:2083–2089. https://doi.org/10.1210/endo.132.5.8477658

Ligrone R, Ducket JG, Renzaglia KS (2000) Conducting tissues and phyletic relationships of bryophytes. Philos Trans R Soc B Biol Sci 355:795–813

Ligrone R, Vaughn KC, Renzaglia KS, Knox JP, Duckett JG (2002) Diversity in the distribution of polysaccharide and glycoprotein epitopes in the cell walls of bryophytes: new evidence for the multiple evolution of water-conducting cells. New Phytol 156:491–508. https://doi.org/10.1046/j.1469-8137.2002.00538.x

Lugtenberg B, Caradus J, Johnson L (2016) Fungal endophytes for sustainable crop production. FEMS Microbiol Ecol 92:fiw194. https://doi.org/10.1093/femsec/fiw194

Márialigeti: The effects of stand structure on ground-fl... – Google Scholar (n.d.) WWW Document. https://scholar.google.com/scholar_lookup?title=The+effects+of+standstructure+on+ground-floor+bryophyte+assemblages+in+temperate+mixed+forests&author=M%C3%A1rialigeti,+S.&author=N%C3%A9meth,+B.&author=Tinya,+F.&author=%C3%93dor,+P.&publication_year=2009&journal=Biodivers.+Conserv.&volume=18&pages=2223%E2%80%932241&doi=10.1007/s10531-009-9586-6. Accessed 1.31.25

McAdam SAM, Duckett JG, Sussmilch FC, Pressel S, Renzaglia KS, Hedrich R, Brodribb TJ, Merced A (2021) Stomata: the holey grail of plant evolution. Am J Bot 108:366–371. https://doi.org/10.1002/ajb2.1619

Moore PA, Smolarz AG, Markle CE, Waddington JM (2019) Hydrological and thermal properties of moss and lichen species on rock barrens: implications for turtle nesting habitat. Ecohydrology 12:e2057. https://doi.org/10.1002/eco.2057

Mosses & Liverworts by Porley, R., and N. Hodgetts | Hardcover | 2005 | HarperCollins | 9780002202121 | BIBLIO (n.d.) WWW Document. https://www.biblio.com/book/mosses-liverworts-porley-r-n-hodgetts/d/1508286525?srsltid=AfmBOoo9GjODNEqZqZKPDehB8u8yQjJDXCx7Hlf01_ty-CzGv5KNfBuI. Accessed 2.1.25

Müller F, Lenz R (2006) Theoretical fundamentals of consistent applications in environmental management. Ecol Indic 6:1–5. https://doi.org/10.1016/j.ecolind.2005.08.001

Müller J, Boch S, Prati D, Socher SA, Pommer U, Hessenmöller D, Schall P, Schulze ED, Fischer M (2019) Effects of forest management on bryophyte species richness in Central European forests. For Ecol Manag 432:850–859. https://doi.org/10.1016/j.foreco.2018.10.019

Nelson: Short-term effects of timber harvest and... – Google Scholar (n.d.) WWW Document. https://scholar.google.com/scholar_lookup?title=Short-term+effects+of+timber+harvest+and+forest+edges+on+ground-layer+mosses+and+liverworts&author=Nelson,+C.R.&author=Halpern,+C.B.&publication_year=2005&journal=Can.+J.+Bot.&volume=83&pages=610%E2%80%93620&doi=10.1139/b05-036. Accessed 1.31.25

Newsham KK, Hopkins DW, Carvalhais LC, Fretwell PT, Rushton SP, O'Donnell AG, Dennis PG (2016) Relationship between soil fungal diversity and temperature in the maritime Antarctic. Nat Clim Chang 6:182–186. https://doi.org/10.1038/nclimate2806

Norden N, Chave J, Caubère A, Châtelet P, Ferroni N, Forget P, Thébaud C (2007) Is temporal variation of seedling communities determined by environment or by seed arrival? A test in a neotropical forest. J Ecol 95:507–516. https://doi.org/10.1111/j.1365-2745.2007.01221.x

Øvstedal DO, Lewis Smith RI(n.d.) Lichens of Antarctica and South Georgia. a guide to their identification and ecology | Biologia Plantarum (WWW Document). https://link.springer.com/article/10.1023/A:1015130808724. Accessed 2.1.25

Paillet Y, Bergès L, Hjältén J, Ódor P, Avon C, Bernhardt-Römermann M, Bijlsma R, De Bruyn L, Fuhr M, Grandin U, Kanka R, Lundin L, Luque S, Magura T, Matesanz S, Mészáros I, Sebastià M, Teresa, Schmidt W, Standovár T, Tóthmérész B, Uotila A, Valladares F, Vellak K, Virtanen R (2010) Biodiversity differences between managed and unmanaged forests: meta-analysis of species richness in Europe. Conserv Biol 24:101–112. https://doi.org/10.1111/j.1523-1739.2009.01399.x

Pires ND, Dolan L (2012) Morphological evolution in land plants: new designs with old genes. Philos Trans R Soc B Biol Sci 367:508–518. https://doi.org/10.1098/rstb.2011.0252

Porras-Alfaro A, Bayman P (2011) Hidden fungi, emergent properties: endophytes and microbiomes. Annu Rev Phytopathol 49:291–315. https://doi.org/10.1146/annurev-phyto-080508-081831

Pressel S, Duckett J (2011) The parasitic ascomycete Mniaecia jungermanniae induces the formation of giant perichaetia and apogamous sporophytes in leafy liverworts. Can J Bot 84:384–392. https://doi.org/10.1139/b06-004

Pressel S, Ligrone R, Duckett J (2008a) Chapter six: the ascomycete Rhizoscyphus ericae elicits a range of host responses in the rhizoids of leafy liverworts: an experimental and cytological analysis. Fieldiana Bot 47:59–72. https://doi.org/10.3158/0015-0746-47.1.59

Pressel S, Ligrone R, Duckett J, Davis C (2008b) A novel ascomycetous endophytic association in the rhizoids of the leafy liverwort family Schistochilaceae (Jungermanniidae, Hepaticopsida). Am J Bot 95:531–541. https://doi.org/10.3732/ajb.2007171

Preussing M, Nebel M, Oberwinkler F, Weiss M (2010) Diverging diversity patterns in the Tulasnella (Basidiomycota, Tulasnellales) mycobionts of Aneura pinguis (Marchantiophyta, Metzgeriales) from Europe and Ecuador. Mycorrhiza 20:147–159. https://doi.org/10.1007/s00572-009-0275-9

Proctor: Poikilohydry and homoihydry: antithesis... – Google Scholar (n.d.) WWW Document. https://scholar.google.com/scholar_lookup?title=Poikilohydry+and+homoihydry:+Antithesis+or+spectrum+of+possibilities?&author=Proctor,+M.C.F.&author=Tuba,+Z.&publication_year=2002&journal=New+Phytol.&volume=156&pages=327%E2%80%93349&doi=10.1046/j.1469-8137.2002.00526.x. Accessed 1.31.25

Pugh G (1980) Strategies in fungal ecology. Trans Br Mycol Soc 75. https://doi.org/10.1016/S0007-1536(80)80188-4

Renzaglia KS, Villarreal JC, Piatkowski BT, Lucas JR, Merced A (2017) Hornwort stomata: architecture and fate shared with 400-million-year-old fossil plants without Leaves1. Plant Physiol 174:788–797. https://doi.org/10.1104/pp.17.00156

Riffo-Donoso V, Osorio F, Fontúrbel FE (2021) Habitat disturbance alters species richness, composition, and turnover of the bryophyte community in a temperate rainforest. For Ecol Manag 496:119467. https://doi.org/10.1016/j.foreco.2021.119467

Rodriguez RJ, White JF Jr, Arnold AE, Redman RS (2009) Fungal endophytes: diversity and functional roles. New Phytol 182:314–330. https://doi.org/10.1111/j.1469-8137.2009.02773.x

Rola K, Plášek V, Rożek K, Zubek S (2021) Effect of tree species identity and related habitat parameters on understorey bryophytes – interrelationships between bryophyte, soil and tree factors in a 50-year-old experimental forest. Plant Soil 466:613–630. https://doi.org/10.1007/s11104-021-05074-w

Rosa LH, Vaz ABM, Caligiorne RB, Campolina S, Rosa CA (2009) Endophytic fungi associated with the Antarctic grass Deschampsia Antarctica Desv. (Poaceae). Polar Biol 32:161–167. https://doi.org/10.1007/s00300-008-0515-z

Rousk K, Degboe J, Michelsen A, Bradley R, Bellenger J-P (2017) Molybdenum and phosphorus limitation of moss-associated nitrogen fixation in boreal ecosystems. New Phytol 214:97–107. https://doi.org/10.1111/nph.14331

Ruisi S, Barreca D, Selbmann L, Zucconi L, Onofri S (2007) Fungi in Antarctica. Rev Environ Sci Biotechnol 6:127–141. https://doi.org/10.1007/s11157-006-9107-y

Sabovljevic M, Vujicic M, Sabovljevic A (2010) Diversity of saproxylic bryophytes in old-growth and managed beech forests in the Central Balkans. Plant Biosyst Int J Deal Asp Plant Biol 144: 234–240. https://doi.org/10.1080/11263500903561015

Shaffer JP, Zalamea P-C, Sarmiento C, Gallery RE, Dalling JW, Davis AS, Baltrus DA, Arnold AE (2018) Context-dependent and variable effects of endohyphal bacteria on interactions between fungi and seeds. Fungal Ecol 36:117–127. https://doi.org/10.1016/j.funeco.2018.08.008

Soliman SS, Trobacher CP, Tsao R, Greenwood JS, Raizada MN (2013) A fungal endophyte induces transcription of genes encoding a redundant fungicide pathway in its host plant. BMC Plant Biol 13:93. https://doi.org/10.1186/1471-2229-13-93

Spitale: Switch between competition and facilitation... – Google Scholar (n.d.) WWW Document. https://scholar.google.com/scholar_lookup?title=Switch+between+competition+and+facilitation+within+a+seasonal+scale+at+colony+level+in+bryophytes&author=Spitale,+D.&publication_year=2009&journal=Oecologia&volume=160&pages=471%E2%80%93482&doi=10.1007/s00442-009-1324-y&pmid=19352717. Accessed 1.31.25

Steel JB, Wilson JB, Anderson BJ, Lodge RHE, Tangney RS (2004) Are bryophyte communities different from higher-plant communities? Abundance relations. Oikos 104:479–486. https://doi.org/10.1111/j.0030-1299.2004.12840.x

Strobel G, Daisy B (2003) Bioprospecting for microbial endophytes and their natural products. Microbiol Mol Biol Rev MMBR 67:491–502. https://doi.org/10.1128/MMBR.67.4.491-502.2003

Sun X, Guo L, Hyde K (2011) Community composition of endophytic fungi in Acer truncatum and their role in decomposition. Fungal Divers 47:85–95. https://doi.org/10.1007/s13225-010-0086-5

Sun M, Li J, Yang C, Schmidt G, Bambacus M, Cahalan R, Huang Q, Xu C, Noble E, Li Z (2012) A web-based geovisual analytical system for climate studies. Future Internet 4:1069–1085. https://doi.org/10.3390/fi4041069

Sun J, Evrin C, Samel SA, Fernández-Cid A, Riera A, Kawakami H, Stillman B, Speck C, Li H (2013a) Cryo-EM structure of a helicase loading intermediate containing ORC–Cdc6–Cdt1–MCM2-7 bound to DNA. Nat Struct Mol Biol 20:944–951. https://doi.org/10.1038/nsmb.2629

Sun S-Q, Wu Y-H, Wang G-X, Zhou J, Yu D, Bing H-J, Luo J (2013b) Bryophyte species richness and composition along an altitudinal gradient in Gongga Mountain, China. PLoS One 8:e58131. https://doi.org/10.1371/journal.pone.0058131

Sussmilch FC, Schultz J, Hedrich R, Roelfsema MRG (2019) Acquiring control: the evolution of stomatal signalling pathways. Trends Plant Sci 24:342–351. https://doi.org/10.1016/j.tplants.2019.01.002

Szövényi P, Waller M, Kirbis A (2019) Evolution of the plant body plan. Curr Top Dev Biol 131:1–34. https://doi.org/10.1016/bs.ctdb.2018.11.005

The microbiology of Antarctic soils – NLM Catalog – NCBI (n.d.) WWW Document. https://www.ncbi.nlm.nih.gov/nlmcatalog/100932917. Accessed 2.1.25

Tosi S, Casado B, Gerdol R, Caretta G (2002) Fungi isolated from Antarctic mosses. Polar Biol 25: 262–268. https://doi.org/10.1007/s00300-001-0337-8

Tuba Z, Slack NG, Stark LR (2011) Bryophyte ecology and climate change (WWW Document). Camb. Core. https://doi.org/10.1017/CBO9780511779701

Tubaki K, Asano I, Eguchi M (1965) Additional species of fungi isolated from the Antarctic materials, JARE 1956–1962 scientific reports. Series E. Polar Section, National Science Museum, Tokyo

Turnau K, Ronikier M, Unrug J (1999) Role of mycorhizal links between plants in establishment of liverworts thalli in natural habitats. Acta Soc Bot Pol 68:63–68. https://doi.org/10.5586/asbp.1999.011

U'Ren JM, Lutzoni F, Miadlikowska J, Arnold AE (2010) Community analysis reveals close affinities between endophytic and endolichenic fungi in mosses and lichens. Microb Ecol 60: 340–353. https://doi.org/10.1007/s00248-010-9698-2

Upson R, Read D, Newsham K (2007) Widespread association between the ericoid Mycorrhizal fungus Rhizoscyphus ericae and a leafy liverwort in the maritime and SubAntarctic. New Phytol 176:460–471. https://doi.org/10.1111/j.1469-8137.2007.02178.x

Vitt: Patterns of bryophyte diversity in peatlands... – Google Scholar (n.d.) [WWW Document]. https://scholar.google.com/scholar_lookup?title=Patterns+of+bryophyte+diversity+in+peatlands+of+continental+western+Canada&author=Vitt,+D.H.&author=Li,+Y.&author=Belland,+R.&publication_year=1995&journal=Bryologist&volume=98&pages=218%E2%80%93227&doi=10.2307/3243306. Accessed 1.31.25

Wang Y, Dai C-C (2011) Endophytes: a potential resource for biosynthesis, biotransformation, and biodegradation. Ann Microbiol 61:207–215. https://doi.org/10.1007/s13213-010-0120-6

Wani ZA, Ashraf N, Mohiuddin T, Riyaz-Ul-Hassan S (2015) Plant-endophyte symbiosis, an ecological perspective. Appl Microbiol Biotechnol 99:2955–2965. https://doi.org/10.1007/s00253-015-6487-3

Weber B, Büdel B, Belnap J (2016) Biological soil crusts: an organizing principle in drylands, Ecological studies. Springer, Cham. https://doi.org/10.1007/978-3-319-30214-0

Wilson D (1995) Endophyte: the evolution of a term, and clarification of its use and definition. Oikos 73:274. https://doi.org/10.2307/3545919

Wynn-Williams DD (1996) Antarctic microbial diversity: the basis of polar ecosystem processes. Biodivers Conserv 5:1271–1293. https://doi.org/10.1007/BF00051979

Xiao B, Bowker MA (2020) Moss-biocrusts strongly decrease soil surface albedo, altering land-surface energy balance in a dryland ecosystem. Sci Total Environ 741:140425. https://doi.org/10.1016/j.scitotenv.2020.140425

Xie F, Li X-B, Zhou J-C, Xu Q-Q, Wang X-N, Yuan H-Q, Lou H-X (2015) Secondary metabolites from aspergillus fumigatus, an endophytic fungus from the liverwort Heteroscyphus tener (Steph.) Schiffn. Chem Biodivers 12:1313–1321. https://doi.org/10.1002/cbdv.201400317

Xiong F, Tentner AR, Huang P, Gelas A, Mosaliganti KR, Souhait L, Rannou N, Swinburne IA, Obholzer ND, Cowgill PD, Schier AF, Megason SG (2013) Specified neural progenitors sort to form sharp domains after noisy shh signaling. Cell 153:550–561. https://doi.org/10.1016/j.cell.2013.03.023

Yergeau E, Vujanovic V, St-Arnaud M (2006) Changes in communities of fusarium and arbuscular mycorrhizal fungi as related to different asparagus cultural factors. Microb Ecol 52:104–113. https://doi.org/10.1007/s00248-006-9047-7

Yu NH, Park S-Y, Kim JA, Park C-H, Jeong M-H, Oh S-O, Hong SG, Talavera M, Divakar PK, Hur J-S (2018) Endophytic and endolichenic fungal diversity in maritime Antarctica based on cultured material and their evolutionary position among Dikarya. Fungal Syst Evol 2:263–272. https://doi.org/10.3114/fuse.2018.02.07

Zhang Y, Rossow WB, Stackhouse PW (2006) Comparison of different global information sources used in surface radiative flux calculation: radiative properties of the near-surface atmosphere. J Geophys Res Atmospheres 111:2005JD006873. https://doi.org/10.1029/2005JD006873

Zhang T, Zhang Y-Q, Liu H-Y, Wei Y-Z, Li H-L, Su J, Zhao L-X, Yu L-Y (2013) Diversity and cold adaptation of culturable endophytic fungi from bryophytes in the Fildes region, King George Island, maritime Antarctica. FEMS Microbiol Lett 341:52–61. https://doi.org/10.1111/1574-6968.12090

Zimmerman NB, Vitousek PM (2012) Fungal endophyte communities reflect environmental structuring across a Hawaiian landscape. Proc Natl Acad Sci 109:13022–13027. https://doi.org/10.1073/pnas.1209872109

赵德先, Dexian Z, 王成, Cheng W, 孙振凯, Sun Z, 郝泽周, Hao Z (2020) 树附生苔藓植物多样性及其影响因素 (WWW Document). https://doi.org/10.5846/stxb201902220322

The Diversity of Endophytic Fungi in the *Lycopodium* Species

Randa Mlik, Slimane Ahmed Benmchih Ayada, Nour Elhouda Mekhadmi, Salim Meddour, Mohamed Kharsi, Ahmed M. Abdel-Azeem, and Makhlouf Sekour

Contents

R. Mlik (✉) · S. A. B. Ayada · M. Kharsi
Saharan Area Laboratory for Modernization and Development of Agriculture, Higher School of Saharan Agriculture, El Oued, Algeria

N. E. Mekhadmi
Department of Biology, SNV Faculty, University of Chahid Hamma Lakhdar, El-Oued, Algeria

Laboratory of Biodiversity and Applications of Biotechnology in the Agriculture Field, University of Chahid Hamma Lakhdar, El-Oued, Algeria

S. Meddour
Laboratoire de Valorisation et Conservation des Ecosystèmes Arides (LVCEA), Faculté des Sciences de la Nature et de la Vie et Sciences de la Terre, Université de Ghardaïa, Ghardaïa, Algeria

A. M. Abdel-Azeem
Botany and Microbiology Department, Faculty of Science, Suez Canal University, Ismailia, Egypt

Research Institute of University of the Bucharest (ICUB), University of Bucharest, Bucharest, Romania

The Centre for Mineral Biogeochemistry, The University of the Free State, Bloemfontein, Free State, South Africa

M. Sekour
Department of Agronomic Sciences, SNV Faculty, University of Kasdi-Merbah, Ouargla, Algeria

A. M. Abdel-Azeem, F. Degola (eds.), *Advances in Lichens and Non-Flowering Plants Fungal Endophytes Research*, Fungal Biology,
https://doi.org/10.1007/978-3-032-01506-8_7

1 Introduction

Endophytic fungi represent a fascinating and diverse group of microorganisms that inhabit plant tissues without causing any apparent harm to their hosts. The association between plants and fungi is extremely common. These fungi are particularly intriguing in the context of *Lycopodium* species, a group of ancient vascular plants known as club mosses. Despite the extensive research on endophytes in angiosperms and conifers, the study of endophytic fungi in lycophytes, such as *Lycopodium*, remains relatively underexplored (Pawłowska et al. 2014; Lehnert et al. 2017). Recent studies have begun to shed light on the diversity and ecological roles of these fungi within *Lycopodium* species. For instance, research conducted in Poland identified a variety of Ascomycota species within the aboveground tissues of *Lycopodium annotinum* and *Lycopodium clavatum*, with *Dothideomycetes* being the most prevalent class (Pawłowska et al. 2014). This study highlights the unique fungal communities associated with different *Lycopodium* species and their potential ecological significance.

Moreover, the symbiotic relationships between *Lycopodium* species and their fungal partners may have deep evolutionary roots. Evidence suggests that these interactions could have played a crucial role in vascular plants' early colonization of land, with specific fungal taxa showing a strong evolutionary concordance with their host plants (Benucci et al. 2020; Fig. 1). This coevolutionary history underscores the importance of understanding the diversity and function of endophytic fungi in *Lycopodium* species. The bioactive potential of these endophytic fungi is also noteworthy. For example, *Colletotrichum alatae*, an endophyte of *Lycopodium clavatum*, has been identified as a producer of antibacterial and antioxidative compounds, demonstrating the potential pharmaceutical applications of these fungi (Santra and Banerjee 2022a, b). Such findings emphasize the need for further exploration of the metabolic capabilities of endophytic fungi in *Lycopodium* species.

In summary, the diversity of endophytic fungi in *Lycopodium* species offers a rich field of study with implications for ecology, evolution, and biotechnology. As research progresses, these fungi will likely reveal even more about their roles in plant health and their potential benefits to human society.

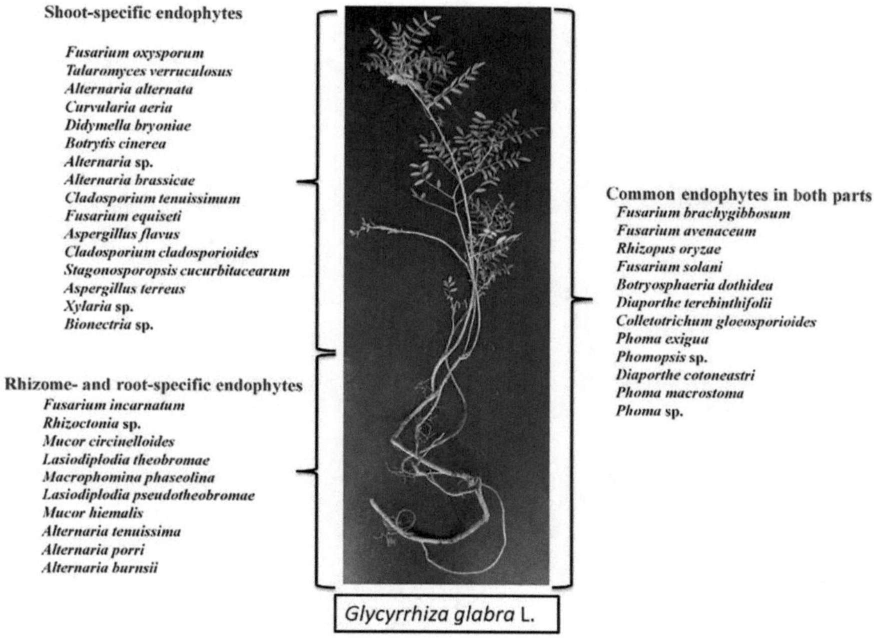

Fig. 1 Endophytic fungi and their colonized specific parts. (Arora et al. 2019)

2 Diversity of Endophytic Fungi

Endophytic fungi are a diverse group of primarily ascomycetous fungi that inhabit the internal tissues of plants without causing any apparent harm. They are found in all major lineages of land plants and across various ecosystems, from the arctic to the tropics (Arnold 2007). The diversity of these fungi is immense, with estimates suggesting over a million species, based on the ratio of vascular plants to fungal species (Sun and Guo 2012). This diversity is taxonomic and functional, as endophytic fungi can significantly affect plant communities by enhancing stress tolerance, increasing biomass, and influencing resource allocation (Rodriguez et al. 2009). For example, 24 culturable endophytic fungal strains were isolated from *Cynomorium songaricum* (Rupr.) (Fig. 2), based on the morphological characteristics and the sequences of the rDNA internal transcribed spacer (ITS) region; the isolates were identified into six genera: *Fusarium*, *Clonostachys*, *Penicillium*, *Alternaria*, *Aspergillus*, and *Madurella* (Jigjiddorj et al. 2024).

Endophytic fungi are taxonomically distinct from other nonpathogenic plant-associated fungi. They exhibit a wide range of ecological interactions with their host plants, which can vary depending on the host species and geographic location (Arnold 2007). For instance, studies on the *Terminalia* species have identified over 47 genera of endophytic fungi, with significant variations in diversity influenced by

Fig. 2 Morphological diversity of endophytic fungi isolated from *Cynomorium songaricum*

factors such as season and location (Toghueo and Boyom 2019). Similarly, research on *Hevea brasiliensis* in Peru revealed a greater diversity of endophytes in sapwood compared to leaves, highlighting spatial heterogeneity in endophyte assemblages (Gazis and Chaverri 2010).

Endophytic fungi are not only diverse but also play crucial roles in plant health and development. They can produce a wide range of secondary metabolites with potential applications in drug discovery and biotechnology. For instance, endophytic fungi from the Cactaceae family have been identified as sources of enzymes and bioactive compounds with antifungal and larvicidal properties (Souza et al. 2023). Similarly, endophytes from *Taxus globosa* have shown a high diversity of novel fungi, suggesting potential for biotechnological applications (Rivera-Orduña et al. 2011).

3 Types of Endophytic Fungi

3.1 Clavicipitaceous Endophytic Fungi

This type of endophytes is a group of fungi that form symbiotic relationships with grasses, living within plant tissues without causing disease symptoms. These endophytes are known for their ability to produce bioactive compounds that can enhance the host plant's resistance to various stresses.

Clavicipitaceous endophytes are known to enhance the resistance of grasses to multiple stresses, including insect herbivores, drought, and poor soil conditions. They achieve this by producing secondary metabolites such as alkaloids, which can deter herbivores and provide competitive advantages to the host plants (Clay 1989; Kuldau and Bacon 2008; Panaccione et al. 2014). These metabolites include ergot alkaloids, lolines, indole-diterpenes, and peramine, which are distributed in different combinations among endophyte taxa (Panaccione et al. 2014).

These endophytes are often transmitted vertically through seeds, ensuring their presence in successive plant generations. This mode of transmission is crucial for maintaining the symbiotic relationship and the associated benefits (Panaccione et al. 2014). The diversity of clavicipitaceous endophytes can vary significantly among different grass species and geographical locations, as seen in studies of *Achnatherum sibiricum* and *Lolium* species (Wilson et al. 1991; Ren et al. 2008; Table 1).

Clavicipitaceous endophytes have potential applications as biocontrol agents due to their ability to confer resistance to insect pests. This could reduce the need for chemical pesticides in managed grasslands, although their use might be limited to non-pasture situations due to potential toxicity to livestock (Clay 1989).

The protective effects of these endophytes are thought to be mediated by the secretion of reactive oxygen species (ROS) and auxin, which enhance the plant's antioxidant systems and suppress cell death, thereby maintaining the symbiotic relationship (Torres et al. 2012). Additionally, the presence of epiphyllous mycelial nets on leaves suggests a role in niche exclusion, potentially defending host plants from pathogens (Moy et al. 2000).

Table 1 Clavicipitaceous endophytic fungi

Aspect	Description	References
Host association	Clavicipitaceous endophytes are primarily associated with grasses, forming mutualistic relationships that enhance plant resistance to various stresses	Clay (1989), Kuldau and Bacon (2008), Panaccione et al. (2014)
Biocontrol potential	These fungi can act as biocontrol agents against insect pests, reducing the need for chemical pesticides in grasslands	Clay (1989)
Stress resistance	They enhance resistance to biotic and abiotic stresses, including drought, pests, and poor soil conditions, by producing secondary metabolites	Kuldau and Bacon (2008), Torres et al. (2012)
Bioactive compounds	Clavicipitaceous endophytes produce bioactive alkaloids such as ergot alkaloids, lolines, and peramine, which protect plants from herbivores	Panaccione et al. (2014)
Transmission	These fungi are often vertically transmitted through seeds, ensuring their presence in successive plant generations	Panaccione et al. (2014)
Diversity	Various morphotypes exist within clavicipitaceous endophytes, with some specific to certain geographic regions or plant species	Ren et al. (2008)
Antifungal activity	Some species, like *Epichloë festucae*, produce compounds with antifungal properties, potentially protecting host plants from other fungal pathogens	Yue et al. (2000)
Symbiotic effects	Endophytes can mediate interactions between host grasses and pathogens, sometimes reducing pathogen infection rates significantly	Perez et al. (2013)

3.2 Nonclavicipitaceous Endophytic Fungi

Nonclavicipitaceous endophytes are a diverse group of fungi that live symbiotically within plant tissues without causing disease. They are distinct from clavicipitaceous endophytes, which are primarily associated with grasses and known for producing protective alkaloids. Nonclavicipitaceous endophytes are more widespread and can be found in a variety of plant species, contributing to plant health and resilience in different ways (Table 2).

Nonclavicipitaceous endophytes are highly diverse, belonging mainly to the phylum Ascomycota, and are prevalent in many plant species (Rodriguez et al. 2009; Kleczewski et al. 2012; Andrade-Linares and Franken 2013). They are categorized into three functional groups based on their colonization patterns, transmission methods, and the benefits they confer on their host plants (Rodriguez et al. 2009). These fungi can enhance plant growth, increase resistance to abiotic and biotic stresses, and produce secondary metabolites that may have agricultural

Table 2 Nonclavicipitaceous endophytic fungi

Aspect	Description	References
Diversity	Nonclavicipitaceous endophytes are highly diverse, primarily belonging to the phylum *Ascomycota*, and are the dominant fungal endophytes in many plants	Rodriguez et al. (2009), Kleczewski et al. (2012), Swamy and Sandhu (2021)
Functional roles	These fungi can enhance plant fitness by increasing stress tolerance, biomass, and nutrient acquisition and by producing bioactive metabolites	Rodriguez et al. (2009), Yuan et al. (2010), Andrade-Linares and Franken (2013), Swamy and Sandhu (2021)
Transmission and colonization	They are horizontally transmitted and can colonize various plant tissues without causing disease symptoms	Yuan et al. (2010), Andrade-Linares and Franken (2013), Verma et al. (2017)
Ecological significance	Nonclavicipitaceous endophytes play a role in plant protection against biotic and abiotic stresses and contribute to plant microbiomes	Yuan et al. (2010), Porras-Alfaro and Bayman (2011), Verma et al. (2017)
Potential applications	They have potential applications in agriculture for improving plant productivity and stress resistance, and in pharmaceuticals for producing novel metabolites	Verma et al. (2017), Swamy and Sandhu (2021), Mishra et al. (2021)

and pharmaceutical applications (Andrade-Linares and Franken 2013; Swamy and Sandhu 2021; Table 2).

These endophytes play a significant role in plant ecophysiology by promoting plant growth and stress tolerance. They can improve biomass production and reduce water consumption, which is beneficial for plant survival in challenging environments (Rodriguez et al. 2009; Kleczewski et al. 2012; Andrade-Linares and Franken 2013). Nonclavicipitaceous endophytes are horizontally transmitted and can establish symbiotic relationships with a wide range of plant species, contributing to the ecological diversity and resilience of plant communities (Andrade-Linares and Franken 2013).

The potential applications of nonclavicipitaceous endophytes are vast, particularly in sustainable agriculture. They can protect crops from various stress factors, improve soil quality, and reduce the need for chemical fertilizers by making nutrients more available to plants (Swamy and Sandhu 2021). Despite their potential, the ecological significance and full range of benefits provided by these fungi are not yet fully understood, highlighting the need for further research to explore their roles and applications in different ecosystems (Rodriguez et al. 2009; Swamy and Sandhu 2021).

4 Principal Genera of Endophytic Fungi

- *Aspergillus*: This genus is frequently cited in studies of endophytic fungi, known for its role in producing secondary metabolites with various applications in medicine and agriculture (Rosa et al. 2010; Torres-Mendoza et al. 2020; Bezerra et al. 2021).
- *Fusarium*: It is a diverse genus with significant genetic and ecological variability. It is known for producing a wide range of bioactive secondary metabolites with antimicrobial, anticancer, and other biological activities (Rosa et al. 2010; Arora et al. 2019; Toghueo 2019).
- *Penicillium*: It is another well-documented genus of endophytic fungi, often noted for its ability to produce proteases and other enzymes, as well as secondary metabolites with potential industrial applications (Torres-Mendoza et al. 2020; Bezerra et al. 2021).
- *Trichoderma*: This genus is recognized for its biocontrol properties and production of secondary metabolites, making it a valuable genus in agricultural biotechnology (Torres-Mendoza et al. 2020).
- *Phomopsis*: Known for its role in producing secondary metabolites, Phomopsis is frequently mentioned in patents related to biotransformation and secondary metabolite production (Torres-Mendoza et al. 2020).
- *Metarhizium* and *Beauveria:* These genera are traditionally recognized as insect pathogenic fungi but also function as endophytes, highlighting their multifunctional lifestyles and potential to promote plant growth (Barelli et al. 2015).
- *Xylaria*: It is a genus within the *Sordariomycetes* class, noted for its prevalence in certain plant hosts and its contribution to the endophytic fungal community (Groff et al. 2024).
- *Cadophora*, *Davidiella*, *Entrophospora*, *Geomyces*, *Gyoerffyella*, *Microdochium*, *Mycocentrospora*, *Phaeosphaeria*: These genera were identified in the endophytic community associated with Antarctic plants, showcasing their adaptability to extreme environments (Rosa et al. 2010).
- Phoma: It is a dominant genus in certain plant hosts, known for its antimicrobial activity and potential in enhancing plant growth (Arora et al. 2019).
- *Bezerromyces* and *Xiliomyces*: These are newly described genera identified in Brazilian cacti, representing novel taxa within the *Dothideomycetes* class (Bezerra et al. 2021) (Fig. 3).

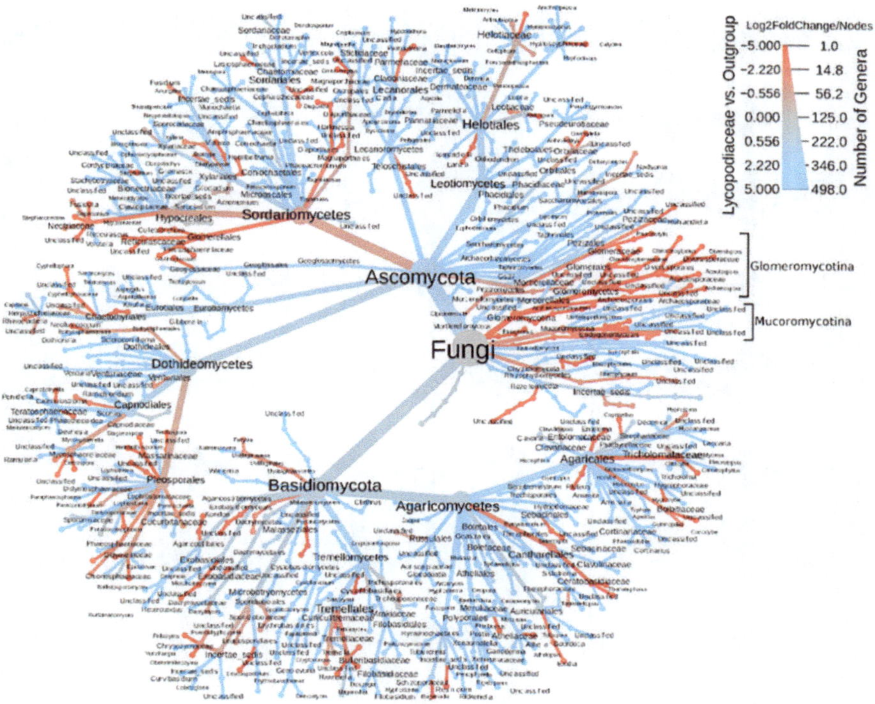

Fig. 3 Heatmap tree of fungal genera detected in the ITS dataset for *Lycopodiaceae* relative to the outgroup (*Selaginella kraussiana*) and representative the number of genera. (Benucci et al. 2020)

5 Endophytic Fungi from *Lycopodium* Species

- Dothideomycetes: This class dominated the fungal endophyte communities in *Lycopodium* species, with specific taxa such as *Phoma brasiliensis* and *Paraconiothyrium lycopodinum* being notably abundant in *L. clavatum* and *L. annotinum*, respectively (Pawłowska et al. 2014).
- Paraboeremia: Identified as an endophytic fungus from *Lycopodium serratum* var. *longipetiolatum*, *Paraboeremia* sp. was noted for producing *Lycopodium* alkaloids, including lycopodine and huperzine A (Ishiuchi et al. 2018).
- Colletotrichum: *Colletotrichum alatae* was isolated from *Lycopodium clavatum* and was recognized for its production of antibacterial and antioxidative compounds (Santra and Banerjee 2022a, b).
- Ascomycota: Various unidentified *Ascomycota* were isolated from *Lycopodium* species, with some taxa being exclusive to specific species like *L. clavatum* (Ishiuchi et al. 2018).
- *Alternaria*, *Fusarium*, *Trichoderma*, *Penicillium*, *Paecilomyces*, and *Phoma*: These genera were identified among the endophytic fungi isolated from *Huperzia*

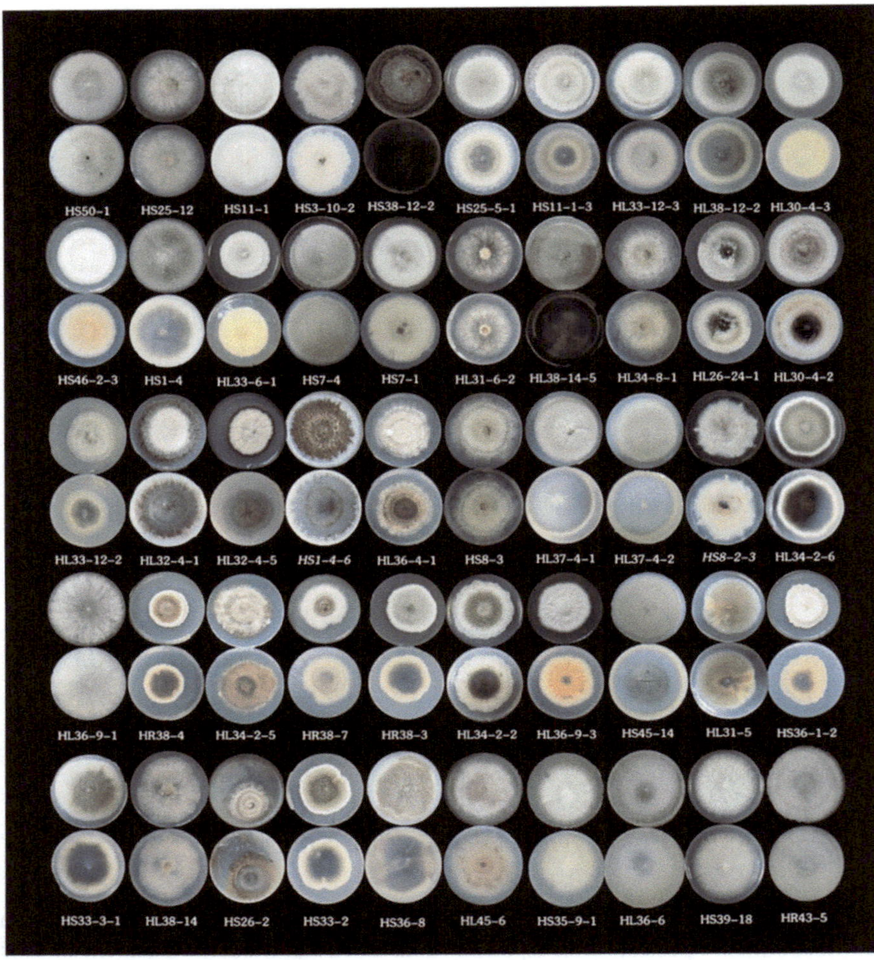

Fig. 4 Colonies of the endophytic fungi from *Huperzia serrate.* (Cui et al. 2021)

serrata, a close relative of *Lycopodium*, indicating a potential overlap in endophytic communities (Le et al. 2019; Cui et al. 2021; Fig. 4).

- *Basidiomycota* and *Ascomycota*: These phyla were found in the rhizobiome of Lycopodiaceae species, including *Lycopodium*, suggesting a diverse fungal community (Benucci et al. 2020).

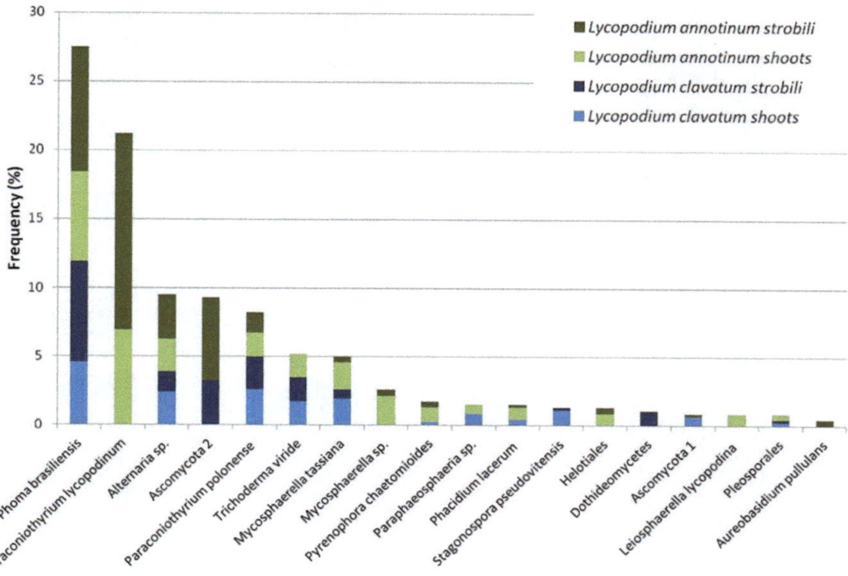

Fig. 5 Endophytic fungi isolated from *Lycopodium clavatum* and *L. annotinum*

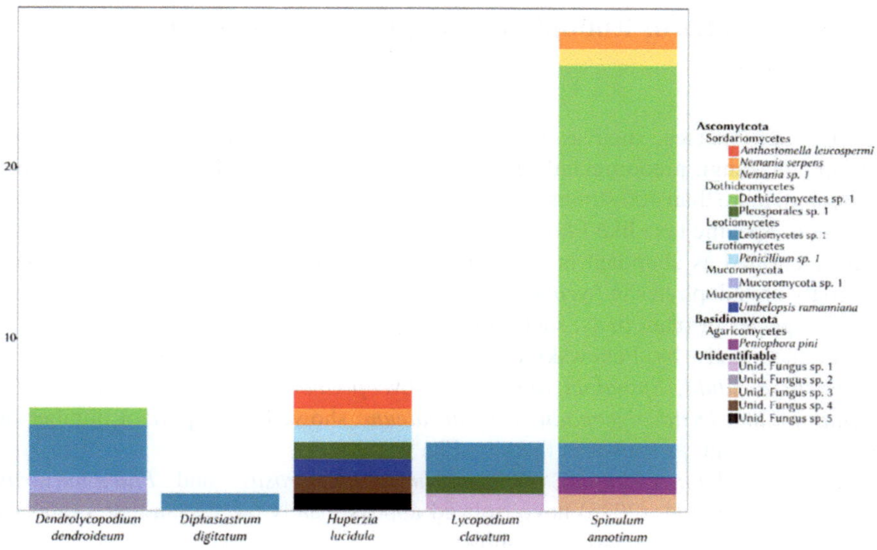

Fig. 6 Endophytic fungal species found in some *Lycopodium* species. (Petlewski 2020)

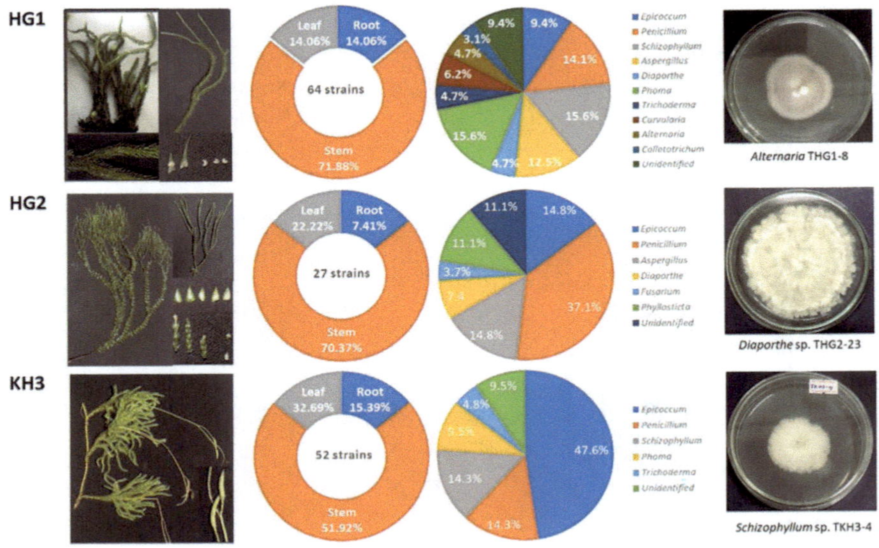

Fig. 7 Isolation and distribution of endophytic fungi recovered from three *Lycopodiaceae* species. (Le et al. 2023)

6 Metabolites of Endophytic Fungi from *Lycopodium* Species

Lycopodium species, such as *L. clavatum* and *L. annotinum*, host a variety of endophytic fungi, predominantly from the *Ascomycete* class. In a study conducted in Poland, more than 400 strains were isolated, with *Dothideomycetes* being the most common. Specific taxa like *Phoma brasiliensis* and *Paraconiothyrium lycopodinum* were identified as abundant in these species (Pawłowska et al. 2014; Fig. 5). The diversity of endophytes in *Lycopodium* is notable, with some taxa being exclusive to certain species or plant parts, such as the strobili.

Another study by Petlewski (2020) in Poland on five *Lycopodium* species, *Huperzia lucidula*, *Spinulum annotinum*, *Lycopodium clavatum*, *Diphasiastrum digitatum*, and *Dendrolycopodium dendroideum*, showed an important number of endophytic fungi, as shown in the next Fig. 6.

In *Lycopodium clavatum*, *Phlegmariurus squarrosus*, and *Phlegmariurus phlegmaria*, Le et al. (2023) have isolated approximately 64, 27, and 52 strains of endophytic fungi, respectively, as shown in Fig. 7.

Fig. 8 Huperzine A chemical structure

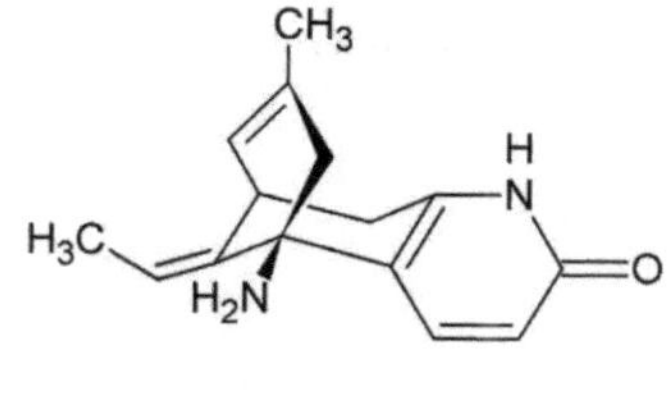

(-)-Huperzine A

6.1 Metabolites of Endophytic Fungi from Lycopodium *Species*

Endophytic fungi associated with *Lycopodium* species produce a diverse range of secondary metabolites, including alkaloids, terpenoids, polyketides, and phenols. These compounds are known for their bioactive properties, such as antimicrobial, antioxidant, and anticancer activities (Keshri et al. 2021; Santra and Banerjee 2022a; Jha et al. 2023). Specifically, *Lycopodium* alkaloids like 12-epi-lycopodine and huperzine A have been identified in endophytic fungi from *Lycopodium serratum*, highlighting the potential for discovering novel compounds from these fungi (Ishiuchi et al. 2018).

The metabolites produced by endophytic fungi exhibit significant biological activities. For instance, *Colletotrichum alatae*, an endophyte from *Lycopodium clavatum*, produces antibacterial and antioxidative compounds effective against various microorganisms, including methicillin-resistant *Staphylococcus aureus* (Santra and Banerjee 2022a). These metabolites have shown potential in inhibiting pathogenic strains and human cancer cell lines, making them promising candidates for drug development (Rustamova et al. 2020; Keshri et al. 2021). The bioactive metabolites from endophytic fungi function by enhancing the host plant's immune system and protecting against pathogens. They achieve this through the production of compounds that mitigate biotic and abiotic stresses (Jha et al. 2023). The metabolites can also possess unique mechanisms of action, such as selective toxicity and novel structural features, which are advantageous for pharmaceutical applications (Ancheeva et al. 2019; Gupta et al. 2020).

Huperzine A (Fig. 8) is a natural compound extracted from the Chinese herb *Huperzia serrata*, known for its potent acetylcholinesterase inhibitory activity. It has been traditionally used in Chinese medicine and is currently being explored for its potential therapeutic effects on neurodegenerative diseases such as Alzheimer's disease (AD) and vascular dementia.

Huperzine A has shown promise in improving cognitive function in patients with Alzheimer's disease. Several studies indicate that it can enhance general cognitive function, global clinical status, and daily living activities, with minimal adverse effects (Zangara 2003; Li et al. 2008; Yang et al. 2013; Damar et al. 2016). However, the quality of the studies varies, and more rigorous trials are needed to confirm these findings (Li et al. 2008; Yang et al. 2013). The evidence for Huperzine A's efficacy

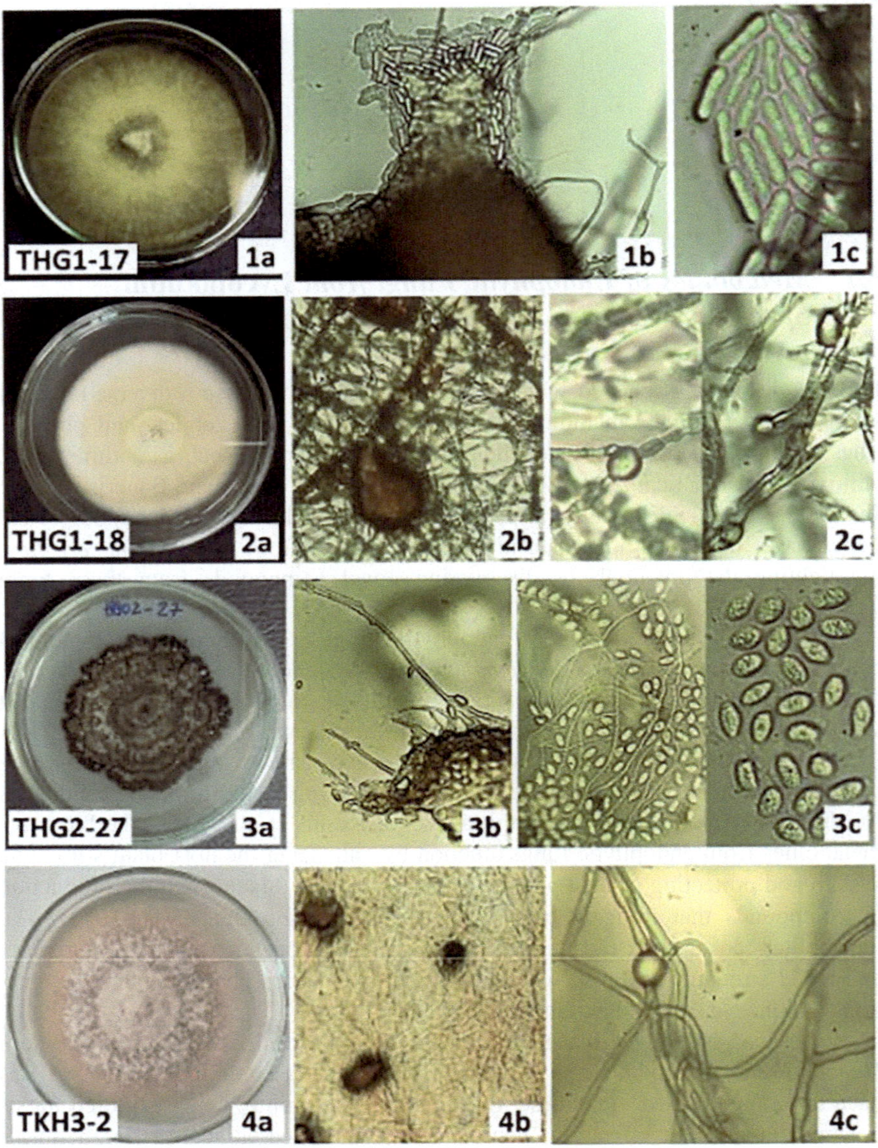

Fig. 9 Morphological characteristics of Hup-producing fungal strains

in treating vascular dementia is less convincing. A small trial did not find significant cognitive improvements, suggesting that more research is needed to establish its effectiveness in this area (Hao et al. 2009). Huperzine A acts as a reversible inhibitor of acetylcholinesterase, preventing the breakdown of acetylcholine and thereby enhancing cholinergic transmission. This mechanism is believed to contribute to its cognitive-enhancing effects (Zangara 2003; Ma et al. 2007; Bredsdorff and Pilegaard 2018). Additionally, Huperzine A has demonstrated neuroprotective properties, potentially offering protection against neuroinflammation and excitotoxicity (Noor et al. 2020; Gupta et al. 2020) (Fig. 9).

7 Conclusion and Perspectives

The diversity of endophytic fungi in *Lycopodium* species reveals a complex and rich symbiotic relationship that is still not fully understood. Studies have shown that *Lycopodium* species host a variety of endophytic fungi, predominantly *Ascomycetes*, with notable species such as *Phoma brasiliensis* and *Paraconiothyrium lycopodinum* being prevalent in different *Lycopodium* species. This diversity is not only significant in terms of ecological interactions but also in the potential bioactive compounds these fungi produce, which have applications in pharmaceuticals, such as antibacterial and antioxidative agents. Furthermore, the coevolutionary history of *Lycopodiaceae* with fungi suggests that these relationships have been crucial for the survival and adaptation of these ancient plant lineages. The presence of diverse fungal communities, including *Basidiomycota* and *Ascomycota*, indicates a long-standing symbiotic relationship that has evolved over time. Moreover, the discovery of unique *Lycopodium* alkaloids produced by endophytic fungi further underscores the biochemical potential of these associations.

Future research should focus on expanding the understanding of the ecological roles and evolutionary dynamics of endophytic fungi in *Lycopodium* species. There is a need to explore the full spectrum of fungal diversity using advanced molecular techniques, as a significant portion of fungal diversity remains poorly classified. Additionally, investigating the functional roles of these fungi in plant health and stress resilience could provide insights into their potential applications in agriculture and biotechnology. Furthermore, the exploration of bioactive compounds from endophytic fungi in *Lycopodium* species should be intensified, given their promising pharmaceutical applications. Understanding the conditions that optimize the production of these compounds could lead to new drug discoveries and enhance the economic value of these symbiotic relationships.

In summary, the diversity of endophytic fungi in *Lycopodium* species offers a rich field of study with significant ecological, evolutionary, and biotechnological implications. Continued research in this area promises to uncover new insights into plant-fungal interactions and their potential benefits.

References

Ancheeva E, Daletos G, Proksch P (2019) Bioactive secondary metabolites from endophytic fungi. Curr Med Chem. https://doi.org/10.2174/0929867326666190916144709

Andrade-Linares D, Franken P (2013) Fungal endophytes in plant roots: taxonomy, colonization patterns, and functions. In: Symbiotic endophytes. Springer, Berlin, Heidelberg, pp 311–334. https://doi.org/10.1007/978-3-642-39317-4_16

Arnold A (2007) Understanding the diversity of foliar endophytic fungi: progress, challenges, and frontiers. Fungal Biol Rev 21:51–66. https://doi.org/10.1016/J.FBR.2007.05.003

Arora P, Wani Z, Ahmad T, Sultan P, Gupta S, Riyaz-Ul-Hassan S (2019) Community structure, spatial distribution, diversity and functional characterization of culturable endophytic fungi associated with *Glycyrrhiza glabra* L. Fungal Biol 123(5):373–383. https://doi.org/10.1016/J.FUNBIO.2019.02.003

Barelli L, Moonjely S, Behie S, Bidochka M (2015) Fungi with multifunctional lifestyles: endophytic insect pathogenic fungi. Plant Mol Biol 90:657–664. https://doi.org/10.1007/s11103-015-0413-z

Benucci G, Burnard D, Shepherd L, Bonito G, Munkacsi A (2020) Evidence for co-evolutionary history of early diverging Lycopodiaceae plants with fungi. Front Microbiol 10. https://doi.org/10.3389/fmicb.2019.02944

Bezerra V, Cardoso S, Fonseca-Bazzo Y, Silveira D, Magalhães P, Souza P (2021) Protease produced by Endophytic fungi: a systematic review. Molecules 26. https://doi.org/10.3390/molecules26227062

Bredsdorff L, Pilegaard K (2018) Concern for adverse effects of huperzine a when sold as an ingredient in food supplements. Toxicol Lett 295. https://doi.org/10.1016/j.toxlet.2018.06.735

Clay K (1989) Clavicipitaceous endophytes of grasses: their potential as biocontrol agents. Fungal Biol 92:1–12. https://doi.org/10.1016/S0953-7562(89)80088-7

Cui L, Noushahi HA, Zhang Y, Liu J, Cosoveanu A, Liu Y, Yan L, Zhang J, Shu S (2021) Endophytic fungal community of *Huperzia serrata*: diversity and relevance to the production of Huperzine a by the plant host. Molecules 26:892. https://doi.org/10.3390/molecules26040892

Damar U, Gersner R, Johnstone J, Schachter S, Rotenberg A (2016) Huperzine a as a neuroprotective and antiepileptic drug: a review of preclinical research. Expert Rev Neurother 16:671–680. https://doi.org/10.1080/14737175.2016.1175303

Gazis R, Chaverri P (2010) Diversity of fungal endophytes in leaves and stems of wild rubber trees (*Hevea brasiliensis*) in Peru. Fungal Ecol 3:240–254. https://doi.org/10.1016/J.FUNECO.2009.12.001

Groff D, Marmentini J, Gaglioti A, Da Silva P, Knob A (2024) Endophytic fungi associated with *Araucaria angustifolia* (Bertol.) Kuntze. Anais da Academia Brasileira de Ciencias 96:e20230251. https://doi.org/10.1590/0001-3765202420230251

Gupta S, Chaturvedi P, Kulkarni M, Van Staden J (2020) A critical review on exploiting the pharmaceutical potential of plant endophytic fungi. Biotechnol Adv 107462. https://doi.org/10.1016/j.biotechadv.2019.107462

Hao Z, Liu M, Liu Z, Lv D (2009) Huperzine A for vascular dementia. Cochrane Database Syst Rev 2:CD007365. https://doi.org/10.1002/14651858.CD007365.pub2

Ishiuchi K, Hirose D, Suzuki T, Nakayama W, Jiang W, Monthakantirat O, Wu J, Kitanaka S, Makino T (2018) Identification of *lycopodium* alkaloids produced by an ultraviolet-irradiated strain of *Paraboeremia*, an Endophytic fungus from *Lycopodium serratum* var. longipetiolatum. J Nat Prod 81 5:1143–1147. https://doi.org/10.1021/acs.jnatprod.7b00627

Jha P, Kaur T, Chhabra I, Panja A, Paul S, Kumar V, Malik T (2023) Endophytic fungi: hidden treasure chest of antimicrobial metabolites interrelationship of endophytes and metabolites. Front Microbiol 14. https://doi.org/10.3389/fmicb.2023.1227830

Jigjiddorj EA, Maidarjav A, Byambasuren B, Nyamgerel D (2024) Identification, antimicrobial and plant growth promoting activities of Endophytic fungi associated with *Cynomorium songaricum* Rupr., a traditional medicinal Plant in Mongolia. Diversity 16:122. https://doi.org/10.3390/d16020122

Keshri P, Rai N, Verma A, Kamble S, Barik S, Mishra P, Singh S, Salvi P, Gautam V (2021) Biological potential of bioactive metabolites derived from fungal endophytes associated with medicinal plants. Mycol Prog 20:577–594. https://doi.org/10.1007/s11557-021-01695-8

Kleczewski N, Bauer J, Bever J, Clay K, Reynolds H (2012) A survey of endophytic fungi of switchgrass (*Panicum virgatum*) in the Midwest, and their putative roles in plant growth. Fungal Ecol 5:521–529. https://doi.org/10.1016/J.FUNECO.2011.12.006

Kuldau G, Bacon C (2008) Clavicipitaceous endophytes: their ability to enhance resistance of grasses to multiple stresses. Biol Control 46:57–71. https://doi.org/10.1016/J.BIOCONTROL.2008.01.023

Le T, Hoang A, Le T, Vo T, Van Quyen D, Chu H (2019) Isolation of endophytic fungi and screening of Huperzine A–producing fungus from *Huperzia serrata* in Vietnam. Sci Rep 9. https://doi.org/10.1038/s41598-019-52481-2

Le TTM, Pham HT, Trinh HTT, Tran HT, Chu HH (2023) Isolation and characterization of novel Huperzine-producing Endophytic fungi from Lycopodiaceae species. J Fungi 9:1134. https://doi.org/10.3390/jof9121134

Lehnert M, Krug M, Kessler M (2017) A review of symbiotic fungal endophytes in lycophytes and ferns: A global phylogenetic and ecological perspective. Symbiosis 71(2):77–89. https://doi.org/10.1007/s13199-016-0454-y

Li J, Wu H, Zhou R, Liu G, Dong B (2008) Huperzine a for Alzheimer's disease. Cochrane Database Syst Rev 2:CD005592. https://doi.org/10.1002/14651858.CD005592.pub2

Ma X, Tan C, Zhu D, Gang D, Xiao P (2007) Huperzine a from *Huperzia* species – an ethnopharmacolgical review. J Ethnopharmacol 113(1):15–34. https://doi.org/10.1016/J.JEP.2007.05.030

Mishra Y, Sharma L, Dhiman M, Sharma M (2021) Endophytic fungal diversity of selected medicinal plants and their bio-potential applications. In: Fungi bio-prospects in sustainable agriculture, environment and Nano-Technology. Academic Press, pp 227–283. https://doi.org/10.1016/b978-0-12-821394-0.00010-x

Moy M, Belanger F, Duncan R, Freehoff A, Leary C, Meyer W, Sullivan R, White J (2000) Identification of epiphyllous mycelial nets on leaves of grasses infected by clavicipitaceous endophytes. Symbiosis 28:291–302

Noor A, Almasri D, Bagalagel A, Abdallah H, Mohamed S, Mohamed G, Ibrahim S (2020) Naturally occurring Isocoumarins derivatives from Endophytic fungi: sources, isolation, structural characterization, biosynthesis, and biological activities. Molecules 25. https://doi.org/10.3390/molecules25020395

Panaccione D, Beaulieu W, Cook D (2014) Bioactive alkaloids in vertically transmitted fungal endophytes. Funct Ecol 28:299–314. https://doi.org/10.1111/1365-2435.12076

Pawłowska J, Wilk M, Śliwińska-Wyrzychowska A, Mętrak M, Wrzosek M (2014) The diversity of endophytic fungi in the above-ground tissue of two *lycopodium* species in Poland. Symbiosis 63:87–97. https://doi.org/10.1007/s13199-014-0291-1

Perez L, Gundel P, Ghersa C, Omacini M (2013) Family issues: fungal endophyte protects host grass from the closely related pathogen *Claviceps purpurea*. Fungal Ecol 6:379–386. https://doi.org/10.1016/J.FUNECO.2013.06.006

Petlewski AR (2020) Exploring Lycopodiaceae endophytes, Dendrolycopodium systematics, and the future of fern model systems (Master's thesis, Cornell University, Ithaca, NY)

Porras-Alfaro A, Bayman P (2011) Hidden fungi, emergent properties: endophytes and microbiomes. Annu Rev Phytopathol 49:291–315. https://doi.org/10.1146/annurev-phyto-080508-081831

Ren A, Gao Y, Wang Y, Zhang X, Wei Y (2008) Diversity of clavicipitaceous fungal endophytes in *Achnatherum sibiricum* in natural grasslands of China. Symbiosis 46:21–31

Rivera-Orduña F, Suarez-Sanchez R, Flores-Bustamante Z, Gracida-Rodríguez J, Flores-Cotera L (2011) Diversity of endophytic fungi of *Taxus globosa* (Mexican yew). Fungal Divers 47:65–74. https://doi.org/10.1007/s13225-010-0045-1

Rodriguez R, White J, Arnold A, Redman R (2009) Fungal endophytes: diversity and functional roles. New Phytol 182(2):314–330. https://doi.org/10.1111/j.1469-8137.2009.02773.x

Rosa L et al (2010) Endophytic fungi community associated with the dicotyledonous plant *Colobanthus quitensis* (Kunth) Bartl. (Caryophyllaceae) in Antarctica. FEMS Microbiol Ecol 73 1:178–189. https://doi.org/10.1111/j.1574-6941.2010.00872.x

Rustamova N, Bozorov K, Efferth T, Yili A (2020) Novel secondary metabolites from endophytic fungi: synthesis and biological properties. Phytochem Rev 19:425–448. https://doi.org/10.1007/s11101-020-09672-x

Santra H, Banerjee D (2022a) Bioactivity study and metabolic profiling of *Colletotrichum alatae* LCS1, an endophyte of club moss *Lycopodium clavatum* L. PLoS One 17. https://doi.org/10.1371/journal.pone.0267302

Santra H, Banerjee D (2022b) Production, optimization, characterization and drought stress resistance by β-Glucan-rich Heteropolysaccharide from an Endophytic fungi *Colletotrichum alatae* LCS1 isolated from clubmoss (*Lycopodium clavatum*). Front Fungal Biol 2. https://doi.org/10.3389/ffunb.2021.796010

Souza J, Fonseca D, De Souza J, Nascimento L, Hughes F, Bezerra J, Góes-Neto A, Ferreira-Silva A (2023) Cactus Endophytic fungi and bioprospecting for their enzymes and bioactive molecules: a systematic review. Chem Biodivers 20. https://doi.org/10.1002/cbdv.202301413

Sun X, Guo L (2012) Endophytic fungal diversity: review of traditional and molecular techniques. Mycology 3:65–76. https://doi.org/10.1080/21501203.2012.656724

Swamy N, Sandhu S (2021) Fungal endophytes: entry, establishment, diversity, future prospects in agriculture. In: Fungi bio-prospects in sustainable agriculture, environment and Nano-Technology. https://doi.org/10.1016/b978-0-12-821394-0.00004-4

Toghueo R (2019) Bioprospecting endophytic fungi from *Fusarium* genus as sources of bioactive metabolites. Mycology 11:1–21. https://doi.org/10.1080/21501203.2019.1645053

Toghueo R, Boyom F (2019) Endophytic fungi from *Terminalia* species: a comprehensive review. J Fungi 5. https://doi.org/10.3390/jof5020043

Torres M, White J, Zhang X, Hinton D, Bacon C (2012) Endophyte-mediated adjustments in host morphology and physiology and effects on host fitness traits in grasses. Fungal Ecol 5:322–330. https://doi.org/10.1016/J.FUNECO.2011.05.006

Torres-Mendoza D, Ortega H, Cubilla-Rios L (2020) Patents on Endophytic fungi related to secondary metabolites and biotransformation applications. J Fungi 6. https://doi.org/10.3390/jof6020058

Verma S, Gond S, Mishra A, Sharma V, Kumar J, Singh D, Kumar A, Kharwar R (2017) Fungal endophytes representing diverse habitats and their role in plant protection. In: Developments in fungal biology and applied mycology. Springer Singapore, Singapore, pp 135–157. https://doi.org/10.1007/978-981-10-4768-8_9

Wilson AD, Clement SL, Kaiser WJ (1991) Survey and detection of endophytic fungi in *Lolium* germ plasm by direct staining and aphid assays. Plant Dis 75:169–173. https://doi.org/10.1094/PD-75-0169

Yang G, Wang Y, Tian J, Liu J (2013) Huperzine a for Alzheimer's disease: a systematic review and meta-analysis of randomized clinical trials. PLoS One 8. https://doi.org/10.1371/journal.pone.0074916

Yuan Z, Zhang C, Lin F (2010) Role of diverse non-systemic fungal endophytes in plant performance and response to stress: progress and approaches. J Plant Growth Regul 29:116–126. https://doi.org/10.1007/s00344-009-9112-9

Yue Q, Miller C, White J, Richardson M (2000) Isolation and characterization of fungal inhibitors from *Epichloë festucae*. J Agric Food Chem 48(10):4687–4692. https://doi.org/10.1021/JF990685Q

Zangara A (2003) The psychopharmacology of huperzine a: an alkaloid with cognitive enhancing and neuroprotective properties of interest in the treatment of Alzheimer's disease. Pharmacol Biochem Behav 75:675–686. https://doi.org/10.1016/S0091-3057(03)00111-4

Fungal Endophytes: A Continuous Interaction with Land Plant Evolution

Lobna Hajji-Hedfi, Okon Godwin Okon, Parwiz Niazi, Takwa Wannassi, Hasadiah Okon Bassey, Abdul Bari Hejran, Antia Ukponobong Efiong, Rym Jaouadi, Laith Khalil Tawfeeq Al-Ani, Mavis Agyeiwaa Acheampong, Nishant Gupta, and Virendra Kumar Yadav

Contents

L. Hajji-Hedfi (✉)
Regional Centre of Agricultural Research of Sidi Bouzid, Sidi Bouzid, Tunisia

Research Laboratory of Agricultural Production Systems and Sustainable Development (LR03AGR02), Carthage University, Tunis, Tunisia
e-mail: lobna.hajji@iresa.agrinet.tn

O. G. Okon
Department of Botany, Akwa Ibom State University, Akwa Ibom, Nigeria

P. Niazi
Department of Biology, Faculty of Education, Kandahar University, Kandahar, Afghanistan

T. Wannassi
Regional Centre of Agricultural Research of Sidi Bouzid, Sidi Bouzid, Tunisia

H. O. Bassey
Department of Biological Sciences, Akwa Ibom State Polytechnic, Akwa Ibom, Nigeria

A. B. Hejran
Department of Biology, Faculty of Education, Helmand University, Lashkar Gah, Afghanistan

A. M. Abdel-Azeem, F. Degola (eds.), *Advances in Lichens and Non-Flowering Plants Fungal Endophytes Research*, Fungal Biology,
https://doi.org/10.1007/978-3-032-01506-8_8

1 Introduction

Fossil fungi and fungal-like structures have been traced back to the early life age (about 1.5–2.4 billion years ago) (Taylor et al. 2015; Bengtson et al. 2017). The Devonian Rhynie chert is, however, the earliest record resource for fungi–plant interactions (Taylor et al. 2003; Berbee and Taylor 2007). The Lower Devonian Rhynie chert is a pivotal geological formation, offering profound insights into the early terrestrial existence of plants, animals, and microorganisms. This extraordinary fossil deposit, situated in Aberdeenshire, Scotland, preserves the intricacies of land-based ecosystems from approximately 407 million years ago (Krings et al. 2018; Strullu-Derrien et al. 2023). Among the microbial fossils, fungi are notably abundant, encompassing representatives from nearly all major fungal lineages, with the notable exception of Basidiomycota (D'Ario et al. 2024; Rungjindamai and Jones 2024). The remarkable preservation of this ancient ecosystem has profoundly

A. U. Efiong
Department of Microbiology, Akwa Ibom State University, Akwa Ibom, Nigeria

R. Jaouadi
Laboratory of Agricultural Production, Higher School of Agriculture of Mograne (ESAM), University of Carthage, Zaghouane, Tunisia

L. K. T. Al-Ani
School of Biology Science, Universiti Sains Malaysia, Minden, Malaysia

M. A. Acheampong
Department of Crop Science, University of Ghana, Accra, Ghana

N. Gupta
Medical Devision, River Engineering Private Limited, Greater Noida, Uttar Pradesh, India

V. K. Yadav
Marwadi University Research Center, Department of Microbiology, Faculty of Sciences, Marwadi University, Rajkot, Gujarat, India

influenced the field of botany, shedding light on early terrestrial life, including plant life cycles, fungal symbiotic associations, soil microbial communities, and arthropod biodiversity. Indeed, the Rhynie cherts offer invaluable insights into the architecture and interactions of primordial land-based communities, which are crucial in understanding their broader implications for Earth's systems. Thought to be no younger than the Middle Old Red Sandstone, subsequent analysis of spore assemblages in the interbedded shales revealed their age to correspond to the Lower Devonian, specifically the middle Pragian to lower Emsian stages (Edwards et al. 2018; Bilański and Kowalski 2022). The ecological composition of the preserved fossils bears striking similarities to the fauna associated with modern Icelandic geothermal springs. The *Rhynie* assemblage likely reflects a well-preserved, largely intact terrestrial animal community from the Early Devonian period (Dunlop and Garwood 2018; Toledo et al. 2021). While the first endophytic fungus was isolated from seeds of ryegrass (*Lolium temulentum*) in 1898 (Vogl 1898), fossil plant tissue records indicated that endophytic fungi have been present for over 400 million years (Krings et al. 2007a, b). In this context, fungal endophytes are fungi that inhabit living plant tissues asymptomatically, without inducing noticeable disease symptoms. In contemporary ecosystems, they are integral to plant health, contributing to improved nutrient uptake, heightened stress resilience, and enhanced resistance to pathogens (Collinge et al. 2022; D'Ario et al. 2024). Fungi are indispensable components of terrestrial ecosystems, yet their diversity and evolutionary trajectory remain incompletely understood. Basal fungal lineages in phylogenetic trees are predominantly unicellular, parasitic, and aquatic, with mycelium-forming species representing most of their diversity. Endophytic fungi have been extensively studied for over 126 years and are now well documented to occur virtually in all plants and environments, including Arctic and Antarctic regions, deserts, oceans, and tropical rainforest (Jin et al. 2021; Wen et al. 2022).

Moreover, the bioactive compounds, beneficial endophytic fungi promote plant growth and increase their tolerance against biotic (pests and diseases) and abiotic environmental stress (Poveda et al. 2022). This has broadened the uses of endophytic fungi with a surge in number of commercial products for the pharmaceutical, phytoremediation, and biopesticide industries. Indeed, research on symbiotic relationships between endophytic fungi and land plants seems to lag bioprospecting and development of their bioactive compounds. Taylor et al. (2004) provided detailed one-to-one associations (saprotrophic, parasitic, and mutualistic) between land plant–fungal interactions from the Rhynie chert. Krings et al. (2007a, b) studied three different fungal endophytes populating 250 rhizoids of the Rhynie chert vascular land plant, *Nothia aphylla*. This chapter, therefore, provides a comprehensive overview of the spatial distribution and mechanism of host response of endophytic fungi and modern plant–fungal interactions.

2 Fossil Fungi

2.1 *Diversity and Fossil Preservation*

An interesting field of paleomycology, fossil fungi are classified according to their geological age, preservation type, and morphological characteristics. As with modern fungi, these ancient fungi are mainly divided into several groups, such as Chytrids, Zygomycetes, Ascomycetes, and Basidiomycetes. Reproductive structures such as spores, hyphae, and fruiting bodies can be preserved in different ways, including compressions, impressions, permineralizations, and amber inclusions are used to identify fossilized fungi. Prototaxites and Paleopyrenomycites are two prominent examples of the Paleozoic era's best-preserved fossil fungi (Krings et al. 2012; Taylor et al. 2015). Based on the development of molecular clocks and phylogenetic analysis, fungi diverged from their closest relatives' animals, as sister taxa, and estimated origins range from 965 million years ago to 1.06 billion years ago, which makes them the most ancient and ecologically diverse organisms on Earth (Berbee and Taylor 2007). Initially, fungi are considered as plants, then they are classified as a separate kingdom based on their mode of nutrition and reproduction. Estimated around 1.5 million fungal species, although only around 100,000 recognized species of fungi were identified. It was revealed that 2.2 to 3.8 million species based on host association and 11.7 to 13.2 million species using high-throughput sequencing were identified (Hyde 2022).

Citing fossilized filamentous structures, estimates suggest they date back to 2.4 billion years ago, discovered in South Africa Precambrian rocks (Bengtson et al. 2017). In fact, restricting the timeline of fungi is very important to understand the development of the Earth's ecosystems.

2.1.1 From Lithic Formations to Paleoenvironments and Biotic Habitats

William Mackie, the pioneer who identified the *Rhynie* locality, conducted exploratory work between 1910 and 1912 while compiling a geological map, which was subsequently published in 1913. During his exploration, Mackie gathered cherty rocks, some containing clasts of altered volcanic debris or silica-cemented sandstones, which are now understood as hydrothermally altered igneous formations and silicified Devonian sandstones. He postulated that the fossiliferous cherts were a product of geyser or hot spring activity coinciding with regional volcanic downturns. Contemporary studies corroborate this hypothesis (Trewin 2003).

(a) **Subsurface lithology—the subterranean conduits of the Rhynie thermal springs**

Geothermal systems typically manifest either acidic (pH 2–4) or alkaline (pH 7–9) vent waters, with a few rare systems exhibiting intermediate pH levels between 4 and 7. The *Rhynie* geothermal spring, shaped by extensional tectonics, aligns with the alkaline model, where the water's chemistry is predominantly chloride-based rather than sulfate-based. Geothermal systems play a significant role in the genesis of mineral deposits through fluid–rock interactions that facilitate rock dissolution, elevating the concentration of metals and metalloids in the ascending heated fluids. At *Rhynie*, minerals such as quartz and adularia precipitate due to interactions with these alkaline-chloride geothermal fluids.

(b) **Superficial lithology—silica sinters, thermal spring microenvironments, and ecological niches**

In geothermal zones, low-pH systems are devoid of sinter formation, whereas alkaline-chloride systems, such as the one at *Rhynie*, are prolific in silica sinter deposits. The *Rhynie* site is characterized by up to 52 sinter beds within a 35-meter sedimentary sequence, indicative of significant silica precipitation within shallow subsurface milieus. The thermal spring waters at *Rhynie* are brackish in nature, and as they cool, silica precipitates in the form of opal-A, enveloping various organisms. This silica encrustation fosters exceptional preservation, capturing both microbial cells and larger organisms.

(c) **Chemical composition and physical characteristics in geothermally impacted aquatic ecosystems**

The *Rhynie chert* is a product of hot-spring sinter formation, wherein opaline silica precipitates from alkaline-chloride thermal waters. Similar geothermal systems, like those in Yellowstone, exhibit comparable conditions, where variables such as water temperature, salinity, and silica concentration promote the precipitation of silica and the formation of sinters. As thermal waters emanate from vent pools, they create surfaces upon which organisms colonize, with cooler regions fostering the encasement of plants and animals in silica. This encasement leads to unparalleled preservation. These processes, in conjunction with environmental stressors, significantly influence both the local ecosystem and the fossilization process at Rhynie (Channing 2018; Wellman 2018; Fay 2015; Gupta et al. 2024).

3 Early Land Plant–Fungal Interactions

It is thought that the fungal–plant association emerged at least 450–460 million years ago, likely involving symbiotic fungi (Priyashantha et al. 2023). The Rhynie chert, a well-known Early Devonian fossil deposit, represents an in situ silicified hot spring environment, marked by small, temporary freshwater pools scattered throughout the landscape. This site is particularly celebrated for its remarkable preservation of fossil fungi and their interactions with various other ecosystem elements (Taylor et al. 2004). While many fungal endophytes are found associated with the plants in the Rhynie chert, a majority are represented by isolated fragments or incomplete stages of their life cycle, which makes detailed analysis challenging. Nevertheless, several instances have been thoroughly documented, revealing a consistent spatial arrangement of the fungi within their plant hosts (Krings et al. 2012).

The early stages of plant evolution were heavily influenced by symbiotic interactions with fungi, which were crucial for the colonization of land. Early land plants, especially bryophytes (liverworts, mosses, and hornworts) and lycophytes, provide valuable insights into the origins and evolutionary history of these symbioses. The establishment of these fungal partnerships was critical for overcoming nutrient limitations in terrestrial habitats, particularly in nutrient-poor soils. As highlighted by Rimington et al. (2020), fungal symbioses were likely indispensable for plants to survive in these challenging environments. These interactions were essential for the evolution of plants, contributing to ecological services such as nutrient acquisition and protection against abiotic stresses. In line with that, Pressel et al. (2010) emphasize the pivotal role of fungi in the early stages of plant evolution, allowing the first land plants to thrive. Moreover, fungi helped early plants adapt to terrestrial environments by aiding in nutrient uptake, especially nitrogen and phosphorus, which are often limited in early terrestrial ecosystems (Field et al. 2012). The symbiotic relationships were particularly important for plants inhabiting challenging environments, such as arid or low-nutrient areas (Bidartondo et al. 2011).

3.1 Diversity of Fungal Endophytes in Early Land Plants

Fossil evidence and molecular phylogenies suggest that arbuscular mycorrhizal fungi (AMF) were the first symbiotic fungi to establish partnerships with land plants, enabling the colonization of terrestrial ecosystems (Rimington et al. 2020). These partnerships were essential for the early plants' success, particularly in nutrient-poor soils where AMF facilitated the acquisition of vital nutrients such as phosphorus (Field et al. 2015). Over time, evolutionary transitions from AMF to other fungal groups, such as Mucoromycotina and Basidiomycota, likely reflected adaptations to changing environmental conditions, including fluctuations in soil nutrient availability and climate (Field et al. 2015; Tedersoo 2017). This evolutionary flexibility suggests that early plants had the capacity to switch symbiotic partners in response to ecological pressures, enhancing their survival and diversification (Bidartondo et al. 2011).

3.2 Ecological Distribution of Fungal Endophytes

The distribution of these fungal endophytes varies widely across plant taxa and is shaped by ecological factors such as soil conditions, moisture, and climate (Brundrett 2009; Field et al. 2015). AMF are predominantly found in tropical and subtropical regions, where nutrient limitations are particularly pronounced (Rimington et al. 2020). Mucoromycotina-like fungi, on the other hand, are more commonly found in temperate zones and in plants that grow in shaded or low-nutrient environments. These fungi contribute significantly to the nutrient cycling in these regions, enhancing plant growth in ecosystems where soil nutrients are scarce (Field et al. 2015). In contrast, Basidiomycota and Ascomycota fungi are often restricted to specific ecological niches, such as peat bogs or montane forests, where specialized conditions foster their growth and symbiotic interactions. These patterns highlight the role of environmental factors in determining the distribution of fungal symbioses. Soil type, moisture availability, and climate conditions strongly influence the types of fungal partnerships established by early plants. Studies by Janowski and Leski (2022) and others further confirm the role of these factors in shaping fungal-plant interactions.

3.3 Functional Roles of Fungal Endophytes

The functional roles of fungal endophytes in early land plants are critical for their survival and growth in nutrient-limited environments. AMF and Mucoromycotina fungi are especially important for enhancing nutrient acquisition, particularly phosphorus and nitrogen and micronutrients essential for plant growth (Field et al. 2012). In return, the fungi received a steady supply of photosynthetically derived sugars, which supported their growth and reproduction. Such interactions likely alleviated nutrient limitations, enabling early plants to colonize and persist in terrestrial environments with poor soil fertility. In addition to nutrient acquisition, these fungal partnerships provide stress tolerance, enabling plants to withstand environmental challenges such as drought and pathogen attack (Rimington et al. 2020). Furthermore, fungal endophytes played a key role in ecosystem engineering by promoting soil formation and enhancing the overall structure of early terrestrial ecosystems (Field et al. 2012). These functional roles are reflected in more recent studies, exploring the specific contributions of mycorrhizal fungi to ecosystem processes such as nutrient cycling and soil aggregation. In fact, fungal activity promoted the formation of soil organic matter, which enhanced water retention and nutrient availability, creating a more hospitable environment for other organisms (Rashid et al. 2016). These interactions played a crucial role in fostering conditions that supported the evolution and diversification of more complex vascular plants, ultimately giving rise to the verdant, forested landscapes observed today (Taylor et al. 2005a, b). Furthermore, the decomposition of fungal structures likely played a role in early carbon cycling, influencing global climate systems.

3.4 Saprotrophic Interactions Between Land Plants and Fungi

Saprotrophism, the decomposition of organic matter by fungi, is one of the most common interactions between land plants and fungi in ancient ecosystems.

However, distinguishing saprotrophic relationships from other plant–fungal interactions, such as mutualism or parasitism, in the fossil record is challenging due to the difficulty in determining fungal life histories and host conditions at the time of colonization (Taylor and Krings 2005; Krings et al. 2009).

Permineralized peat from the Carboniferous, such as coal balls, and silicified peat from the Permian and Triassic also provide evidence of saprotrophic fungi, with fossilized spore-like structures and sporocarps (fungal reproductive structures) found within decaying plant remains, indicating that fungi played a crucial role in plant matter degradation and the carbon cycle in early terrestrial ecosystems (Davis and Leisman 1962; Krings et al. 2010). Despite their ecological significance, saprotrophic fungi are underrepresented in the fossil record, likely due to preservation biases and methodological challenges in studying fossilized fungi (Taylor et al. 2011). Additionally, distinguishing whether fungal colonization occurred before or after plant tissue decay remains problematic (Zhou and Hyde 2001; Hyde and Soytong 2008). Nevertheless, fungal saprotrophism has played a fundamental role in ancient ecosystems by decomposing organic matter and recycling nutrients. Although still poorly understood in the geological past, these interactions were likely as complex as those observed in modern ecosystems (Dighton 2003; Purin and Rillig 2008). Continued research into fossilized fungi and their associations with plants will help uncover the evolutionary history of these essential decomposers.

4 Fungal Community in the Rhynie Chert

4.1 Geological and Ecological Context of the Rhynie Chert

The *Rhynie chert* of the Lower Devonian age (Wellman 2006), located in Aberdeenshire, Scotland, represents a significant geological formation that offers critical insights into the early evolution of continental plant, animal, and microbial life in ancient freshwater and terrestrial environments. This chert is particularly renowned for its exceptional preservation of microbial diversity, encompassing bacteria, cyanobacteria, eukaryotic algae, fungi, and various other microorganisms. Fungi, in particular, exhibit remarkable diversity, with a wide range of reproductive structures, mycelial filaments, and mycelial cords found embedded within the chert. These fungal associations, including parasitic, mycorrhizal, and saprotrophic interactions, underscore fungi's essential role within the Rhynie ecosystem.

The *Rhynie chert* Lagerstätte, situated in northern Aberdeenshire, Scotland, was formed within an alluvial plain adjacent to geothermal wetland systems, with alkaline hot springs crucial in its preservation. Fossils are primarily found within chert lenses that accumulate due to intermittent flooding by silica-rich waters or groundwater. The biota preserved in the Rhynie chert dates from the Early Pragian to the Earliest Emsian period, approximately 411.5 ± 1.3 million years ago, as we mentioned before. Fossils were examined using transmitted-light microscopy and digital imaging, with specimens currently housed in several paleontological collections (Saladin et al. 2017; Krings 2021; Reinhardt et al. 2021; Loron et al. 2023).

4.2 Fungal Diversity in the Rhynie Chert

4.2.1 Vegetative Structures

Fragments of fungal filaments, mycelia, and hyphae are commonly encountered within the *Rhynie chert*. These include septate and nonseptate hyphae, mycelial cords, and occasional structures reminiscent of nematophytes. Mycelial cords, composed of hyphae arranged in parallel orientation, are particularly noteworthy, as they likely served an ecological function akin to modern fungal mycelia, assisting in transporting nutrients and water.

4.2.2 Reproductive Structures

The *Rhynie chert* is distinguished by the presence of small, diverse fungal reproductive units, which are often detached from their original systems, posing challenges to their taxonomic classification. Certain reproductive forms, such as chlamydospores and glomeromycotinan spores, exhibit distinctive characteristics, including multilayered walls or germination shields, suggesting affiliations with specific fungal groups, such as *Palaeomyces* or glomeromycotinan fungi. Other reproductive structures, such as those associated with *Zwergimyces vestitus* and *Mycocarpon rhyniense*, display varying degrees of hyphal investment, which aid in their differentiation from one another. These investment structures bear resemblance to those found in modern fungi, particularly glomeromycotinan and mucoromycotinan species.

4.2.3 Fungal Interactions and Symbioses

The *Rhynie chert* offers rich evidence of diverse fungal interactions with terrestrial plants, fungi, charophytes, animals, and cyanobacteria. By drawing comparisons with modern ecosystems, these fossilized associations provide valuable insights into the nutritional interrelationships that existed in ancient times.

(i) *Fungi–Terrestrial Plants:* The most prominent fungal interaction identified in the *Rhynie chert* is the endomycorrhizal relationship in *Aglaophyton majus*, involving the fungus *Glomus rhyniensis* (*Glomeromycotina*). This fungus forms an extramatrical mycelium that penetrates the plant's stomata, spreading throughout the cortex and forming vesicles, glomoid spores, and arbuscules. Occasionally, a filamentous cyanobacterium is also observed within the plant, although its relationship with the mycorrhizal system remains undetermined. The discovery of endomycorrhiza in *A. majus* offers significant insights into the early evolution of mycorrhizal networks. Other plants from the *Rhynie chert*, such as *Rhynia gwynne-vaughanii* and *N. aphylla*, also exhibit fungal associations, potentially including endomycorrhizal connections. Additionally, fungi like *Palaeoglomus boullardii* and *Palaeoendogone gwynne-vaughaniae* suggest that both Glomeromycotina and Mucoromycotina fungi formed mutualistic relationships with these early plants.

(ii) *Fungi–Fungi:* The *Rhynie chert* also reveals an array of interfungal associations, such as mycelial penetration within large fungal spores and hyphae invading glomeromycotinan vesicles. Chytrid-like organisms parasitize fungal spores, often forming epibiotic zoosporangia and rhizomycelia that extend into the spore lumen. *Globicultrix nugax*, a polycentric chytrid-like fungus, is notably found within glomeromycotinan spores. Other fungal species show host responses characterized by the development of inwardly directed pegs or papillae that encapsulate the invading fungal hyphae.

(iii) *Fungi–Charophytes:* Chytrid-like fungi, such as *Milleromyces rhyniensis*, parasitize the charophyte *Palaeonitella cranii*, forming spheroidal zoosporangia with discharge tubes. Other parasitic chytrids, such as *Lyonomyces pyriformis* and *Krispiromyces discoides*, induce hypertrophic responses in host cells, akin to the reactions observed in modern *Chara* species under chytrid parasitism.

(iv) *Fungi–Animals:* A monocentric chytrid, *Cultoraquaticus trewinii*, resembling those parasitizing *P. cranii* and fungal spores, has been documented. The zoosporangia of this chytrid are found alongside resting eggs of the crustacean *Lepidocaris rhyniensis*, suggesting that chytrids played a significant role in early aquatic food webs. Additionally, fungal hyphae and spores have been discovered in coprolites, providing direct evidence of fungi within ancient food chains.

(v) *Fungi–Cyanobacteria:* A cyanolichen-like symbiosis, termed *Winfrenatia reticulata*, is found in the Rhynie chert. This association consists of parallel hyphal layers featuring a unique pattern of ridges and depressions on the surface, indicative of a complex ecological relationship between cyanobacteria and fungi (Hoysted et al. 2023; Krings et al. 2007a, b; Tiwari and Bae 2022; Schardl et al. 2004; Selosse et al. 2006; Chen et al. 2018).

5 Spatial Distribution of Fungal Endophytes

The spatial distribution of fungal endophytes is nonrandom and is determined by multiple factors, including host species, tissue type, environmental gradients, and ecological interactions.

Decades of research have shown that fungal endophytes are widely distributed across different plant groups and environmental conditions (Singh et al. 2017). Their distribution can vary in plants due to the chemical nature of host (plant) tissues, and environmental factors such as salinity, temperature, and seasonal variations. The symbiotic relationship between plants and endophytic fungi typically does not cause visible disease symptoms. This relationship is key in improving plant resilience to stress, reducing disease susceptibility, enhancing biomass, and influencing plant biochemical pathways and functional traits (Rodriguez et al. 2009; Sun et al. 2012; Harrison and Griffin 2020).

The study of fungal endophyte distribution is essential for understanding their ecological roles, including their contributions to plant health, nutrient cycling, and ecosystem resilience. This chapter provides an overview of the spatial distribution patterns of fungal endophytes at different scales and discusses the underlying factors that drive these patterns.

5.1 Factors Affecting the Spatial Distribution of Fungal Endophytes

Their spatial distribution is shaped by a complex interplay of biotic and abiotic factors, influencing their community structure, diversity, and ecological roles.

5.1.1 Host Plant Phenotype and Genotype

The identity and traits of host plant species are key factors influencing the distribution of fungal endophytes. Various plants can support different endophyte communities due to their specific biochemical characteristics, which shape the microhabitats for colonization. For example, certain grasses may host endophytes that align with their unique physiological features, while woody plants may support a different range of fungi (Arnold and Lutzoni 2007). Additionally, genetic diversity within plant species plays a role in the variation of endophyte communities. Different genotypes can create diverse chemical environments that either foster or hinder fungal colonization (Saikkonen et al. 2010).

The host plant species is crucial in determining the endophytic community. Distinct plant species can support unique endophyte communities due to differences in physiology, biochemistry, and immune responses. Plants produce secondary metabolites that can either attract or repel specific fungi, thus affecting the diversity and composition of endophytes (Arnold and Lutzoni 2007). Genetic variations within plant species can also result in differences in endophyte communities, as various genotypes may provide unique niches or chemical environments (Saikkonen et al. 2010).

5.1.2 Environmental Factors

They include abiotic factors such as temperature, humidity, soil type, and light availability which play significant roles in shaping the distribution of fungal endophytes. Moisture levels are particularly crucial; higher humidity is often associated with increased endophyte diversity as many fungi thrive in moist conditions (Rodriguez et al. 2009). Temperature variations can also impact fungal growth rates and community structures, leading to distinct endophyte populations across different climatic zones. For example, tropical ecosystems generally exhibit higher endophyte diversity than temperate ones due to more stable and favorable conditions for fungal growth. Soil type is also one of the main environmental factors that may have a significant effect on the endophytic communities. Soil is considered as one of the main sources of endophytes in roots and leaves (Compant et al. 2010). Furthermore, soil type affects host plant physiology, phenology, and biochemical characteristics, which in turn could influence endophytes. Most studies have reported that soil physical and chemical properties are the main environmental factors influencing diversity of mycorrhizal fungi and dark septate root endophytes (Pandey et al. 2016; Oehl et al. 2010; Sun et al. 2012).

5.1.3 Geographic and Biogeographic Factors

Geographic location profoundly affects the spatial distribution of fungal endophytes. Biogeographic history, including past climate changes and continental drift, has shaped endophytes' distribution patterns and evolutionary trajectories by creating isolated populations and unique evolutionary pressures (Huang et al. 2020). Research has shown that endophyte communities can vary significantly between regions, reflecting the local ecology and evolutionary history. For instance, studies have indicated that endophyte diversity is typically greater in tropical regions than temperate areas, likely due to the higher biodiversity and more favorable climatic conditions in these regions (Huang et al. 2020).

Different geographical distributions also exist in the same kind of plants. Early in 1994, Fisher et al. (1994) detected that endophytic fungal communities in leaves of *Quercus ilex* from England, Majorca, and Switzerland were significantly different. Vega et al. (2010) also detected those endophytic fungi diversity was significantly different in coffee plants from Colombia, Hawaii, Mexico, and Puerto Rico. Investigation of endophytic fungi communities of *Taxus chinensis* var. *mairei*, distributed in Jiangxi, Zhejiang, and Chongqing regions of China, gave similar results (Wu et al. 2013). Semiarid and arid lands cover approximately 33% of the Earth's surface area (Ali et al. 2018), but the diversity, community composition and functional traits of endophytic fungi living within desert plants are poorly investigated compared with other terrestrial habitats. Desert plants endure physiological and ecological challenges such as lack of water availability, extreme temperatures, intense solar radiation, nutrient-poor soils, etc. (Massimo et al. 2015).

5.1.4 Plant Age and Developmental Stage

The age of the host plant can influence the composition and abundance of its endophytic community. Younger plants may be more susceptible to colonization by a wider range of endophytes, while mature plants often host more stable communities (Kerry 2000). As plants develop, the microenvironments within plant tissues change, affecting the availability of niches for fungal colonization. This dynamic can lead to shifts in endophyte populations throughout the plant's life cycle (Suryanarayanan et al. 2012). Endophytic fungal communities have different distributions in different tissues of a single tree species. Endophytic fungi of the same plant species in the same area are similar, but for some fungi, the abundance and distribution differ with the age and the tissue of plants (Arnold and Herre 2003; Mei et al. 2014).

Normally, the species and abundance of endophytic fungi increase as the host ages. There are two main reasons for this: on the one hand, fungi spread by air and rainfall afford older trees the opportunity for repeat infection; on the other hand, changes in plant physiological status and tree bark structure in aging trees create new access in the plant tissue, allowing fungal invasion (Fisher et al. 1994).

The distribution of endophytic fungal communities can also be affected by crown height and canopy cover (Arnold and Herre 2003). Commonly, the species diversity and abundance of endophytic fungi are higher in samples taken from densely wooded sites than in samples taken from more open sites (Petrini et al. 1982). Generally, the colonization rate of endophytic fungi is significantly higher in the stems than in the leaves.

5.1.5 Biotic Interactions

Fungal endophytes do not exist in isolation, they interact with other microorganisms, including bacteria, other fungi, and even viruses. These interactions can be competitive, antagonistic, or mutualistic, each influencing endophyte distribution. For example, bacterial communities in the rhizosphere can inhibit or facilitate fungal colonization, thus shaping endophyte communities (Huang et al. 2020). Symbiotic relationships, such as those with mycorrhizal fungi, can also affect endophyte dynamics distribution by altering nutrient availability and host plant health.

5.1.6 Anthropogenic Effects

Human activities, such as agriculture, deforestation, and urbanization, have significant impacts on the spatial distribution of fungal endophytes. Agricultural practices, particularly monoculture systems, can lead to a decline in plant diversity, resulting in simplified endophyte communities (Bardgett and Van Der Putten 2014). Additionally, land-use changes can alter local environmental conditions, further affecting endophyte diversity and distribution. The introduction of nonnative plant species can disrupt existing endophyte communities and create new ecological dynamics (Fang et al. 2023).

5.1.7 Climate Change

Climate change is reshaping ecosystems globally and presents new challenges for fungal endophytes. Alterations in temperature and precipitation patterns can lead to shifts in suitable habitats for various endophyte species, potentially resulting in changes in community composition and distribution. As climate conditions fluctuate, endophytes may be forced to adapt to new environments, which could affect their interactions with host plants and other microorganisms (Giauque and Hawkes 2013).

5.2 *Spatial Distribution Within Plant Tissues*

The distribution of fungal endophytes within plant tissues is highly heterogeneous. Different tissues, such as roots, stems, leaves, and seeds, exhibit distinct fungal communities.

5.2.1 Root Endophytes

Root tissues are a hotspot for fungal endophytes due to their direct interaction with the soil microbiome. Arbuscular mycorrhizal fungi and dark septate endophytes are commonly found in roots, aiding in nutrient acquisition and stress tolerance (Smith and Read 2008). Root endophyte communities are influenced by soil composition, pH, and plant root exudates (Kowalski et al. 2015).

5.2.2 Leaf Endophytes

Leaves include a diverse array of endophytes, primarily Ascomycetes. The composition of leaf endophyte communities is influenced by factors such as light availability, leaf age, and microclimatic conditions (Arnold et al. 2000). For instance, younger leaves often exhibit lower fungal diversity compared to older, senescing leaves (Helander et al. 2013).

5.2.3 Stem and Seed Endophytes

Stem tissues often have lower fungal diversity compared to roots and leaves, possibly due to limited nutrient availability and structural barriers (Hardoim et al. 2015). Seeds, on the other hand, can host vertically transmitted fungal endophytes that play critical roles in seed germination and early plant development (Truyens et al. 2015).

5.2.4 Spatial Distribution Across Plant Species

The host plant species is a significant determinant of fungal endophyte diversity and composition. Host specificity can range from broad (generalist endophytes) to narrow (specialist endophytes). For example, Arnold and Lutzoni (2007) demonstrated that plant species within the same family often host similar endophyte communities, suggesting phylogenetic constraints on endophyte colonization.

5.3 Geographical and Environmental Influences on Distribution

Geographical location and environmental conditions strongly influence fungal endophyte communities.

5.3.1 Biogeographical Patterns

Endophyte diversity tends to be higher in tropical regions compared to temperate zones, correlating with increased plant biodiversity (Arnold et al. 2000). Unique fungal taxa have been documented in extreme environments, such as arctic tundras and desert ecosystems, where they contribute to plant adaptation to harsh conditions (Fig. 1).

5.3.2 Environmental Gradients

Abiotic factors such as temperature, humidity, and soil characteristics shape fungal endophyte communities. For instance, arid environments favor xerophilic endophytes, while humid environments support a greater abundance of mesophilic fungi (Newsham 2011).

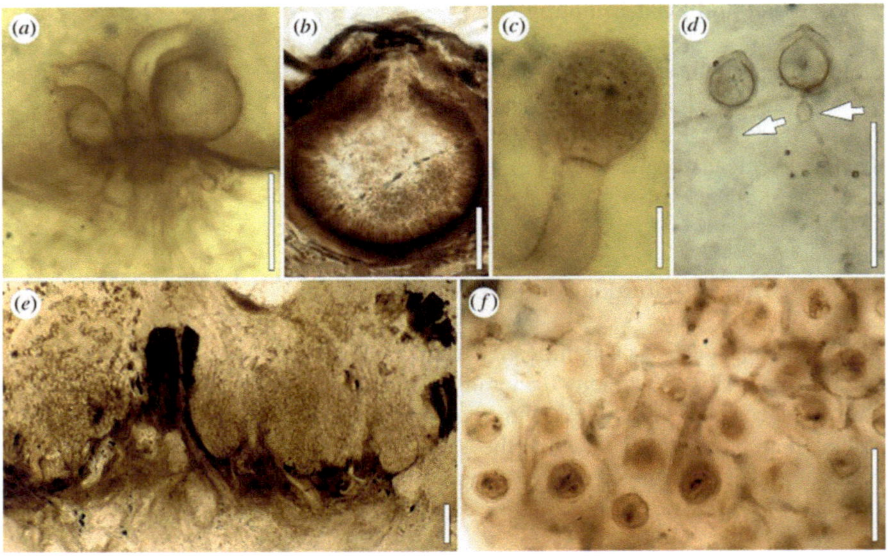

Fig. 1 Fungal associations and interactions within the Lower Devonian Rhynie chert: (**a**) *Trewinomyces annulifer* proliferating from the surface of a land plant axis (BSPG 1965 I 336, scale bar = 50 μm). (**b**) Longitudinal section of the perithecium of *Paleopyrenomycites devonicus* in *A. mackiei* (courtesy of H. Kerp & H. Hass, P3411, scale bar = 100 μm). (**c**) Zoosporangium of *Paleopyrenomycites milleri* (P2054, scale bar = 10 μm). (**d**) Monocentric chytrid fungi on *Paleopyrenomycites cranii* exhibiting endobiotic apophysis and rhizomycelium (BSPG 2016 VII 5, scale bar = 50 μm). (**e**) Thallus of *W. reticulata* with hyphal cavities containing cyanobacteria (P1323, scale bar = 500 μm). (**f**) Hyphal network of *W. reticulata* enveloping cyanobacterial cells (P1386, scale bar = 15 μm). (Krings et al. 2017)

5.4 Mechanisms Driving Spatial Distribution

Several mechanisms influence the spatial distribution of fungal endophytes (Tables 1, 2, 3, and 4):

- *Horizontal transmission*: Fungal endophytes can be horizontally transmitted through spores dispersed by wind, water, or insect vectors (Card et al. 2016).
- *Vertical transmission*: Some endophytes are inherited from parent plants through seeds, ensuring their presence in subsequent generations (Hodgson et al. 2014).
- *Environmental filtering*: Abiotic factors act as filters, allowing only certain fungal taxa to thrive in specific environments (Geml et al. 2022).

The spatial distribution of fungal endophytes is a dynamic and multifaceted phenomenon influenced by host plant characteristics, environmental gradients, biogeographical factors, developmental stages, biotic interactions, human activities, and climate change. These symbiotic microorganisms exhibit remarkable diversity across plant tissues, with roots, leaves, stems, and seeds hosting distinct fungal communities. Root tissues are dominated by nutrient-acquiring fungi like arbuscular

Table 1 Early land plant–fungal interactions

Interaction type	Description	Impact on plants	References
Arbuscular mycorrhizal (AM) symbiosis	Mutualistic association between early land plants and Glomeromycotan fungi	Improved nutrient acquisition, stress tolerance	Field et al. (2015), Taylor et al. (2005a, b)
Mucoromycotina symbiosis	Early mutualistic fungi alternative to AM in some ancient plant lineages	Phosphate acquisition, enhanced growth	Field et al. (2015)
Endophytic fungi	Fungal colonization of plant tissues, often conferring benefits like stress resistance	Protection against pathogens, hormone regulation	Krings et al. (2009), Priyashantha et al. (2023)
Saprotrophic fungi	Decomposers breaking down organic matter, contributing to nutrient cycling	Increased soil nutrient availability by breaking down decomposing organic matter	Zhou and Hyde (2001)
Parasitic fungi	Fungi exploiting plant hosts, often pathogenic in nature	Physiological stress, weakened growth	Purin and Rillig (2008), Krings et al. (2010)

mycorrhizal fungi and dark septate endophytes, while leaves host ascomycetes that contribute to disease resistance and stress tolerance. Host-specificity and genetic diversity within plant species further shape endophyte communities by providing unique chemical and structural niches. Environmental factors such as soil composition, humidity, temperature, and light availability play critical roles in determining fungal endophyte diversity and community structure. Geographically, tropical ecosystems exhibit the highest diversity, driven by stable climates and abundant plant biodiversity, whereas arid and temperate regions support specialized and resilient fungal communities adapted to extreme conditions. The developmental stage of plants also influences endophyte colonization, with older plants hosting more stable and diverse communities due to repeated exposure and physiological changes over time. Biotic interactions, including competition and mutualism with other microorganisms, significantly impact fungal endophyte dynamics, while anthropogenic activities such as agriculture and deforestation alter community structures, often reducing diversity. Additionally, climate change poses new challenges, driving shifts in fungal distribution and potentially reshaping plant-endophyte relationships. Understanding the spatial distribution of fungal endophytes is critical for unlocking their ecological and functional potential. These fungi contribute to plant health, stress tolerance, and ecosystem stability, offering promising applications in sustainable agriculture, forestry, and environmental remediation. Future research should focus on integrating advanced molecular techniques, ecological modeling, and functional studies to further elucidate the intricate relationships between fungal endophytes, their host plants, and the environment. The future of fungal endophyte

Table 2 Synopsis of fungal taxa and fungus-like organisms described from the Rhynie chert (Krings et al. 2018)

Taxon	Suggested systematic affinities	Occurrence/substrate
Fungi		
Cultoraquaticus trewini	Chytridiomycota	Epibiotic on large, spheroidal structures of uncertain affinity. Intermixed with crustacean resting eggs in freshwater environments
Globicultrix nugax	Chytridiomycota	Endobiotic in large fungal (probably glomeromycotinan) spores
Glomites rhyniensis	Mucoromycota (Glomeromycotina)	In axes of *A. majus*
Glomites sporocarpoides	Mucoromycota (Glomeromycotina)	In axes of *R. gwynne-vaughanii* and *A. majus*
Helmutella devonica	Mucoromycota incertae sedis	Free-floating in chert matrix
Illmanomyces corniger	Chytridiomycota	Epibiotic on fungal (probably glomeromycotinan) spores
K. discoides	Chytridiomycota	Extramatrical, associated with *P. cranii*
Kryphiomyces catenulatus	Inconclusive, perhaps Chytridiomycota	Endobiotic in large fungal (probably glomeromycotinan) spores
L. pyriformis	Chytridiomycota	Embedded in the surface coating of *P. cranii* cell walls
M. rhyniensis	Chytridiomycota	Endobiotic in cells of the green alga *P. cranii*
M. rhyniense	Mucoromycota incertae sedis	Free-floating in chert matrix
Mycokidstonia sphaerialoides	Mucoromycota (Glomeromycotina)	Free-floating in chert matrix
P. gwynne-vaughaniae	Mucoromycota (Mucoromycotina)	In rhizomes of *H. lignieri*
P. boullardii	Mucoromycota (Glomeromycotina)	In aerial axes of *H. lignieri*
Palaeomyces agglomeratus	Mucoromycota incertae sedis	In aerial axes of *R. gwynne-vaughanii* and surrounding chert matrix
Palaeomyces asteroxyli		In intact and decayed tissue of *Asteroxylon mackiei*
Palaeomyces gordonii		In axes of *A. mackiei*, free-floating in chert matrix
Palaeomyces horneae		In rhizomes and aerial axes of *H. lignieri*
Palaeomyces simpsonii		In decayed axes of *R. gwynne-vaughanii*
Palaeozoosporites renaultii	Zoosporic fungi incertae sedis, perhaps Blastocladiomycota	Endobiotic in rhizomes of *A. mackiei*

(continued)

Table 2 (continued)

Taxon	Suggested systematic affinities	Occurrence/substrate
Paleoblastocladia milleri	Blastocladiomycota	Epidictic on partially degraded axes of *A. majus*
P. devonicus	Ascomycota	In aerial axes and lateral portions of *A. mackiei*
Perexiflasca tayloriana	Inconclusive, perhaps Chytridiomycota	In intact and degraded land plant tissue, on fungal spores and hyphae, free-floating in chert matrix
Perithecial ascomycetes	Ascomycota	Cortex of aerial stems and rhizomes of *Asteroxylon*
P. asteroxylicola	Ascomycota incertae sedis (Dikarya)	Aerial axes and leaf-like appendages of *Asteroxylon mackiei*
Rhizophydites shutei	Chytridiomycota	On glomeromycotan acaulospores
Scepasmatocarpion fenestrulatum	Inconclusive, perhaps Mucoromycota or Ascomycota	Free-floating in microbial mats
Scutellosporites devonicus	Mucoromycota (Glomeromycotina)	In aerial axes of *A. mackiei*
T. annulifer	Inconclusive, perhaps Chytridiomycota or Blastocladiomycota	Epibiotic on partially degraded land plant axes
Zwergimyces vestitus	Mucoromycota incertae sedis	In intact and degraded land plant tissues, in chert litter layers
Lichen-like associations		
Winfrenatia reticulata	Mycobiont: Mucoromycota photobiont: Cyanobacteria	Inconclusive, probably on hard terrestrial substrate
Eoentophysalis	Cyanobacteria	Colony-forming, similar to modern Entophysalis
Peronosporomycetes		
Frankbaronia polyspora	Peronosporomycetes (Oomycota)	Free-floating in microbial mats and litter layers
Frankbaronia velata		Free-floating in microbial mats
Hassiella monospora		Free-floating in chert matrix
Nematophytes		
Nematophyton taiti	Nematophyta	Inconclusive, probably on hard substrate
Nematoplexus rhyniensis	Nematophyta	Inconclusive, probably on hard substrate

research is promising and multidimensional. By integrating interdisciplinary approaches and focusing on practical applications, researchers can harness the potential of these microorganisms to address pressing global challenges in agriculture, ecosystem management, and environmental conservation.

Table 3 Distribution of endophytes across plant tissues

Tissue type	Dominant fungal endophytes	Functions	References
Roots	Dark septate endophytes and arbuscular mycorrhizal fungi	Nutrient acquisition, drought tolerance	Smith and Read (2008)
Leaves	Ascomycetes	Disease resistance, growth promotion	Arnold et al. (2000)
Stems	Fungi from *Phoma* and *Fusarium*	Structural integrity, stress resilience	Hardoim et al. (2015)
Seeds	*Acremonium*, *Fusarium*	Enhanced germination under stress	Truyens et al. (2015)

Table 4 Environmental influences on fungal endophyte distribution

Environment	Dominant fungal groups	Adaptations	References
Tropical forests	Ascomycetes	Host-specific, high diversity	Arnold et al. (2000)
Temperate forests	Generalist ascomycetes	Cold tolerance, lower diversity	Arnold and Lutzoni (2007)
Deserts	Dark septate endophytes	Drought resistance, low water requirements	Newsham (2011)
Arctic tundra	Basidiomycetes, psychrophilic fungi	Cold adaptation, nutrient recycling	Rivkina et al. (2000)

6 Mechanisms of Host Response Pathways Between Fungal Endophyte and Plants

Land plants have devised coping mechanisms to cohabitate with endophytic microorganisms, including fungi. These coping mechanisms deployed right from recognition of the invading microbe/s to penetration, invasion, and their colonization have been reported on early and modern land plants dating back over 400 million years (Krings et al. 2007a, b).

6.1 Host Recognition and Response to Penetration by Endophytic Fungi

Majority of endophytic fungi are brought in contact with their hosts from the prevailing environment of their host plants, while others remain in plants following vertical transmission in seeds of parent materials (Hodgson et al. 2014; Ji et al. 2022). Chemical cues released by plants facilitate their detection by fungal endophytes (Besserer et al. 2006). For example, the well-known Chinese herb, *Gastrodia elata,* first establishes a relationship with its endophytic fungi, such as *Armillaria mellea*, by secreting strigolactones to attract the fungus, which promotes its colonization, consequently increasing the likelihood of it coming into contact with the host plant (Hua et al. 2024).

Following contact with the host, penetration of the fungal endophytes can be achieved through wounds, natural openings, or enzymatic breakdown of the plant cell wall aided by cuticle-degrading enzymes (e.g., cellulase and pectinase), mechanical force, and turgor pressure (Suryanarayanan et al. 2012; Chow and Ting 2019). The penetration of fungal endophyte triggers host's pattern recognition receptors (PRRs) to detect microbial-associated molecular patterns of the invading organism (Poveda et al. 2022). This extracellular recognition stimulates plant pattern-triggered immunity (PTI), resulting in an increase in cytosolic calcium levels as the first line of defense (Ji et al. 2022; Jha et al. 2023). Preexisting physical structures of plants, such as the cell wall as well as the cuticle, which covers natural openings such as hydathodes, leaves, and trichomes are also useful weapons against penetration by the invading microorganism (Ji et al. 2022). Nonetheless, contrary reports where *Trichoderma* spp. could colonize trichomes, leaves, and stem tissues of cocoa endophytically are noted (*Theobroma cacao*) (Bailey et al. 2009).

6.2 *Response of Host Plant to Invasion and Colonization by Endophytic Fungi*

While the first line of defense by the host plant (PT1) is intended to limit microbial invasion, many endophytic microbes, including fungi, secrete effector proteins and secondary metabolites (mainly alkaloids and terpenoids) to reduce or avoid them (Alam et al. 2021; Grabka et al. 2022). Thus, fungi that escape PTI could establish endophytically within the host plant. Consequently, an effector-triggered immunity (ETI), which is a stronger defense mechanism of the plant is activated, leading to the following: (1) reactive oxygen species (ROS) generation to promote oxidative cross-linking of cell wall polymers to fortify them against degradation and (2) accumulation of nitric oxides (precursors for plant defense hormones), and synthesis of secondary metabolites such as phenolics and phytoalexins (to act against the invaders). This in turn induces the production of plant hormones such as ethylene, jasmonic acid, and salicylic acid, in addition to callose deposition and stomatal closure (Bedini et al. 2018. Yuan et al. 2021a, b).

Krings et al. (2007a, b) studied petrographic thin sections of three fungal endophytes of the Early Devonian plant, *N. aphylla,* from the Rhynie chert and proposed the following as host responses following fungal infection (i) "bulging of infected rhizoids; (ii) formation of encasement layers around intercellular hyphae; and (iii) separation of infected from noninfected tissues by secondary thickening of cell walls." These observations are not different from those of modern plants under attack by microbes, including fungi (Grabka et al. 2022; Bard et al. 2024).

6.3 Adaptations to Maintain Symbiosis of Host Plants and Endophytic Fungi

Symbiosis, the interaction between dissimilar organisms, has been studied for over a century, revealing that most plants in natural ecosystems form symbiotic relationships with *mycorrhizal fungi* and/or *fungal endophytes*. These symbionts significantly impact plant ecology, fitness, and evolution, shaping plant communities and influencing associated organisms like bacteria, nematodes, and insects. Fossil records show that plants have been interacting with fungi for over 400 million years (Niazi 2024), playing a crucial role in land life evolution. Fungal endophytes differ from mycorrhizal fungi by residing entirely within plant tissues. There are two main groups: *Clavicipitaceous endophytes (C-endophytes)*, which infect grasses, and non-Clavicipitaceous endophytes *(NC-endophytes)*, found in a wide range of plants. While much research focuses on C-endophytes due to their agricultural importance, NC-endophytes are diverse and have distinct ecological roles (Rodriguez et al. 2009).

Despite the robust defense mechanisms mounted by the host plant/s, the endophytic fungi equally elicit coping mechanisms to remain symbiotic with the host. Armed with effector proteins, they interfere with plant signaling pathways, including jasmonic acid and salicylic acid, giving rise to tolerance instead of resistance (Ji et al. 2022). In the well-studied *Epichloë*-grass association, the symbiotic relationship between the endophytic symbiont and the host plant is generally maintained as follows: The symbiont ceases production of toxic proteins or secondary metabolites to avoid host defense responses; The intercalary hyphal growth is controlled; hyphae colonizes intercellular spaces of host aerial tissues only, do produce specialized feeding structures, rarely occur in host vascular tissues, develop only when leaves are growing, resulting in equal hyphal biomass in young and old leaves (Christensen et al. 2002; Scott et al. 2018).

Additionally, secondary metabolites produced by endophytic microbes are also well documented to prevent invasion and establishment of their counterpart plant pathogenic microbes (possibly via antibiosis) (Baiyee et al. 2019; Rodrigo et al. 2022). In the long run, these interactions result in improvements in growth, tolerance to biotic and abiotic stress, and the overall fitness of the plant (Rodriguez et al. 2009). In recent years, entomopathogens such as *Beauveria bassiana* and *Metarhizium anisopliae* and other species of both genera, traditionally known to be insect pathogens only, are now well documented to be capable of living as endophytes on virtually all agricultural crops (Vega et al. 2009), while exhibiting their insecticidal activities (Akello et al. 2008; Klieber and Reineke 2016). The endophytic potentials of fungal entomopathogens are gaining attention as an additional marketing tool for these biocontrol agents against UV radiation, which is their key abiotic environmental constraint in the field. In addition to their antiinsect herbivory benefits, endophytic fungal entomopathogens have been shown to play other roles in promoting plant growth, such as reducing the severity of nutrient deficiencies or transporting nitrogen from mycosed cadavers to plants (Behie et al. 2012; Raya-Díaz et al. 2017) providing protection against plant pathogens through antibiosis (Ownley et al. 2004) improving plant growth parameters (Liao et al. 2014; Lopez and Sword 2015), and producing plant growth regulators (e.g., indole acetic acid). As research unravels other nontraditional roles of endophytic fungi, in plants and their bioactive compounds, additional comprehensive studies will be needed to fully understand their mode of action, as well as host responses, especially for endophytic fungi used in biological control of agricultural pests and bioremediation.

7 Conclusions

The study of fungal endophytes and those preserved in the *Rhynie chert* provides a remarkable window into the symbiotic dynamics of Early Devonian ecosystems, approximately 407 million years ago. These ancient interactions reveal the profound role fungi played in facilitating the territorialization of plants, offering key insights into the evolutionary and ecological underpinnings of mutualism. The fossilized evidence demonstrates that *fungal endophytes* were integral to host plants, inhabiting rhizoids, stems, and cortical tissues, where they formed specialized structures reminiscent of modern arbuscular mycorrhizal associations. These relationships likely enhanced nutrient acquisition, bolstered host defense mechanisms, and contributed to the adaptability of plants in nutrient-poor, fluctuating Devonian environments. The adaptations observed within these symbiotic systems, including hyphal networks and vesicle-like storage structures, underscore the evolutionary antiquity of plant-fungal mutualism. These findings illuminate how co-evolutionary dynamics between fungi and plants laid the groundwork for more complex ecosystems and biotic interactions. Additionally, the intimate proximity and biotrophic interactions captured in the fossil record challenge traditional views of early terrestrial ecosystems as simplistic, instead portraying them as dynamic, interconnected networks where symbiosis played a foundational role in ecological resilience and productivity. Advances in methodologies such as high-resolution microscopy, synchrotron radiation, and chemical analyses have been instrumental in uncovering the morphological and biochemical properties of these ancient fungi. By bridging paleontological evidence with modern molecular and ecological approaches, researchers have reconstructed the evolutionary trajectory of these symbiotic systems, highlighting their relevance to contemporary ecological and evolutionary theories. Furthermore, the implications of these findings extend beyond Devonian landscapes, raising important questions about the role of fungi in shaping early biogeochemical cycles and influencing the diversification of terrestrial life. The *Rhynie chert fungal endophytes* exemplify the complexity of early life and the intricate interplay between organisms that shaped Earth's evolutionary history. Their study not only enhances our understanding of ancient ecosystems but also informs future research into the continuity of symbiotic mechanisms. Comparative analyses with extant fungi and contemporaneous fossil ecosystems, coupled with advanced imaging and isotopic techniques, could further unravel the metabolic exchanges and functional roles of these ancient symbionts. In summation, the *fungal endophytes* of the *Rhynie chert* are a cornerstone for understanding the origins of symbiosis and the evolutionary processes that underpin terrestrial ecosystems. Their fossilized legacy bridges the gap between ancient and modern mutualisms, illustrating the enduring significance of symbiotic relationships in driving ecological stability and evolutionary innovation. In addition, the study of fungal endophytes is rapidly evolving, driven by advancements in molecular biology, ecological modeling, and biotechnological applications. While significant progress has been made in understanding the spatial distribution and ecological roles of fungal endophytes, numerous gaps in knowledge remain, paving the way for future research. By exploring the profound interconnectedness of life through deep time, this research enriches our comprehension of the complex interplay between fungi and plants, both past and present.

References

Akello J, Dubois T, Coyne D, Kyamanywa S (2008) Endophytic *Beauveria bassiana* in banana (*Musa* spp.) reduces banana weevil (*Cosmopolites sordidus*) fitness and damage. Crop Prot 27(11):1437–1441. https://doi.org/10.1016/j.cropro.2008.07.003

Alam B, Lǐ J, Gě Q, Khan MA, Gōng J, Mehmood S et al (2021) Endophytic fungi: from symbiosis to secondary metabolite communications or vice versa? Front Plant Sci 12:791033. https://doi.org/10.3389/fpls.2021.791033

Ali BH, Mostafa A, Usama R, Soad E, Magdi AE (2018) Effect of Thermomyces fungal endophyte isolated from extreme hot desert-adapted plant on heat stress tolerance of cucumber. Appl Soil Ecol 124:155–162

Arnold AE, Herre EA (2003) Canopy cover and leaf age affect colonization by tropical fungal endophytes: ecological pattern and process in *Theobroma cacao* (Malvaceae). Mycologia 95: 388–398

Arnold AE, Lutzoni F (2007) Diversity and host specificity of foliar fungal endophytes: are there more than two sides to the story? Fungal Divers 35(1):93–103

Arnold AE, Mejía LC, Kyllo D, Rojas E, Maynard Z, Robbins N, Herre EA (2000) Fungal endophytes limit pathogen damage in a tropical tree. Proc Natl Acad Sci 100(26):15649–15654

Bailey BA, Strem MD, Wood D (2009) *Trichoderma* species form endophytic associations within *Theobroma cacao* trichomes. Mycol Res 113(12):1365–1376. https://doi.org/10.1016/j.mycres.2009.09.004

Baiyee B, Ito SI, Sunpapao A (2019) *Trichoderma asperellum* T1 mediated antifungal activity and induced defense response against leaf spot fungi in lettuce (Lactuca sativa L.). Physiol Mol Plant Pathol 106:96–101. https://doi.org/10.1016/j.pmpp.2018.12.009

Bard NW, Cronk QCB, Davies TJ (2024) Fungal endophytes can modulate plant invasion. Biol Rev Camb Philos Soc 99(5):1652–1671. https://doi.org/10.1111/brv.13085

Bardgett RD, Van Der Putten WH (2014) Belowground biodiversity and ecosystem functioning. Nature 515(7528):505–511

Bedini A, Mercy L, Schneider C, Franken P, Lucic-Mercy E (2018) Unraveling the initial plant hormone signaling, metabolic mechanisms and plant defense triggering the endomycorrhizal symbiosis behavior. Front Plant Sci 17(9):1800. https://doi.org/10.3389/fpls.2018.01800

Behie SW, Zelisko PM, Bidochka MJ (2012) Endophytic insect-parasitic fungi translocate nitrogen directly from insects to plants. Science 336(6088):1576–1577. https://doi.org/10.1126/science.1222289

Bengtson S, Rasmussen B, Ivarsson M, Muhling J, Broman C, Marone F, Bekker A (2017) Fungus-like mycelial fossils in 2.4-billion-year-old vesicular basalt. Nat Ecol Evol 1(6):0141. https://doi.org/10.1038/s41559-017-0141

Berbee ML, Taylor JW (2007) Rhynie Chert: a window into a lost world of complex plant-fungus interactions. New Phytol 174(3):475–479. https://doi.org/10.1111/j.1469-8137.2007.02080.x

Besserer A, Puech-Pagès V, Kiefer P, Gomez-Roldan V, Jauneau A, Roy S et al (2006) Strigolactones stimulate arbuscular mycorrhizal fungi by activating mitochondria. PLoS Biol 4(7):e226. https://doi.org/10.1371/journal.pbio.0040226

Bidartondo MI, Read DJ, Trappe JM, Merckx V, Ligrone R, Duckett JG (2011) The dawn of symbiosis between plants and fungi. Biol Lett 7:574–577

Bilański P, Kowalski T (2022) Fungal endophytes in Fraxinus excelsior petioles and their in vitro antagonistic potential against the ash dieback pathogen Hymenoscyphus fraxineus. Microbiol Res 257:126961. https://doi.org/10.1016/j.micres.2022.126961

Brundrett MC (2009) Mycorrhizal associations and other means of nutrition of vascular plants: understanding the global diversity of host plants by resolving conflicting information and developing reliable means of diagnosis. Plant Soil 320:37–77

Card S, Johnson L, Teasdale S, Caradus J (2016) Deciphering endophyte behaviour: the link between endophyte biology and efficacious biological control agents. FEMS Microbiol Ecol 92(8). https://doi.org/10.1093/femsec/fiw114

Channing A (2018) A review of active hot-spring analogues of Rhynie: environments, habitats and ecosystems. Philos Trans R Soc Lond Ser B Biol Sci 373(1739):20160490. https://doi.org/10.1098/rstb.2016.0490

Chen ECH, Morin E, Beaudet D, Noel J, Yildirir G, Ndikumana S, Charron P, St-Onge C, Giorgi J, Krüger M, Marton T, Ropars J, Grigoriev IV, Hainaut M, Henrissat B, Roux C, Martin F, Corradi N (2018) High intraspecific genome diversity in the model arbuscular mycorrhizal symbiont Rhizophagus irregularis. New Phytol 220(4):1161–1171. https://doi.org/10.1111/nph.14989

Chow YY, Ting ASY (2019) Influence of fungal infection on plant tissues: FTIR detects compositional changes to plant cell walls. Fungal Ecol 37:38–47

Christensen MJ, Bennett RJ, Schmid J (2002) Growth of Epichloë/Neotyphodium and p-endophytes in leaves of Lolium and Festuca grasses. Mycol Res 106(1):93–106. https://doi.org/10.1017/S095375620100510X

Collinge DB, Jensen B, Jørgensen HJ (2022) Fungal endophytes in plants and their relationship to plant disease. Curr Opin Microbiol 69:102177. https://doi.org/10.1016/j.mib.2022.102177

Compant S, Van Der Heijden MG, Sessitsch A (2010) Climate change effects on beneficial plant–microorganism interactions. FEMS Microbiol Ecol 73:197–214. https://doi.org/10.1111/j.1574-6941.2010.00900.x

D'Ario M, Lane B, Fioratti Junod M, Leslie A, Mosca G, Smith RS (2024) Hidden functional complexity in the flora of an early land ecosystem. New Phytol 241(2):937–949. https://doi.org/10.1111/nph.19228

Davis B, Leisman GA (1962) Further observations on Sporocarpon and allied genera. Bull Torrey Bot Club 89:97–109

Dighton J (2003) Fungi in ecosystem processes. In: Dekker M (ed) Fungi in ecosystem processes, 2nd edn. CRC Press, New York

Dunlop JA, Garwood RJ (2018) Terrestrial invertebrates in the Rhynie chert ecosystem. Philos Trans R Soc Lond Ser B Biol Sci 373(1739):20160493. https://doi.org/10.1098/rstb.2016.0493

Edwards D, Kenrick P, Dolan L (2018) History and contemporary significance of the Rhynie Cherts-our earliest preserved terrestrial ecosystem. Philos Trans R Soc Lond Ser B Biol Sci 373(1739):20160489. https://doi.org/10.1098/rstb.2016.0489

Fang K, Yang L, Li X, Zeng Y, Wang F, Li T, Zhao W, Zhang B (2023) Native plants change the endophyte assembly and growth of an invasive plant in response to climatic factors. Appl Environ Microbiol 89(10):e01093–e01023. https://doi.org/10.1128/aem.01093-23

Fay MF (2015) Flora of Great Britain and Ireland. Volume 2: Capparaceae-Rosaceae. https://doi.org/10.1111/boj.12311

Field KJ, Cameron DD, Leake JR, Tille S, Bidartondo MI, Beerling DJ (2012) Contrasting arbuscular mycorrhizal responses of vascular and non-vascular plants to a simulated Palaeozoic CO2 decline. Nat Commun 3:835

Field KJ, Pressel S, Duckett JG, Rimington WR, Bidartondo MI (2015) Symbiotic options for the conquest of land. Trends Ecol Evol 30(8):477–486. https://doi.org/10.1016/j.tree.2015.05.007

Fisher PJ, Petrini O, Petrini LE, Sutton BC (1994) Fungal endophytes from the leaves and twigs of *Quercus ilex* L. from England, Majorca and Switzerland. New Phytol 127(1):133–137

Geml J, Leal CM, Nagy R, Sulyok J (2022) Abiotic environmental factors drive the diversity, compositional dynamics and habitat preference of ectomycorrhizal fungi in Pannonian Forest types. Front Microbiol 13:1007935. https://doi.org/10.3389/fmicb.2022.1007935

Giauque H, Hawkes CV (2013) Climate affects symbiotic fungal endophyte diversity and performance. Am J Bot 100(7):1435–1444. https://doi.org/10.3732/ajb.1200568

Grabka R et al (2022) Fungal endophytes and their role in agricultural plant protection against pests and pathogens. Plants 11(3):384. https://doi.org/10.3390/plants11030384

Gupta M et al (2024) Lead toxicity in plants: mechanistic insights into toxicity, physiological responses of plants and mitigation strategies. Plant Signal Behav 19(1):2365576. https://doi.org/10.1080/15592324.2024.2365576

Hardoim PR, Van Overbeek LS, Van Elsas JD (2015) Properties of bacterial endophytes and their proposed role in plant growth. Trends Microbiol 23(10):749–761

Harrison JG, Griffin EA (2020) The diversity and distribution of endophytes across biomes, plant phylogeny and host tissues: how far have we come and where do we go from here? Environ Microbiol 22(6):2107–2123. https://doi.org/10.1111/1462-2920.14968

Helander M, Ahlholm J, Sieber TN, Hinneri S, Saikkonen K (2013) Endophyte-mediated resistance against pathogens in foliage and roots. FEMS Microbiol Ecol 83(3):377–389

Hodgson S, De Cates C, Hodgson J et al (2014) Vertical transmission of fungal endophytes is widespread in forbs. Ecol Evol 4(8):1199–1208. https://doi.org/10.1002/ece3.953

Hoysted GA, Field KJ, Sinanaj B, Bell CA, Bidartondo MI, Pressel S (2023) Direct nitrogen, phosphorus and carbon exchanges between Mucoromycotina 'fine root endophyte' fungi and a flowering plant in novel monoxenic cultures. New Phytol 238(1):70–79. https://doi.org/10.1111/nph.18630

Hua Z, Teng X, Huang J, Zhou J, Zhao Y, Huang L, Yuan Y (2024) The Armillaria response to *Gastrodia elata* is partially mediated by strigolactone-induced changes in reactive oxygen species. Microbiol Res 278:127536. https://doi.org/10.1016/j.micres.2023.127536

Huang LQ, Niu YC, Su L, Deng H, Lyu H (2020) The potential of endophytic fungi isolated from cucurbit plants for biocontrol of soilborne fungal diseases of cucumber. Microbiol Res 231:126369. https://doi.org/10.1016/j.micres.2019.126369

Hyde KD (2022) The numbers of fungi. Fungal Divers 114:1. https://doi.org/10.1007/s13225-022-00507-y

Hyde KD, Soytong K (2008) The fungal endophyte dilemma. Fungal Divers 33:163–173

Janowski D, Leski T (2022) Factors in the distribution of Mycorrhizal and soil fungi. Diversity 14:1122. https://doi.org/10.3390/d14121122

Jha P, Kaur T, Chhabra I, Panja A, Paul S, Kumar V, Malik T (2023) Endophytic fungi: hidden treasure chest of antimicrobial metabolites interrelationship of endophytes and metabolites. Front Microbiol 14:1227830. https://doi.org/10.3389/fmicb.2023.1227830

Ji X, Xia Y, Zhang H, Cui JL (2022) The microscopic mechanism between endophytic fungi and host plants: from recognition to building stable mutually beneficial relationships. Microbiol Res 261:127056. https://doi.org/10.1016/j.micres.2022.127056

Jin LR, Yang L, Li WJ, Xu D, Yang NN, Li GQ, Wan P (2021) Diversity and biocontrol potential of culturable endophytic fungi in cotton. Front Microbiol 12:698930. https://doi.org/10.3389/fmicb.2021.698930

Kerry BR (2000) Fungal endophyte: the role in plant health. Agric Sci 33(2):45–51

Klieber J, Reineke A (2016) The entomopathogenic *Beauveria bassiana* has epiphytic and endophytic activity against the tomato leaf miner *Tuta absoluta*. J Appl Entomol 140(8):580–589. https://doi.org/10.1111/jen.12287

Kowalski KP, Bacon C, Bickford W, Braun H, Clay K, Leduc-Lapierre M et al (2015) Advancing the science of microbial symbiosis to support invasive species management: a case study on Phragmites in the Great Lakes. Front Microbiol 6:95

Krings M, Harper CJ, Taylor EL (2017) Fungi and fungal interactions in the Rhynie chert: a review of the evidence, with the description of Perexiflasca tayloriana gen. et sp. nov.†. Philos Trans R Soc Lond B Biol Sci 373(1739):20160500. https://doi.org/10.1098/rstb.2016.0500

Krings M (2021) The Rhynie chert land plant Aglaophyton majus harbored cyanobacteria in necrotic local lesions. N Jb Geol Paläont Abh 300:279–289. https://doi.org/10.1127/njgpa/2021/0990

Krings M, Taylor TN, Hass H, Kerp H, Dotzler N, Hermsen EJ (2007a) Fungal endophytes in a 400-million-yr-old land plant: infection pathways, spatial distribution, and host responses. New Phytol 174(3):648–657. https://doi.org/10.1111/j.1469-8137.2007.02008.x

Krings M, Taylor TN, Hass H, Kerp H, Dotzler N, Hermsen EJ (2007b) An alternative mode of early land plant colonization by putative endomycorrhizal fungi. Plant Signal Behav 2(2): 125–126. https://doi.org/10.4161/psb.2.2.3970

Krings M, Dotzler N, Taylor TN, Gaktier J (2009) A late Pennsylvanian fungal leaf endophyte from grand-Croix, France. Rev Palaeobot Palynol 156:449–453

Krings M, Dotzler N, Taylor TN et al (2010) Microfungi from the upper Visean (Mississippian) of Central France: structure and development of the sporocarp Mycocarpon cinctum nov. sp. Zitteliana 50:127–135

Krings M, Taylor TN, Dotzler N (2012) Fungal endophytes as a driving force in land plant evolution: evidence from the fossil record. In: Biocomplexity of plant–fungal interactions, pp 5–27. https://doi.org/10.1002/9781118314364.ch1

Krings M, Harper CJ, Taylor EL (2018) Fungi and fungal interactions in the Rhynie Chert: a review of the evidence, with the description of *Perexiflasca tayloriana* gen. et sp. nov. Philos Trans R Soc Lond Ser B Biol Sci 373(1739):20160500. https://doi.org/10.1098/rstb.2016.0500

Liao X et al (2014) The plant beneficial effects of *Metarhizium* species correlate with their association with roots. Appl Microbiol Biotechnol 98(16):7089–7096. https://doi.org/10.1007/s00253-014-5788-2

Lopez DC, Sword GA (2015) The endophytic fungal entomopathogens *Beauveria bassiana* and *Purpureocillium lilacinum* enhance the growth of cultivated cotton (*Gossypium hirsutum*) and negatively affect survival of the cotton bollworm (*Helicoverpa zea*). Biol Control 89:53–60. https://doi.org/10.1016/j.biocontrol.2015.03.010

Loron CC, Rodriguez Dzul E, Orr PJ, Gromov AV, Fraser NC, McMahon S (2023) Molecular fingerprints resolve affinities of Rhynie chert organic fossils. Nat Commun 14(1):1387. https://doi.org/10.1038/s41467-023-37047-1

Massimo NC, Nandi Devan MM, Arendt KR, Wilch MH, Riddle JM, Furr SH, Steen C, U'Ren JM, Sandberg DC, Arnold AE (2015) Fungal endophytes in aboveground tissues of desert plants: infrequent in culture, but highly diverse and distinctive symbionts. Microb Ecol 70(1):61–76

Mei L, Zhu M, Zhang DZ, Wang YZ, Guo J, Zhang HB (2014) Geographical and temporal changes of foliar fungal endophytes associated with the invasive plant *Ageratina adenophora*. Microb Ecol 67(2):402–409

Newsham KK (2011) A meta-analysis of plant responses to dark septate root endophytes. New Phytol 190(3):783–793

Niazi P (2024) Isolation and characterization of a (Surfactin-like molecule) produced by Bacillus subtilis: antagonistic impact on root-knot nematodes. Scient Res Commun 4(2). https://doi.org/10.52460/src.2024.010

Oehl F, Laczko E, Bogenrieder A, Stahr K, Bösch R, Van der Heijden M, Sieverding E (2010) Soil type and land use intensity determine the composition of arbuscular mycorrhizal fungal communities. Soil Biol Biochem 42:724–738. https://doi.org/10.1016/j.soilbio.2010.01.006

Ownley BH, Pereira RM, Klingeman WE, Quigley NB, Leckie BM (2004) *Beauveria bassiana*, a dual-purpose biocontrol organism, with activity against insect pests and plant pathogens. In: Lartey RT, Cesar AJ (eds) Emerging concepts in plant health management. Research Signpost, Trivandrum, pp 255–269

Pandey SS, Singh S, Babu CS, Shanker K, Srivastava NK, Shukla AK, Kalra A (2016) Fungal endophytes of *Catharanthus roseus* enhance vindoline content by modulating structural and regulatory genes related to terpenoidindole alkaloid biosynthesis. Sci Rep 25(6):265–283

Petrini O, Stone J, Carroll FE (1982) Endophytic fungi in evergreen shrubs in western Oregon: a preliminary study. Can J Bot 60(6):789–796

Poveda J, Baptista P, Sacristán S, Velasco P (2022) Beneficial effects of fungal endophytes in major agricultural crops. Front Plant Sci 13:1061112. https://doi.org/10.3389/fpls.2022.1061112

Pressel S, Bidartondo MI, Ligrone R, Duckett JG (2010) Fungal symbioses in bryophytes: new insights in the twenty first century. Phytotaxa 9:238–253

Priyashantha KH, Dai DQ, Bhat DJ, Stephenson SL, Promputtha I, Kaushik P, Tibpromma S, Karunarathna SC (2023) Plant–fungi interactions: where it goes? Biology 12:809

Purin S, Rillig MC (2008) Parasitism of arbuscular mycorrhizal fungi: reviewing the evidence. FEMS Microbiol Lett 279:8–14

Rashid MI, Mujawar LH, Shahzad T, Almeelbi T, Ismail IMI, Oves M (2016) Bacteria and fungi can contribute to nutrients bioavailability and aggregate formation in degraded soils. Microbiol Res 183:26–41

Raya-Díaz S et al (2017) Entomopathogenic fungi-based mechanisms for improved Fe nutrition in sorghum plants grown on calcareous substrates. PLoS One 12(10):e0185903. https://doi.org/10.1371/journal.pone.0185903

Reinhardt D, Roux C, Corradi N, Di Pietro A (2021) Lineage-specific genes and cryptic sex: parallels and differences between Arbuscular Mycorrhizal fungi and fungal pathogens. Trends Plant Sci 26(2):111–123. https://doi.org/10.1016/j.tplants.2020.09.006

Rimington WR, Duckett JG, Field KJ, Bidartondo MI, Pressel S (2020) The distribution and evolution of fungal symbioses in ancient lineages of land plants. Mycorrhiza 30:23–49

Rodrigo S, García-Latorre C, Santamaria O (2022) Metabolites produced by fungi against fungal phytopathogens: review, implementation and perspectives. Plants 11(1):81. https://doi.org/10.3390/plants11010081

Rivkina EM, Friedmann EI, McKay CP, Gilichinsky DA (2000) Metabolic activity of permafrost bacteria below the freezing point. Appl Environ Microbiol 66(8):3230–3233. https://doi.org/10.1128/AEM.66.8.3230-3233.2000

Rodriguez RJ, White JF Jr, Arnold AE, Redman ARA (2009) Fungal endophytes: diversity and functional roles. New Phytol 182(2):314–330. https://doi.org/10.1111/j.1469-8137.2009.02773.x

Rungjindamai N, Jones EBG (2024) Why are there so few Basidiomycota and basal fungi as endophytes? A review. J Fungi (Basel, Switzerland) 10(1):67. https://doi.org/10.3390/jof10010067

Saikkonen K, Wäli PR, Helander M (2010) Genetic compatibility determines endophyte-grass combinations. PLoS One 5(6):e11395. https://doi.org/10.1371/journal.pone.0011395

Saladin B, Leslie AB, Wüest RO, Litsios G, Conti E, Salamin N, Zimmermann NE (2017) Fossils matter: improved estimates of divergence times in Pinus reveal older diversification. BMC Evol Biol 17(1):95. https://doi.org/10.1186/s12862-017-0941-z

Schardl CL, Leuchtmann A, Spiering MJ (2004) Symbioses of grasses with seedborne fungal endophytes. Annu Rev Plant Biol 55:315–340. https://doi.org/10.1146/annurev.arplant.55.031903.141735

Scott B, Green K, Berry D (2018) The fine balance between mutualism and antagonism in the *Epichloë festucae* grass symbiotic interaction. Curr Opin Plant Biol 44:32–38. https://doi.org/10.1016/j.pbi.2018.01.010

Selosse MA, Richard F, He X, Simard SW (2006) Mycorrhizal networks: des liaisons dangereuses? Trends Ecol Evol 21(11):621–628. https://doi.org/10.1016/j.tree.2006.07.003

Singh DK, Sharma VK, Kumar J, Mishra A, Verma SK, Sieber TN, Kharwar RN (2017) Diversity of endophytic mycobiota of tropical tree Tectona grandis Linn. F.: spatiotemporal and tissue type effects. J Scient Rep 7(1):1–4

Smith SE, Read DJ (2008) Mycorrhizal symbiosis, 3rd edn. Academic Press

Strullu-Derrien C, Goral T, Spencer ART, Kenrick P, Catherine Aime M, Gaya E, Hawksworth DL (2023) A fungal plant pathogen discovered in the Devonian Rhynie Chert. Nat Commun 14(1): 7932. https://doi.org/10.1038/s41467-023-43276-1

Sun X, Ding Q, Hyde KD, Guo LD (2012) Community structure and preference of endophytic fungi of three woody plants in a mixed forest. Fungal Ecol 5(5):624–632

Suryanarayanan TS, Thirunavukkarasu N, Govindarajulu MB, Gopalan V (2012) Fungal endophytes: an untapped source of biocatalysts. Fungal Divers 54(1):19–30. https://doi.org/10.1007/s13225-012-0168-7

Taylor TN, Krings M (2005) Fossil microorganisms and land plants: associations and interactions. Symbiosis 40:119–135

Taylor TN, Klavins SD, Krings M, Taylor EL, Kerp H, Hass H (2003) Fungi from the Rhynie Chert: a view from the dark side. Earth Environ Sci Trans R Soc Edinb 94(4):457–473. https://doi.org/10.1017/S026359330000081X

Taylor TN, Klavins SD, Krings M, Taylor M (2004) Fungi from the Rhynie Chert: a view from the dark side. Trans R Soc Edinb Earth Sci 94:457–473

Taylor TN, Hass H, Kerp H, Krings M, Hanlin RT (2005a) Perithecial ascomycetes from the 400-million-year-old Rhynie chert: an example of ancestral polymorphism. Mycologia 97(1): 269–285. https://doi.org/10.1080/15572536.2006.11832862

Taylor TN, Kerp H, Hass H (2005b) Life history biology of early land plants: deciphering the gametophyte phase. Proc Natl Acad Sci USA 102(16):5892–5897. https://doi.org/10.1073/pnas.0501985102

Taylor TN, Krings M, Dotzler N, Galtier J (2011) The advantage of thin sections over acetate peels in the study of late Paleozoic fungi and other microorganisms. PALAIOS 26:239–244

Taylor TN, Krings M, Taylor EL (2015) Chapter 3 – how old are the fungi? In: Taylor TN, Krings M, Taylor EL (eds) Fossil Fungi. Academic Press, pp 27–39

Tedersoo L (2017) Global biogeography and invasions of ectomycorrhizal plants: past, present and future. In: Tedersoo L (ed) Biogeography of mycorrhizal symbiosis. Springer International, Cham, pp 469–531

Tiwari P, Bae H (2022) Endophytic fungi: key insights, emerging prospects, and challenges in natural product drug discovery. Microorganisms 10(2):360. https://doi.org/10.3390/microorganisms10020360

Toledo S, Bippus AC, Atkinson BA, Bronson AW, Tomescu AMF (2021) Taxon sampling and alternative hypotheses of relationships in the euphyllophyte plexus that gave rise to seed plants: insights from an early Devonian radiatopsid. New Phytol 232(2):914–927. https://doi.org/10.1111/nph.17511

Trewin NH (2003) History of research on the geology and palaeontology of the Rhynie area, Aberdeenshire, Scotland. Trans R Soc Edinb Earth Sci 94(04):285–297. https://doi.org/10.1017/s0263593300000699

Truyens S, Weyens N, Cuypers A, Vangronsveld J (2015) Bacterial seed endophytes: generators of plant growth-promoting functions. Plant Soil 368(1):1–20

Vega FE, Goettel MS, Blackwell M, Chandler D, Jackson MA, Keller S, Koike M, Maniania NK, Monzon A, Ownley BH, Pell JK, Rangel D, Roy HE (2009) Fungal entomopathogens: new insights on their ecology. Fungal Ecol 2(4):149–159. https://doi.org/10.1016/j.funeco.2009.05.001

Vega FE, Simpkins A, Aime MC, Posada F, Peterson SW, Rehner SA, Infante F, Castillo A, Arnold AE (2010) Fungal endophyte diversity in coffee plants from Colombia, Hawai'i, Mexico and Puerto Rico. Fungal Ecol 3(3):122–138

Vogl AE (1898) Mehl und die anderen mehlprodukte der cerealien und leguminosen. Nahrungsm Unters Hyg Warenk 12:25–29

Wellman CH (2006) Spore assemblages from the lower Devonian 'lower old red Sandstone' deposits of the Rhynie outlier, Scotland. Earth Environ Sci Trans R Soc Edinb 97(2): 167–211. https://doi.org/10.1017/S0263593300001449

Wellman CH (2018) Palaeoecology and palaeophytogeography of the Rhynie chert plants: further evidence from integrated analysis of *in situ* and dispersed spores. Philos Trans R Soc Lond Ser B Biol Sci 373(1739):20160491. https://doi.org/10.1098/rstb.2016.0491

Wen J, Okyere SK, Wang S, Wang J, Xie L, Ran Y, Hu Y (2022) Endophytic fungi: an effective alternative source of plant-derived bioactive compounds for pharmacological studies. J Fungi 8(2):205. https://doi.org/10.3390/jof8020205

Wu L, Han T, Li W, Jia M, Xue L, Rahman K, Qin L (2013) Geographic and tissue influences on endophytic fungal communities of *Taxus chinensis* var. *mairei* in China. Curr Microbiol 66(1): 40–48

Yuan M, Jiang Z, Bi G, Nomura K, Liu M, Wang Y, Cai B, Zhou JM, He XY, Xin ZF (2021a) Pattern-recognition receptors are required for NLR-mediated plant immunity. Nature 592(7852):105–109. https://doi.org/10.1038/s41586-021-03316-6

Yuan M, Ngou BPM, Ding P, Xin ZF (2021b) PTI-ETI crosstalk: an integrative view of plant immunity. Curr Opin Plant Biotechnol 62:10203. https://doi.org/10.1016/j.pbi.2021.102030

Zhou DQ, Hyde KD (2001) Host-specificity, host-exclusivity and host-recurrence in saprobic fungi. Mycol Res 105:1449–1457

Discovery of Endo-fungi Within Water Ferns: Community Analyses Based on High-Throughput Sequencing

Hawraa F. H. Al-Abedi, Maha Abdel Kareem Taher, Nawar Ali Jasim, Raid D. Thanoon, Suhail Gawdat Fadhil, Israa Ibrahim Khalil, Semaa F. H. Al-Abedi, and Manal Muwaffaq Mohammed

Contents

H. F. H. Al-Abedi (✉) · R. D. Thanoon · S. F. H. Al-Abedi · M. M. Mohammed
Department of Biology, Collage of Education for Pure Sciences, University of Al-Hamdaniya, Al- Hamdaniya, Iraq
e-mail: Hawraafaisal@uohamdaniya.edu.iq

M. A. K. Taher
Laser and Photonics Research Center, University of Al-Hamdaniya, Al- Hamdaniya, Iraq

N. A. Jasim
Department of Pathology & Poultry Diseases, College of Veterinary Medicine, University of Tikrit, Tikrit, Iraq

S. G. Fadhil
Mosul Medical Technical Institute Northern Technical University (NTU), Mosul, Iraq

I. I. Khalil
Department of Microbiology, College of Veterinary Medicine, University of Mosul, Mosul, Iraq

A. M. Abdel-Azeem, F. Degola (eds.), *Advances in Lichens and Non-Flowering Plants Fungal Endophytes Research*, Fungal Biology,
https://doi.org/10.1007/978-3-032-01506-8_9

1 Introduction

Endophytes encompass a remarkably diverse group of microorganisms, mainly fungi and bacteria, that establish close, often symbiotic relationships with plants. The term "endophyte," rooted in the Greek words "endo" (within) and "phytón" (plant), was introduced by Anton de Bary in 1866 to describe fungi residing inside plant tissues, as opposed to epiphytes, which inhabit plant surfaces (Dutta et al. 2014). Endophytic fungi form a diverse and heterogeneous community found in virtually all types of plants, thriving across diverse climatic and ecological zones. They have been identified in various plant forms, including trees, shrubs, grasses, herbaceous plants, ferns, mosses, and even marine vegetation (Chowdhary et al. 2012). Estimates suggest that the number of endophytic fungal species could approach 1 million, extrapolated from a proposed 6:1 fungi-to-plant species ratio. However, this figure may be significantly underestimated, as only about 5% of fungal species have been formally described to date (Abhijeet Singh 2014). Moreover, the 6:1 ratio has faced scrutiny, with some research indicating it could be substantially higher—potentially up to five times greater in tropical and subtropical regions, where fungal diversity is exceptionally rich (Sun and Guo 2012; Chávez et al. 2015).

Initially, fungal endophytes were defined as those fungi inhabiting plant tissues without causing visible harm or disease to their hosts. Over time, this definition has evolved and is likely to continue adapting as research in endophyte biology progresses (Gamboa et al. 2003; Gakuubi et al. 2021). Clear distinctions between fungal endophytes, mycorrhizal fungi, and phytopathogens are not always straightforward. For instance, certain fungal strains may act as pathogens in some plants while displaying neutral or beneficial effects in others (Hardoim et al. 2015). Moreover, some fungi remain latent within plant tissues as asymptomatic pathogens, only exhibiting pathogenic traits under favorable conditions. Another ambiguous aspect lies in determining whether plant pathogens that have lost their virulence should be considered endophytes or categorized otherwise (Kloepper et al. 2013). Furthermore, defining fungal endophytes solely based on pathogenicity excludes unculturable fungi whose pathogenic potential might still be unclear (Kim et al. 2013).

Most fungal endophytes belong to the Ascomycetes group, with occasional representation from Basidiomycetes and Zygomycetes (Meshram and Gupta 2019). Their diversity can vary widely depending on factors such as environmental conditions, host plant genotypes and physiology, seasons, and human influence.

Even within the same geographic location, the diversity and abundance of endophytic fungi can differ significantly among various host plants and even among different parts of a single plant (Banerjee 2011). Tropical and subtropical regions are known to harbor much higher fungal diversity compared to other regions. Additionally, seasonal changes play an important role in shaping fungal distribution patterns (Materatski et al. 2019).

This chapter delves into the origins, proliferation, and ecological invasions of endophytic fungi while exploring how host plants recognize these invaders. It examines the intricate interactions between endophytic fungi and their plant hosts within varying ecological contexts.

2 Classification of Fungal Endophytes

The classification of fungal endophytes is complex due to their taxonomic diversity and polyphyletic nature. Various schemes have been developed to categorize fungal endophytes based on different aspects of their biology (Bamisile et al. 2018). Early classification approaches considered factors such as phylogeny, ecological functions, and life history strategies. This led to the division of fungal endophytes into two primary groups: clavicipitaceous endophytes (C-endophytes) and nonclavicipitaceous endophytes (NC-endophytes). Clavicipitaceous endophytes, also known as grass endophytes or class 1 endophytes, predominantly include species from the family Clavicipitaceae (Hypocreales; Ascomycota) (Rodriguez et al. 2009). This group comprises 27 recognized genera, encompassing both free-living and symbiotic species. In contrast, nonclavicipitaceous endophytes form a diverse and polyphyletic group with taxonomically undefined members. These are commonly associated with vascular and non-vascular plants, with most belonging to the Dikarya subkingdom (Ascomycota or Basidiomycota) (Uzma et al. 2018).

As research in fungal endophyte biology has expanded, additional criteria for their classification have emerged. These classifications take into account factors such as host range, modes of transmission and infection, reproduction strategies, colonization sites within the host, and the nature of interactions with their hosts (Kaul et al. 2012). For example, fungal endophytes have been categorized by their host plant types, leading to studies focusing on specific plant associations. Examples include endophytes from medicinal plants, mangroves, coniferous trees, and grasses (Demers et al. 2018).

When classified by modes of transmission, fungal endophytes fall into two main categories: vertically transmitted and horizontally transmitted. Vertically transmitted endophytes are passed between generations of a host, typically via seeds (Tanaka et al. 2012). Horizontally transmitted endophytes, on the other hand, spread between individuals in a population through spores or vegetative propagules (Daley et al. 2017).

Tissue colonization preferences provide another basis for classification. Research indicates that fungal endophytes often demonstrate specificity in colonizing particular plant tissues rather than exhibiting systemic colonization (Bamisile et al. 2018). Root endophytic fungi are predominantly isolated from plant roots; foliar endophytes, by contrast, are commonly found in stems and leaves (Chadha et al. 2014).

Other classifications consider reproductive modes (sexual or asexual) or infection expression, distinguishing between symptomatic and asymptomatic fungal endophytes (Jia et al. 2020). This multifaceted classification framework underscores the biological complexity of fungal endophytes and the diverse interactions they maintain with their host plants (Terhonen et al. 2019).

3 Isolation of Fungal Endophytes: Rationale for Plant Selection and Prioritization

Nearly all higher plant species in natural ecosystems support the presence of one to several hundred species of endophytic fungi (Meshram and Gupta 2019; Gakuubi et al. 2021). Given this abundance of potential fungal sources, effective and innovative strategies are essential for selecting host plants for fungal isolation. Such prioritization increases the likelihood of uncovering uncharacterized fungi, which can, in turn, enhance the discovery of novel bioactive molecules. While no formalized guidelines have been established for selecting plant sources for fungal endophyte isolation, various researchers have proposed bioprospecting strategies. These strategies, combined with statistical and comparative analyses of the growing body of literature, can help identify the most promising host plants for exploration. One widely cited framework suggests four key criteria for prioritizing plant selection in fungal endophyte isolation (Strobel and Daisy 2003).

1. Plants from distinctive ecological niches with unique biological traits and survival mechanisms
2. Medicinal plants and those with documented ethnobotanical significance
3. Plants endemic to specific regions
4. Plants located in biodiversity hotspots

The first criterion emphasizes plants from specialized environments or those exhibiting extraordinary survival strategies, as such adaptations may facilitate associations with rare microbial endophytes. For instance, mangrove plants represent a notable example. These species thrive under extreme conditions including high salinity, fluctuating seawater levels, intense heat, and continual water movement. Adaptations to these challenges have fostered mangroves' association with diverse fungal endophytes capable of producing an extensive array of bioactive secondary metabolites (Shearer et al. 2007; Deshmukh et al. 2018).

Another important strategy involves focusing on medicinal plants (Rana et al. 2020a, b). Particular attention is placed on plants utilized in traditional medicine or those with known ethnopharmacological applications across diverse local communities. These plants can be identified through direct engagement with indigenous communities, ethnobotanical surveys, or existing literature reviews. Investigating whether their medicinal properties may stem from associated fungal endophytes provides a promising avenue for discovery (Tiwari 2014). In recent years, the search for bioactive compounds among fungal endophytes of medicinal plants has gained momentum due to discoveries like fungal strains capable of independently producing host-specific bioactive compounds (Venieraki et al. 2017). A well-documented case involves paclitaxel (commonly known as Taxol), a potent anticancer agent originally sourced from the bark of *Taxus brevifolia*. However, the limited yield of paclitaxel in *T. brevifolia* rendered its extraction unsustainable (Gakuubi et al. 2021). This challenge was addressed through the groundbreaking discovery of paclitaxel-producing endophytic fungi like *Taxomyces andreanae*, followed by additional findings of similar fungi in other Taxus and non-Taxus species (Roopa et al. 2015).

Endemic plants represent another promising category for sourcing endophytic fungi. After inhabiting distinct geographic areas over extended periods, these plants often develop associations with unique fungal communities capable of synthesizing complex and bioactive molecules (Tan et al. 2018). The isolation of fungi from these plant populations can therefore enhance the chances of discovering new and valuable strains.

Finally, prioritizing plants in biodiversity hotspots offers a particularly fruitful approach. Such regions, particularly in tropical and subtropical zones, harbor unparalleled plant diversity (Meshram and Gupta 2019). Plants within these ecosystems face intense biotic and abiotic pressures, including competition for resources and high selection pressures due to environmental factors (Hodkinson et al. 2019). Over time, these challenging conditions may drive the evolution of unique and as-yet uncharacterized endophytic fungi, further expanding opportunities for impactful discoveries.

4 Water Fern

Ferns, as the second-largest group of vascular plants, are placed between lower and higher plants in pteridophytes and comprised more than 12,000 known species (Razaghi and Abdel-Azeem 2024). These earliest vascular plants are nonflowering and more similar to the seed-bearing plants (Prado and Sylvestre 2010). Ferns are abundantly distributed across tropical and subtropical ecosystems with high humidity, especially in Southeast Asia and the neotropical region (Mehltreter et al. 2010) and a few species in Arctic and Antarctic regions (Smith 2014). Almost more than half of the fern species have been reported from China with 2600 species (Lu 2007), followed by India with 1200 species (Chandra 2000), and Malaysia with more than 1000 species (Razaghi and Abdel-Azeem 2024). Ferns are divided into four categories according to the purpose of their usage from rhizomes, stems, croziers, leaves, young fronds, and shoots including edible ferns, cattle bed ferns, ornamental ferns, and medicinal ferns. Based on the survey of fern flora in (Chandra 2000) and Thapa (2016). The water fern, belonging to the genus *Ceratopteris* within the family Pteridaceae, comprises a small group of aquatic fern species. Currently, at least four species are identified: the broadleaf water sprite (*Ceratopteris cornuta*), the floating antler fern or water horn fern (*Ceratopteris pteridoides*), the triangle water fern (*Ceratopteris richardii*), and the water sprite (*Ceratopteris thalictroides*). These plants are widely distributed across tropical and warm temperate regions globally, and several are popularly cultivated as aquarium plants. While they sometimes take root in mud, they more commonly float on the surface of shallow waters found in environments such as ditches, lakes, and slow-moving backwaters (Yatskievych et al. 2018).

Individual plants feature dimorphic leaves, producing a rosette of lettuce-like vegetative leaves that are lobed or divided. These leaves often exhibit inflated petioles, an adaptation for buoyancy. Along their margins, small meristematic buds develop, which give rise to asexual "daughter plants"—vegetative offshoots that grow into free-living clones of the parent plant. In contrast to the vegetative leaves, the fertile leaves are much more finely dissected. The narrow segments of these leaves have their undersides densely covered with sporangia, specialized structures for spore production. The sporangia are partially shielded by the recurved edges of the leaf segments, and the spores they produce are globose and trilete in form (Yatskievych et al. 2018).

Ceratopteris species have gained significant importance as a tool for education and genetic research due to their uniquely fast life cycle compared to many other plants. Mature spores germinate within 3–4 days, quickly developing through the gametophyte (haploid) phase. The complete life cycle from a mature sporulating plant to the next generation of sporulating plants can be accomplished in about 4 months, whereas most terrestrial ferns require a year or more to achieve the same. This rapid development makes water ferns an excellent model for studying genetic inheritance patterns.

Numerous genetic strains of *Ceratopteris*, each with a distinct mutation, have been cultivated and preserved in specialized culture collections. Detailed protocols for developing new mutant lines have also been established. These strains and mutation guidelines are made accessible to researchers and educators under the trade name C-Fern, further enhancing their utility in scientific and teaching applications (Yatskievych et al. 2018).

5 Fern Application in Medicine

Dates back more than 2000 years throughout the world in Europe, North and South America, Asia, and Africa (Van and Wink 2017; Razaghi and Abdel-Azeem 2024). Ethnomedicinal uses of ferns have been mainly documented in traditional medicine system in China (Chang et al. 2011) and other indigenous systems of medicine among other ethnic groups around the world (Pan et al. 2014). The highest diversity of fern species distributed in the Asian countries compared to other continents might be due to the large number of reports and surveys conducted in such regions indicating the popularity of herbal medicine and use of ferns in the traditional medicines in these regions. It is estimated that there are more than 400 medicinal fern taxa distributed in 121 genera in 36 families with promising therapeutic values. *Culcita macrocarpa*, *C. Presl*, and *Blechnum eburneum* Christ have been reported as near threatened and vulnerable plants, respectively (Muhammad et al. 2020), which might be due to overexploitation of resources, prevalence of invasive species, growth of cities, and forest fragmentation and loss of habitats for ferns (Paciencia and Prado 2005). Pteridaceae, Polypodiaceae, Dryopteridaceae, Aspleniaceae, Thelypteridaceae, and Cyatheaceae are among the most medicinally used families. Fern taxa with potential uses are mostly founded in the genera *Adiantum, Asplenium, Dryopteris, Equisetum*, and *Pteris* to treat gastrointestinal ailments, stomach ailments, diseases related to accessory digestive organs and small or large intestine, skin issues, gynecological, and infertility disorders, respiratory diseases, urogenital diseases, and musculoskeletal disorder (Muhammad et al. 2020). Pharmacological effects of fern plants are related to the production of diverse phytochemicals including alkaloids (Dong et al. 2012), flavonoids (Xia et al. 2014), polyphenols, and terpenoids (Socolsky et al. 2007), and steroids (Ho et al. 2012). *Angiopteris evecta* as a member of Marattiaceae can be a valuable source of alkaloid, angiopteroside, flavonoid, monoterpenoid, polyphenol, saponin, steroid, triterpenoid, and quinone, which has anticancer, antiviral properties and can treat headache, wounds, broken bones, and fractures (Baskaran et al. 2018). Phenols and flavonoids derived from Asplenium nidus (family Aspleniaceae) have anti-lice, antibacterial, antiviral activity and can be used for treatment of fever, mouth disorders, childbirth disorders, tuberculosis, chest pain, elephantiasis, skin disorders (Modolo and Foglio 2020). *Adiantum trapeziforme* (family Pteridaceae) contains compounds like ascorbic acid, peroxidase, polyphenol oxidase and shows antibacterial activity (Quattrocchi 2012). *Blechnum orientale* (family Blechnaceae), by having flavonoids, tannins, and terpenoids has a strong antioxidant and anticancer capacity (Baskaran et al. 2018). Whole plants of *Ceratopteris thalictroides* (family Pteridaceae), especially leaves contain various bioactive compounds including alkaloids, phenols, flavonoids, saponins, and tannins which has been shown to have antiinflammatory and anticancer effects and used for skin disease treatment (Baskaran et al. 2018; Joshi et al. 2019). Benzoic acid, flavonoids, and terpenoids identified in the roots of *Drynaria rigidula* (family Polypodiaceae) make this fern potent in anti-inflammatory and antiviral activities (Nugraha et al. 2013). Caffeic acid derivatives identified in the methanol extract of rhizomes of *Abacopteris penangiana* (family Thelypteridaceae) show antioxidant activity against neurotoxicity in mice (Fu et al. 2013). This medicinal fern has been traditionally used by Chinese locals for the treatment of blood stasis by promoting blood circulation, dampness, swelling in the tissues and inflammation (Administration Bureau of National Chinese Traditional Medicine, 1998/Chinese Materia Medica). *A. penangiana* also contains flavonoids components with antiinflammatory activity (Lei et al. 2011), cytotoxic activity, antioxidative activity (Zhao et al. 2007), and neurotoxicity in PC12 cells via dopamine levels (Wei et al. 2013). *Macrothelypteris* is another genus in Thelypteridaceae with massive uses in Chinese medicine. Protoapigenone is a flavonoid compound isolated from *M. torresiana* with potential cytotoxicity against various tumor cells (Chang et al. 2008). Other flavonoids and total polyphenols identified in this fern have been shown to be effective in renal protection, cancer chemoprevention, and cancer treatment. Various bioactive compounds including alkaloids, antheridiogen, phenols, flavonoids, and tannins have been identified in leaves, roots, and rhizomes of *Lygodium circinatum* (family Lygodiaceae), as an evergreen fern used for the treatment of snakebites, conjunctivitis, and eye problems (Baskaran et al. 2018). *Lygodium microphyllum* is another species in the same fern genus which contains alkaloid, flavonoid, isoquercetin, and steroid with antioxidant and antimicrobial activities (Kuncoro et al. 2018).

6 How Endophytic Fungi Interact with Their Water Ferns

Endophytic fungi establish various relationships with their host plants through mutualism, antagonism, and neutralism. The population structure of endophytic fungi is largely affected under varying ecological pressures, for example, the genetic background, nutrient level, and ecological habitats of host plants. In turn, they confer certain benefits such as induced growth, increased resistance upon attack by pathogens and/or herbivores (Rodriguez et al. 2009), as well as accumulation of bioactive components (Khare et al. 2018). The bioactive compounds produced by endophytic fungi associated with host plants in a mutual interrelation are essential to increase the adaptability of both endophytic fungi and their host plants to biotic stresses: herbivore attacks, pests and phytopathogens (Azevedo 1998 cited by Razaghi and Abdel-Azeem 2024), and abiotic stresses: dry seasons, salinity, extreme temperatures, and others (Khare et al. 2018). These fungi can induce the production of various biologically active secondary metabolites, which can be utilized for medicinal purposes (Rodriguez et al. 2009). However, the colonization of these fungi is not accidental, as they are attracted by specific chemicals produced by the host plants. Additionally, the secondary metabolites produced by the host plant, such as saponin and essential oils can act as a resistance mechanism against pathogens. As a result, the secondary metabolites enhance barriers for the endophytic fungi colonization. To overcome this problem, endophytic fungi must secrete corresponding detoxifying enzymes such as cellulase, laccaseactas, xylanase, and protease to break down these secondary metabolites before they enter the defense system of the resident host plants (Firáková et al. 2007). Once in the host plant's tissue, the fungi can remain in a dormant state either for the entire life cycle of the host plant (neutral) or for a long period of time (mutually or antagonistically) until favorable environmental conditions or changes in the host plant's ontogenetic state allow them to thrive (Sieber 2007). Ferns and lycophytes have developed a wide spectrum of antagonistic and mutualistic relationships with fungi and animals. While some of these interactions, such as endomycorrhizae, are old and may have coexisted with their host plants for a long time, other interactions may have originated more recently, such as some herbivorous insects that have switched from seed plants to ferns (Mehltreter et al. 2012). More than 80% of sporophytes possess endomycorrhizae, while for a few species, fern-ericoid mycorrhizae and ectendomycorrhizae have been reported. In the gametophytic stage, mycorrhizae are obligate in older fern and lycophyte lineages but are facultative or can be absent in more modern lineages (Mehltreter et al. 2012). Interactions with parasitic, symbiotic and neutral endophytic fungi that infect aerial parts of the ferns seem to be as common as in seed plants, while the proportion of interactions with insects seems to be three to seven times lower than in seed plants. Most ferns have few specific biochemical defense mechanisms in comparison with seed plants, yet ferns and seed plants sustain similar levels of herbivore damage (Mehltreter et al. 2012). A study by Olmo-Ruiz and Arnold 2014 showed that ferns are an ancient and diverse lineage of vascular plants that differ morphologically, chemically, and in growth habits from the angiosperms with which they co-occur. We used a culture-based approach coupled with phylogenetic analyses to characterize the incidence, diversity, and composition of fungal endophyte assemblages in ferns, with a focus on healthy aboveground tissues of seven species of eupolypods at La Selva, Costa Rica. Endophytes were isolated from every individual plant and were similarly abundant and diverse in frond blades and stalks, in different vegetation types, in epiphytic versus terrestrial species, and between sampling years. However, abundance, diversity, and community structure differed significantly among fern species, and composition differed markedly between sampling years. Phylogenetic classification using separate and combined datasets revealed that as for many Neotropical angiosperms, the majority (95%) of endophyte taxa were Ascomycota, with particular dominance by Sordariomycetes, Eurotiomycetes, and Dothideomycetes. However, our data suggest higher phylogenetic richness and stronger host affinities in fern associated endophytes relative to those studied in angiosperms thus far (Götz et al. 2006).

7 Unique Traits of Endophytic Colonization and the Exploration of Endophytic Biodiversity

The range of endophytic fungi linked to a plant can differ greatly based on varying environmental factors (Vega et al. 2010; Baron and Rigobelo 2022). Even within plants of the same species, the composition of fungal endophytes in their tissues can differ significantly, influenced by variations in the physiological state of each individual plant (Aly et al. 2011). The age of the plant is an important factor in shaping the fungi endophytic commune profiles (Sieber 2007). While some fungal endophytes are capable of colonizing a broad range of plant species, others display high specificity, being limited to only a few host plants. This specificity can also extend to the particular plant part they inhabit (Bamisile et al. 2018). Furthermore, fungi that are transmitted vertically tend to exhibit associations with plants that are more mutualistic in nature, whereas those that are transmitted horizontally are more often antagonistic (Aly et al. 2011). The genetic communication between the endophyte and the host plant for the establishment of the interaction is a complex and poorly understood process that involves the selective expression of fungal genes responsible for the production of enzymes and secondary metabolites that aid in colonization (Yan et al. 2019). The approximation of the germinative tube of the endophyte to the root causes the loss of apical dominance of the root and the formation of a hyphal penetration apparatus (aspersorium), which enters the root cortex with hyphae of infection, thereby starting the colonization process (Khan et al. 2015). These events promote a balanced activation of plant defense genes, and when the fungus reaches the inner cortex, the hyphae penetrate the plant cell wall and continue the colonization of adjacent tissues (endoderm, pericycle, xylem, phloem) of the roots and of the soil (Yan et al. 2019). Various factors highlight the intricate relationship between plants and their fungal endophytes. These include the absence of defensive reaction against the fungi and the ability of some endophytes to synthesize metabolites similar to those produced by the host plant (as noted by Kusari et al. 2012), and the synchronous reproductive processes of the plant and fungal hosts, particularly evident during vertical transmission via seeds (Baron and Rigobelo 2022).

Analyzing the fungal endophytic communities in plants presents significant challenges, though science has progressively advanced in this area. Detecting these microorganisms with plant tissue remains challenging due to the rarity of observable hyphae and a lack of distinct features (Rashmi et al. 2019). Traditionally, researchers have studied endophytic communities by isolating fungi from surface-sterilized plant tissues, focusing on strains located internally. This approach typically involves the use of conventional culture media, with certain modifications to enhance fungal growth. For example, doubling the water content in the medium can prevent osmotic shock and promote exploratory hyphal growth. Additionally, incorporating plant extracts into culture media has proven beneficial (Murphy et al. 2018).

Despite its usefulness, cultivation-based characterization of fungal endophyte diversity is limited by biotic and abiotic influences (Chen et al. 2020). Consequently, molecular tools have revolutionized studies of endophytic fungal communities in various plant species. Advanced methods, such as mass DNA sequencing, eliminate the need for cultivation and enable the identification of numerous previously uncharacterized endophytic taxa (Deyett and Rolshausen 2020). In terms of their distribution and biodiversity, many aspects of fungal endophytes remain poorly understood.

Endophytic fungi have already been recovered from a wide range of habitats, including arctic environments, hot deserts, and mangrove, temperate, and tropical forests (Arnold 2008). As reviewed by Chadha et al. (2015) and Lugtenberg et al. (2016), the characterization of the diversity and the distribution of fungal endophytes across large geographical areas is still in the beginning, and only some general aspects can be affirmed, such as that the diversity of endophytic fungi is higher in the tropics than in higher latitudes. Moreover, tropical environments host a greater diversity of endophytic species, which are typically confined to a limited number of classes (Baron and Rigobelo 2022).

8 From Biocontrol to Enhancing Biofertilization and Stress Resilience: The Role of Endophytic Fungi

Endophytic fungi play a multifaceted role, ranging from biocontrol to enhancing biofertilization and improving the stress tolerance of plants. These fungi inhabit plant tissues, securing a sheltered environment rich in essential nutrients that supports their survival and growth while reducing competition with other microorganisms. In return for this protected habitat, endophytic fungi boost plants fitness through various mechanisms, as outlined by Chitnis et al. (2020). The advantages they provide can be both direct and indirect, though drawing clear distinctions between these two types of benefits often proves difficult (Berg 2009). Direct mechanisms primarily involve nutrient acquisition and the production of phytohormones, whereas indirect mechanisms include increased tolerance to biotic and abiotic stresses, such as resistance to pathogens, which further promote plant growth (Souza and dos Santos 2017).

Given the challenges associated with conventional plant breeding in certain crops and the ongoing restrictions on genetically modified organisms (GMOs) due to uncertainties about their long-term effects on human, animal, and environmental health (Chadha et al. 2015), natural solutions hold significant promise for advancing sustainable agriculture. The use of endophytic fungi is still in its early stages. However, their potential to improve plant health is becoming increasingly evident. The following section will explore the indirect and direct mechanisms by which these microorganisms contribute, along with future applications in agricultural practices (Latz et al. 2018). Direct benefits of endophytic fungal interactions include enhanced nutrient uptake and increased production of phytohormones in host plants. These improvements lead to greater biomass, expanded root development, increased plant height, reproductive success, and overall yield. Due to these advantages, endophytic fungi can effectively be regarded as natural biofertilizers (Bamisile et al. 2018).

In fact, endophytic fungi enhance the absorption of essential macronutrients, including phosphorus, nitrogen, potassium, and magnesium, as well as micronutrients like zinc, iron, and copper, from the soil and organic matter. In doing so, they contribute to an increased nutrient supply for their plant hosts (Rana et al. 2020a; b).

Additionally, endophytic fungi have the capability to produce important phytohormones such as auxin, gibberellin, and cytokinin. Although the possibility for phytohormone production by these fungi remains largely underexplored, especially regarding gibberellins, these molecules play critical roles as chemical signals and messengers that regulate plant growth under diverse environmental conditions

(Khan et al. 2015). Among the auxins synthesized by fungi, indole-3-acetic acid is the most prominent. Auxins are key regulators of plant development, showcasing various beneficial effects on shoot and root growth. These include influencing tropic responses, cell division and elongation, vascular tissue differentiation, and initiating root formation processes (Jaroszuk-Ściseł et al. 2014).

Among the benefits of indirectly interacting with fungal endophytes, we can acknowledge:

(i) **Activation of Systemically Acquired Resistance (SAR)**

Fungal endophytes play a significant role in enhancing a plant's self-defense mechanisms by triggering induce systemically resistance (ISR) pathway. These pathways often overlap with acquired systemically resistances (ASR), which contribute to improved plant growth (Busby et al. 2016) and provide protection against pests and pathogens (Chadha et al. 2015). The process begins when plants recognize molecules associated with pathogens, referred to as pathogen-associated molecular patterns (PAMPs) (Poveda et al. 2020).

(ii) **Production of Antibiotics and Secondary Metabolites**

Beyond stimulating plants to produce their own defense molecules, endophytic fungi act as a rich source of compounds beneficial to their host. These fungi excel in producing substances with anti-pathogenic and anti-herbivore properties, include alkaloid, phenol, steroid, quinol, peptide, polyketide, terpenoid, flavonoid, chlorinate component, and volatile organic component (VOCs) (Kaddes et al. 2019). Additionally, study have documented their product of antiviral, antibacterial, antifungal, and insecticidal compounds (Chowdappa et al. 2020).

The use of synthetic chemical compounds in agriculture has raised concerns due to their harmful effects on human health, animals, and the environment. Many of these substances are now restricted or banned, making their future commercialization unlikely (Kaddes et al. 2019). Consequently, efforts in research and industry are focusing on developing biological alternatives to promote more sustainable agricultural practices (Kaddes et al. 2019).

(iii) **Production of Siderophores**

Iron is a crucial micronutrient for all living cells. Siderophores are small molecules with iron-chelating properties produced by certain microorganisms, include endophytic fungi, to link ferric ions in the rhizospheres. For example, studies by Suebrasri et al. (2020) detected siderophore production in endophytic strains such as *Macrophomina phaseolina* SS1R10, *Macrophomina phaseolina* SS1L10, and *Trichoderma koningii* ST-KKU1, suggesting that fungal siderophores play a significant role in promoting the growth of sunchoke plants. Similarly, recombinant strains of *Trichoderma harzianum* colonizing beans (*Phaseolus vulgaris*) have been shown to produce siderophores (Eslahi et al. 2020; Card et al. (2016) the specific roles and mechanisms of siderophore production by fungal endophytes, particularly in inducing systemic resistance (ISR), are still not fully understood (Fouda et al. 2015).

(iv) **Protection Against Biotic and Abiotic Stress**

Agricultural practices and global environmental changes increasingly expose plants to challenge growth conditions, necessitating enhanced support for crop resilience. This has led to growing interest in the potentially of endophytic fungi as a mean to protect plants from biotic and abiotic stresses. Endophytic fungi have demonstrated the ability to assist plants in coping with a variety of stressors, including drought, extreme temperatures, salinity, and heavy metal toxicity (Khan et al. 2015).

These fungi support plant adaptation to stress through multiple mechanisms. During oxidative stress, for instance, plants activate antioxidant enzymes such as catalase and peroxidase, which facilitate the product of reactive oxygen species (ROS). ROS accumulation can damage membranes, leading to lipid peroxidation. Although the underlying processes are not fully understood, endophytic fungi enhance plant tolerant to ROS by mitigating lipid peroxides. The challenge posed by abiotic stressors such as drought, heat, and salinity is often linked to electrolyte leakage from cell membranes and alterations in lipid composition under stress conditions. Endophytic fungi can modify the lipid composition of cell membranes to prevent such leakage (Khan et al. 2015; Yan et al. 2019). Advances in research on plant-microbe interactions have revealed that plant genomes interact directly with beneficial microorganisms, opening up fresh avenues for sustainable agricultural practices (Card et al. 2016).

9 Conventional Methods Employed in Endophyte Research

Traditional, endophytic fungi in plant tissue can be recognized by two basic techniques, that is, direct observation and cultivation-dependent methods. In the direct observation method, endophytic fungal structures within living plant tissues are directly examined under a light and electron microscope, which can show all endophytic microbiota within the plant tissue, particularly biotrophic fungi that cannot be cultured on standard growth media (Lucero et al. 2011). However, most endophytic fungi within plant tissue have only a hyphal structure, and therefore cannot be identified to any taxonomic category according to morphology due to lack of spore-producing structures and sexual or asexual spores. In addition, endophytic isolates cannot be obtained as microbial resources for further use with the direct observation method. Therefore, this is not commonly used in endophyte diversity studies (Deckert et al. 2001).

Unlike direct observation methods, cultivation-dependent techniques remain a common approach in studies exploring endophyte diversity (Vieira et al. 2011). These techniques are essential for isolating endophytic fungi, facilitating more in-depth investigations, whether focusing on their characterization, population dynamics, species diversity, or their application as inoculants to enhance plant growth and health. They are also valuable for screening purposes aimed at

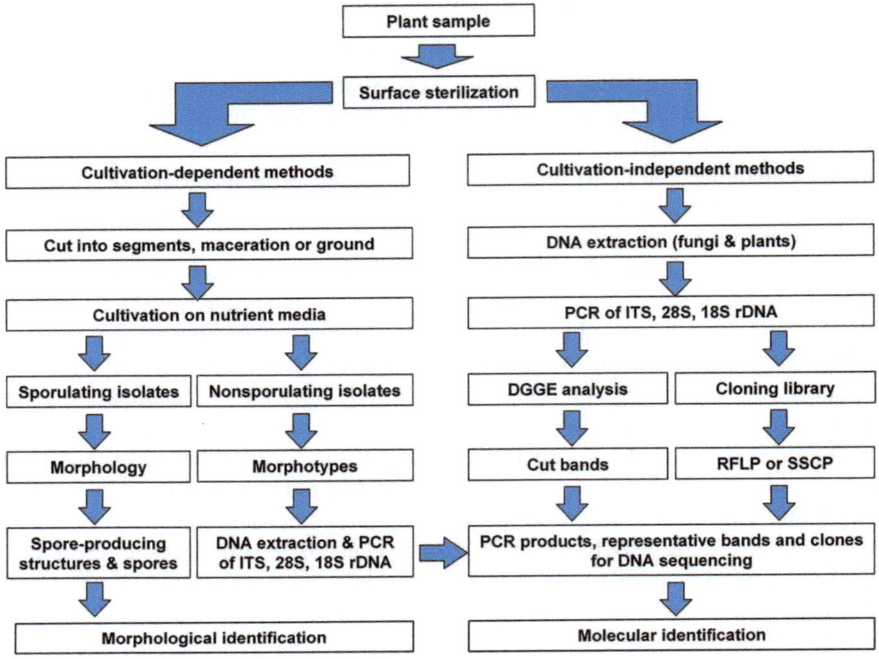

Fig. 1 Endophytic fungi types: reviewing traditional and molecular techniques. (Sun et al. 2011; Sun and Guo 2012)

identifying new biologically active secondary metabolites (Tejesvi et al. 2011). Within this context, the isolation process plays a pivotal role and serves as a crucial step when work with endophytes fungi. Live plant tissue undergoes a series of surface sterilization processes to remove all organisms from the plant surface. Only endophytic fungi are isolated by culturing plant samples on nutrient plates. Cultivation-dependent techniques general involve:(1). thorough washing of plant tissues under running water to remove adhering soil particles, debris, and important epiphytes; (2). surface sterilization of plant tissues to kill any microorganisms on the hosted surface, using different protocols for different tissue types (Hallmann et al. 2006); (3). isolation of endophytic fungi from the sample on nutrient agar; (4). purification and sporulation under different culture conditions; (5). identification of endophytic fungi based on morphological characteristics of the culture (Fig. 1) (Sun et al. 2011).

The effectively isolate technique are method-dependent processes but are suitable for the fast retrieval of a large number of endophytic fungal species from plant tissue. However, the plant-derived endophytic fungal community is directly affected by the surface sterilization technique, culture conditions, and whether the isolate sporulates. The activity of the isolation process must be tailored to the specific tissue and fungus of the plant. In practice, if complete surface sterilization is confirmed, the isolated strain can be considered an endophyte. If no fungal growth occurs, press the surface-sterilized plant tissue onto nutrient medium or culture an aliquot of the previous rinse on nutrient medium. To investigate fungal diversity, sterilized plant tissue is cut into small pieces (about 5 mm in diameter), macerated or ground, and then cultured on nutrient agar to detect both fast and slow-growing fungi.. Isolation should be performed using a variety of media, such as standard MEA (malt extract agar) and PDA (potato dextrose agar), as well as lower media containing plant tissues or extracts. In addition to the amount and size of tissue sampled from the plant, these factors can influence the observed fungal diversity and community composition. Gamboa et al. (2002) demonstrate that in several tropical plants, the number of preserved endophytic fungi increased as the size of the cultured tissue fragment decreased.

10 Advanced Molecular Approaches in Endophyte Studies

The development of molecular biology has opened up new perspectives for studying the diversity of endophytes. The application of molecular biology techniques, such as DNA fingerprinting and sequencing methods, has the potential to overcome the obstacles of traditional culture methods.

10.1 Molecular Identification of Sterile Mycelia

In traditionally culture-dependent methods, fungal isolates can be identified by their morphological characteristics when they form spores on culture medium. Although various methods have been developed to promote spore formation (Sun and Guo 2012), many isolates (54% of the total) do not form spores in culture (Sun et al. 2011). Since conventionally classifications of fungi relied heavily on reproductive structures, it is not possible to provide taxonomic names for these nonspore strains. To understand the considerable diversity of these sterile mycelia, they are usually classified into “morphotypes” based on cultural characteristics like texture, growth rate, and similar colony color (Sun et al. 2011). However, the division of types into differently morphed morphotypes does not reflect the phylogeny of species because morphotypes are not true taxonomic units. To identify and comprehend the diversity of endophytic mycelia sterilia, molecular methods are essential. In a study examining endophytic fungi from the fronds of *L. chinensis* in Hong Kong, a significant portion of the total isolates (16.5%) did not produce spores. It remained classified as Mycelia sterile (Guo et al. 2000). These nonsporulating isolates were categorized into 19 distinct morphotypes based on their cultural morphology and identified across various genera (*Xylaria*, *Mycosphaerella*, and *Diaporthe*), families (Clypeosphaeriaceae and Pleosporaceae), and order (Xylariales) through ITS sequencing analysis. Sun et al. (2011) classified 221 nonsporulating endophyte strains into 56 morphotypes, further organizing these into 37 types based on similarity in ITS sequence and phylogenetic analysis. Likewise, in research involving endophytic fungi from *Pinus tabulaeformis* in China, a considerable number of the total isolates (11%) were also found to remain as mycelia sterile (Wang and Guo 2007). The classification of these nonsporulation isolates resulted in the identification of 74 morphotypes based on their cultural morphology, which were subsequently categorized into 64 types through ITS sequencing analysis (Wang et al. 2005). Additionally, 18 sterile strains classified within the white morphotype were recognized as belonging to *Entoleuca*, *Rosellinia*, and *Nemania* of Xylariaceae, as well as members of Rhytismataceae, using ITS sequencing analysis. This indicated that certain sterile isolates exhibiting similar cultural traits are actually distantly related types (Guo et al. 2003). The study was conducted on endophytic mycota linked to *Vitis vinifera* in central Spain. The nonsporulation strain (78) was identified at both the generic and species levels by comparing its ITS sequencing with those available in GenBank (González and Tello 2011). As a result, the molecular method can enhance our understanding of fungal diversity within natural ecosystems, especially for non-sporulating isolates that cannot be classified taxonomically through morphological characteristics and are overlooked in diversity assessments using conventional techniques.

10.2 Molecular Identification of Isolate Community

Endophytic fungal communities encompass a diverse array of fungi lineages, including Zygomycota, Basidiomycota, and Ascomycota. Due to this diversity, mycologists face considerable challenges in classifying endophytic fungi to specific species or genera based solely on morphology traits. Additionally, the process of complete identification is highly time-intensive. As a result, DNA sequencing techniques, coupled with morphological analysis, have become widely adopted for exploring endophyte diversity, particularly in ecological studies.

For instance, Hoff et al. (2004) identified 27 fungal genera spanning Ascomycota, Zygomycota, and Basidiomycota from the roots of *Pseudotsuga menziesii* and Pinus ponderosa through a combination of morphology and ITS sequence analysis. Similarly, Arnold et al. (2008) isolated 59 distinct ITS genotypes from the asymptomatic foliage of *Pinus taeda*, which corresponded to 24 and 37 individual groups based on 90% and 95% sequencing similarity thresholds, respectively. Ghimire et al. (2011) utilized ITS sequence analysis to identify an endophytic strain of Panicum virgatum located in native tallgrass prairies, revealing a broad spectrum of fungi species belonging to at least 18 taxonomic orders. Their study also highlighted the significantly greater fungal species diversity in shoot tissues compared to root tissues of switchgrass plants. Conversely, Su et al. (2010) observed higher species diversity among endophytic fungi in the roots compared to the leaves of *Stipa grandis* in the Inner Mongolian steppe of China. Studies on endophytes have utilized 18S and 28S genes to classify endophytic fungi at higher taxonomic ranks. For instance, analysis of 18S gene sequencing placed 71 representative strains (out of 257) isolated from bamboo species *Phyllostachys* and *Sasa* into the classes Sordariomycetes and Dothideomycetes. Further identification at lower taxonomic levels was achieved using ITS sequencing data, with some strains potentially representing novel types (Morakotkarn et al. 2007). Likewise, 47 distinct genotypic groups, defined by 90% ITS sequencing similarity, were identified among 280 representative strains sourced from *Picea mariana*, *Huperzia selago*, and *Dryas integrifolia* in boreal forests and Arctic tundra. Phylogenetic analysis combining 18S and 28S gene data revealed that these genotypes belonged to the classes Chaetothyriomycetidae, Leotiomycetes, Sordariomycetes, Ascomycota, Dothideomycetes within Pezizomycetes (Higgins et al. 2007).

In another study, 59 morphologically unidentified strain isolates from the healthy stem and pod of *Theobroma cacao* were identified using 28S gene sequences (Crozier et al. 2006). Most isolates were members of Basidiomycota, particularly corticoid and polyporoid taxa. While some isolates represented rare genera like *Byssomerulius*, the majority belonged to the cosmopolitan genus *Coprinellus* (Agaricales). Similarly, the phylogenetic diversity of fungal endophytes in *Pinus halepensis* was examined through ITS and 28S gene analyses (Botella and Diez 2011). This study found Dothideomycetes to be the dominant group and confirmed the presence of endophytic stages for several pathogens previously linked to *P. halepensis* reject at Spain. Molecular sequence technique thus provides faster and more comprehensive insights into fungal diversity within natural ecosystems compared to traditional morphology-based methods, even when isolates do not sporulate in culture.

10.3 Molecular Methods for Detecting Endophytes Within Plant Tissues

Traditional isolation techniques often fail to capture the full diversity of endophytic fungi, primarily because some species either do not grow or grow too slowly to compete with faster-growing ones on artificial media. To address this limitation and reduce technical bias, molecular techniques have been adopted to detect these fungal direct within hosted tissue. The molecular approach typically involved a series of steps:

1. Extraction of total genomic DNA from fungi and plants using surface-sterilized plant tissue
2. Amplifications of specific DNA fragments (28S, ITS, and 18S genes) from the total DNA using fungus-specific primers
3. Separation of PCR product by denaturing gradient gel electrophoresis (DGGE), followed by excision of distinct DGGE bands, each representing difference types
4. Cloning of the PCR products into plasmid vectors (pGEM-T) and transferring them to *E. coli* DH5α cells

5. Screening of positive clones for distinct types using DNA fingerprint technique (PCR-RFLP)
6. Sequencing of representative clones based on unique fingerprint patterns and DGGE band profiles
7. Identification of sequences at various taxonomic levels through phylogenetic analyses and comparison with similar sequences

This molecular workflow provides a more comprehensive understanding of fungal diversity by circumventing the limitations associated with culture-dependent methodologies.

A prior phylogenetic study on Picea identified a contaminant ITS sequence amplified from plant tissue as belonging to *Hormonema dematioide*, a widespread foliar endophyte of conifers (Camacho et al. 1997). Similarly, a wide range of fungal ITS sequences was directly amplified from genomic DNA derived from Heterosmilax japonica tissue (Gao et al. 2005). Some of these sequences corresponded to well-known fungal genera such as *Guignardia*, *Glomerella*, *Cladosporium*, *Mycosphaerella*, *Phomopsis*, *Botryosphaeria*, and *Aureobasidium*. Others, such as YJ4–70, YJ4–9, and YJ4–61, were more closely related to the encultured environment sample and might signify novel fungal types (Gao et al. 2005).

This present study directly amplified the ITS region of five fungi from the total DNA extract of the fronds of *L. chinensis*. Of the cloned sequence, four matched common isolate fungal, including *Mycosphaerella* (anamorph Cladosporium), *Glomerella* (anamorph Colletotrichum), and members of the Herpotrichiellaceae family within the Ascomycetes. The remaining sequence, classified within Basidiomycota, could not be detected using traditionally cultivation-dependent methods (Guo et al. 2001).

Similarly, endophytic fungal rDNA fragments (ITS and 28S) amplified from surface-sterilized needles of 12 *Pinus taeda* trees (Arnold and Lutzoni 2007) revealed that cloned endophytic fungi spanned multiple lineages of Basidiomycota and Ascomycota. Among these, Leotiomycetes and Dothideomycetes within Ascomycota were both previously known to be common isolates from P. taeda through culture methods. Interestingly, no Sordariomycetes were identified among the clone sequences despite their prevalence in culture-based studies. A similar discrepancy in diversity and community composition among endophytic fungi was observed in the leaves of Rhododendron tomentosum when comparing cultivation-dependent methods with molecular techniques (Tejesvi et al. 2011).

Denaturing gradient gel electrophoresis (DGGE) has proven effective for separating closely related DNA sequences based on their differential mobility in a gradient of denaturant. This method has been successfully utilized to document endophytic fungi communities by excising and sequencing distinct bands, allowing researchers to identify community members through database searches and phylogenetic analyses. For instance, Duong et al. (2006) identified 14 Operational Taxonomic Units (OTUs) from *Magnolia liliifera* using DGGE combined with sequencing. These findings were completely distinct from those uncovered through traditional, cultivation-dependent methods (Promputtha et al. 2005). However, while DGGE offers several advantages, it also has limitations. Typically, shorter DNA fragments (<500 bp) yield better resolution within band profiles; however, this compromises the taxonomic information required for identifying fungi at the genus or species level (Duong et al. 2006). Additionally, even the most advanced stain techniques are often insufficient to detect the complete diversity present in a sample, especially rarer fungal community members (Anderson and Cairney 2004). Future studies should consider using primers that are more universally applicable to fungal species and capable of providing improved phylogenetic resolution at finer taxonomic levels (Dumbrell et al. 2011).

The simple sequence repeat (SSR) marker technique offers an inexpensive, rapid, and straightforward alternative. It shares some similarities with randomly amplified polymorphic DNA (RAPD) analyses, but employs long primers (around 18 nucleotides) and requires more stringent amplification conditions, such as higher annealing temperatures. Microsatellite-containing genomic regions evolve and mutate faster than other genomic areas (Burgess et al. 2001). Consequently, using longer primers and stricter conditions produces high-reproducing SSR markers that are far more robust than traditional RAPD markers (Liang et al. 2005). In endophyte research, Groppe and Boller (1997) designed specific primer pairs targeting microsatellite-flanked regions and successfully detected rDNA fragments of endophytic Epichloë species in infected *Bromus erectus* tissue. No fragment was amplified from DNA isolated from uninfected plant material or unrelated fungal species from the same grass. These findings underscore how molecular techniques reveal fungal diversity and community compositions that diverge significantly from those identified by cultivation-based methods.

10.4 Advances in High-Throughput Sequencing for Endophyte Research

High-throughput sequencing, specifically pyrosequencing, has emerged as a cutting-edge tool for metagenomics and metagenetic analysis, offering a compelling alternative for studying fungal communities in natural environments. This method is particularly advantageous due to its speed, relatively low cost, elimination of the cloning step, and capacity for high output approximately 100 times greater than that of traditional Sanger sequence (Margulies et al. 2005). Pyrosequencing has been effectively applied to explore fungal diversity in various contexts, including clinically fungal (Gharizadeh et al. 2004), phyllo sphere fungal (Jumpponen and Jones 2009), wood-inhabiting fungal (Ovaskainen et al. 2010), soil fungal (Jumpponen et al. 2010), freshwater fungi (Monchy et al. 2011), and mycorrhizal fungal (Zhang et al. 2024).

Compared to Sanger sequencing, pyrosequencing provides several clear benefits, such as reduced costs, faster processing times, higher sample throughput, improved species recovery, and a more comprehensive, less biased profile of microbial community composition in natural ecosystems (Jumpponen et al. 2010). Despite these advantages, certain technical biases can influence the representation of fungal diversity in pyrosequencing datasets (Tedersoo et al. 2010). For instance, the DNA extraction technique significantly impacts both the number of sequences obtained and the diversity of species identified, likely due to varying capacities to break down fungal cell walls and eliminate inhibitors. Additionally, the choice of primers plays a critical role in determining the number of fungal sequences recovered (Lucero et al. 2011).

To address PCR biases in environmental studies, DNA extract protocols are optimized to improve outcomes. Primers should be designed to encompass all target types, multiple PCR replicates are conduct, and the number of PCR cycles is typically reduced to 20 to 25 to minimize amplification errors (Tedersoo et al. 2010). Additionally, the DNA fragments obtained through pyrosequencing are often limited in length, usually ranging from 350 to 500 base pairs, which can hinder taxonomic resolution. However, advancements in sequencing technology, such as single-molecule real-time sequencing by Pacific Biosciences, have reportedly achieved an average read length of 2700 bases. In conventional pyro sequence datasets, singletons represent the largest source of error, as most are artificial and exhibit a significantly higher frequency of insertion compared to naturally occurring and interspecific variation (Tedersoo et al. 2010). As such, future research will require automated species identification methods accompanied by robust error assessment frameworks (Ovaskainen et al. 2010).

The pyrosequence technique has greatly improved the ability to characterize fungal diversity compared to the traditional Sanger sequencing method. For instance, a study by Gillevet et al. (2009) examined the number of fungal ITS sequences obtained from *Spartina alterniflora* in a salt marsh using both clone sequencing and pyrosequencing. Their findings showed that pyrosequence not only provided deeper coverage by identifying a greater number of contains but also revealed entire clades that traditionally methods failed to detect. Pyrosequence was first applied in the study of endophytic fungi from *Atriplex canescens* and *A. torreyi* var. *griffithsii* (Lucero et al. 2011).

High-throughput sequence marks a remarkable technologically advancement with the potential to transform mycology. However, there is a pressing need for a unified framework to standardize the process and interpreted of results within a taxonomic case. Notably, Nilsson et al. (2011) advocated for standardizing the description and publication of next-generation sequence datasets for fungi communities, offering significant benefits for future research on fungal diversity and ecology.

10.5 DNA Barcoding Approaches for Identifying Endophytic Fungi

The DNA barcode system utilizes a short, efficient, and standardized gene region to identify species (Hebert et al. 2003). For successful species identification, the interspecific distance of the DNA barcode region must surpass the inter- and intraspecific distance. Identification becomes more straightforward when a sequence is unique to a particular species and remains consistent within that species (Letourneau et al. 2010). Ideally, a single locus would serve as the universal DNA barcode region across all organism groups and kingdom. To date, the DNA barcode has proven effective in the animal kingdom, notably employing a 648-base-pair region of the cytochrome c oxidase 1 (*CO1*) gene (Hajibabaei et al. 2006).

In mycology, various DNA gene fragments have been explored for fungal DNA barcoding (Zhao et al. 2011). While the *CO1* gene was initially considered, findings revealed the presence of multiple introns (ranging from 1 to 7), with varying lengths (134 base pairs to 3.1 kilobases), making it unsuitable as a fungal barcode (Vialle et al. 2009). The authors examined 979 internal transcribed spacer (ITS) sequences from 88 *Hypocrea* species (anamorph Trichoderma) and developed a DNA barcoding system for species identification (The TrichOKEY v1.0) (Druzhinina et al. 2005). This system successfully identifies 51 strains isolated from soil to their respective known species. The ITS region, as a candidate DNA barcode, has been extensively applied in the identification of fungal species. Examples include its use in Cortinarius (Frøslev et al. 2007), Amanita (Zhang et al. 2010), Melampsora and Chrysomyxa (Vialle et al. 2009), Mucorales (Schwarz et al. 2006), the Tricholoma scalpturatum complex (Jargeat et al. 2010), lichenized fungal (Kelly et al. 2011), aquatic hyphomycete species (Seena et al. 2010), and ectomycorrhizal fungal in naturally forests (Wang et al. 2011).

The ITS region serves as a widely utilized DNA barcode for fungi, offering several undeniable advantages. These include a high success rate of amplification across various fungal lineages, the use of universal primers, an optimal fragment length, and access to a substantial number of existing databases (Seifert 2009). Despite these strengths, the ITS barcode has notable limitations. Among them are the varying inter- and intra-specific distances across different fungal groups (Zhao et al. 2011), which complicates the establishment of a unified ITS divergence threshold for species identification (Sun et al. 2011). On the other hand, Arnold et al. (2007) demonstrated that ITS genotypes grouped based on 90% sequence similarity, aligned with species categorized by 28S rDNA in 72 Ascomycota and Basidiomycota, 145 culture endophytic fungi, and 33 environmentally PCR samples.

Still, several prior studies highlight ITS's insufficiency for delimiting some fungal species, particularly in genera or species complexes that evolve rapidly or display extensive diversity (Gazis et al. 2011). Moreover, the reference sequences available are limited; less than 1% of the estimated 1.5 million fungal species are represented in databases like EMBL, BOLD, and NCBI GenBank, despite daily contributions of new fungal DNA sequences (Vilgalys 2003). Additionally, misidentifications in published sequences pose another challenge, with approximately 20% of named sequences potentially associated with organisms that have been incorrectly identified. This issue further restricts the reliability of sequence-based fungal identification methods (Jumpponen et al. 2010).

A workshop focused on identifying the most suitable gene for fungal DNA barcoding took place from April 19 to 20, 2011, in Amsterdam. The event brought together 55 mycologists from 18 countries, many of whom were part of the fungi work group of the Consortium for the Barcode of Life (CBOL). Preliminary analysis involving six preselected gene sequences, which represent all major Eumycota lineages, was reviewed (details available at http://www.fungalbarcoding.org). Among these, the ITS region emerged as the strongest candidate for a fungal DNA barcode marker, achieving a 72% identification success rate across all fungi. The workshop concludes with a structure plan to submit a formal proposal to CBOL, advocating ITS as the primary fungal DNA barcode. Subsequently, ITS was officially endorsed as the primary DNA barcoding method for fungi at the fourth International Barcode of Life Conference, held in Adelaide, Australia, from November 28 to December 3, 2011. In endophyte research, ITS remains the most widely utilized DNA barcode for molecular identification. It has proven valuable in studies of diversity and ecology, although it presents some limitations in distinguishing closely related species (Sun et al. 2011).

11 Conclusion

Increasing research suggests that endophytic fungi have a significant impact on plant physiology, development, and metabolism, making them an essential focus in plant science. Despite this, the intricate micromolecular mechanisms of these interactions remain poorly understood due to their complexity. This chapter explores the recent advances in understanding the symbiotic dynamics between endophytic fungi and plants. It highlights key aspects such as signal generation, plant defense mechanisms, and fungal strategies that help establish and maintain balanced relationships between fungi and their host plants, ultimately forming what can be described as a "community of shared future." These interactions occur in a synergistic manner, with both fungi and plants playing essential roles in fostering stable symbiosis.

The findings contribute to a deeper understanding of species interactions and offer practical applications for addressing real-world challenges. DNA barcoding is set to become a crucial tool in identifying fungal species, although current sequence repositories face limitations due to inconsistent taxonomy and insufficient annotation in public DNA databases. To address this, taxonomic experts across the fungal group should prioritize sequencing well-identified and thoroughly annotating specimens from herbaria worldwide to build a reliable and correctly named sequence database. A common set of primers and technologies that allow the amplification and sequencing of all fungi at both the primary and secondary barcode loci; This step will significantly enhance the future application of DNA barcoding in the study of endophytic fungi.

Endophytic fungi are widespread, existing in all known plants across diverse ecosystems. However, geographic variations in their diversity, community composition, and host or tissue preferences remain underexplored. To gain a better understanding of fungal endophyte ecology, more comprehensive data on key parameters of endophyte symbiosis are needed on both regional and continental scales, encompassing various ecosystems.

References

Abhijeet Singh YM (2014) Understanding the biodiversity and biological applications of endophytic fungi: a review. J Microb Biochem Technol s8

Aly AH, Debbab A, Proksch P (2011) Fungal endophytes: unique plant inhabitants with great promises. Appl Microbiol Biotechnol 90(6):1829–1845

Anderson IC, Cairney JWG (2004) Diversity and ecology of soil fungal communities: increased understanding through the application of molecular techniques. Environ Microbiol 6:769–779

Arnold A (2008) Hidden within our botanical richness, a treasure trove of fungal endophytes, vol 32. Plant Press, pp 13–15

Arnold AE, Lutzoni F (2007) Diversity and host range of foliar fungal endophytes: are tropical leaves biodiversity hotspots? Ecology 88(3):541–549

Azevedo JL (1998) Microrganismos endofíticos. Ecologia Microbiana:117–137

Bamisile BS, Dash CK, Akutse KS, Keppanan R, Wang LD (2018) Fungal endophytes: beyond herbivore management. Front Microbiol Mar 9:11

Banerjee D (2011) Endophytic fungal diversity in tropical and subtropical plants. Res J Microbiol 2011(6):54–62

Baron NC, Rigobelo EC (2022) Endophytic fungi: a tool for plant growth promotion and sustainable agriculture. Mycol Int J Fungal Biol 13(1):39–55

Baskaran XR, Geo Vigila AV, Zhang SZ, Feng SX, Liao WB (2018) A review of the use of pteridophytes for treating human ailments. J Zhejiang Univ Sci B 19:85–119

Berg G (2009) Plant-microbe interactions promoting plant growth and health: perspectives for controlled use of microorganisms in agriculture. Appl Microbiol Biotechnol 84(1):11–18

Botella L, Diez JJ (2011) Phylogenic diversity of fungal endophytes in Spanish stands of Pinus halepensis. Fungal Divers 47:9–18

Burgess T, Wingfield MJ, Wingfield BW (2001) Simple sequence repeat markers distinguish among morphotypes of Sphaeropsis sapinea. Appl Environ Microbiol 67:354–362

Busby PE, Ridout M, Newcombe G (2016) Fungal endophytes: modifiers of plant disease. Plant Mol Biol 90(6):645–655

Camacho FJ, Gernandt DS, Liston A, Stone JK, Klein AS (1997) Endophytic fungal DNA, the source of contamination in spruce needle DNA. Mol Ecol 6:983–987

Card S, Johnson L, Teasdale S, Caradus J (2016) Deciphering endophyte behaviour: the link between endophyte biology and efficacious biological control agents. FEMS Microbiol Ecol 92(8):19

Chadha N, Mishra M, Prasad R (2014) Varma a. root endophytic fungi: research update. J Biol Life Sci 5:135

Chadha N, Mishra M, Rajpal K, Bajaj R, Choudhary DK, Varma A (2015) An ecological role of fungal endophytes to ameliorate plants under biotic stress. Arch Microbiol 197:869–881

Chandra S (2000) The ferns of India: enumeration, synonyms & distribution. International Book Distributors, Dehra Dun, p 459

Chang HL, Su JH, Yeh YT, Lee YC, Chen HM, Wu YC, Yuan SS (2008) Protoapigenone, a novel flavonoid, inhibits ovarian cancer cell growth in vitro and in vivo. Cancer Lett 267:85–95

Chang HC, Gupta SK, Tsay HS (2011) Studies on folk medicinal fern: an example of 'Gu-sui-Bu'. In: Kumar A, Fernández H, Revilla MA (eds) Working with ferns: issues and applications. Springer, New York, pp 285–304

Chávez R, Fierro F, García-Rico RO, Vaca I (2015) Filamentous fungi from extreme environments as a promising source of novel bioactive secondary metabolites. Front Microbiol 6

Chen WH, Wu SJ, Sun XL, Feng KM, Rahman K, Tan HY, Han T (2020) High-throughput sequencing analysis of endophytic fungal diversity in cynanchum sp. S Afr J Bot 134:349–358

Chitnis VR, Suryanarayanan TS, Nataraja KN, Prasad SR, Oelmuller R, Shaanker RU (2020) Fungal endophyte-mediated crop improvement: the way ahead. Front Plant Sci 11:10

Chowdappa S, Jagannath S, Konappa N, Udayashankar AC, Jogaiah S (2020) Detection and characterization of antibacterial siderophores secreted by endophytic fungi from cymbidium aloifolium. Biomolecules 10(10):18

Chowdhary K, Kaushik N, Coloma AG, Raimundo CM (2012) Endophytic fungi and their metabolites isolated from Indian medicinal plant. Phytochem Rev 11(4):467–485

Crozier J, Thomas SE, Aime MC, Evans HC, Holmes KA (2006) Molecular characterization of fungal endophytic morphospecies isolated from stems and pods of theobromine cacao. Plant Pathol 55:783–791

Daley DK, Brown KJ, Badal S (2017) Pharmacognosy: fundamentals, applications and strategy. In: Fungal metabolites. Elsevier Inc, Amsterdam, pp 413–421

Deckert RJ, Melville LH, Peterson RL (2001) Structural features of a Lophodermium endophyte during the cryptic life-cycle phase in the foliage of Pinus strobus. Mycol Res 105:991–997

Demers D, Knestrick M, Fleeman R, Tawfik R, Azhari A, Souza A, Vesely B, Netherton M, Gupta R, Colon B et al (2018) Exploitation of mangrove endophytic fungi for infectious disease drug discovery. Mar Drugs 16:376

Deshmukh SK, Gupta MK, Prakash V, Sudhakara Reddy M (2018) Mangrove-associated fungi: a novel source of potential anticancer compounds. J Fungi 4:101

Deyett E, Rolshausen PE (2020) Endophytic microbial assemblage in grapevine. FEMS Microbiol Ecol 96(5):fiaa053

Dong LB, Yang J, He J, Luo HR, Wu XD, Deng X, Peng LY, Cheng X, Zhao QS (2012) Lycopalhine a, a novel sterically congested lycopodium alkaloid with an unprecedented skeleton from Palhinhaea cernua. Chem Commun 48:9038–9040

Druzhinina IS, Kopchinskiy AG, Komon M, Bissett J, Szakacs G, Kubicek CP (2005) An oligonucleotide barcode for species identification in Trichoderma and Hypocrea. Fungal Genet Biol 42:813–828

Dumbrell AJ, Ashton PD, Aziz N, Feng G, Nelson M, Dytham C, Fitter AH, Helgason T (2011) Distinct seasonal assemblages of arbuscular mycorrhizal fungi revealed by massively parallel pyrosequencing. New Phytol 190:794–804

Duong LM, Jeewon R, Lumyong S, Hyde KD (2006) DGGE coupled with ribosomal DNA gene phylogenies reveal uncharacterized fungal phylotypes. Fungal Divers 23:121–138

Dutta D, Puzari KC, Gogoi R, Dutta P (2014) Endophytes: exploitation as a tool in plant protection. Braz Arch Biol Technol 57:621–629

Eslahi N, Kowsari M, Motallebi M, Zamani MR, Moghadasi Z (2020) Influence of recombinant Trichoderma strains on growth of bean (Phaseolus vulgaris L) by increased root colonization and induction of root growth related genes. Sci Hortic 261:108932

Firáková S, Šturdíková M, Múčková M (2007) Bioactive secondary metabolites produced by microorganisms associated with plants. Biologia 62:251–257

Fouda AH, Hassan SED, Eid AM, Ewais EED (2015) Biotechnological applications of fungal endophytes associated with medicinal plant Asclepias sinaica (Bioss.). Ann Agric Sci 60:95–104

Frøslev TG, Jeppesen TS, Laessoe T, Kjoller R (2007) Molecular phylogenetics and delimitation of species in Cortinarius section Calochroi (Basidiomycota, Agaricales) in Europe. Mol Phylogenet Evol 44:217–227

Fu W, Du G, Liu D, Ruan JL (2013) Neuroprotective effect of a caffeic acid derivative from Abacopteris penangiana. Pharm Biol 51:376–382

Gakuubi MM, Munusamy M, Liang ZX, Bee Ng S (2021) Fungal endophytes: a promising frontier for discovery of novel bioactive compounds. J Fungi (Basel) 7(10):786

Gamboa MA, Laureano S, Bayman P (2002) Measuring diversity of endophytic fungi in leaf fragments: does size matter? Mycopathologia 156:41–45

Gamboa MA, Laureano S, Bayman P (2003) Measuring diversity of endophytic fungi in leaf fragments: does size matter? Mycopathologia 2003:41

Gao XX, Zhou H, Xu DY, Yu CH, Chen YQ, Qu LH (2005) High diversity of endophytic fungi from the pharmaceutical plant, Heterosmilax japonica Kunth revealed by cultivation-independent approach. FEMS Microbiol Lett 249:255–266

Gazis R, Rehner S, Chaverri P (2011) Species delimitation in fungal endophyte diversity studies and its implications in ecological and biogeographic inferences. Mol Ecol 20:3001–3013

Gharizadeh B, Norberg E, Löffler J, Jalal S, Tollemar J, Einsele H, Klingspor L, Nyrén P (2004) Identification of medically important fungi by the pyrosequencing (TM) technology. Mycoses 47:29–33

Ghimire SR, Charlton ND, Bell JD, Krishnamurthy YL, Craven KD (2011) Biodiversity of fungal endophyte communities inhabiting switchgrass (Panicum virgatum L.) growing in the native tallgrass prairie of northern Oklahoma. Fungal Divers 47:19–27

Gillevet PM, Sikaroodi M, Torzilli AP (2009) Analyzing salt-marsh fungal diversity: comparing ARISA fingerprinting with clone sequencing and pyrosequencing. Fungal Ecol 2:160–167

González V, Tello ML (2011) The endophytic mycota associated with Vitis vinifera in Central Spain. Fungal Divers 47:29–42

Götz M, Nirenberg H, Krause S, Wolters H, Draeger S, Buchner A, Lottmann J, Berg G, Smalla K (2006) Fungal endophytes in potato roots studied by traditional isolation and cultivation-independent DNA-based methods. FEMS Microbiol Ecol 58:404–413

Groppe K, Boller T (1997) PCR assay based on a microsatellite-containing locus for detection and quantification of Epichloë endophytes in grass tissue. Appl Environ Microbiol 63:1543–1550

Guo LD, Hyde KD, Liew ECY (2000) Identification of endophytic fungi from Livistona chinensis (Palmae) using morphological and molecular techniques. New Phytol 147:617–630

Guo LD, Hyde KD, Liew ECY (2001) Detection and identification of endophytic fungi within frond tissues of Livistona chinensis based on rDNA sequence. Mol Phylogenet Evol 20:1–13

Guo LD, Huang GR, Wang Y, He WH, Zheng WH, Hyde KD (2003) Molecular identification of white morphotype strains of endophytic fungi from Pinus tabulaeformis. Mycol Res 107:680–688

Hajibabaei M, Janzen DH, Burns JM, Hallwachs W, Hebert PDN (2006) DNA barcodes distinguish species of tropical Lepidoptera. Proc Natl Acad Sci USA 103:968–971

Hallmann J, Berg G, Schulz B (2006) Isolation procedures for endophytic microorganisms. In: Schulz BJE, Boyle CJC, Sieber TN (eds) Microbial root endophytes. Springer, Berlin, pp 299–319

Hardoim PR, Overbeek LSV, Berg G, Pirttilä AM, Compant S, Coampisano A, Döring M, Sessitsch A (2015) The hidden world within plants: ecological and evolutionary considerations for defining functioning of microbial endophytes. Microbiol Mol Biol Rev 2015:293

Hebert PDN, Cywinska A, Ball SL, Waard JR (2003) Biological identifications through DNA barcodes. Proc R Soc Lond B 270:313–321

Higgins KL, Arnold AE, Miadlikowska J, Sarvate SD, Lutzoni F (2007) Phylogenetic relationships, host affinity, and geographic structure of boreal and arctic endophytes from three major plant lineages. Mol Phylogenet Evol 42:543–555

Ho R, Girault JP, Raharivelomanana P, Lafont R (2012) E-and Z-isomers of new phytoecdysteroid conjugates from French Polynesian Microsorum membranifolium (Polypodiaceae) fronds. Molecules 17:11598–11606

Hodkinson TR, Doohan FM, Saunders MJ, Murphy BR (2019) Endophytes for a growing world. Cambridge University Press, Cambridge, pp 3–22

Hoff JA, Klopfenstein NB, McDonald GI, Tonn JR, Kim MS, Zambino PJ, Hessburg PF, Rogers JD, Peever TL, Carris LM (2004) Fungal endophytes in woody roots of Douglas-fir (Pseudotsuga menziesii) and ponderosa pine (Pinus ponderosa). For Pathol 34:255–271

Jargeat P, Martos F, Carriconde F, Gryta H, Moreau PA, Gardes M (2010) Phylogenetic species delimitation in ectomycorrhizal fungi and implications for barcoding: the case of the Tricholoma scalpturatum complex (Basidiomycota). Mol Ecol 19:5216–5230

Jaroszuk-Ściseł J, Kurek E, Trytek M (2014) Efficiency of indoleacetic acid, gibberellic acid and ethylene synthesized in vitro by fusarium culmorum strains with different effects on cereal growth. Biologia 69(3):281–292

Jia Q, Qu J, Mu H, Sun H, Wu C (2020) Foliar endophytic fungi: diversity in species and functions in forest ecosystems. Symbiosis 80:103–132

Joshi B, Tewari S, Srivastava A, Mishra RK (2019) Ecology and ethnobotany of ceratopteris thalictroides (L.) Brongn. In: Horticulture and ecology-studies & management. Discovery Publishing House Pvt. Ltd., New Delhi

Jumpponen A, Jones KL (2009) Massively parallel 454 sequencing indicates hyperdiverse fungal communities in temperate Quercus macrocarpa phyllosphere. New Phytol 18:438–444

Jumpponen A, Jones KL, Blair J (2010) Vertical distribution of fungal communities in tallgrass prairie soil. Mycologia 102:1027–1041

Kaddes A, Fauconnier ML, Sassi K, Nasraoui B, Jijakli MH (2019) Endophytic fungal volatile compounds as solution for sustainable agriculture. Molecules 24(6):16

Kaul S, Gupta S, Ahmed M, Dhar MK (2012) Endophytic fungi from medicinal plants: a treasure hunt for bioactive metabolites. Phytochem Rev 11:487–505

Kelly LJ, Hollingsworth PM, Coppins BJ, Ellis CJ, Harrold P, Tos J, Yahr R (2011) DNA barcoding of lichenized fungi demonstrates high identification success in a floristic context. New Phytol 191:288–300

Khan AL, Hussain J, Al-Harrasi A, Al-Rawahi A, Lee IJ (2015) Endophytic fungi: resource for gibberellins and crop abiotic stress resistance. Crit Rev Biotechnol 35(1):62–74

Khare E, Mishra J, Arora NK (2018) Multifaceted interactions between endophytes and plant: developments and prospects. Front Microbiol 9:2732

Kim CK, Eo JK, Eom AH (2013) Diversity and seasonal variation of endophytic fungi isolated from three conifers in Mt. Taehwa, Korea. Mycobiology 41:82–85

Kloepper JW, McInroy JA, Liu K, Hu CH (2013) Symptoms of fern distortion syndrome resulting from inoculation with opportunistic endophytic fluorescent pseudomonas spp. PLoS One 8

Kuncoro H, Farabi K, Rijai L, Julaeha E, Supratman U, Shiono Y (2018) Flavonoid compounds from Krokot herbs (Lygodium microphyllum) and their antioxidants activity against DPPH. J Math Fund Sci 50:192–202

Kusari S, Hertweck C, Spiteller M (2012) Chemical ecology of endophytic fungi: origins of secondary metabolites. Chem Biol 19(7):792–798

Latz MAC, Jensen B, Collinge DB, Jorgensen HJL (2018) Endophytic fungi as biocontrol agents: elucidating mechanisms in disease suppression. Plant Ecol Divers 11(5–6):555–567

Lei YF, Chen JL, Wei H, Xiong CM, Zhang YH, Ruan JL (2011) Hypolipidemic and anti-inflammatory properties of Abacopterin a from Abacopteris penangiana in high-fat diet-induced hyperlipidemia mice. Food Chem Toxicol 49:3206–3210

Letourneau A, Seena S, Marvanová L, Bärlocher F (2010) Potential use of barcoding to identify aquatic hyphomycetes. Fungal Divers 40:51–64

Liang Y, Guo LD, Ma KP (2005) Population genetic structure of an ectomycorrhizal fungus amanita manginiana in a subtropical forest over two years. Mycorrhiza 15:137–142

Lu S (2007) Pteridology. Higher Education Press, Beijing

Lucero ME, Unc A, Cooke P, Dowd S, Sun SL (2011) Endophyte microbiome diversity in micropropagated Atriplex canescens and Atriplex torreyi var. griffithsii. PLoS One 6:e17693

Lugtenberg BJJ, Caradus JR, Johnson LJ (2016) Fungal endophytes for sustainable crop production. FEMS Microbiol Ecol 92(12):17

Margulies M, Egholm M, Altman WE, Attiya S, Bader JS, Bemben LA, Berka J, Braverman MS, ChenYJ Chen ZT (2005) Genome sequencing in microfabricated high-density picolitre reactors. Nature 437:376–380

Materatski P, Varanda C, Carvalho T, Dias AB, Campos MD, Rei F, Félix MR (2019) Spatial and temporal variation of fungal endophytic richness and diversity associated to the phyllosphere of olive cultivars. Fungal Biol 123:66–76

Mehltreter K, Walker LR, Sharpe JM (2010) Fern ecology. Cambridge University Press, Cambridge

Mehltreter K, Rand WL, Sharpe JM (2012) Interactions of ferns with fungi and animals. In: Fern ecology. Cambridge University Press, pp 220–254

Meshram V, Gupta M (2019) Endophytic fungi: a quintessential source of potential bioactive compounds. Cambridge University Press, Cambridge

Modolo LV, Foglio MA (2020) Brazilian medicinal plants. CRC Press, Boca Raton

Monchy S, Sanciu G, Jobard M, Rasconi S, Gerphagnon M, Chabé M, Cian A, Meloni D, Niquil N, Christaki U (2011) Exploring and quantifying fungal diversity in freshwater lake ecosystems using rDNA cloning/sequencing and SSU tag pyrosequencing. Environ Microbiol 13:1433–1453

Morakotkarn D, Kawasaki H, Seki T (2007) Molecular diversity of bamboo-associated fungi isolated from Japan. FEMS Microbiol Lett 266:10–19

Muhammad M, Ismail ZS, Schneider H, Hawkins JA (2020) Medicinal use of ferns: an ethnobotanical review. Sains Malaysiana 49:1003–1014

Murphy BR, Doohan FM, Hodkinson TR (2018) From concept to commerce: developing a successful fungal endophyte inoculant for agricultural crops. J Fungi 4(1):11

Nilsson RH, Tedersoo L, Lindah BD, Kjøller R, Carlsen T, Quince C, Abarenkov K, Pennanen T, Stenlid J, Bruns T (2011) Towards standardization of the description and publication of next-generation sequencing datasets of fungal communities. New Phytol 191:314–318

Nugraha AS, Haritakun R, Keller PA (2013) Constituents of the Indonesian epiphytic medicinal plant Drynaria rigidula. Nat Prod Commun 8:703–705

Olmo-Ruiz MD, Arnold AE (2014) Interannual variation and host affiliations of endophytic fungi associated with ferns at La Selva, Costa Rica. Mycologia 106(1):8–22

Ovaskainen O, Nokso-Koivista J, Hottola J, Rajala T, Pennanen T, Ali-Kovero H, Miettinen O, Oinonen P, Auvinen P, Paulin L (2010) Identifying wood-inhabiting fungi with 454 sequencing – what is the probability that BLAST gives the correct species? Fungal Ecol 3:274–283

Paciencia MLB, Prado J (2005) Effects of forest fragmentation on pteridophyte diversity in a tropical rain forest in Brazil. Plant Ecol 180:87–104

Pan SY, Litscher G, Gao SH, Zhou SF, Yu ZL, Chen HQ, Zhang SF, Tang MK, Sun JN, Ko KM (2014) Historical perspective of traditional indigenous medical practices: the current renaissance and conservation of herbal resources. Evid Based Complement Alternat Med 27:1–20

Poveda J, Abril-Urias P, Escobar C (2020) Biological control of plant-parasitic nematodes by filamentous fungi inducers of resistance: trichoderma, mycorrhizal and endophytic fungi. Front Microbiol 11:14

Prado J, Sylvestre LS (2010) As samambaias e licofitas do Brasil. In: Forzza RC (ed) Catalogo de plantas e fungos do Brasil. Instituto de Pesquisas Jardim Botanico, Rio de Janeiro

Promputtha I, Jeewon R, Lumyong S, McKenzie EHC, Hyde KD (2005) Ribosomal DNA fingerprinting in the identification of non sporulating endophytes from Magnolia liliifera (Magnoliaceae). Fungal Divers 20:167–186

Quattrocchi UFLS (2012) CRC world dictionary of medicinal and poisonous plants: common names, scientific names, eponyms, synonyms, and etymology. CRC Press, Boca Raton

Rana KL, Kour D, Kaur T, Devi R, Negi C, Yadav AN, Singh K, Saxena AK (2020a) Endophytic fungi from medicinal plants: biodiversity and biotechnological applications. In: Kumar A, Radhakrishnan EK (eds) Microbial endophytes. Woodhead Publishing, Cambridge, pp 273–305
Rana KL, Kour D, Kaur T, Devi R, Yadav AN, Yadav N, Dhaliwal HS, Saxena AK (2020b) Endophytic microbes: biodiversity, plant growth-promoting mechanisms and potential applications for agricultural sustainability. Antonie Van Leeuwenhoek 113(8):1075–1107
Rashmi M, Kushveer JS, Sarma VV (2019) A worldwide list of endophytic fungi with notes on ecology and diversity. Mycosphere 10(1):798–1079
Razaghi P, Abdel-Azeem AM (2024) Diversity of endophytic fungi hosted medicinal ferns: biotechnological potentials and possible applications
Rodriguez RJ, White JF, Arnold AE, Redman RS (2009) Fungal endophytes: diversity and functional roles: Tansley review. New Phytol 182:314–330
Roopa G, Madhusudhan MC, Sunil KCR, Lisa N, Calvin R, Poornima R, Zeinab N, Kini KR, Prakash HS, Geetha N (2015) Identification of Taxol-producing endophytic fungi isolated from Salacia oblonga through genomic mining approach. J Genet Eng Biotechnol 13:119–127
Schwarz P, Bretagne S, Gantier JC, Garcia-Hermoso D, Lortholary O, Dromer F, Dannaoui E (2006) Molecular identification of Zygomycetes from culture and experimentally infected tissues. J Clin Microbiol 44:340–349
Seena S, Pascoal C, Marvanová L, Cássio F (2010) DNA barcoding of fungi: a case study using ITS sequences for identifying aquatic hyphomycete species. Fungal Divers 44:77–87
Seifert KA (2009) Progress towards DNA barcoding of fungi. Mol Ecol Res 9(1):83–89
Shearer CA, Descals E, Kohlmeyer B, Kohlmeyer J, Marvanová L, Padgett D, Porter D, Raja HA, Schmit JP, Thorton HA et al (2007) Fungal biodiversity in aquatic habitats. Biodivers Conserv 16:49–67
Sieber TN (2007) Endophytic fungi in forest trees: are they mutualists? Fungal Biol Rev 21:75–89
Smith RL (2014) A fern cultured from Antarctic glacier detritus. Antarct Sci 26:341–344
Socolsky C, Asakawa Y, Bardon A (2007) Diterpenoid glycosides from the bitter fern Gleichenia quadripartita. J Nat Prod 70:1837–1184
Souza BD, dos Santos TT (2017) Endophytic fungi in economically important plants: ecological aspects, diversity and potential biotechnological applications. J Bioenergy Food Sci 4(2): 113–126
Strobel G, Daisy B (2003) Bioprospecting for microbial endophytes and their natural products. Microbiol Mol Biol Rev 2003(67):491–502
Su YY, Guo LD, Hyde KD (2010) Response of endophytic fungi of Stipa grandis to experimental plant function group removal in Inner Mongolia steppe, China. Fungal Divers 43:93–101
Suebrasri T, Harada H, Jogloy S, Ekprasert J, Boonlue S (2020) Auxin-producing fungal endophytes promote growth of sunchoke. Rhizosphere 16:9
Sun X, Guo LD (2012) Endophytic fungal diversity: review of traditional and molecular techniques. Mycol Int J Fungal Biol 3(1):65–76
Sun X, Guo LD, Hyde KD (2011) Community composition of endophytic fungi in Acer truncatum and their role in decomposition. Fungal Divers 47:85–95
Tan X, Zhou Y, Zhou X, Xia X, Wei Y, He L, Tang H, Yu L (2018) Diversity and bioactive potential of culturable fungal endophytes of Dysosma versipellis; a rare medicinal plant endemic to China. Sci Rep 8:1–9
Tanaka A, Takemoto D, Chujo T, Scott B (2012) Fungal endophytes of grasses. Curr Opin Plant Biol 15:462–468
Tedersoo L, Nilsson RH, Abarenkov K, Jairus T, Sadam A, Saar I, Bahram M, Bechem E, Chuyong G, Kõljalg U (2010) Pyrosequencing and sanger sequencing of tropical mycorrhizal fungi provide similar results but reveal substantial methodological biases. New Phytol 188:291–301
Tejesvi MV, Kajula M, Mattila S, Pirttilä AM (2011) Bioactivity and genetic diversity of endophytic fungi in Rhododendron tomentosum Harmaja. Fungal Divers 47:97–107

Terhonen E, Blumenstein K, Kovalchuk A, Asiegbu FO (2019) Forest tree microbiomes and associated fungal endophytes: functional roles and impact on forest health. Forests 10:42

Thapa N (2016) Studies on the pteridophytic flora of Darjeeling Hills. Doctoral dissertation, University of North Bengal

Tiwari K (2014) The future products: endophytic fungal metabolites. J Biodivers Bioprospect Dev 2014(2):1–7

Uzma F, Mohan CD, Hashem A, Konappa NM, Rangappa S, Kamath PV, Singh BP, Mudili V, Gupta VK, Siddaiah CN et al (2018) Endophytic fungi-alternative sources of cytotoxic compounds: a review. Front Pharmacol 9:1–37

Van BEW, Wink M (2017) Medicinal plants of the world: an illustrated scientific guide to important medicinal plants and their uses. Timber Press, Portland

Vega FE, Simpkins A, Aime MC, Posada F, Peterson SW, Rehner SA, Arnold AE (2010) Fungal endophyte diversity in coffee plants from Colombia, Hawaii, Mexico and Puerto Rico. Fungal Ecol 3(3):122–138

Venieraki A, Dimou M, Katinakis P (2017) Endophytic fungi residing in medicinal plants have the ability to produce the same or similar pharmacologically active secondary metabolites as their hosts. Hell Plant Prot J 10:51–66

Vialle A, Feau N, Allaire M, Didukh M, Martin F, Moncalvo JM, Hamelin RC (2009) Evaluation of mitochondrial genes as DNA barcode for Basidiomycota. Mol Ecol Resour 9:99–113

Vieira PDS, de Souza Motta CM, Lima D, Torres JB, Quecine MC, Azevedo JL, de Oliveira NT (2011) Endophytic fungi associated with transgenic and non-transgenic cotton. Mycology 2:91–97

Vilgalys R (2003) Taxonomic misidentification in public DNA database. New Phytol 160:4–5

Wang Y, Guo LD (2007) A comparative study of endophytic fungi in needles, bark, and xylem of Pinus tabulaeformis. Can J Bot 85:911–917

Wang Y, Guo LD, Hyde KD (2005) Taxonomic placement of sterile morphotypes of endophytic fungi from Pinus tabulaeformis (Pinaceae) in Northeast China based on rDNA sequences. Fungal Divers 20:235–260

Wang Q, Gao C, Guo LD (2011) Ectomycorrhizae associated with Castanopsis fargesii (Fagaceae) in a subtropical forest, China. Mycol Progress 10:323–332

Wei H, Wu G, Chen J, Zhang X, Xiong C, Lei Y, Chen W, Ruan J (2013) (2 S)-5, 2′, 5′-trihydroxy-7- methoxyflavanone, a natural product from Abacopteris penangiana, presents neuroprotective effects in vitro and in vivo. Neurochem Res 38:1686–1694

Xia X, Cao J, Zheng Y, Wang Q, Xiao J (2014) Flavonoid concentrations and bioactivity of flavonoid extracts from 19 species of ferns from China. Ind Crop Prod 58:91–98

Yan L, Zhu J, Zhao XX, Shi JL, Jiang CM, Shao DY (2019) Beneficial effects of endophytic fungi colonization on plants. Appl Microbiol Biotechnol 103(8):3327–3340

Yatskievych G, Petruzzello M, Hosch WL (2018) Water fern. Fact-checked by The Editors of Encyclopaedia Britannica. Britannica

Zhang P, Chen ZH, Xiao B, Tolger B, Bao HY, Yang ZL (2010) Lethal amanitas of East Asia characterized by morphological and molecular data. Fungal Divers 42:119–133

Zhang J, Li X, Yin Y, Luo S, Wang D, Zheng H, Liu Y (2024) High-throughput sequencing-based analysis of the composition and diversity of the endophyte community in roots of *Stellera chamaejasme*. Sci Rep 14:8607

Zhao Z, Jin J, Ruan J, Zhu C, Lin C, Fang W, Cai Y (2007) Antioxidant flavonoid glycosides from aerial parts of the fern Abacopteris penangiana. J Nat Prod 70:1683–1686

Zhao P, Luo J, Zhuang WY (2011) Practice towards DNA barcoding of the nectriaceous fungi. Fungal Divers 46:183–191

Diversity of Aquatic Hyphomycetes as Root Endophytes Hosted Pteridophytic Plants

Sara A. Gezaf, Fatma A. Abo Nouh, Teroj A. Mohamed, Mohamed A. Abdel-Azeem, Lobna Hajji-Hedfi, Amal Belaid, Salma S. Higazy, Youssef M. Salah, Bola N. Aziz, Amany R. Ali, Norhan A. Soliman, Fatma A. Altamany, Hager M. Fathy, Roaa S. Abdel-Monem, Kholoud A. Elsayed, Alshaimaa S. Abdel-Salam, Ola F. Thabet, Mardian A. Adel, Amal M. Elshazly, Emad M. Ahmed, Mennatullah M. Allam, and Ahmed M. Abdel-Azeem

Contents

S. A. Gezaf (✉)
Botany and Microbiology Department, Faculty of Science, Arish University, Arish, Egypt

F. A. Abo Nouh
Botany and Microbiology Department, Faculty of Science, Suez Canal University, Ismailia, Egypt

Natural and Medical Sciences Research Center, University of Nizwa, Nizwa, Oman

T. A. Mohamed
Dental Basic Sciences Department, College of Dentistry, University of Duhok, Duhok, Iraq

M. A. Abdel-Azeem
Pharmacognosy Department, Faculty of Pharmacy, Sinai University – Arish Branch, Arish, Egypt

L. Hajji-Hedfi
Regional Centre of Agricultural Research of Sidi Bouzid, CRRA, Sidi Bouzid, Tunisia

Research Laboratory of Agricultural Production Systems and Sustainable Development (LR03AGR02), Carthage University, Tunis, Tunisia

A. Belaid
Biology Department, Faculty of Science, Selcuk University, Konya, Turkey

S. S. Higazy · Y. M. Salah · B. N. Aziz · A. R. Ali · N. A. Soliman · F. A. Altamany · H. M. Fathy · R. S. Abdel-Monem · K. A. Elsayed · A. S. Abdel-Salam · O. F. Thabet · M. A. Adel · A. M. Elshazly · E. M. Ahmed · M. M. Allam
Botany and Microbiology Department, Faculty of Science, Suez Canal University, Ismailia, Egypt

A. M. Abdel-Azeem, F. Degola (eds.), *Advances in Lichens and Non-Flowering Plants Fungal Endophytes Research*, Fungal Biology,
https://doi.org/10.1007/978-3-032-01506-8_10

1 Introduction

Endophytes are microorganisms that colonize the internal tissues of plants without inducing pathological symptoms in the host (Petrini 1991; Hyde and Soytong 2008; Gezaf et al. 2022; Mossa et al. 2024). The term "endophyte," derived from the Greek words "endon" (inside) and "phyton" (plant) (Abdel-Azeem and Salem 2012; Dutta et al. 2014; Razaghi and Abdel-Azeem 2024), was first coined by de Bary in 1866 (Melo et al. 1998) to describe microorganisms residing within plants without causing disease. These endophytic fungi, which have co-evolved with terrestrial plants for over 400 million years (Krings et al. 2007; Chadha et al. 2015), colonize living plant tissues asymptomatically for at least a portion of their life cycle, without detrimental effects on the host (Fouda et al. 2015). This endophytic fungal-plant association represents a crucial component of plant micro-ecosystems, significantly influencing host plant growth and development (Razaghi and Abdel-Azeem 2024).

Fungal endophytes are ecologically categorized into three groups according to Schultz: mycorrhizal, balansia-like (pasture), and non-pasture endophytes (Faeth and Fagan 2002). A separate classification, based on phylogeny and life history, distinguishes two major groups of fungal endophytes: clavicipitaceous and non-clavicipitaceous. Non-clavicipitaceous endophytes are further subdivided into three functional groups based on their colonization patterns. The first group, comprising both ascomycetes and basidiomycetes, colonizes both aerial and subterranean plant tissues. The second and third groups exhibit tissue specificity, colonizing

A. M. Abdel-Azeem
Botany and Microbiology Department, Faculty of Science, Suez Canal University, Ismailia, Egypt

Research Institute of University of the Bucharest (ICUB), University of Bucharest, Bucharest, Romania

The Centre for Mineral Biogeochemistry, The University of the Free State, Bloemfontein, Free State, South Africa exclusively above-ground or below-ground plant parts, respectively. While the first group demonstrates localized diversity, the second (ascomycetes and basidiomycetes) and third (including mycorrhizal fungi and dark septate endophytes) groups possess characteristics facilitating extensive colonization across diverse host ranges. For instance, members of the second group can colonize over 20 species within a single tropical leaf (Arnold et al. 2003), whereas the third group, notably dark septate endophytes (DSE), can colonize up to 600 plant species across broad geographical areas (Jumpponen 2001). The association between filamentous fungi and plants, specifically their endophytic relationship, is an ancient one, with fossil and molecular data placing its origins in the Devonian period (Taylor et al. 1995; Tehler et al. 2003). Contemporary research in this field is driven by three key goals: first, to broaden our understanding of endophyte diversity by documenting cryptic species; second, to discover novel metabolites with potential applications in both medicine and agriculture; and third, to investigate the influence of endophytic fungi on plant fitness under varying geographical conditions (Arnold et al. 2000; Suryanarayanan et al. 2009; Newsham 2011).

The presence of endophytic fungi has been reported in diverse plant species, spanning from mosses and ferns to conifers, grasses, and angiosperms (Sati and Belwal 2005). These endophytes, inhabiting a unique ecological niche, play a crucial role in shaping the distribution, ecology, physiology, and biochemistry of their host plants (Sridhar and Seetharam 1995). Beyond their ecological roles, endophytic hyphomycetes offer a promising source of biotechnologically and agriculturally relevant products (Bills and Polishook 1992; Sati and Belwal 2005). Furthermore, a wide array of fungal species, notably aquatic hyphomycetes, are known to colonize the roots of aquatic plants (Bärlocher 2006; Ghate and Sridhar 2017).

Aquatic hyphomycetes were initially characterized by Prof. C.T. Ingold in 1942, as they(Sati and Belwal 2005) complete their life cycle on submerged substrates in well-aerated water, producing distinct spore types (Sati and Belwal 2005). Additionally, Waid (1954) made the pioneering observation of these aquatic hyphomycetes on root surfaces for the first time. (Fisher and Petrini 1989) identified a distinct group of aquatic hyphomycetes, referred to as endophytic hyphomycetes. This classification was later substantiated by Fisher et al. (1991), who provided experimental evidence of their presence in the plant roots of aquatic environments. Their study involved the examination of the bark and xylem of *Alnus glutinosa* roots. Further research by various mycologists has also reported the occurrence of endophytic aquatic hyphomycetes on different host plants, including studies conducted by Sridhar and Bärlocher (1992) and Sati and Belwal (2009).

Aquatic hyphomycetes constitute a vital component of the microbial community in freshwater lotic ecosystems, fulfilling a critical role in the decomposition of allochthonous organic matter (detritus) and the subsequent transfer of energy through the food web (Baerlocher 2005). Molecular phylogenetic analyses have demonstrated that aquatic hyphomycetes represent a polyphyletic group comprising up to 300 species distributed among 5 classes within the Pezizomycota subphylum

(Belliveau and Bärlocher 2005; Baschien et al. 2013). Beyond their typical aquatic habitats, these fungi have colonized a range of semiaquatic and terrestrial environments, including leaf litter, soil, roots, and the canopy. This adaptation likely compensates for the loss of propagules due to the unidirectional flow of streams (Sridhar and Bärlocher 1992; Seena and Monroy 2016; Chauvet et al. 2016; Ghate and Sridhar 2017). In addition to leaf litter and woody debris serving as detrital substrates in stream ecosystems, the exposed live roots of riparian tree species within stream environments provide a suitable habitat for colonization by aquatic hyphomycetes (Chauvet et al. 2016). To date, approximately 20 species have been identified as root endophytes in three angiosperm species (*Acer spicatum* Lam., *Alnus glutinosa* (L.) Gaertner, *Betula papyrifera* Marsh.) and one gymnosperm species (*Picea glauca* (Moench) Voss) (Sridhar and Bärlocher 1992; Ghate and Sridhar 2017). All of these tree species naturally occur along streams and rivers in temperate regions. The present study aims to expand these observations to include two plantation crops (coffee and rubber trees) and four fern species found near two streams within the Western Ghat forests of India. Aquatic hyphomycetes are widely distributed in the streams of the Western Ghats, India (Sridhar et al. 1992). Studies have shown that mid-altitude streams host a greater diversity of aquatic hyphomycetes compared to those in mountainous regions, the foothills of the Western Ghats, and stream reaches along the west coast (Sridhar et al. 1992; Ghate and Sridhar 2017).

Angiosperms, the seed-bearing plants, currently constitute the majority of global flora. However, during the late Paleozoic era (approximately 280–230 million years ago), pteridophytes, a group of seedless vascular plants, were the dominant terrestrial plant life (Krishnan and Rekha 2021). Pteridophytes represent one of the most ancient plant groups and are commonly referred to as "vascular cryptogams." They are also known as "ferns and fern allies." It is believed that these plants first emerged during the Silurian and Devonian periods and later became the dominant vegetation during the Carboniferous period. Pteridophytes played a crucial role in the colonization of terrestrial habitats during these periods. Their high reproductive capacity and relatively simple genetic structure enabled them to rapidly establish themselves along seashores, riverbanks, and areas with minimal disturbances (Kenrick and Crane 1997). In the past, large fern trees with secondary growth were widespread, but today, tree ferns are relatively scarce. Currently, pteridophytes make up the second-largest plant group. Ferns and fern allies are typically found in moist and shaded areas. Pteridophytes are typically found in tropical, subtropical, and moist deciduous forest ecosystems. Their distribution is wide-ranging, extending from sea level to high elevations in mountainous terrain (Krishnan 2021). They exhibit a high degree of sensitivity to microclimatic variations. Consequently, even minor disturbances to climatic or other environmental factors can lead to localized extinctions. Therefore, their continued existence is largely contingent upon the preservation of their forest habitats. Beyond their terrestrial forest habitats, pteridophytes also exhibit diverse growth forms, including epiphytes on trees, lithophytes within rock crevices, and hydrophytes fully submerged in aquatic environments. India, as a subcontinent with diverse habitats, encompasses a wide range of climatic conditions, from the arid Thar Desert in the northwest to the extremely humid environment of Mawsynram in the northeast (Krishnan 2021). In the present chapter, aquatic hyphomycetes as endophyte of the Pteridophytic plant roots.

2 Aquatic Hyphomycetes: Characteristics and Diversity

Aquatic fungi exhibit significant morphological, ecological, and phylogenetic diversity, resulting in their ubiquitous presence across freshwater and marine ecosystems. Following their initial description by Ingold (1942) on decaying leaves in running water, current knowledge encompasses over 100,000 species within more than 3000 species (Bärlocher and Boddy 2016). Despite their known presence and diversity, the functional roles of aquatic fungi in ecosystem processes remain understudied (Grossart and Rojas-Jimenez 2016; Grossart et al. 2019). In freshwater ecosystems, aquatic fungi represent a functionally significant group of organisms, notably contributing to energy transfer to higher trophic levels through the processing of detritus (Sridhar and Bärlocher 2000). In freshwater ecosystems, the main source of energy comes from organic matter derived from adjacent terrestrial habitats, such as riparian vegetation, as well as from terrestrial runoff. Aquatic fungi play a vital role in carbon sequestration within these environments, significantly influencing biogeochemical and nutrient cycles. Their impact is largely determined by their diversity and the enzymatic activities they perform (Chandrashekar et al. 1991; Solé et al. 2012; Heeger et al. 2021). Aquatic fungi enhance the nutritional value of organic matter by enriching it with proteins, enzymes, lipids, and bioactive compounds (Krauss et al. 2011). This enrichment of plant detritus makes it a more attractive food source for invertebrates and vertebrates. Within freshwater lotic ecosystems, aquatic hyphomycetes represent the most diverse and functionally indispensable group of organisms (Duarte et al. 2016). These fungi serve as the principal agents of plant litter decomposition, thereby playing a crucial role in carbon and nutrient cycling and energy transfer to higher trophic levels. The distribution of individual species is subject to the influence of both geographic factors, such as latitude and altitude, and local environmental conditions, including the physical and chemical properties of the stream or river (Duarte et al. 2016). While originally studied primarily in temperate regions of North America and Europe (Bärlocher 1992), aquatic hyphomycetes are now understood to have a wide geographic distribution (Ghate and Sridhar 2017), with peak occurrence observed in cooler mid-latitudes (Barreto et al. 2023). Although they inhabit diverse climatic and geographic settings, the evolutionary pressures within lotic systems (e.g., water flow, substrate availability, and physicochemical factors) appear to exert a consistent selective force. Most aquatic hyphomycetes are classified within the Ascomycota phylum (Shearer et al. 2007). Within this phylum, Leotiomycetes contains a large number of species, while the remaining species are distributed among several other classes, including Sordariomycetes, Dothideomycetes, Orbiliomycetes, and Pezizomycetes (Campbell et al. 2009). The taxonomic classification of numerous aquatic hyphomycete species remains unresolved due to the absence of observed teleomorphs or molecular data. These polyphyletic fungi are characterized by the production of two primary conidial morphs: stauroconidia (predominantly tetraradiate or branched spores) and scolecoconidia (sigmoid, curved, or nearly straight spores). In addition to spore morphology, the mechanisms of conidiogenesis have traditionally been employed in their systematic classification (Franco-Duarte et al. 2022). The convergent evolution of conidial morphology in aquatic hyphomycetes, driven by shared environmental pressures, poses a challenge to phylogenetic studies, as spore shape cannot be reliably used to infer evolutionary relationships (Campbell et al. 2006). In recent years, fungal taxonomy and systematics have shifted toward using nucleotide sequence comparisons of select genes as a primary tool, sometimes replacing and sometimes supplementing phenotypic characters (Johnston et al. 2019). DNA sequence analysis is increasingly employed in the study of aquatic hyphomycetes. These fungi function as primary colonizers, converting plant detritus into a palatable biomass for aquatic invertebrates, thereby facilitating energy transfer to higher trophic levels (Baerlocher and Sridhar 2014). The downstream transport of substrates and spores in lotic systems raises the question of how these fungi maintain populations within a given stream reach. Several potential mechanisms may contribute to their persistence: (1) colonization of long-lasting woody substrates; (2) the presence of teleomorphs in some species, enabling aerial dispersal of ascospores; (3) dispersal of colonized leaves or conidia by animals; and (4) the documented occurrence of these fungi in terrestrial habitats, such as plant pathogens or endophytes (Bärlocher 1992).

2.1 Habitats

The interchange of water between the surface and subsurface of streams establishes the hyporheic zone, offering a suitable environment for aquatic hyphomycetes (Bärlocher 2006). This dynamic process, influenced by the sediment's structure and porosity (Zilliox 1994; Findlay 1995), creates regions of downwelling and upwelling (Crenshaw et al. 2002; Chauvet et al. 2016). Floods and sediment transport cause leaf litter burial (Naegeli et al. 1995; Chauvet et al. 2016), which significantly contributes to the organic matter content of the hyporheic zone (Metzler and Smock 1990; Cornut et al. 2012).

Bärlocher and Murdoch (1989) were the first to identify aquatic hyphomycete conidia in hyporheic zones, despite the presence of fungal hyphae having been previously reported (Barlocher and Murdoch 1989; Ellis et al. 1998). While hyporheic conditions may limit fungal leaf conditioning compared to surface waters (Essafi et al. 1994), and research on fungal activity is somewhat contradictory (Chauvet et al. 2016), studies have shown a connection between aquatic hyphomycetes and vascular plant detritus (Sudheep and Sridhar 2012).

Reduced dissolved oxygen levels in the hyporheic zone may lead to a decline in fungal biomass and sporulation (Medeiros et al. 2009), potentially creating favorable conditions for specific species such as *Flagellospora curvula* (Medeiros et al. 2009). Although their presence is reduced, aquatic hyphomycetes may still play a crucial role in leaf decomposition within the hyporheic zone, largely due to the lower activity of invertebrates (Cornut et al. 2010). The morphology of conidia, which influences both dispersal and attachment, may restrict their mobility in this environment (Danger et al. 2012). Additionally, sediment porosity and spore size are critical factors that influence dispersal (Cornut et al. 2014). The physical screening of spores, along with environmental factors such as low oxygen levels, plays a significant role in shaping fungal diversity within the hyporheic zone (Cornut et al. 2010). Additionally, fungal enzymatic activity and decomposition processes may be influenced by these conditions, leading to potential alterations in their functionality (Cornut et al. 2010; Danger et al. 2012).

Molecular techniques are essential for identifying and documenting fungal taxa, including those that do not produce spores (Duarte et al. 2016). Traditional approaches for detecting aquatic hyphomycetes include submerging detritus (Webster 1977), concentrating conidia (Sridhar and Bärlocher 1993), and promoting conidial release (Sridhar and Bärlocher 1993). Endophyte isolation and culture techniques are also employed for studying fungal communities (Sokolski et al. 2006). Molecular methods, particularly next-generation sequencing, are crucial for comprehensive assessments of fungal diversity and activity (Bärlocher 2016). However, field observations and pure culture studies continue to offer valuable insights, as demonstrated by *Articulospora tetracladia* and *Spirosphaera dimorpha*, which exhibit different morphological forms depending on their habitat (Marvanová and Bärlocher 1998).

The diverse aquatic ecosystems of the Western Ghats and Himalayas, both biodiversity hotspots, have been the main areas of focus for research on aquatic hyphomycetes in India. Studies have explored these fungi in a range of habitats, from streamlets and rivers to waterfalls and occasional dams, across different altitudinal gradients. The presence of aquatic hyphomycetes has even been documented in thermal sulfur springs within the Western Ghats (Chandrashekar et al. 1991; Rajashekhar and Kaveriappa 1996). A recent survey in Madhya Pradesh revealed 45 species (29 genera) inhabiting various substrates, including water, foam, detritus, stem flow, and live roots, with some species showing endophytic or stem flow-associated growth patterns (Chaudhari et al. 2016). In addition to water samples, studies have analyzed various substrates, including leaf and woody litter, conifer needles, foam, scum, sediments, and damp terrestrial litter. Endophytic aquatic hyphomycetes have also been found in the live roots of riparian vegetation in the Western Ghats and Himalayas (Sati and Belwal 2005). Studies of aquatic hyphomycetes have expanded beyond streams to encompass various terrestrial habitats, including leaf litter, stemflow, throughfall, decaying epiphytes, tree holes, and crown humus in the Western Ghats and southwest coast (Sridhar and Kaveriappa 1988). A recent study in southwest Karnataka identified 31 species of aquatic and aero-aquatic hyphomycetes on damp leaf litter (Sridhar et al. 2020). Despite their presence in these terrestrial settings, their lower abundance compared to aquatic habitats suggests potentially distinct functional roles.

2.2 Distribution

Analysis of published research reveals a geographically uneven distribution of documented aquatic hyphomycete diversity within the Indian subcontinent. Karnataka exhibits the highest reported diversity (37%), followed by Uttarakhand (21%), Maharashtra (20%), Andhra Pradesh (9%), Madhya Pradesh (5%), Goa (3%), Gujarat (2%), Tamil Nadu (2%), Kerala (1%), and Assam and Himachal Pradesh (<1%). This distribution pattern corroborates prior checklists of freshwater mitosporic fungi (Patil and Borse 2015) and is reflected in a recent comprehensive checklist (Sridhar 2021). A recent checklist identifies approximately 191 aquatic hyphomycete species (66 genera) in the Indian subcontinent. This represents a substantial portion (up to 57%) of the globally estimated 335 morphospecies (Duarte et al. 2016), suggesting a rich diversity within India. This diversity is likely driven by the subcontinent's varied tropical, subtropical, and temperate climates, coupled with its ten distinct biogeographic zones (Singh and Chaturvedi 2017). However, it is important to note that many ecoregions remain undersampled, suggesting that the true extent of diversity may be even higher. Although aquatic mitosporic fungi have been categorized into four groups (Goh and Hyde 1996), a more focused approach, concentrating on typical aquatic hyphomycetes with specific conidial morphologies (staurosporous, scolecosporous, helicosporous, and conventional cylindrical/oval/spherical), better reflects adaptations to lotic environments. This approach excludes transient or immigrant fungal species, as well as those present due to substrate transfer or terrestrial runoff, thus facilitating more targeted research. The occurrence of certain aquatic hyphomycetes in tree canopies and damp terrestrial litter, especially in the consistently wet Western Ghats (with 4–6 months of wet conditions), is likely a consequence of these persistently moist conditions. This observation supports the findings of Shearer et al. (2007), who proposed that species inhabiting tropical aquatic environments may also be found in adjacent rainforest habitats due to the high levels of humidity (Shearer et al. 2007).

The widespread distribution of aquatic hyphomycetes is supported by data from 352 publications documenting 335 morphospecies across diverse substrates (Duarte et al. 2016). Research efforts have historically focused on temperate regions of the Northern Hemisphere, particularly Europe, where Ingold's foundational studies originated (Ingold 1942). Subsequent research has expanded to include North America, tropical regions (Venezuela, Brazil, Puerto Rico), and Asia (especially India). In contrast, research on aquatic hyphomycetes is limited in Africa (excluding South Africa) and the polar regions (Shearer et al., 2007). This disparity in sampling effort creates bias in our current understanding of their global biogeography (Duarte et al. 2016). Ecologically, species richness typically increases with sampling area (Zinger et al. 2014). Using the number of publications per region as a proxy for sampling effort, and applying a rectangular hyperbola model, we estimate that roughly 29 publications per region are required to capture 50% of the true species richness, while 275 publications would be needed to reach 90%. Of the geographic regions examined, only the Northern Temperate/Tropical (Western Hemisphere) and European/Tropical Asian (Eastern Hemisphere) regions have attained the 50% estimated species richness level. The absence of any region reaching the 90% threshold underscores the preliminary nature of current knowledge regarding global aquatic hyphomycete diversity. Despite the limitations in sampling effort, this substantial dataset enables the exploration of several key ecological concepts related to aquatic hyphomycetes. These include species-area relationships, distance-decay patterns, the impact of latitude on species richness, and the influence of geographic region and climate on community structure. Because traditional similarity indices are highly sensitive to sample size, particularly when dealing with numerous rare species (as in this dataset), Chao's abundance-based index was employed for all species turnover analyses (Chao et al. 2005).

2.3 Diversity

While many studies in Indian aquatic habitats have focused on species richness, a more comprehensive approach is needed. Including data on frequency of occurrence, conidial output, and substrate preference would significantly enhance our understanding. This broader perspective would allow for the calculation of diversity indices (evenness, rarefaction), identification of core-group and keystone fungal species, and the detection of cryptic species (Sridhar 2021). Furthermore, incorporating basic water quality parameters would add valuable context to these studies. Finally, integrating molecular methods would reveal the full extent of fungal colonization within substrates, uncovering species often missed by traditional techniques. Such a multifaceted approach would provide crucial insights into seasonal dynamics, community succession, decomposition processes, organic matter mass loss, food web interactions, and energy flow patterns within these vital ecosystems (Sridhar 2021).

Studies have shown a greater diversity of aquatic hyphomycetes in Western Ghats streams compared to foothill, coastal, and plain regions (Sudheep and Sridhar 2013). Research by S.C. Sati and colleagues (Duarte et al. 2016; Sridhar 2021) suggests that the diversity in Western Ghats streams may be comparable to that found in Himalayan streams. The elevated diversity of aquatic hyphomycetes observed in Himalayan streams is attributed to the presence of dense riparian vegetation and an abundance of available substrate. Within the Western Ghats, an altitudinal gradient in species richness has been documented, with the highest diversity occurring in mid-altitude streams (20 spp.), followed by a progressive decline in mountain streams (10 spp.), foothill streams (14 spp.), and coastal streams (12.5 spp.) (Sridhar 2021). The decline in aquatic hyphomycete species richness observed in foothill and coastal regions is likely a consequence of reduced forest cover and altered water quality resulting from human activities, such as agricultural chemical runoff and sewage input. This hypothesis is strengthened by the findings of Rajashekhar and Kaveriappa (2003), who documented a strong correlation between vegetation and aquatic hyphomycete communities. Beyond temperature, other water quality parameters affected by human interference also appear to negatively influence both the richness and diversity of these fungal communities (Rajashekhar and Kaveriappa 2003). Nair and Bhat (2016) assessed seasonal variation in aquatic hyphomycete species richness and diversity within three streams located in Goan wildlife sanctuaries (altitude: 123–280 m asl). Their findings indicated a correlation between species density and factors such as rainfall, riparian vegetation, and substrate availability. The absence of comparable investigations across other Indian ecoregions limits the scope for broader comparative analyses (Nair and Bhat 2016).

In addition to the more commonly studied substrates (water, foam, and leaf litter), investigations of aquatic hyphomycetes in the Western Ghats have encompassed submerged woody litter (Sudheep and Sridhar 2011). Analysis of woody litter from high-altitude streams yielded a total of 30 aquatic hyphomycete species, with a higher diversity observed in the bark (20 spp.) compared to the cambium (18 spp.) (Sudheep and Sridhar 2011). Naturally deposited woody litter collected from the Kali River and Kadra Dam supported 15 and 11 species, respectively (Sudheep and Sridhar 2011). Submersion experiments utilizing *Anacardium occidentale* and *Terminalia paniculata* in the Kaiga stream and Kadra Dam demonstrated a slightly higher species richness on *T. paniculata* (17 spp. in stream, 16 spp. in dam) compared to *A. occidentale* (17 spp. in stream, 14 spp. in dam) (Sudheep and Sridhar 2013). Aquatic hyphomycetes have been found as endophytes in live riparian tree roots, in addition to their presence on leaf litter and woody debris (Ghate and Sridhar 2017). Research has documented 22 endophytic species in the Western Ghats and 29 in the Himalayas (Ghate and Sridhar 2017). The diversity of these root endophytes mirrors patterns seen in other substrates, with higher diversity in mid-altitude streams (Ghate and Sridhar 2017). A study in Madhya Pradesh identified six endophytic species (Chaudhari et al. 2016), further emphasizing the ecological significance of riparian roots for aquatic hyphomycetes. Analysis of various substrates in India reveals a clear preference for submerged leaf litter by aquatic hyphomycetes, as shown by Patil and Borse (2015) and additional research on woody litter (Sudheep and Sridhar 2011). The order of preference, from highest to lowest species richness, is: submerged leaf litter, woody litter, foam, water, and roots.

3 Pteridophytic Plants: Diversity and Adaptations

Pteridophytes, vascular cryptogams that reproduce via spores, are divided into two groups: Lycophytes and Monilophytes (Schneider and Schuettpelz 2016). While both possess vascular bundles, a trait shared with seed plants (Pigg and Rothwell 2001; Bondada and Tu 2006), a key difference lies in their stem stele structure.

Lycophytes lack leaf gaps, whereas Monilophytes possess them (Mir et al. 2015). Pteridophytes typically possess a protostele, a type of stele where the xylem occupies the central region and is encircled by phloem (Sofiyanti et al. 2019). Of the approximately 12,000 identified pteridophyte species globally (Della and Falkenberg 2019), these plants occupy diverse habitats, including coastal regions. India alone is home to over 1200 pteridophyte species, encompassing both ferns and fern allies (Arjun et al. 2021). Pteridophytes are recognized for their evolutionary significance, stemming from their classification as an ancient plant lineage. This is further emphasized by the substantial number of relict and endemic species within the group (Patil et al. 2016). Pteridophytes are regarded as an evolutionary bridge between lower nonvascular plants and higher vascular plants, including gymnosperms and angiosperms, due to their well-developed vascular system. Their life cycle is characterized by an independent alternation of generations between the sporophyte and gametophyte phases. Pteridophytes are commonly classified into ferns and fern allies, though this distinction is not strictly scientific (Arjun et al. 2021). Ferns, a more recently evolved group appearing in the Carboniferous period (325 million years ago), possess a well-developed sporophyte (spore-bearing generation) characterized by rhizomes, roots, and large leaves, but lack flowers and fruits. In contrast, fern allies, originating in the Silurian period (395 million years ago), are characterized by small (microphyllous) leaves with a single vein in their stele. Spore arrangement also differs: ferns have sporangia located superficially (on the underside of the leaf) or marginally, while fern allies typically produce spores in terminal strobili (sori) at branch tips (Arjun et al. 2021).

The potential of pteridophytes, particularly their edible and medicinal properties, has been extensively studied in India. Researchers have documented considerable ethnobotanical knowledge and medicinal uses of these plants (Benjamin and Manickam 2007; Arjun et al. 2021; Irfan et al. 2021). Ethnobotanical research has documented the utilization of 60 pteridophyte species for medicinal purposes by tribal communities residing in the Western Ghats (Benjamin and Manickam 2007). In addition to their medicinal significance, numerous pteridophytes are cultivated as ornamentals, both domestically and in outdoor settings, owing to the aesthetic appeal of their fronds (Sathiyaraj 2015). The pharmacological properties of these plants have been the subject of numerous reports (Arjun et al. 2021). Within the Nilgiris region, 20 pteridophyte species have been identified as possessing ornamental potential(Arjun et al. 2021). Many pteridophyte species, particularly those inhabiting the diverse and dense forests of the tropics, are yet to be discovered. The ongoing destruction of these tropical habitats is therefore a major concern. While pteridophytes are adaptable to various niches, their greatest diversity is concentrated in these warm, wet regions (Kramer and Viane 1990). Approximately 43% of pteridophyte species in India are classified as threatened, according to a threat assessment (Chandra et al. 2008). The Western Ghats, a biodiversity hotspot, boasts a rich assemblage of wild flora and fauna. The Kerala portion of this region, with its consistently high rainfall (around 3000 mm annually), is particularly diverse in plant life. A recent survey documented 283 species of ferns and fern allies in Kerala (Arjun et al. 2021). Kerala's climate provides optimal conditions for the growth and flourishing of ferns and fern allies. This region of the Western Ghats also harbors numerous endemic fern species, including the tree fern *Cyathea nilgirensis* (Dudani et al. 2014). Several studies have investigated the fern flora of the Western Ghats (Antony 2013). Notably, the world's smallest fern, *Ophioglossum malviae,* was recently discovered in the northern part of this region (Patel and Reddy 2018). Habitat destruction and climate change pose significant threats to fern populations. Many fern species exhibit narrow ecological amplitudes, requiring specific microclimatic conditions for optimal growth and development. Eravikulam National Park, characterized by a diverse flora including angiosperms, gymnosperms, and pteridophytes, was formerly managed as game reserves. The park's unique floral and faunal composition, sustained by its specialized shola-grassland ecosystem, led to its designation as Kerala's first national park in 1978 (Arjun et al. 2021). Ferns play a significant role both ecologically and economically, as evidenced by numerous studies (Batke et al. 2021). Ecologically, they function as primary producers in food chains and contribute substantially to the nitrogen cycle. Economically, ferns hold considerable potential as traded commodities due to their ornamental and medicinal properties. Research on ferns in Indonesia has explored both their geographic distribution, as seen in the Selamat study (Batke et al. 2021), and their taxonomic relationships, exemplified by work on the Polypodiaceae family (Batke et al. 2021). The rapid climate variability across short altitudinal gradients, coupled with unique geographical features, makes forests and mountains ideal locations for studying climate change impacts on species distributions. These factors, along with altitude, precipitation, and plant abundance, all play a role in shaping community composition (Kessler et al. 2016; Rogora et al. 2018). Ferns (Pteridophyta) and lycophytes are particularly susceptible to the combined effects of rising temperatures and declining precipitation, both of which are projected under future climate change scenarios. Terrestrial and epiphytic species within these groups are likely to exhibit differing responses to these altered conditions (Mandl et al. 2010). Given the significant and sensitive nature of climate's role in ecosystems, the globally distributed and diverse pteridophytes have been a focus of considerable research exploring global altitudinal distribution patterns (Sánchez González et al. 2010).

4 Aquatic Hyphomycetes as Endophytes

Aquatic hyphomycetes have been found in the surprising niche of plant endophytes, residing within plant tissues, typically without causing harm (Wilson 1995). These endophytic relationships can range from neutral to beneficial or slightly detrimental. While their presence in roots near or in water has been frequently shown through cultivation (Bärlocher 2006; Sati and Belwal 2009), molecular barcoding has revealed their wider distribution, including in ectomycorrhizal roots, liverworts, mycorrhizal orchid roots, and ericaceous plants, even at some distance from water (Russell and Bulman 2005; Murat et al. 2005; Tedersoo et al. 2007; Stark et al. 2009; Chauvet et al. 2016). Aquatic hyphomycetes also exist as endophytes within trees and ferns (Fisher et al. 1991; Suryanarayanan et al. 2009; Chauvet et al. 2016), a fact further substantiated by next-generation sequencing (e.g., the detection of *Tetracladium* and *Gyoerffyella* in *Bistorta vivipara* roots; Davey et al. 2015). For instance, *Dwayaangamcolodena, Tripospermum camelopardus*, and *Tr. myrti* are frequently found as endophytes in spruce needles (*Picea mariana*) (Sokolski et al. 2006), and several *Gyoerffyella* strains have been isolated from the ectomycorrhizal roots of *P. abies* (Vohník et al. 2013). Seena and Monroy (2016) provide a review of potentially endophytic aquatic hyphomycetes. A key piece of evidence supporting their dual ecology is the isolation of *Dwayaangam* strains from surface-sterilized spruce needles using both direct culture and a flow chamber mimicking running water (Sokolski et al., 2006), which yielded consistent molecular barcodes and asexual spores. This dual lifestyle may explain the appearance of aquatic hyphomycetes on submerged leaves without added inoculum (Bandoni 1972), prompting Bandoni (1972) to propose their endophytic nature in aerial plant tissues. Phylogenetic analysis of *Tetracladium* and related genera suggests a common ancestor capable of both endophytic and aquatic lifestyles (Seena and Monroy 2016). The evolutionary trajectory (from saprotrophy to endophytism, or vice versa) remains unresolved but may be clarified through further phylogenetic analyses. The contribution of endophytic strains to canopy populations also warrants investigation. While the dual saprotrophic and endophytic niche may appear unusual, it has been observed in other fungal taxa (Davis et al. 2003; Wedin et al. 2004; Porras-Alfaro and Bayman 2011; Behie et al. 2012), underscoring the limitations of rigidly defined fungal niches. Genomic studies of fungi such as *Sebacinales* and *Tulasnellales* (Zuccaro et al. 2011, 2014; Kohler et al. 2015) have revealed a combination of hydrolytic enzyme genes (typical of saprotrophs) and effector protein genes (characteristic of biotrophs). Sequencing additional endophytic aquatic hyphomycete strains is essential to ascertain whether they exhibit similar genomic characteristics (Seena and Monroy 2016).

Fisher and Petrini (1989) provided the initial evidence of an endophytic stage in two aquatic hyphomycete species isolated from terrestrial *Alnus glutinosa* roots. Colonization rates were low relative to the common endophyte *Cylindrocarpon destructans*. A subsequent study (Fisher et al. 1991), comparing submerged and terrestrial alder roots, found a higher prevalence of endophytic aquatic hyphomycetes, particularly in the bark, of submerged roots. In addition to typical aquatic hyphomycetes, *Fusarium* and *Cylindrocarpon* species were also identified. Cluster analysis indicated distinct endophyte populations in root samples from aquatic and soil environments, suggesting a greater influence of the external environment on root endophytes compared to leaf endophytes (Fisher and Petrini 1993). Following the initial discovery, three new species (*Fontanospora fusiramosa* and two *Filosporella* species) were isolated from submerged alder roots (Bärlocher 2006; Sridhar and Bärlocher 1992) expanded the known host range to include spruce (*Picea glauca*), birch (*Betula papyrifera*), and maple (*Acer spicatum*), again observing a higher concentration of endophytes in the bark. While they successfully induced sporulation by aerating root fragments, this method required longer aeration periods and yielded fewer spores compared to submerged branches or leaves.

Nevertheless, the considerable root biomass and rapid turnover in streams (Waid 1954) suggest that roots represent a substantial secondary resource for aquatic hyphomycetes. Their endophytic lifestyle may confer a competitive advantage in the utilization of root detritus (Fisher and Petrini 1993). In spruce roots, the highest concentration of aquatic hyphomycetes in the xylem was found in 4- to 5-year-old segments (Sridhar and Bärlocher 1993). The total fungal biomass, measured using ergosterol, ranged from 0.002% to 0.2% of the root biomass, significantly lower than that found on decaying leaves (Gessner et al. 2003). Seetharm et al. (1996) also detected aquatic endophytes in the submerged roots of two plantation crops and four fern species in India, with a higher prevalence observed in the bark. Notably, the release of conidia from fern roots was considerably greater than that from other plant species examined. In mangrove swamps, Ananda and Sridhar (2002) discovered aquatic hyphomycetes, including *Mycocentrospora acerina* and *Triscelophorus acuminatus*, within the prop roots and pneumatophores of several mangrove species, alongside both marine and terrestrial fungi. Across 8 studies, a total of approximately 35 aquatic hyphomycete species, representing about 10% of all described species, have been documented from the submerged roots of 13 plant species, encompassing angiosperms, gymnosperms, and ferns. These submerged roots likely offer a stable refuge for aquatic hyphomycetes, contributing to their survival in lotic environments (Ananda and Sridhar 2002).

5 Endophytic Interactions Between Aquatic Hyphomycetes and Pteridophytic Plants

Pteridophytes, unlike angiosperms, have a high proportion of epiphytes, with up to 13,000 species of epiphytic ferns reported in tropical and subtropical ecosystems. While Lehnert et al. (2017) have reviewed symbiotic fungal associations (AM, EM, and DSE fungi) with pteridophytes, research on endophytic fungi in these plants remains limited (Sati and Belwal 2009). Seetharm et al. (1996) examined the roots of four riparian ferns (*Angiopteris evecta*, *Cristela dentata*, *Diplazium esculentum*, and *Macrothelypteris torresiana*) growing in streams along India's west coast and the Western Ghats (Seetharam et al. 1996). They found that up to ten aquatic hyphomycete species colonized these ferns, with three species (*Triscelophorus acuminatus*, *T. monosporus*, and an unidentified species) common to all four fern species. Kumaresan et al. (2013) studied endophytic fungi in 5 pteridophyte species from India's east coast, identifying 40 fungal species (331 isolates). *Colletotrichum* sp. was commonly observed, with *Pteris* exhibiting the highest fungal species richness and *Selaginella* the highest number of isolates. Sati and Belwal (2009) reported 11 aquatic hyphomycete species as endophytes in the roots of several riparian pteridophytes from two locations in Kumaun, Himalaya. *Botrychium* roots were found to be colonized by up to seven endophytic hyphomycete species. In southern India, 14 hyphomycetes, 3 coelomycetes, and 1 non-sporulating endophytic fungus were isolated from leaflets, stolons, runners, and roots of the aquatic medicinal fern *Marsilea minuta* (Nyayiru Kannaian et al. 2018). Although ferns are recognized for producing secondary metabolites with a wide range of pharmacological applications, including cytotoxicity, hepatoprotective, antihyperglycemic, leishmanicidal, trypanocidal, antinociceptive, antiinflammatory, immunomodulatory, and chemopreventive activities (Cao et al. 2017). Research on their endophytic fungal communities is limited (Asiandu et al. 2021). There is a particular scarcity of studies focused on identifying the diversity of fungal endophytes within medicinal ferns and characterizing their potential to synthesize diverse bioactive compounds (Chen et al. 2015; Thi Minh Le et al. 2019; Asiandu et al. 2019). For instance, Sati and Belwal (2009) and Sati and Belwal (2005) isolated fungal species from 9 genera of aquatic hyphomycetes (*Alatospora*, *Anguillospora*, *Campylospora*, *Clavariopsis*, *Heliscus*, *Lunulospora*, *Pestalotiopsis*, *Tetrachaetum*, and *Tetracladium*) from the roots of the pteridophytes *Equisetum* sp., *Botrychium* sp., and other ferns. (Kumaresan et al. 2013) isolated forty endophytic fungi from five pteridophyte species: *Adiantum* sp., *Gleichenia linearis*, *Lygodium flexuosum*, *Pteris* sp., and *Selaginella* sp. These isolates belonged to various genera, including *Aspergillus*, *Aureobasidium*, *Botryodiplodia*, *Cladosporium*, *Colletotrichum*, *Drechslera*, *Fusarium*, *Penicillium*, *Pestalotiopsis*, *Phialophora*, *Phomopsis*, *Phyllosticta*, *Sporormiella*, and *Trichoderma*. The limited research on fungal endophyte-fern interactions (including isolation, diversity, ecology, and metabolite production) has spurred interest in discovering novel fungal species associated with ferns and reviewing their bioactive metabolites. This chapter summarizes other endophytic fungi isolated from medicinal ferns.

5.1 Adiantum *spp.*

Ramesha et al. (2020) isolated the endophytic fungus *Nigrospora sphaerica* from healthy leaves of *Adiantum philippense*, an ethnomedicinal fern used to treat various ailments, including burning sensations, erysipelas, epileptic seizures, diarrhea, bronchitis, coughs, and elephantiasis (Ramesha et al. 2020). Analysis using TLC-bioautography and spectroscopic techniques identified phomalactone, a potent antimicrobial agent, produced by *N. sphaerica*. Phomalactone, identified from *N. sphaerica*, exhibits a range of biological activities, including antifungal activity against *Aspergillus fumigatus* (Razaghi and Abdel-Azeem 2024), nematocidal activity (Khambay et al. 2000), immunomodulatory activity (Krivobok et al. 1994), insecticidal activity (Krasnoff and Gupta 1994), and phytotoxic activity (Fukushima et al., 1998), and is effective against both human and phytopathogenic bacteria and fungi. *N. sphaerica* is also known to produce other antimicrobial compounds, such as nigrosporolides (Razaghi and Abdel-Azeem 2024), nigrosporins (Tanaka et al. 1997), lactones (Fukushima et al. 1998), epoxydons and pyrones, diterpenes, diketopiperazines (Razaghi and Abdel-Azeem 2024), and nigrosporolides (Zhang et al. 2009). *Rhizopus oryzae*, an endophyte found in *Adiantum capillus-veneris*, has demonstrated potential in enhancing thermal stress tolerance in crops like sunflower and soybean by increasing chlorophyll content, shoot and root lengths, and fresh and dry biomass (Hussain et al. 2020).

5.2 Asplenium nidus

Asplenium nidus (Aspleniaceae) has been shown to produce bioactive flavonoids with tumoricidal, antibacterial, and antioxidant properties (Jarial et al. 2018). Therefore, endophytic fungi isolated from this fern may serve as a source of similar bioactive compounds. Table 1 lists some endophytic species isolated from *A. nidus* (Jarial et al. 2018).

5.3 Botrychium *spp.*

In Nepal's traditional medicine, various *Botrychium* species (e.g., *B. lanuginosum*, *B. lunaria*, and *B. multifidum*) are commonly used to treat a range of ailments (Ojha and Devkota 2021; Razaghi and Abdel-Azeem 2024). *Botrychium ternatum*, found in China, Korea, and Japan, is also used in folk medicine to address conditions such as dizziness, headaches, coughs, and fevers. These therapeutic effects may be

Table 1 Lists some endophytic species associated with medicinal ferns

Fern species	Fungal endophytes	Isolate sources	References
Asplenium nidus.	*Fusarium* sp., *Colletotrichum* spp., *Aspergillus* sp., and *chaetomium* sp.	Leaves, Stem, Root	Suhartina et al. (2018), Razaghi and Abdel-Azeem (2024)
Equisetum sp. *Equisetum arvense.*	*Anguillospora longissima* *Teratosphaeria* sp.	Roots, Stem	Sati and Singh (2014), Singh and Sati (2017)
Pteris sp.	*Acremonium* sp., *Aspergillus tereus*, *Aureobasidium* sp., *Botryodiplodia* sp., *Colletotrichum* spp., *Fusarium* sp., and *Phomopsis* sp.	Shoots	Kumaresan et al. (2013)
Elaphoglossum sp.	*Phomatospora berkeleyi*, *Chrysosporium* sp., *Scopulariopsis candida*, *Verticillium* cf. *psalliotae* and *Phoma* sp.	Leaves, Stem, Root	Razaghi and Abdel-Azeem (2024)
Elaphoglossum spp.	dark septate endophytes (DSE)	Root	Lehnert et al. (2009)
Elaphoglossum doanense.	*Apioclypea* sp., *Colletotrichum*, *Ophioceras*, *Paramicrothyrium*, *Annulohypoxylon* and *Anthostomella*	Leaves, Petiole, Branch, and Root	Rojas-Jimenez and Tamayo (2021)

attributed to the presence of kaempferol glycosides and four quercetin glycosides within this fern (Cao et al. 2017). The endophytic fungus *Phyllosticta capitalensis* has been isolated from *Botrychium* sp. in the USA (Wikee et al. 2013). Notably, *P. capitalensis* is known to produce compounds like meroterpene analogues and polyketides, which exhibit antimicrobial activity against several human pathogens (Xu et al. 2020).

5.4 Dryopteris *spp.*

Dryopteris plants (Dryopteridaceae) are abundant in secondary metabolites, such as glycosides, steroids, alkaloids, phenols, terpenes, flavonoids, and tannins, which contribute to their antibacterial, antifungal, nematicidal, and antioxidant properties (Soare et al. 2012; Egorova et al. 2021; Razaghi and Abdel-Azeem 2024). Ismail et al. (2021) isolated the endophytic fungus *Aspergillus violaceofuscus* from *Dryopteris filix* roots and leaves. The culture filtrate of *A. violaceofuscus* enhanced total chlorophyll content, plant height, and biomass in *Glycine max* and *Helianthus annuus* seedlings under high temperature conditions. This heat stress tolerance was likely due to the presence of secondary metabolites (flavonoids, phenolics, indole-3-acetic acid (IAA), and salicylic acid (SA)) in the fungal culture filtrate, which improved lipid, protein, and sugar concentrations, and enhanced ROS scavenging antioxidant systems (flavonoids, proline, phenolics, AAO, and CAT) in the crops. Similar results were obtained by Ismail et al. (2021) with the endophytic fungus *Penicillium glabrum*, isolated from *Dryopteris blanfordii*, under thermal stress.

5.5 Elaphoglossum *spp.*

Elaphoglossum, a medicinal fern now classified within the Dryopteridaceae family, is a rich source of secondary metabolites, including prenylated acylphloroglucinols like elaphogayanin A-B, elaphopilosin C-E, lindbergins A-F, and yungensins A-F (Cao et al. 2017). Lehnert et al. (2009) observed dark septate endophytes (DSE, Ascomycota) colonizing the roots of several *Elaphoglossum* species (*E. engelii*, *E. erinaceum*, *E. glossophyllum*, *E. guamanianum*, *E. petiolosum*, *E. quitense*, *E. yatesii*, and *Elaphoglossum* sp.) in their study of mycorrhizal status in neotropical ferns, suggesting a mycorrhizal association similar to ericoid mycorrhiza (Lehnert et al. 2009). Rojas-Jimenez and Rojas-Jimenez and Tamayo (2021) identified diverse fungal endophytes from various parts of *Elaphoglossum doanense* in Costa Rican tropical forests, including *Apioclypea* sp. (Amphisphaeriales), *Colletotrichum* (Glomerellales), *Ophioceras* (Magnaporthales), *Paramicrothyrium* (Microthyriales), *Annulohypoxylon* (Xylariales), and *Anthostomella* (Xylariales) (Rojas-Jimenez and Tamayo 2021; Razaghi and Abdel-Azeem 2024).

5.6 Equisetum *spp.*

Molecular and microscopic staining confirmed the presence of dark septate endophytes (DSE), also known as fine root endophytes (FRE), in six *Equisetum* species (*E. arvense*, *E. sylvaticum*, *E. fluviatile*, *E. hyemale*, *E. palustre*, and *E. telmateia*). Endophytic fungi identified from members of the Equisetaceae family included DSE (comprising over 80% of the observed endophytes), unidentified fungi from the Helotiales order, *Phialocephala fortinii*, *Cladophialophora chaetospira*, *Tetracladium* spp., and arbuscular mycorrhiza (AM) Glomeromycotina. These endophytic fungi appear to function similarly to mycorrhizae, colonizing plants and facilitating nutrient exchange. DSE contribute to nutrient acquisition, particularly by accessing 15 N-enriched soil organic compounds, while receiving organic carbon compounds produced through plant photosynthesis (Giesemann et al. 2020). *Teratosphaeria* sp. AK1128, an endophytic fungal strain belonging to the Teratosphaeriaceae family (Dothideomycetes, Ascomycota), was isolated from the stem of *Equisetum arvense*. This fungus produces ten bioactive compounds, including naphtho-γ-pyrone dimers, teratopyrones A–C, naphtho-γ-pyrones (aurasperone B, aurasperone C, aurasperone F, nigerasperone A, and fonsecin), and diketopiperazines (asperazine and isorugulosuvine). Notably, nigerasperone A exhibited cytotoxic activity against several cancer cell lines, including human metastatic prostate cancer, human non–small-lung cancer, human central nervous system glioma, human breast cancer, human metastatic breast cancer, and normal human lung fibroblast cells (Xu et al. 2020). *Anguillospora longissima*, a freshwater hyphomycete endophytic fungus, has also been reported from the roots of riparian *Equisetum* sp. in India. The bioactivity, specifically antimicrobial potential, of *A. longissima* was evaluated using preliminary and secondary antibacterial assays. The fungus inhibited the growth of bacterial strains, likely due to the production of antibiotic substances, suggesting that *A. longissima* represents a potential source of natural products and secondary metabolites (Sati and Singh 2014). *Anguillospora longissima* has demonstrated the potential to inhibit plant pathogenic fungi. Studies using well diffusion and dual culture techniques have shown its antagonistic activity against various pathogens. In dual culture assays, *A. longissima* inhibited *Tilletia indica* (79.76% maximum inhibition) and *Fusarium oxysporum* (35.73% minimum inhibition). Using well diffusion, it inhibited *Colletotrichum falcatum* (62.96% maximum inhibition) and *Pyricularia oryzae* (17.85% minimum inhibition) (Singh and Sati 2019). Furthermore, Singh and Sati (2017) found that *Anguillospora longissima* effectively solubilizes phosphate by producing organic acids. These acids can dissolve mineral phosphate or chelate iron (Fe) and aluminum (Al) ions, making the phosphate available to plants (Singh and Sati 2017; Razaghi and Abdel-Azeem 2024).

6 Conclusions

Pteridophytes, an ancient and diverse group of plants, play vital roles in various ecosystems. They contribute to soil stability, water regulation, and serve as a food source for some organisms. Many pteridophyte species possess medicinal properties, producing bioactive compounds with therapeutic potential. Endophytic fungi associated with pteridophytes, including aquatic hyphomycetes in aquatic ferns, offer a promising source of similar bioactive substances, potentially revolutionizing drug discovery and reducing pressure on wild fern populations. These endophytes also contribute to plant health and nutrient cycling, further enhancing the ecological importance of pteridophytes. Investigating the diversity and functions of these endophytic communities is crucial for understanding pteridophyte biology and unlocking their potential for various biotechnological and pharmaceutical applications. Further research into these symbiotic relationships can lead to sustainable utilization of these valuable plant resources.

References

Abdel-Azeem A, Salem F (2012) Biodiversity of laccase producing fungi in Egypt. Mycosphere 3: 900–920

Ananda K, Sridhar K (2002) Diversity of endophytic fungi in the roots of mangrove species on the west coast of India. Can J Microbio 48:871–878

Antony R (2013) Pteridophyte Fora of Agasthyamala, Western Ghats

Arjun MS, Antony R, Ali AA et al (2021) Diversity of Pteridophyte Flora in Rajamala, Eravikulam National Park, Kerala, India. Asian J Environ Ecol 28–36. https://doi.org/10.9734/ajee/2021/v15i430235

Arnold A et al (2000) Are tropical fungal endophytes hyperdiverse? Ecol Lett 3:267–274. https://doi.org/10.1046/j.1461-0248.2000.00159.x

Arnold AE, Mejía LC, Kyllo D et al (2003) Fungal endophytes limit pathogen damage in a tropical tree. Proc Natl Acad Sci 100:15649–15654. https://doi.org/10.1073/pnas.2533483100

Asiandu A, Widjajanti H, Nurnawati E (2019) Exploration of endophytic fungi of dragon scale's fern (Pyrrosia piloselloides (L.) M.G. Price) as an antibacterial sources. Bioval Biol Res J 5. https://doi.org/10.24233/BIOV.5.2.2019.149

Asiandu A, Wahyudi A, Ramadhan F, Widjajanti H (2021) Bioprospecting of non-mycorrhizal endophytic fungi associated with ferns and mosses. Curr Res Environ Appl Mycol 11:416–437. https://doi.org/10.5943/cream/11/1/30

Baerlocher F (2005) Freshwater fungal communities. In: The fungal community: its organization and role in the ecosystem, pp 39–59

Baerlocher F, Sridhar K (2014) Association of animals and fungi in leaf decomposition. pp 413–441

Bandoni RJ (1972) Terrestrial occurrence of some aquatic hyphomycetes. Can J Bot 50:72–297

Bärlocher F (1992) Community organization. In: Bärlocher F (ed) The ecology of aquatic hyphomycetes. Springer, Berlin, Heidelberg, pp 38–76

Bärlocher F (2006) Fungal endophytes in submerged roots. In: Schulz BJE, Boyle CJC, Sieber TN (eds) Microbial root endophytes. Springer, Berlin, Heidelberg, pp 179–190

Bärlocher F (2016) Aquatic hyphomycetes in a changing environment. Fungal Ecol 19:14–27. https://doi.org/10.1016/j.funeco.2015.05.005

Bärlocher F, Boddy L (2016) Aquatic fungal ecology—how does it differ from terrestrial? Fungal Ecol 19:5–13. https://doi.org/10.1016/j.funeco.2015.09.001

Barlocher F, Murdoch JH (1989) Hyporheic biofilms – a potential food source for interstitial animals. Hydrobiologia 184:61–67. https://doi.org/10.1007/BF00014302

Barreto GG, Hepp LU et al (2023) The cooler the better: increased aquatic hyphomycete diversity in subtropical streams along a neotropical latitudinal gradient. Fungal Ecol 62:101223. https://doi.org/10.1016/j.funeco.2022.101223

Baschien C, Tsui CK-M, Gulis V et al (2013) The molecular phylogeny of aquatic hyphomycetes with affinity to the Leotiomycetes. Fungal Biol 117:660–672. https://doi.org/10.1016/j.funbio.2013.07.004

Batke S, Reyes Chávez J, Quail M et al (2021) Nowhere to escape -diversity and community composition of ferns and lycophytes on the highest mountain in Honduras. J Trop Ecol 37. https://doi.org/10.1017/S0266467421000122

Behie SW, Zelisko PM, Bidochka MJ (2012) Endophytic insect-parasitic fungi translocate nitrogen directly from insects to plants. Science 336:1576–1577. https://doi.org/10.1126/science.1222289

Belliveau MJR, Bärlocher F (2005) Molecular evidence confirms multiple origins of aquatic hyphomycetes. Mycol Res 109:1407–1417. https://doi.org/10.1017/s0953756205004119

Benjamin A, Manickam VS (2007) Medicinal pteridophytes from the Western Ghats. Indian J Trad Knowl 6:611–618

Bills G, Polishook JD (1992) Recovery of endophytic fungi from Chamaecyparis thyoides. Sydowia 44:1–12

Bondada B, Tu C (2006) Surface structure and anatomical aspects of Chinese brake fern (Pteris vittata; Pteridaceae). Brittonia 58:217–228

Campbell J, Shearer C, Marvanová L (2006) Evolutionary relationships among aquatic anamorphs and teleomorphs: Lemonniera, Margaritispora, and Goniopila. Mycol Res 110:1025–1033. https://doi.org/10.1016/j.mycres.2006.04.012

Campbell J, Marvanová L, Gulis V (2009) Evolutionary relationships between aquatic anamorphs and teleomorphs: Tricladium and Varicosporium. Mycol Res 113:1322–1334. https://doi.org/10.1016/j.mycres.2009.09.003

Cao H, Chai T-T, Wang X et al (2017) Phytochemicals from fern species: potential for medicine applications. Phytochem Rev 16:379–440. https://doi.org/10.1007/s11101-016-9488-7

Chadha N, Mishra M, Rajpal K et al (2015) An ecological role of fungal endophytes to ameliorate plants under biotic stress. Arch Microbiol 197:869–881. https://doi.org/10.1007/s00203-015-1130-3

Chandra S, Fraser-Jenkins C, Kumari A, Srivastava A (2008) A summary of the status of threatened Pteridophytes of India. Taiwania 53:170–209. https://doi.org/10.6165/tai.2008.53(2).170

Chandrashekar KR, Sridhar KR, Kaveriappa KM (1991) Aquatic hyphomycetes of a Sulphur spring. Hydrobiologia 218:151–156. https://doi.org/10.1007/BF00006787

Chao A, Chazdon RL, Colwell RK, Shen T-J (2005) A new statistical approach for assessing similarity of species composition with incidence and abundance data. Ecol Lett 8:148–159. https://doi.org/10.1111/j.1461-0248.2004.00707.x

Chaudhari S, Patil V, Borse B (2016) List of freshwater mitosporic fungi of Madhya Pradesh. Int J Res Biosci Agric Technol 4:109–105

Chauvet E, Cornut J, Sridhar K et al (2016) Beyond the water column: aquatic hyphomycetes outside their preferred habitat. Fungal Ecol 19:112–127. https://doi.org/10.1016/j.funeco.2015.05.014

Chen Y-M, Yang Y-H, Li X-N et al (2015) Diterpenoids from the Endophytic fungus Botryosphaeria sp. P483 of the Chinese herbal medicine Huperzia serrata. Molecules 20:16924–16932. https://doi.org/10.3390/molecules200916924

Cornut J, Elger A, Lambrigot D et al (2010) Early stages of leaf decomposition are mediated by aquatic fungi in the hyporheic zone of woodland streams. Freshw Biol 55:2541–2556. https://doi.org/10.1111/j.1365-2427.2010.02483.x

Cornut J, Elger A, Greugny A et al (2012) Coarse particulate organic matter in the interstitial zone of three French headwater streams. Ann Limnol Int J Lim 48:303–313. https://doi.org/10.1051/limn/2012019

Cornut J, Chauvet E, Mermillod-Blondin F et al (2014) Aquatic Hyphomycete species are screened by the Hyporheic zone of woodland streams. Appl Environ Microbiol 80:1949–1960. https://doi.org/10.1128/AEM.03024-13

Crenshaw CL, Valett HM, Tank JL (2002) Effects of coarse particulate organic matter on fungal biomass and invertebrate density in the subsurface of a headwater stream. J N Am Benthol Soc 21:28–42. https://doi.org/10.2307/1468297

Danger M, Cornut J, Elger A, Chauvet E (2012) Effects of burial on leaf litter quality, microbial conditioning and palatability to three shredder taxa. Freshw Biol 57:1017–1030. https://doi.org/10.1111/j.1365-2427.2012.02762.x

Davey M, Blaalid R, Vik U et al (2015) Primary succession of istorta vivipara (L.) Delabre (Polygonaceae) root-associated fungi mirrors plant succession in two glacial chronosequences. Environ Microbiol 17:2777–2790. https://doi.org/10.1111/1462-2920.12770

Davis EC, Franklin JB, Shaw AJ, Vilgalys R (2003) Endophytic Xylaria (Xylariaceae) among liverworts and angiosperms: phylogenetics, distribution, and symbiosis. Am J Bot 90:1661–1667. https://doi.org/10.3732/ajb.90.11.1661

de Melo IS, Azevedo JL et al (1998) Ecologia microbiana

Della PA, Falkenberg D (2019) Pteridophytes as ecological indicators: an overview. Hoehnea 46. https://doi.org/10.1590/2236-8906-52/2018

Duarte S, Bärlocher F, Pascoal C, Cássio F (2016) Biogeography of aquatic hyphomycetes: current knowledge and future perspectives. Fungal Ecol 19:169–181. https://doi.org/10.1016/j.funeco.2015.06.002

Dudani SN, Mahesh MK, Chandran MDS, Ramachandra TV (2014) Cyathea nilgirensis Holttum (Cyatheaceae: Pteridophyta): a threatened tree fern from Central Western Ghats, India. J Threat Taxa 6:5413–5416. https://doi.org/10.11609/JoTT.o3343.3413-5413-6

Dutta D, Puzari K, Gogoi R, Dutta P (2014) Endophytes: exploitation as a tool in plant protection. Braz Arch Biol Technol 57:621–629. https://doi.org/10.1590/S1516-8913201402043

Egorova A, Gatiyatullina A, Terenzhev D et al (2021) Toxic action of substances from male fern Dryopteris filix-mas (L.) Schott (1834) on free-living soil nematode Caenorhabditis elegans Maupas (1900). E3S Web Conf 254:09011. https://doi.org/10.1051/e3sconf/202125409011

Ellis BK, Stanford JA, Ward JV (1998) Microbial assemblages and production in alluvial aquifers of the Flathead River, Montana, USA. J N Am Benthol Soc 17:382–402. https://doi.org/10.2307/1468361

Essafi K, Chergui H, Pattee E, Mathieu J (1994) The breakdown of dead leaves buried in the sediment of a permanent stream in Morocco. Arch Hydrobiol 130:105–112. https://doi.org/10.1127/archiv-hydrobiol/130/1994/105

Faeth SH, Fagan WF (2002) Fungal endophytes: common host plant symbionts but uncommon Mutualists1. Integr Comp Biol 42:360–368. https://doi.org/10.1093/icb/42.2.360

Findlay S (1995) Importance of surface-subsurface exchange in stream ecosystems: the hyporheic zone. Limnol Oceanogr 40:159–164. https://doi.org/10.4319/lo.1995.40.1.0159

Fisher PJ, Petrini O (1989) Two aquatic hyphomycetes as endophytes in Alnus glutinosa roots. Mycol Res 92:367–368. https://doi.org/10.1016/S0953-7562(89)80081-4

Fisher P, Petrini O (1993) Ecology, biodiversity and physiology of endophytic fungi. Curr Top Bot Res 1:271–279

Fisher PJ, Petrini O, Webster J (1991) Aquatic hyphomycetes and other fungi in living aquatic and terrestrial roots of Alnus glutinosa. Mycol Res 95:543–547. https://doi.org/10.1016/S0953-7562(09)80066-X

Fouda AH, Hassan SE-D, Eid AM, Ewais EE-D (2015) Biotechnological applications of fungal endophytes associated with medicinal plant Asclepias sinaica (Bioss.). Ann Agric Sci 60:95–104. https://doi.org/10.1016/j.aoas.2015.04.001

Franco-Duarte R, Fernandes I, Gulis V et al (2022) ITS rDNA barcodes clarify molecular diversity of aquatic Hyphomycetes. Microorganisms 10:1569. https://doi.org/10.3390/microorganisms10081569
Fukushima T, Tanaka M, Gohbara M, Fujimori T (1998) Phytotoxicity of three lactones from Nigrospora sacchari. Phytochemistry 48:625–630. https://doi.org/10.1016/S0031-9422(97)01023-6
Gessner M, Barlocher F, Chauvet E (2003) Qualitative and quantitative analyses of aquatic hyphomycetes in streams, Tsui CKM, Hyde KD. Yunnan, Fungal Diversity Research
Gezaf S, Hamedo H, Ibrahim A, Mossa M (2022) Mycosynthesis of silver nanoparticles by endophytic fungi: mechanism, characterization techniques and their applications. Microbial Biosystems 7:48–65. https://doi.org/10.21608/mb.2023.185718.1066
Ghate SD, Sridhar KR (2017) Endophytic aquatic hyphomycetes in roots of riparian tree species of two Western Ghat streams. Symbiosis 71:233–240. https://doi.org/10.1007/s13199-016-0435-6
Giesemann P, Eichenberg D, Stöckel M et al (2020) Dark septate endophytes and arbuscular mycorrhizal fungi (Paris-morphotype) affect the stable isotope composition of 'classically' non-mycorrhizal plants. Funct Ecol 34:2453–2466. https://doi.org/10.1111/1365-2435.13673
Goh TK, Hyde KD (1996) Biodiversity of freshwater fungi. J Ind Microbiol Biotechnol 17:328–345. https://doi.org/10.1007/BF01574764
Grossart H-P, Rojas-Jimenez K (2016) Aquatic fungi: targeting the forgotten in microbial ecology. Curr Opin Microbiol 31:140–145. https://doi.org/10.1016/j.mib.2016.03.016
Grossart H-P, Van den Wyngaert S, Kagami M et al (2019) Fungi in aquatic ecosystems. Nat Rev Microbiol 17:339–354. https://doi.org/10.1038/s41579-019-0175-8
Heeger F, Bourne EC, Wurzbacher C et al (2021) Evidence for lignocellulose-decomposing enzymes in the genome and transcriptome of the aquatic Hyphomycete Clavariopsis aquatica. Journal of Fungi 7:854. https://doi.org/10.3390/jof7100854
Hussain A, Mehmood A, Qadir M et al (2020) Thermal stress alleviating potential of endophytic fungus Rhizopus oryzae inoculated to sunflower (Helianthus annuus L.) and soybean (Glycine max L.). Pak J Bot 52
Hyde K, Soytong K (2008) The fungal endophyte dilemma. Fungal Divers 33:163–173
Ingold CT (1942) Aquatic hyphomycetes of decaying alder leaves. Trans Br Mycol Soc 25:339. https://doi.org/10.1016/S0007-1536(42)80001-7
Irfan M, Jan G, Murad W et al (2021) Ethnomedicinal and traditional uses of the ferns of Khyber Pakhtunkhwa, Pakistan. Braz J Biol 84:e250256. https://doi.org/10.1590/1519-6984.250256
Ismail I, Hamayun M, Hussain A et al (2021) Penicillium Glabrum acted as a heat stress relieving endophyte in soybean and sunflower. Pol J Environ Stud 30. https://doi.org/10.15244/pjoes/128579
Jarial R, Thakur S, Sakinah M et al (2018) Potent anticancer, antioxidant and antibacterial activities of isolated flavonoids from Asplenium nidus. J King Saud Univ Sci 30:185–192. https://doi.org/10.1016/j.jksus.2016.11.006
Johnston PR, Quijada L, Smith CA et al (2019) A multigene phylogeny toward a new phylogenetic classification of Leotiomycetes. IMA Fungus 10:1. https://doi.org/10.1186/s43008-019-0002-x
Jumpponen A (2001) Dark septate endophytes – are they mycorrhizal? Mycorrhiza 11:207–211. https://doi.org/10.1007/s005720100112
Kenrick P, Crane PR (1997) The origin and early evolution of plants on land. Nature 389:33–39. https://doi.org/10.1038/37918
Kessler M, Karger DN, Kluge J (2016) Elevational diversity patterns as an example for evolutionary and ecological dynamics in ferns and lycophytes. J Syst Evol 54:617–625. https://doi.org/10.1111/jse.12218
Khambay BPS, Bourne JM, Cameron S et al (2000) A nematicidal metabolite from Verticillium chlamydosporium. Pest Manag Sci 56:1098–1099
Kohler A, Kuo A, Nagy LG et al (2015) Convergent losses of decay mechanisms and rapid turnover of symbiosis genes in mycorrhizal mutualists. Nat Genet 47:410–415. https://doi.org/10.1038/ng.3223

Kramer K, Viane R (1990) Reprint from the families and genera of vascular plants. pp 52–56
Krasnoff SB, Gupta S (1994) Identification of the antibiotic phomalactone from the entomopathogenic fungusHirsutella thompsonii var.synnematosa. J Chem Ecol 20:293–302. https://doi.org/10.1007/BF02064437
Krauss G-J, Solé M, Krauss G et al (2011) Fungi in freshwaters: ecology, physiology and biochemical potential. FEMS Microbiol Rev 35:620–651. https://doi.org/10.1111/j.1574-6976.2011.00266.x
Krings M, Taylor TN, Hass H et al (2007) Fungal endophytes in a 400-million-yr-old land plant: infection pathways, spatial distribution, and host responses. New Phytol 174:648–657. https://doi.org/10.1111/j.1469-8137.2007.02008.x
Krishnan A, Rekha K (2021) Diversity of pteridophytes in western ghats- a review. Plant Arch 21
Krivobok S, Thomasson F, Seigle-Murandi F, Steiman R, Bottex-Gauthier C (1994) 6-Allyl-5,6-dihydro-5hydroxypyran-2-one, a lactone produced by a new Drechslera species: specified 1H and 13C NMR assignments, mutagenic and immunomodulating testings. Pharmazie 49:605–607
Kumaresan V, Veeramohan R, Bhat M et al (2013) Fungal endophyte assemblages of some Pteridophytes from Mahe, India. World J Sci Technol 2013:7–10
Lehnert M, Kottke I, Setaro S et al (2009) Mycorrhizal associations in ferns from southern Ecuador. Am Fern J 99. https://doi.org/10.1640/0002-8444-99.4.292
Lehnert M, Krug M, Kessler M (2017) A review of symbiotic fungal endophytes in lycophytes and ferns – a global phylogenetic and ecological perspective. Symbiosis 71:77–89. https://doi.org/10.1007/s13199-016-0436-5
Mandl N, Lehnert M, Kessler M, Gradstein SR (2010) A comparison of alpha and beta diversity patterns of ferns, bryophytes and macrolichens in tropical montane forests of southern Ecuador. Biodivers Conserv 19:2359–2369. https://doi.org/10.1007/s10531-010-9839-4
Marvanová L, Bärlocher F (1998) Hyphomycetes from Canadian streams. IV. Spirosphaera dimorpha sp. nov. 68
Medeiros AO, Pascoal C, Graça M (2009) Diversity and activity of aquatic fungi under low oxygen conditions. Freshw Biol 54:142–149. https://doi.org/10.1111/j.1365-2427.2008.02101.x
Metzler GM, Smock LA (1990) Storage and dynamics of subsurface detritus in a sand-bottomed stream. Can J Fish Aquat Sci 47:588–594. https://doi.org/10.1139/f90-067
Mir SA, Mishra AK, Pala SA et al (2015) Ferns and fern allies of district Shopian, Kashmir Valley, India. Biodivers J Biol Diver 16. https://doi.org/10.13057/biodiv/d160105
Mossa M, Gezaf S, Ibrahim A, Hamedo H (2024) Microbial biosystems preliminary screening of endophytic fungi hosted some wild plants in North Sinai for biogenic production of silver nanoparticles. Microbial Biosystems 8. https://doi.org/10.21608/mb.2024.341385
Murat C, Vizzini A, Bonfante P, Mello A (2005) Morphological and molecular typing of the below-ground fungal community in a natural tuber magnatum truffle-ground. FEMS Microbiol Lett 245:307–313. https://doi.org/10.1016/j.femsle.2005.03.019
Naegeli MW, Hartmann U, Meyer EI, Uehlinger U (1995) POM-dynamics and community respiration in the sediments of a floodprone prealpine river (Necker, Switzerland). Arch Hydrobiol 133:339–347. https://doi.org/10.1127/archiv-hydrobiol/133/1995/339
Nair SK, Bhat DJ (2016) Diversity and seasonal appearance of aquatic fungi in three streams of Western Ghat forests of Goa, India
Newsham KK (2011) A meta-analysis of plant responses to dark septate root endophytes. New Phytol 190:783–793. https://doi.org/10.1111/j.1469-8137.2010.03611.x
Nyayiru Kannaian UP, Natarajan A, Poomagal D et al (2018) Fungal endophytes of an aquatic weed Marsilea minuta Linn. Curr Res Environ Appl Mycol 8. https://doi.org/10.5943/cream/8/1/7
Ojha R, Devkota HP (2021) Edible and medicinal Pteridophytes of Nepal: a review. Ethnobot Res Appl 22:1–16
Patel M, Reddy MN (2018) Discovery of the world's smallest terrestrial Pteridophyte. Sci Rep 8: 5911. https://doi.org/10.1038/s41598-018-24135-2

Patil V, Borse B (2015) Checklist of freshwater mitosporic fungi of India. Int J Bioassays 4:4090–4099

Patil S, Lavate R, Rawat V, Dongare M (2016) Diversity and distribution of Pteridophytes from Satara District, Maharashtra (India). Plant Sci Today 3:149–156. https://doi.org/10.14719/pst.2016.3.2.216

Petrini O (1991) Fungal endophytes of tree leaves. In: Andrews JH, Hirano SS (eds) Microbial ecology of leaves. Springer, New York, pp 179–197

Pigg KB, Rothwell GW (2001) Anatomically preserved Woodwardia virginica (Blechnaceae) and a new filicalean fern from the middle Miocene Yakima canyon flora of Central Washington, USA. Am J Bot 88:777–787

Porras-Alfaro A, Bayman P (2011) Hidden fungi, emergent properties: endophytes and microbiomes. Annu Rev Phytopathol 49:291–315. https://doi.org/10.1146/annurev-phyto-080508-081831

Rajashekhar M, Kaveriappa KM (1996) Studies on the aquatic hyphomycetes of a sulfur spring in the Western Ghats, India. Microb Ecol 32:73–80. https://doi.org/10.1007/BF00170108

Rajashekhar M, Kaveriappa KM (2003) Diversity of aquatic hyphomycetes in the aquatic ecosystems of the Western Ghats of India. Hydrobiologia 501:167–177. https://doi.org/10.1023/A:1026239917232

Ramesha KP, Mohana NC, Nuthan BR et al (2020) Antimicrobial metabolite profiling of Nigrospora sphaerica from Adiantum philippense L. J Genetic Eng Biotechnol 18:66. https://doi.org/10.1186/s43141-020-00080-4

Razaghi P, Abdel-Azeem A (2024) Diversity of endophytic fungi hosted medicinal ferns: biotechnological potentials and possible applications. Microbial Biosyst 9:105–128. https://doi.org/10.21608/mb.2024.366970

Rogora M, Frate L, Carranza ML et al (2018) Assessment of climate change effects on mountain ecosystems through a cross-site analysis in the Alps and Apennines. Sci Total Environ 624:1429–1442. https://doi.org/10.1016/j.scitotenv.2017.12.155

Rojas-Jimenez K, Tamayo G (2021) Fungal endophytes and their bioactive compounds in tropical forests of Costa Rica. In: Neotropical endophytic fungi: diversity, ecology, and biotechnological applications. Springer International Publishing, Cham, pp 109–130

Russell J, Bulman S (2005) The liverwort Marchantia foliacea forms a specialized symbiosis with arbuscular mycorrhizal fungi in the genus glomus. New Phytol 165:567–579. https://doi.org/10.1111/j.1469-8137.2004.01251.x

Sánchez González A, Zúñiga E, Díez JD (2010) Richness and distribution patterns of ferns and lycopods in Los Mármoles National Park, Hidalgo, Mexico. J Torrey Bot Soc 137:373–379. https://doi.org/10.3159/10-RA-002.1

Sathiyaraj G (2015) Ethnomedicinal importance of fern and fern allies traditionally used by tribal people of Palani Hills (Kodaikanal), Western Ghats, South India. J Med Herbs Ethnomed 4–9. https://doi.org/10.5455/jmhe.2015-07-08

Sati SC, Belwal M (2005) Aquatic hyphomycetes as endophytes of riparian plant roots. Mycologia 97:45–49. https://doi.org/10.1080/15572536.2006.11832837

Sati SC, Belwal M (2009) Diversity of aquatic Hyphomycetes as root endophytes on Pteridophytic plants in Kumaun Himalaya. J Am Sci 5

Sati SC, Singh L (2014) Bioactivity of root Endophytic freshwater Hyphomycetes Anguillospora longissima (Sacc. & Syd.) Ingold. Sci World J 2014:707368. https://doi.org/10.1155/2014/707368

Schneider H, Schuettpelz E (2016) Systematics and evolution of lycophytes and ferns: editorial. J Syst Evol 54:561–562. https://doi.org/10.1111/jse.12231

Seena S, Monroy S (2016) Preliminary insights into the evolutionary relationships of aquatic hyphomycetes and endophytic fungi. Fungal Ecol 19:128–134. https://doi.org/10.1016/j.funeco.2015.07.007

Seetharam R, Sridhar K, Baerlocher F (1996) Endophytic aquatic hyphomycetes of roots of plantation crops and ferns from India. Sydowia 48

Shearer CA, Descals E, Kohlmeyer B et al (2007) Fungal biodiversity in aquatic habitats. Biodivers Conserv 16:49–67. https://doi.org/10.1007/s10531-006-9120-z

Singh J, Chaturvedi R (2017) Diversity of ecosystem types in India: a review. Proc Indian Natl Sci Acad 83:569–594

Singh L, Sati SC (2017) Bio-prospecting of root Endophytic freshwater fungi Anguillospora longissima and Cylindrocarpon aquaticum as phosphate Solubilization potential. J Pure Appl Microbiol 11:1929–1937. https://doi.org/10.22207/JPAM.11.4.34

Singh L, Sati SC (2019) Antagonistic activity of isolated root endophytic freshwater fungus Anguillospora longissima against pathogenic fungi. Natl Acad Sci Lett 43:1–5. https://doi.org/10.1007/s40009-019-00818-w

Soare LC, Ferdes M, Stefanov S et al (2012) Antioxidant activity, polyphenols content and antimicrobial activity of several native Pteridophytes of Romania. Notulae Botanicae Horti Agrobotanici Cluj-Napoca 40:53–57. https://doi.org/10.15835/nbha4016648

Sofiyanti N, Iriani D, Fitmawati F, Marpaung AA (2019) Morphology, palynology, and stipe anatomy of four common ferns from Pekanbaru, Riau Province, Indonesia. Biodivers J Biol Diver 20:327–336. https://doi.org/10.13057/biodiv/d200138

Sokolski S, Piché Y, Chauvet É, Bérubé JA (2006) A fungal endophyte of black spruce (Picea mariana) needles is also an aquatic hyphomycete. Mol Ecol 15:1955–1962. https://doi.org/10.1111/j.1365-294X.2006.02909.x

Solé M, Müller I, Pecyna MJ et al (2012) Differential regulation by organic compounds and heavy metals of multiple laccase genes in the aquatic Hyphomycete Clavariopsis aquatica. Appl Environ Microbiol 78:4732–4739. https://doi.org/10.1128/AEM.00635-12

Sridhar KR (2021) Five decades of research on the freshwater Hyphomycetes in India. In: Satyanarayana T, Deshmukh SK, Deshpande MV (eds) Progress in mycology: an Indian perspective. Springer, Singapore, pp 491–529

Sridhar KR, Bärlocher F (1992) Aquatic Hyphomycetes in spruce roots. Mycologia 84:580–584. https://doi.org/10.2307/3760325

Sridhar KR, Bärlocher F (1993) Aquatic hyphomycetes on leaf litter in and near a stream in Nova Scotia, Canada. Mycol Res 97:1530–1535. https://doi.org/10.1016/S0953-7562(09)80229-3

Sridhar KR, Bärlocher F (2000) Initial colonization, nutrient supply, and fungal activity on leaves decaying in streams. Appl Environ Microbiol 66:1114–1119. https://doi.org/10.1128/AEM.66.3.1114-1119.2000

Sridhar K, Kaveriappa KM (1988) Occurrence and survival of aquatic hyphomycetes in brackish and seawater. Arch Hydrobiol 113:153–160. https://doi.org/10.1127/archiv-hydrobiol/113/1989/153

Sridhar K, Seetharam R (1995) Endophytes – a crucial issue. Curr Sci 69:570–571

Sridhar KR, Chandrashekar KR, Kaveriappa KM (1992) Research on the Indian subcontinent. In: Bärlocher F (ed) The ecology of aquatic Hyphomycetes. Springer, Berlin, Heidelberg, pp 182–211

Sridhar K, Nagesh H, Kodandoor S (2020) Assemblage and diversity of asexual fungi in 10 terrestrial damp leaf litters: comparison of two incubation techniques. Asian J Mycol 3:362–375. https://doi.org/10.5943/ajom/3/1/10

Stark C, Babik W, Durka W (2009) Fungi from the roots of the common terrestrial orchid Gymnadenia conopsea. Mycol Res 113:952–959. https://doi.org/10.1016/j.mycres.2009.05.002

Sudheep NM, Sridhar KR (2011) Diversity of lignicolous and Ingoldian fungi on woody litter from the river Kali (Western Ghats, India). Mycology 2:98–108. https://doi.org/10.1080/21501203.2011.554905

Sudheep NM, Sridhar KR (2012) Aquatic hyphomycetes in hyporheic freshwater habitats of Southwest India. Limnologica 42:87–94. https://doi.org/10.1016/j.limno.2012.02.001

Sudheep N, Sridhar K (2013) Fungal colonization and decomposition of submerged woody litter in river Kali of the Western Ghats, India. Curr Res Environ Appl Mycol 3:160–180

Suhartina, Kandou FEF, Singkoh MFO (2018) Isolasi dan Identifikasi Jamur Endofit Pada Tumbuhan Paku Asplenium nidus. Jurnal MIPA 7:24–28. https://doi.org/10.35799/jm.7.2.2018.20640

Suryanarayanan TS, Thirunavukkarasu N, Govindarajulu MB et al (2009) Fungal endophytes and bioprospecting. Fungal Biol Rev 23:9–19. https://doi.org/10.1016/j.fbr.2009.07.001

Tanaka M, Fukushima T, Tsujino Y, Fujimori T (1997) Nigrosporins a and B, new phytotoxic and antibacterial metabolites produced by a fungus Nigrospora oryzae. Biosci Biotechnol Biochem 61:1848–1852. https://doi.org/10.1271/bbb.61.1848

Taylor TN, Remy W, Hass H, Kerp H (1995) Fossil Arbuscular mycorrhizae from the early Devonian. Mycologia 87:560–573. https://doi.org/10.2307/3760776

Tedersoo L, Pellet P, Kõljalg U, Selosse M-A (2007) Parallel evolutionary paths to mycoheterotrophy in understorey Ericaceae and Orchidaceae: ecological evidence for mixotrophy in Pyroleae. Oecologia 151:206–217. https://doi.org/10.1007/s00442-006-0581-2

Tehler A, Little DP, Farris JS (2003) The full-length phylogenetic tree from 1551 ribosomal sequences of chitinous fungi, fungi*. Mycol Res 107:901–916. https://doi.org/10.1017/S0953756203008128

Thi Minh Le T, Thi Hong Hoang A, Thi Bich Le T et al (2019) Isolation of endophytic fungi and screening of Huperzine A-producing fungus from Huperzia serrata in Vietnam. Sci Rep 9: 16152. https://doi.org/10.1038/s41598-019-52481-2

Vohník M, Mrnka L, Lukešová T et al (2013) The cultivable endophytic community of Norway spruce ectomycorrhizas from microhabitats lacking ericaceous hosts is dominated by ericoid mycorrhizal Meliniomyces variabilis. Fungal Ecol 6:281–292. https://doi.org/10.1016/j.funeco.2013.03.006

Waid JS (1954) Occurence of aquatic hyphomycetes upon the root surfaces of beech grown in woodland soils. Trans Br Mycol Soc 37:420–421. https://doi.org/10.1016/S0007-1536(54)80025-8

Webster J (1977) Seasonal observations on 'aquatic' hyphomycetes on oak leaves on the ground. Trans Br Mycol Soc 68:108–111. https://doi.org/10.1016/S0007-1536(77)80162-9

Wedin M, Döring H, Gilenstam G (2004) Saprotrophy and lichenization as options for the same fungal species on different substrata: environmental plasticity and fungal lifestyles in the Stictis–Conotrema complex. New Phytol 164:459–465. https://doi.org/10.1111/j.1469-8137.2004.01198.x

Wikee S, Lombard L, Crous P et al (2013) Phyllosticta capitalensis, a widespread endophyte of plants. Fungal Divers 60. https://doi.org/10.1007/s13225-013-0235-8

Wilson D (1995) Endophyte: the evolution of a term, and clarification of its use and definition. Oikos 73:274–276. https://doi.org/10.2307/3545919

Xu Y-M, Arnold AE, Ren U et al (2020) Teratopyrones A-C, dimeric Naphtho-γ-Pyrones and other metabolites from Teratosphaeria sp. AK1128, a fungal endophyte of Equisetum arvense. Molecules 25:5058. https://doi.org/10.3390/molecules25215058

Zhang Y, Mu J, Feng Y et al (2009) Broad-Spectrum antimicrobial epiphytic and Endophytic fungi from marine organisms: isolation, bioassay and taxonomy. Mar Drugs 7:97–112. https://doi.org/10.3390/md7020097

Zilliox L (1994) Chapter 3 – porous media and aquifer systems. In: Gibert J, Danielopol DL, Stanford JA (eds) Groundwater ecology. Academic Press, San Diego, pp 69–96

Zinger L, Boetius A, Ramette A (2014) Bacterial taxa–area and distance–decay relationships in marine environments. Mol Ecol 23:954–964. https://doi.org/10.1111/mec.12640

Zuccaro A, Lahrmann U, Güldener U et al (2011) Endophytic life strategies decoded by genome and transcriptome analyses of the mutualistic root symbiont Piriformospora indica. PLoS Pathog 7:e1002290. https://doi.org/10.1371/journal.ppat.1002290

Zuccaro A, Lahrmann U, Langen G (2014) Broad compatibility in fungal root symbioses. Curr Opin Plant Biol 20:135–145. https://doi.org/10.1016/j.pbi.2014.05.013

Arbuscular Mycorrhiza (AM) of Pteridophytes and Gymnosperms: Diversity and Plant-Microbe Interaction

Hira Bashir and Munazza Kiran

Contents

H. Bashir
Department of Botany, Women University Mardan, Khyber Pakhtunkhwa, Pakistan
e-mail: drhirabashir@wumardan.edu.pk

M. Kiran (✉)
Department of Botany, Division of Science & Technology, University of Education, Lahore, Punjab, Pakistan
e-mail: munazza.kiran@ue.edu.pk

A. M. Abdel-Azeem, F. Degola (eds.), *Advances in Lichens and Non-Flowering Plants Fungal Endophytes Research*, Fungal Biology,
https://doi.org/10.1007/978-3-032-01506-8_11

1 Introduction

Arbuscular mycorrhiza (AM) is a mutualistic symbiotic association between arbuscular mycorrhizal fungi (AMF) and plants, including pteridophytes (ferns and allied plants) and gymnosperms (cone-bearing plants). This association plays a crucial role in nutrient uptake and plant growth, making it an essential component of terrestrial ecosystems. Understanding the diversity of AMF and their interactions with these plant groups is fundamental for comprehending the functioning of plant-microbe interactions in natural and agricultural ecosystems. AMF are found in an immensely wide variety of host plants as obligatory symbiotic fungi. They have lately been reclassified based on DNA sequence data into a distinct fungal phylum, the *Glomeromycota* (Schüβler et al. 2001). It seems that the fungi had their ancestry possibly more than 1000 million years ago and that AM symbiosis is also very ancient. As far as nutrient uptake is concerned, AM fungi were perhaps the first and the most important in the colonization of land by plants (Remy et al. 1994). They remain a major determining factor in plant interactions in ecosystems to date.

The name "arbuscular" is a derivative of unique structures, the arbuscules, which can be seen within the cortical cells of several plant roots as well as some mycothalli colonized by AM fungi. Along with storage vesicles situated within or between the plant cells, these specific structures have been thought to be diagnostic for AM symbioses. AM fungi are believed to have a monophyletic origin in the Ordovician, about 480 million years ago (Redecker et al. 2013). It is commonly found in most taxa of land plants and effectively all ecological niches. Most land plants are involved in facultative symbiotic associations, i.e., they benefit from AM fungi, but are also able to live without them, although at significant fitness costs. However, some of the plant species became obligate parasites on the AM fungus, i.e., they became fully reliant on fungal nutrients and lost their photosynthetic ability (mycoheterotrophs). On the other side of the scale, some plant species belonging to families such as the *Brassicaceae* and *Chenopodiaceae* became asymbiotic, i.e., they lost their ability to make successful interactions with AM fungi and evolved alternative approaches to meet their nutritional requirements (Fig. 1) (Boller and Wiemken 1992).

Arbuscular mycorrhizal symbiotic interaction is said to be mainly promiscuous involving more than 100,000 plant taxa and a few hundred AM fungal morphotypes, which have long been considered as the equivalent of species. However, due to the presence of surprisingly few distinguishing morphological features of AM fungi (broadly speaking, associated with spores), and because of their fundamentally asexual mode of propagation, the traditional species concept is quite elaborative within the context of AM fungi. AM fungi have by no means been shown to shape sexual tiers or to mate, however, they are able to undergo hyphal fusions (anastomoses) and alternate genetic fabric, thereby reshuffling their genomes and producing new genetic range within the absence of classical meiosis and recombination (Dučić et al. 2009). Anastomosis depends on genetic relatedness; for this reason, this option

Fig. 1 AM fungi inside the plant root cells

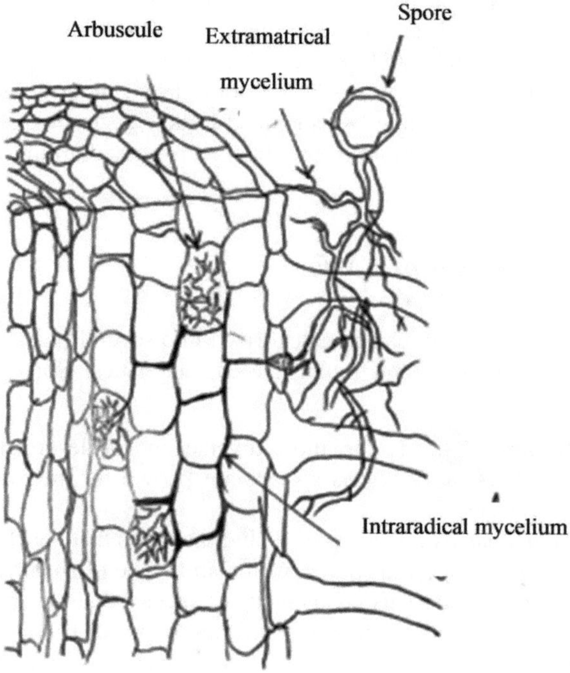

ought to doubtlessly be used as an extra criterion to define taxonomic units besides spore morphotypes.

2 Diversity of AMF

AMF are a group of soil-dwelling fungi that form a symbiotic relationship with plant roots. AMF belongs to the phylum *Glomeromycota* and forms a mutualistic symbiosis with a wide range of host plants. AMF diversity is crucial for maintaining ecosystem balance, promoting plant growth, and enhancing soil fertility. Several studies have shown that AMF colonizes the roots of pteridophytes and gymnosperms, forming characteristic intracellular structures called arbuscules. These arbuscules act as sites for nutrient exchange between the fungus and the plant.

The diversity of AMF in pteridophytes and gymnosperms is not as well-studied as in angiosperms (flowering plants). However, a few recent studies have begun to shed light on the presence and distribution of AMF in these plant groups. For example, studies have reported the presence of AMF in various fern species, such as *Athyrium filix-femina* (L.) Roth, *Dryopteris filix-mas* (L.) Schott, and *Polystichum acrostichoides* (Michx.) Schott. Similarly, AMF colonization has been observed in gymnosperms, including *Pinus* spp. L. and *Picea* spp. L. The occurrence of AM association in pteridophytes has received little attention, and only 180 species of ferns have been analyzed for mycorrhizal interactions since 1960. In contrast, approximately 6000 species of angiosperms and gymnosperms have been examined in this regard (Trappe 1987). Prior to 1960, about 420 species of ferns were observed for AM, excluding Newman and Reddle's survey of specimens. In the study, more than 75% of samples were obtained from herbarium specimens (Boullard 1957).

2.1 Types of AMF

AMF diversity can be categorized into several types based on their morphological, physiological, and molecular characteristics. Some of the main types of AMF include:

Glomus: This is the most diverse and widespread genus of AMF, with over 80 species. *Glomus* species are known for their ability to form arbuscules and vesicles.

Scutellospora: This genus includes species that are characterized by the presence of scutellosporoid spores, which are a type of spore that is unique to this genus.

Acaulospora: This genus includes species that are characterized by the presence of accaulosporoid spores, which are a type of spore that is unique to this genus.

Entrophospora: This genus includes species that are characterized by the presence of entrophosporoid spores, which are a type of spore that is unique to this genus.

Gigaspora: This genus includes species that are characterized by the presence of gigasporoid spores, which are a type of spore that is unique to this genus.

2.2 Diversity of AMF in Different Ecosystems

AMF diversity varies widely across different ecosystems, including:

Tropical ecosystem: Tropical ecosystems tend to have higher AMF diversity due to the presence of a wide range of plant species.

Temperate ecosystems: Temperate ecosystems tend to have lower AMF diversity compared to tropical ecosystems.

Agricultural ecosystems: Agricultural ecosystems tend to have lower AMF diversity due to the use of intensive farming practices.

Soil types: AMF diversity also varies across different soil types, with some soils supporting a wider range of AMF species than others.

3 Plant-Microbe Interactions and Intracellular Mechanisms of AM Fungi

The association between pteridophytes, gymnosperms, and AMF is mutually beneficial. The fungal partner enhances the plant's nutrient uptake , particularly phosphorus, which is often limited in natural ecosystems. In return, the plant supplies the fungus with carbohydrates produced through photosynthesis. This exchange of nutrients not only promotes plant growth but also contributes to soil fertility and ecosystem nutrient cycling. Furthermore, AMF are known to enhance plant resistance against various biotic and abiotic stresses, including pathogen attacks, drought, and heavy metal toxicity. They can also improve plant tolerance to salinity and nutrient deficiencies. These beneficial effects have significant implications for the cultivation and conservation of pteridophytes and gymnosperms, especially in nutrient-poor and degraded environments (Fig. 2).

A high degree of adaptation and genetic/metabolic coordination between mycorrhizal partners is suggested by the very long evolutionary history of AM symbiosis, which spans more than 400 million years (Schüssler and Walker 2010; Redecker et al. 2013), as well as the involvement of plant-derived and fungal signaling molecules that promote AM. The root-borne signal, strigolactone, is secreted to encourage AM fungal activity; it is, in fact, necessary for AM production. A signal transduction system known as the Common Symbiosis Signaling Pathway (CSSP),

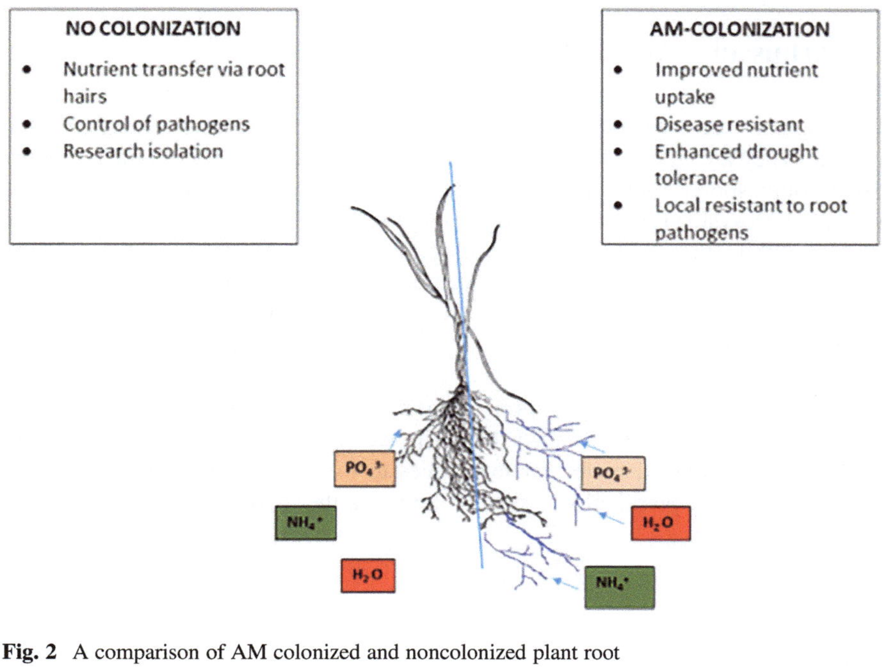

Fig. 2 A comparison of AM colonized and noncolonized plant root

which is shared with root nodule symbiosis, is activated by lipochito-oligosaccharides that AM fungi subsequently release (Harper et al. 2015). The involvement of a bidirectional exchange of symbiosis signals challenges our current understanding of communication between partners, particularly in the context of the extremely low host specificity in AM, as it would require either a large number of alternative signals for each potential partner or a small number of signals that a broad range of potential partners can recognize.

A large body of microscopic evidence demonstrates that the contact has a very high degree of cellular coordination at later stages, even though fundamental concerns about recognition and infection remain unanswered. The production of an infectious structure (pre-penetration apparatus; PPA) that permits cellular invasion and the development of intracellular arbuscules, which act as a nutritional interface between cells are two of the most striking instances between the partners (Harper et al. 2015). PPAs are assumed to be a precondition for AM fungal infection of host roots and to need communication through the CSSP, even if the molecular-genetic basis of PPA development is unclear (Ferrol et al. 2016). The development of AM is accompanied by a fundamental reprogramming of the host cells, which involves the activation of hundreds of genes, some of which are expressed primarily or exclusively in cells with arbuscules. Although it is believed that these genes are necessary for the fungus to fit inside cells. Their molecular and cellular activities have only rarely been clarified in relation to the coordination of symbiotic processes (Ferrol et al. 2016).

4 Arbuscular Mycorrhizae of Pteridophytes

Pteridophytes are a diverse group of plants that includes ferns, horsetails, and club mosses. They are characterized by the presence of vascular tissue and the reproduction through spores, rather than seeds. Pteridophytes play important roles in various terrestrial ecosystems, occupying a range of habitats including forests, wetlands, and deserts. While they have distinct ecological and morphological characteristics, pteridophytes share common features in their association with mycorrhizal fungi. Although the arbuscular mycorrhizae are commonly found in angiosperms, their occurrence in pteridophytes has been a subject of debate. Early studies suggested that pteridophytes are largely non-mycorrhizal or form associations with other types of mycorrhizal fungi. However, more recent research has consistently demonstrated the presence of arbuscular mycorrhizae in various pteridophyte species. These studies have employed molecular techniques to examine the presence of specific fungal markers in pteridophyte roots, confirming the association with arbuscular mycorrhizal fungi.

4.1 Arbuscular Mycorrhizal Fungi and Pteridophytes

Pteridophytes have unique life cycles and are of ancient origin. They play a crucial role in the emergence and orderly evolution of vascular plants. The fossil rhizomes of *Rhynia* and *Asteroxylon*, as described by Kidston & Lang, which were widespread in the Devonian to Carboniferous periods of the Paleozoic era, contained mycorrhizae formed between pteridophytes and fungi, and these mycorrhizae were thought to be the earliest arbuscular mycorrhizae (Hass et al. 1994; Remy et al. 1994). The arbuscular mycorrhizal status of modern pteridophytes is still unclear despite having endured throughout evolution for 300 to 400 million years. Boullard (1957) conducted a systematic examination of the mycorrhizae of pteridophytes, surveying the mycorrhizae of more than 420 species (mainly sporophytes and a few gametophytes). According to Boullard, all pteridophytes had "endophytic symbioses" except for the aquatic ferns (*Marsileaceae*, *Salviniaceae*, and *Isoetaceae*). It is highly regrettable that Boullard did not examine the many kinds of endophytic symbioses and the endophytic fungus, and that he obtained 75% of his material (roots and rhizomes) from dried herbarium specimens rather than from fresh plants. Cooper (1976) came to a similar conclusion to Boullard after examining the vesicular-arbuscular mycorrhizae (VAM) of 101 species of pteridophytes in New Zealand. When researching VAM in southern Ontario ferns and fern-allies, Berch and Kendrick (1982) discovered that there was no VAM in the sporophytes of fern-allies, but that all eusporangiates in the family *Filicineae* were arbuscular mycorrhizal partners. Arbuscular mycorrhizae were less common in the *Filicineae* than what Boullard (1957) or Cooper (1976) observed. In Hawaii, Gemma et al. (1992) examined the mycorrhizae of 89 species of ferns. They discovered 66 species to be mycorrhizal; however, the *Dicksoniaceae*, *Dryopteridaceae*, and *Lindsaeaceae* leptosporangiates had the largest percentage. Mycorrhiza was quite infrequent in the eusporangiates. It appears that the arbuscular mycorrhizal status of pteridophytes found in diverse regions varies, and the distinct species' mycorrhizal status is yet unknown.

5 Ecological Significance of Arbuscular Mycorrhizae in Pteridophytes

Arbuscular mycorrhizae play crucial roles in nutrient acquisition and plant performance in many ecosystems. In pteridophytes, these mycorrhizal associations have several ecological implications:

1. **Nutrient Acquisition:** Like in angiosperms, arbuscular mycorrhizae in pteridophytes facilitate the uptake of essential nutrients, particularly phosphorus, from nutrient-deficient soils. The extensive fungal hyphae network can explore a larger soil volume, enhancing nutrient availability for the host plant.

2. **Drought Tolerance:** Pteridophytes growing in arid and semi-arid environments often face water limitations. Arbuscular mycorrhizae have been shown to enhance drought tolerance in pteridophytes by improving water uptake and reducing water stress-induced damage.
3. **Plant Growth and Fitness:** The association with arbuscular mycorrhizal fungi can positively influence the growth, biomass accumulation, and reproductive success of pteridophytes. These fungi contribute to the overall health and fitness of the plants, allowing them to compete effectively in their respective habitats.
4. **Soil Stabilization:** Pteridophytes, including ferns, often colonize disturbed areas or soil with poor fertility. Arbuscular mycorrhizal fungi aid in the establishment and recovery of vegetation by improving soil structure, nutrient cycling, and organic matter decomposition.

5.1 Arbuscular Mycorrhizal Status of Pteridophytes in Different Evolutionary Stages

The number of species, frequency, and distribution of the *Filicineae* (eusporangiates and leptosporangiates) in extant pteridophytes all outnumber fern-allies. Bower (1959) discovered that ferns with synchronous sporangia production and maturation in the sorus (*Simplices*) predominated in the Paleozoic period by comparing the anatomy of *Filicineae* with the geological records (fossils) of pteridophytes. Along with the *Simplices*, the ferns of the *Gradatae* family, which dominated the Mesozoic period, have sporangia that develop progressively from the center to the base of the sorus. The production and maturation of the sorus is neither simultaneous nor gradual in more recent ferns (*Mixtae*), but rather blended from the center to the base. Ching (1978) listed pteridophyte families into three groups: *Helminthostachyaceae* to *Lygodiaceae* can be placed in the *Simplices*, *Hymenophyllaceae* to *Hypolepidaceae* in the *Gradatae*, and the families after *Pteridiaceae* in the *Mixtae*. The pteridophytes surveyed for this study can be categorized according to their arbuscular mycorrhizal status using the *Simplices*, *Gradatae*, and *Mixtae* to represent three stages in the vertical evolution process of the *Filicineae*. Arbuscular mycorrhizae were less common in each of the three evolutionary groups as they progressed from *Simplices* to *Mixtae*. While the percentage of non-mycorrhizal species gradually increased in each of these three categories (35%, 56%, and 74%, respectively), the proportion of continuously arbuscular mycorrhizal plants was 56% *Simplices*, 36% *Gradatae*, and 12% *Mixtae*. There were only minor differences in the proportion of the interim facultative arbuscular mycorrhizal type among the three classes. The *Filiniceae*'s apparent evolutionary transition from mycorrhizal to facultative to nonmycorrhizal greatly supports the theory that land plants evolved from mycotrophs. The further phylogenetic study of plants has benefited from the use of this evolutionary development (Trappe 1987; Gemma et al. 1992). Additionally, because pteridophytes originated before angiosperms and have continued to evolve in a non-mycorrhizal path, they have a lower incidence of arbuscular mycorrhiza than angiosperms do.

6 Systematic Survey

6.1 *Lycopsida*

Gametophytes

The gametophytes of *Lycopodium* L. species within *Lycopodiaceae* were investigated to determine whether the species are partially surface-living or completely subterranean. The first study based on sequencing was conducted on two gametophytes of Lycopodium, and the identified fungal association belonged to the family Glomeraceae (Redecker et al. 2013). A second molecular-based study was conducted on both the internal transcribed spacer (ITS) region and the larger subunit (LSU) region of nrDNA, and the study resulted in the identification of the AM fungus in the gametophyte of *Lycopodium alpinum* L. (Horn et al. 2013).

Sporophytes

The symbiotic associations appeared to be sporadic in thin wiry roots of families *Lycopodiaceae* and *Selaginellaceae*, as observed through light microscopic surveys. Morphologically, these fungi are found to be AM with large trunk hyphae, fine hyphal coils, and/or arbuscules and vesicles. Similarly, the thick and fleshy roots of *Isoetes* L. are found to be far better candidates for mycorrhizal association compared to narrow counterparts of *Lycopodium* and *Selaginella* P. Beauv. Three light microscopic studies reported the presence of AM fungi in the roots of two submerged species viz. *Isoetes lacustris* L. and *I. echinospora* A. Braun, in Europe, as well as in several terrestrial plant roots in India (Radhika and Rodrigues 2007) suggesting a broader occurence of AM fungi. In general, the AM associations are observed to be less frequent in ferns (Rimington et al. 2014).

6.2 *Monilophytes*

Gametophytes

Electron microscopic studies of subterranean fern gametophytes revealed that *Botrychium* Sw., *Ophioglossum* L., *Psilotum* Sw., and *Tmesipteris* Bernh. have exclusive intracellular symbionts comprising hyphal coils with arbuscules and vesicles, indicating typical features of *Glomeromycota*. The DNA-based studies confirmed that fungi in *Botrychium* (Winther and Friedman 2007), *Psilotum* (Winther and Friedman 2009; Rimington et al. 2014), and *Tmesipteris* (Rimington et al. 2014) are members of *Glomeraceae*. In contrast, fungal associations are not observed in endosporic gametophytes in heterosporous ferns and lycophytes. Furthermore, the symbionts are absent in gametophytes of *Equisetum,* although their multicellular cushions are found attached to the substratum and are pre-adapted structurally for fungal colonization. This absence is likely linked to ecological factors. All the associated liverworts also lack fungal partners. Most of the studies that were carried out on wild ferns have focused on ecology and reproductive biology and have found it difficult to identify them at the species or genus level. It also contributed to the lack of data on fungi. Overall, two features appeared to be helpful for fungal associations. One of them is rhizoids that are major entry points for AM fungi and bona fide symbionts are invariably present on ventral cell layers in central cushion region (Ogura-Tsujita et al. 2013, 2016). In addition to *Glomeraceae*, the members of *Gigasporaceae*, *Acaulosporaceae*, *Claroideoglomeraceae*, and *Archaeosporales* fungal partners are unknown in ferns previously, but widespread in thalloid liverworts and hornworts. Recent morphological, microscopic, and molecular studies have indicated similar associations in a range of genera, including *Adiantum* L., *Pellaea* Link, *Dryopteris* (L.) Schott, *Nephrolepis* Schott, and *Pteris* L.

Sporophytes
Fungi appear to be ubiquitously present in all taxa having fleshy roots, occupying several layers of cortical cells. However, in *Ophioglossum,* root hairs are absent. Ultrastructural investigations on *Ophioglossum, Psilotum, Tmesipteris*, and the marattioid fern (*Ptisana* Murdock) indicated the host-fungal relationship observed to be the same in both generations of the same taxon. Typical AM structures found in these five genera confirmed the AM associations in them (Rimington et al. 2014). Other electron microscopic examinations have shown that ascomycetes are found in *Loxsomopsis* Christ (Lehnert et al. 2009). However, none of the micrographs showed that ascomycete symbionts are surrounded by healthy host cytoplasm, thus question arise for the existence of fern mycorrhization. It is highly unlikely that pteridophytes form mutualistic associations with ascomycetes.

7 Overview of Arbuscular Mycorrhizal Fungi (AMF) in Gymnosperms

Gymnosperms comprise a diverse group of plants that includes conifers, cycads, ginkgo, and gnetophytes. They are characterized by their cone-bearing reproductive structures and the production of uncovered seeds. Gymnosperms are among the most endangered species of living things on Earth, with 40% of their species facing a high risk of extinction, which is approximately twice the latest estimate for all plants (i.e., 21.4%) (Brummitt et al. 2015). The dilemma of which species (and places) should receive priority in conservation programs is one that frequently arises, but it does so even more so in groupings like gymnosperms, where there are so many threatened species. It has been suggested that the best method for establishing and implementing this prioritization process in conservation is to use the metric of phylogenetic diversity (Faith 1992), which is based on evolutionary history. Gymnosperms play crucial roles in many terrestrial ecosystems, ranging from temperate forests to Arctic tundra, and often dominate in nutrient-poor and harsh environments. Mycorrhizae are symbiotic associations between plant roots and fungi that play a significant role in nutrient acquisition and plant growth. Understanding their mycorrhizal associations, particularly arbuscular mycorrhizae, can shed light on their nutrient acquisition strategies and ecological adaptations. Historically, gymnosperms were considered to be non-mycorrhizal or to form associations with other types of mycorrhizal fungi such as ectomycorrhizae. However, recent studies using molecular techniques have confirmed the presence of arbuscular mycorrhizal associations in several gymnosperm species. These studies have employed gene sequencing methods to target specific fungal markers in the roots, confirming the presence of arbuscular mycorrhizal fungi. The occurrence of this type of mycorrhizae in gymnosperms is variable, with some species showing a high prevalence of arbuscular mycorrhizal associations while others exhibit a lower frequency or absence of these associations.

Among the various types of mycorrhizae, arbuscular mycorrhizae (AM) have been extensively studied in relation to angiosperms. However, the occurrence and ecological significance of AM in gymnosperms (a group of non-flowering seed-producing plants) have received relatively less attention. On the other hand, it has always been believed that ectomycorrhizal fungus typically colonizes gymnosperms. AMF has been reported to have colonized several members of family *Pinaceae* (Smith et al. 1998; Wagg et al. 2008; Cázares and Trappe 1993). For instance, AMF vesicles have been found in the roots of *Pseudotsuga menziesii* (Mirb.) Franco (Wagg et al. 2011). In addition, several authors have noted the presence of AMF vesicles and hyphae in the roots of five additional *Pinaceae* species, belonging to the genera *Tsuga* Carrière, *Pinus*, and *Abies* L. (Chilvers et al. 1987; Dučić et al. 2009).

Here, we will explore the arbuscular mycorrhizal associations in gymnosperms and their importance in nutrient uptake and ecosystem functioning.

7.1 Arbuscular Mycorrhizal Fungi and Gymnosperms

Arbuscular mycorrhizal fungi (AMF) currently form symbioses with around 80% of vascular plants and have a long fossil history (Remy et al. 1994; Taylor et al. 2015; Walker et al. 2018) globally, including the five main conifer species (Smith and Read 2010; Spatafora et al. 2016). AMF colonizes the roots of the *Taxaceae* in contemporary conifers, *Araucariaceae*, *Cupressaceae*, *Sciadopityaceae*, and *Podocarpaceae* (Dickie and Holdaway 2011). Members of the *Podocarpaceae* and *Araucariales* clade, including *Sciadopitys* Siebold & Zuccarini, harbor AMF (arbuscular mycorrhizal fungi) in their young fine roots and in small, spherical structures called nodules. These nodules are a characteristic feature of this group of conifers (Spratt 1912; Mcgee et al. 1999; Russell et al. 2002; Dickie and Holdaway 2011; Padamsee et al. 2016). Although arbuscular mycorrhizal relationships with conifers and related species are known from the ancient record (Taylor et al. 2015; Strullu-Derrien et al. 2014; Walker et al. 2018), there is a dearth of information on these associations. Voltzialean conifers from the Triassic of Antarctica (Schwendemann et al. 2011; Harper et al. 2015), Carboniferous *Cordaites* from the Massif Central, France, and a taxodiaceous *Cupressaceae* from the Middle Eocene of Canada, all contain structurally preserved AM.

Nodular roots associated with the *Araucariales* have been found in the Triassic voltzialean *Notophytum krauselli* (Schwendemann et al. 2011; Harper et al. 2015) as well as in the Cretaceous of Antarctica, Australia, and India (Cantrill and Douglas 1988; Cantrill and Falcon-Lang 2001; Banerji and Ghosh 2002). The final record is the only one that describes the anatomical preservation of the rootlets by AM fungal invasion.

Among hot spring environments, the Middle-Late Jurassic Baha Laura Group chert deposits of the Deseado Massif in Santa Cruz Province, Southern Patagonia, Argentina, include exceptionally preserved organisms. Based on their structural similarities to the hot springs of Yellowstone National Park in the United States, the best-preserved Jurassic geothermal paleosurfaces can be considered positively comparable (Channing and Edwards 2009a,b). These are the only geothermal fossiliferous deposits known during the Mesozoic that have ecosystem-level preservation. A few localities have undergone preliminary analyses of the fossiliferous content and the reconstruction of the paleoenvironment (Channing and Edwards 2003; Guido et al. 2010; García Massini et al. 2012, 2013; Sagasti et al. 2016). In addition to the plant that is still present, fungi and fungus-like vegetative and reproductive structures are prevalent (Guido et al. 2010; García Massini et al. 2012, 2013; Sagasti et al. 2016, 2019), and frequently found in conjunction with one another. This offers the chance to discuss various plant-fungal interactions.

Only a few sporadic coenocytic hyphae are found in the outer cortex, close to the entry points beneath the epidermis, whereas AM fungal formations are restricted to the interior layers of the cortex. Hyphae are also found in the regenerated cortex of the nodules when resurgent development in the host nodules is present, which compares favorably to what has been documented in extant *Araucariales* (Russell et al. 2002; Dickie and Holdaway 2011). The present *Araucariales* likewise show a similar overall distribution of AM fungi in host nodules (McGee et al. 1999; Russell et al. 2002; Dickie and Holdaway 2011). Similar findings were made regarding fossil mycorrhizal plants, where the infected root tissues had the same zonal distribution of AM fungus.

7.2 Ecological Significance of Arbuscular Mycorrhizae in Gymnosperms

The presence of arbuscular mycorrhizae in gymnosperms carries several ecological implications:

1. **Nutrient Acquisition:** Gymnosperms, particularly in nutrient-poor habitats, benefit from the arbuscular mycorrhizal associations, as these fungi enhance the uptake of essential nutrients like phosphorus from the soil. The extensive fungal hyphal network extends beyond the root zone, thereby increasing the nutrient-absorbing capacity of the plant.

2. **Drought Resistance:** Gymnosperms are often found in arid and semi-arid regions where water availability is limited. Arbuscular mycorrhizal fungi play a crucial role in improving the drought resistance of gymnosperms by enhancing water uptake and promoting plant water-use efficiency.
3. **Plant Fitness and Growth:** The association with arbuscular mycorrhizal fungi positively influences the growth, biomass accumulation, and overall plant fitness in gymnosperms. These fungi contribute to the nutrient status and overall health of the plants, enabling them to thrive and compete successfully in their respective environments.
4. **Ecosystem Functioning:** Gymnosperms, as foundation species in many ecosystems, significantly contribute to key ecological processes. Arbuscular mycorrhizal associations in gymnosperms enhance soil nutrient cycling, carbon sequestration, and overall ecosystem productivity. They also have implications for soil stability.
5. The cortical tissues of conifer root nodules include coenocytic hyphae from glomeromycotan fungus that produce coils, ramified arbuscules, and spores. It was noted that just a few scattered coenocytic hyphae are found in the outer cortex, adjacent to the entry locations beneath the epidermis, while the AM fungal formations are isolated to the interior layers of the cortex. Hyphae are also found in the regenerated cortex of the nodules when resurgent development in the host nodules is present, which compares favorably to what has been documented in extant *Araucariales* (Baylis et al. 1963; Russell et al. 2002; Dickie and Holdaway 2011). In present *Araucariales*, a similar overall distribution of AM fungi is also seen in host nodules (McGee et al. 1999; Russell et al. 2002; Dickie and Holdaway 2011). Similar results were obtained in the infected root tissues of prior records of fossil mycorrhizal plants (Remy et al. 1994; Phipps and Taylor 1996; Strullu-Derrien et al. 2009, 2014; Harper et al. 2015). Arbuscular mycorrhizae can be classified classically into two morphological kinds: Arum type, which is defined by intracellular coiling hyphae with arbuscules growing along the coil, and Paris type, which is characterized by short side branches that give birth to arbuscules almost at right angles (Taylor et al. 2015; Walker et al. 2018). The Paris morphological type is congruent with the characteristics of the fungi described here. The morphological types, however, have been shown to vary continuously depending on various variables such as host plant tissues, taxa involved, and environment (Dickson 2004; Van Aarle et al. 2005), making this character merely descriptive and of relative use for the classification of AM fungi. Additionally, the differences in these morphologies have been considered in fossil comparisons (Harper et al. 2015, 2016; Walker et al. 2018).

8 Systematic Affinities of the Root Nodules in Gymnosperms and the Fungi

The root nodules have variable patterns of right-angle emergence from the parental root along the main axis, with a spherical to spheroidal three-dimensional morphology, and a central vascular cylinder surrounded by an endodermis, followed by several layers of cortex and an epidermis. The existing *Araucariales*, which contain the *Podocarpaceae* and *Araucariaceae*, share these characteristics. The La Matilde Formation has additional reproductive and vegetative organs that can be attributed to the *Araucariaceae* and *Podocarpaceae*, such as wood, leafy branches, and cones (Gnaedinger 2007; Falaschi et al. 2011; Kloster and Gnaedinger 2018). It is not possible to further elucidate the systematic affinities of the nodules based on the preserved morphological characteristics, especially in light of the morphological heterogeneity that is known to exist in both extant and fossil *Podocarpaceae* and *Araucariaceae* (Spratt 1912; Dickie and Holdaway 2011). It is still plausible that many taxa are present, given the presence of both podocarpaceous and araucariaceous organs in the chert blocks under study, as well as the lack of significant morphological variation among the mycorrhizal nodules. Based on phylogenetic analysis of combined molecular and morphological data, many families within the old *Glomeromycota* have been identified (Schüβler et al. 2001; Spatafora et al. 2016): *Paraglomeraceae, Archaeosporaceae, Geosiphonaceae, Ambisporaceae, Glomeraceae, Claroideoglomeraceae, Pacisporaceae, Gigasporaceae, Pacisporaceae, Sacculospporaceae, Acaulosporaceae,* and *Diversisporaceae* (Redecker et al. 2013; http://www.amf-phylogeny.com). The fungi described here are unrelated to other groups due to differences in spore type creation, spore wall construction, vesicle existence, and even ecology. Specific morphological traits lead us to believe that these fossils have relationships with the *Paraglomeraceae*. Arbuscules linked to slender trunks (4 m in diameter), the formation of intra-rhizal spores, and intracellular hyphae that regularly coil are distinguishing features that suggest similarities with the *Paraglomeraceae*. However, the wall of *Paraglomus* J.B. Morton & D. Redecker spores is made up of three layers, but the wall of our specimens is made up of only two layers. However, according to Morton and Redecker (2001), the outer layer of *Paraglomus* spores is said to disintegrate at senescence, resulting in a two-layered spore wall that is comparable to that seen by the fossils shown here. The *Glomeraceae* often generate straight hyphae, but the fossil most frequently displays coiled hyphae. On the other hand, the fossil's relationships with the *Glomeraceae* are also probable, differing primarily in the absence of vesicles. The fossil shares several characteristics with *Diversisporaceae* and *Ambisporaceae*, despite the fact that these two families also produce acaulosporoid spores, which are not present in the fossil. The Triassic of Antarctica contains the earliest evidence of nodular roots invaded by arbuscular mycorrhizal fungus (Schwendemann et al. 2011; Harper et al. 2015). In this instance, many fungi (including hyphal coils, arbuscules, extra-rhizal hyphae, vesicles, and spores (Schwendemann et al. 2011; Harper et al. 2015) have invaded *Notophytum krauselli*'s rootlets and root nodules. However, the fungal structures in the nodules and secondary rootlets are distinct from one another, indicating that several arbuscular mycorrhizal symbioses are present in various organs of the same plant root system (Harper et al. 2015). Additional instances of conifers with root nodules are from several Southern Hemisphere Cretaceous locations (Cantrill and Douglas 1988; Cantrill and Falcon-Lang 2001; Banerji and Ghosh 2002).

These nodules were discovered as imprints, compressions, and molds, but there is no anatomical preservation to detect any fungal associations. Based on the relationship with other organs, many affinities were proposed for these Cretaceous conifer nodule-bearing roots, including affiliations with the *Podocarpaceae, Araucariaceae, Sciadopityaceae*, and possibly *Cupressaceae*. This novel discovery of arbuscular mycorrhizae in root nodules from the Jurassic of Patagonia, with connections to the *Araucariales*, closes a gap in the geological history of AM fungi and adds to the body of knowledge regarding the morphological diversity and ecological preferences of *Glomeromycetes* across time.

9 Conclusions

Arbuscular mycorrhizae have proven to be ecologically relevant in pteridophytes, despite the earlier notion that these plants do not form associations with these fungi. The diversity and interactions between pteridophytes, gymnosperms, and AMF are important components of plant-microbe interactions in terrestrial ecosystems. Further research is needed to better understand the specific AMF taxa associated with these plant groups and their ecological roles. This knowledge will be essential for harnessing the potential of AMF in sustainable agriculture, ecosystem restoration, and conservation efforts.

References

Banerji J, Ghosh AK (2002) Mutualism/symbiosis from the early cretaceous (intertrappeans) of Rajmahal Basin, Jharkhand. India Curr Sci 83(9):1073–1074

Baylis GTS, McNabb RFR, Morrison TM (1963) The mycorrhizal nodules of podocarps. Trans Br Mycol Soc 46(3):378–384. https://doi.org/10.1016/S0007-1536(63)80031-5

Berch SM, Kendrick B (1982) Vesicular-arbuscular mycorrhizae of southern Ontario ferns and fern-allies. Mycologia 74(5):769–776. https://doi.org/10.1080/00275514.1982.12021584

Boller T, Wiemken A (1992) Arbuscular mycorrhiza in ferns and lycophytes: are the nonseed vascular understood as AM. Plant Soil 144(1):175–182

Boullard B (1957) Mycotrophy in pteridophytes. Its frequency, its characteristics, its meaning. Botanist 41:5–185

Bower FO (1959) Origin of a land flora; a theory based upon the facts of alternation. Hafner Publishing, p 727

Brummitt NA, Bachman SP, Griffiths-Lee J, Lutz M, Moat JF, Farjon A et al (2015) Green plants in the red: a baseline global assessment for the IUCN sampled red list index for plants. PLoS One 10(8):e0135152. https://doi.org/10.1371/journal.pone.0135152

Cantrill DJ, Douglas JG (1988) Mycorrhizal conifer roots from the lower cretaceous of the Otway Basin. Victoria Aust J Bot 36(3):257–272. https://doi.org/10.1071/BT9880257

Cantrill DJ, Falcon-Lang HJ (2001) Cretaceous (late Albian) coniferales of Alexander Island, Antarctica. 2. Leaves, reproductive structures and roots. Rev Palaeobot Palynol 115:119–145. https://doi.org/10.1016/S0034-6667(01)00053-7

Cázares E, Trappe JM (1993) Vesicular endophytes in roots of the *Pinaceae*. Mycorrhiza 2:153–156. https://doi.org/10.1007/BF00210584

Channing A, Edwards D (2003) Experimental taphonomy: silicification of plants in Yellowstone hot-spring environments. Earth and Environ Sci Trans Royal Soc Edin 94(4):503–521. https://doi.org/10.1017/S0263593300000845

Channing A, Edwards D (2009a) Silicification of higher plants in geothermally influenced wetlands: Yellowstone as a lower Devonian Rhynie analog. PALAIOS 24(8):505–521. https://doi.org/10.2110/palo.2008.p08-131r

Channing A, Edwards D (2009b) Yellowstone hot spring environments and the palaeoecophysiology of Rhynie chert plants: towards a synthesis. Plant Ecol Diverst 2(2):111–143. https://doi.org/10.1080/17550870903349359

Chilvers GA, Lapeyrie FF, Horan DP (1987) Ectomycorrhizal vs endomycorrhizal fungi within the same root system. New Phytol 107(2):441–448. https://doi.org/10.1111/j.1469-8137.1987.tb00195.x

Ching RC (1978) The Chinese fern families and genera: systematic arrangement and historical origin. Acta Phytotax Sin 16:16–37

Cooper KM (1976) A field survey of mycorrhizas in New Zealand ferns. New Zealand J of Bot 14(2):169–181. https://doi.org/10.1080/0028825X.1976.10428891

Dickie IA, Holdaway RJ (2011) Podocarp roots, mycorrhizas, and nodules. In: Turner BL, Cernusak LA (eds) Ecology of the *Podocarpaceae* in tropical forests, Smithsonian Contributions to Botany 95, pp 175–187

Dickson S (2004) The arum-Paris continuum of mycorrhizal symbioses. New Phytol 163:187–200

Dučić T, Berthold D, Langenfeld-Heyser R, Beese F, Polle A (2009) Mycorrhizal communities in relation to biomass production and nutrient use efficiency in two varieties of Douglas fir (*Pseudotsuga menziesii* var. *menziesii* and var. *glauca*) in different forest soils. Soil Biol Biochem 41(4):742–753. https://doi.org/10.1016/j.soilbio.2009.01.013

Faith DP (1992) Conservation evaluation and phylogenetic diversity. Biol Conserv 61(1):1–10. https://doi.org/10.1016/0006-3207(92)91201-3

Falaschi P, Grosfeld J, Zamuner AB, Foix N, Rivera SM (2011) Growth architecture and silhouette of Jurassic conifers from La Matilde formation, Patagonia, Argentina. Palaeogeogr Palaeoclimatol Palaeoecol 302:122–141. https://doi.org/10.1016/j.palaeo.2010.12.005

Ferrol N, Pérez-Tienda J, Azcón-Aguilar C (2016) Arbuscular mycorrhizas in pteridophytes and gymnosperms: a review. Mycorrhiza 26(7):313–326. https://doi.org/10.1002/9783527823468.ch21

Garcia Massini J, Channing A, Guido DM, Zamuner AB (2012) First report of fungi and fungus-like organisms from Mesozoic hot springs. PALAIOS 27(1):55–62. https://doi.org/10.2110/palo.2011.p11-076r

García Massini J, Channing A, Guido DM (2013) Diversidad paleobotánica en depósitos de origen geotermal del Jurásico (Santa Cruz, Argentina). In: V Simposio Argentino del Jurásico

Gemma JN, Koske RE, Flynn T (1992) Mycorrhizae in Hawaiian pteridophytes: occurrence and evolutionary significance. Amer J Bot 79(8):843–852. https://doi.org/10.1002/j.1537-2197.1992.tb13665.x

Gnaedinger S (2007) *Podocarpaceae* woods (*Coniferales*) from middle Jurassic La Matilde formation, Santa Cruz province, Argentina. Rev Palaeobot Palynol 147:77–93. https://doi.org/10.1016/j.revpalbo.2007.06.003

Guido DM, Channing A, Campbell KA, Zamuner A (2010) Jurassic geothermal landscapes and fossil ecosystems at San Agustín, Patagonia. Argentina J Geol Soc 167(1):11–20. https://doi.org/10.1144/0016-76492009-109

Harper CJ, Taylor TN, Krings M, Taylor EL (2015) Arbuscular mycorrhizal fungi in a voltzialean conifer from the Triassic of Antarctica. Rev Palaeobot Palynol 215:76–84. https://doi.org/10.1016/j.revpalbo.2015.01.005

Harper CJ, Taylor TN, Krings M, Taylor EL (2016) Structurally preserved fungi from Antarctica: diversity and interactions in late Palaeozoic and Mesozoic polar forest ecosystems. Antarct Sci 28(3):153–173. https://doi.org/10.1017/S0954102016000018

Hass H, Taylor TN, Remy W (1994) Fungi from the lower Devonian Rhynie chert: my Coparasitism. Ame J Bot 81(1):29–37. https://doi.org/10.1002/j.1537-2197.1994.tb15405.x

Horn K, Franke T, Unterseher M, Schnittler M, Beenkeb L (2013) Morphological and molecular analyses of fungal endophytes of achlorophyllous gametophytes of *Diphasiastrum alpinum* (*Lycopodiaceae*). Ame J Bot 100:2158–2174. https://doi.org/10.3732/ajb.1300011

Kloster AC, Gnaedinger SC (2018) Coniferous wood of *Agathoxylon* from the La Matilde formation, (middle Jurassic), Santa Cruz. Argentina J Pale 92(4):546–567. https://doi.org/10.1017/jpa.2017.145

Lehnert M, Kottke I, Setaro S, Pazmino LF, Kessler M (2009) Mycorrhizal associations in ferns from southern Ecuador. Amer Fern J 99:292–306. https://doi.org/10.1640/0002-8444-99.4.292

McGee PA, Bullock S, Summerell BA (1999) Structure of mycorrhizae of the Wollemi pine (*Wollemia noblis*) and related *Araucariaceae*. Aust J Bot 47(1):85–95. https://doi.org/10.1071/BT97064

Morton JB, Redecker D (2001) Two new families of *Glomales*, *Archaeosporaceae* and *Paraglomaceae*, with two new genera *Archaeospora* and *Paraglomus*, based on concordant molecular and morphological characters. Mycologia 93(1):181–195

Ogura-Tsujita Y, Sakoda A, Ebihara A, Yukawa T, Imaichi R (2013) Arbuscular mycorrhiza formation in cordate gametophytes of two ferns, *Angiopteris lygodiifolia* and *Osmunda japonica*. J Plant Res 126:41–50. https://doi.org/10.1007/s10265-012-0511-9

Ogura-Tsujita Y, Hirayama Y, Sakoda A, Suzuki A, Ebihara A, Morita N et al (2016) Arbuscular mycorrhizal colonization in field-collected terrestrial cordate gametophytes of pre-polypod leptosporangiate ferns (*Osmundaceae, Gleicheniaceae, Plagiogyriaceae, Cyatheceae*). Mycorrhiza 26:41–50. https://doi.org/10.1007/s00572-015-0648-1

Padamsee M, Johansen RB, Stuckey SA, Williams SE, Hooker JE, Burns BR, Bellgard SE (2016) The arbuscular mycorrhizal fungi colonising roots and root nodules of New Zealand kauri *Agathis australis*. Fungal Biol 120(5):807–817. https://doi.org/10.1016/j.funbio.2016.01.015

Phipps CJ, Taylor TN (1996) Mixed arbuscular mycorrhizae from the Triassic of Antarctica. Mycologia 88(5):707–714. https://doi.org/10.1080/00275514.1996.12026707

Radhika KP, Rodrigues BF (2007) Arbuscular mycorrhizae in association with aquatic and marshy plant species in Goa, India. Aqua Bot 86:291–294. https://doi.org/10.1016/j.aquabot.2006.10.009

Redecker D, Schüßler A, Stockinger H, Stürmer SL, Morton JB, Walker C (2013) An evidence-based consensus for the classification of arbuscular mycorrhizal fungi (Glomeromycota). Mycorrhiza 23:515–531. https://doi.org/10.1007/s00572-013-0486-y

Remy W, Taylor TN, Hass H, Kerp H (1994) Four hundred-million-year-old vesicular arbuscular mycorrhizae. Proc Natl Acad Sci USA 91:11841–11843. https://doi.org/10.1073/pnas.91.25.11841

Rimington WR, Pressel S, Duckett JG, Bidartondo MI (2014) Fungal diversity in early vascular plants: reopening a closed book? New Phytol 205:1394–1398

Russell AJ, Bidartondo MI, Butterfield BG (2002) The root nodules of the *Podocarpaceae* harbour arbuscular mycorrhizal fungi. New Phytol 156(2):283–295. https://doi.org/10.1046/j.1469-8137.2002.00504.x

Sagasti AJ, Massini JG, Escapa IH, Guido DM, Channing A (2016) *Millerocaulis zamunerae* sp. nov.(Osmundaceae) from Jurassic, geothermally influenced, wetland environments of

Patagonia, Argentina. Alcheringa: An Aust J Palaeo 40(4):456–474. https://doi.org/10.1080/03115518.2016.1210851

Sagasti AJ, Massini JG, Escapa IH, Guido DM (2019) Multitrophic interactions in a geothermal setting: arthropod borings, actinomycetes, fungi and fungal-like microorganisms in a decomposing conifer wood from the Jurassic of Patagonia. Palaeogeogr Palaeoclimatol Palaeoecol 514:31–44. https://doi.org/10.1016/j.palaeo.2018.09.004

Schüssler A, Walker C (2010) The *Glomeromycota*: a species list with new families and new Gener

Schüßler A, Schwarzott D, Walker C (2001) A new fungal phylum, the *Glomeromycota*: phylogeny and evolution. Mycol Res 105(12):1413–1421. https://doi.org/10.1017/S0953756201005196

Schwendemann AB, Decombeix AL, Taylor TN, Taylor EL, Krings M (2011) Morphological and functional stasis in mycorrhizal root nodules as exhibited by a Triassic conifer. Pro Nat Aca Sci 108(33):13630–13634. https://doi.org/10.1073/pnas.1110677108

Smith SE, Read DJ (2010) Mycorrhizal symbiosis. Academic press

Smith JE, Johnson KA, Cázares E (1998) Vesicular mycorrhizal colonization of seedlings of Pinaceae and Betulaceae after spore inoculation with *glomus intraradices*. Mycorrhiza 7:279–285. https://doi.org/10.1007/s005720050193

Spatafora JW, Chang Y, Benny GL, Lazarus K, Smith ME, Berbee ML, Stajich JE (2016) A phylum-level phylogenetic classification of zygomycete fungi based on genome-scale data. Mycol 108(5):1028–1046. https://doi.org/10.3852/16-042

Spratt ER (1912) The formation and physiological significance of root nodules in the *Podocarpineae*. Annals Bot 26(103):801–814

Strullu-Derrien C, Rioult JP, Strullu DG (2009) Mycorrhizas in upper carboniferous Radiculites-type cordaitalean rootlets. New Phytol 182:561–564

Strullu-Derrien C, Kenrick P, Pressel S, Duckett JG, Rioult JP, Strullu DG (2014) Fungal associations in *Horneophyton ligneri* from the Rhynie Chert (c. 407 million year old) closely resemble those in extant lower land plants: novel insights into ancestral plant–fungus symbioses. New Phytol 203(3):964–979. https://doi.org/10.1111/nph.12805

Taylor TN, Krings M, Taylor EL (2015) Fossil Fungi. Academic Press

Trappe JM (1987) Phylogenetic and ecologic aspects of mycotrophy in the angiosperms from an evolutionary standpoint. Ecophy VA myco Plants:5–25

Van Aarle IM, Cavagnaro TR, Smith SE, Smith FA, Dickson S (2005) Metabolic activity of glomus intraradices in arum and Paris type arbuscular mycorrhizal colonization. New Phytol 166(2): 611–618. https://doi.org/10.1111/j.1469-8137.2005.01340.x

Wagg C, Pautler M, Massicotte HB, Peterson RL (2008) The co-occurrence of ectomycorrhizal, arbuscular mycorrhizal, and dark septate fungi in seedlings of four members of the *Pinaceae*. Mycorrhiza 18:103–110. https://doi.org/10.1007/s00572-007-0157-y

Wagg C, Antunes PM, Peterson RL (2011) Arbuscular mycorrhizal fungal phylogeny-related interactions with a non-host. Symbiosis 53:41–46. https://doi.org/10.1007/s13199-011-0107-5

Walker C, Harper CJ, Brundrett MC, Krings M (2018) Looking for arbuscular mycorrhizal fungi in the fossil record: an illustrated guide. In *Transformative paleobotany* Academic Press. pp481–pp517

Winther JL, Friedman WE (2007) Arbuscular mycorrhizal symbionts in *Botrychium* (*Ophioglossaceae*). Amer J Bot 94:1248–1255

Winther JL, Friedman WE (2009) Phylogenetic affinity of arbuscular mycorrhizal symbionts in *Psilotum nudum*. J Plant Res 122:485–496

Exploring Endophytic Fungal Biodiversity in Different Coniferous Species

Fatma A. Abo Nouh, Sara A. Gezaf, Hager M. Fathy, Roaa S. Abdel-Monem, Kholoud A. Elsayed, Ola F. Thabet, Mardian A. Adel, Alshaimaa S. Abdel-Salam, Amal M. Elshazly, Emad M. Ahmed, Mennatullah M. Allam, Teroj A. Mohamed, Mohamed A. Abdel-Azeem, Lobna Hajji-Hedfi, Salma S. Higazy, Youssef M. Salah, Bola N. Aziz, Amany R. Ali, Norhan A. Soliman, Fatma A. Altamany, and Ahmed M. Abdel-Azeem

F. A. Abo Nouh (✉)
Botany and Microbiology Department, Faculty of Science, Suez Canal University, Ismailia, Egypt

Natural and Medical Sciences Research Center, University of Nizwa, Nizwa, Oman

S. A. Gezaf
Botany and Microbiology Department, Faculty of Science, Arish University, Arish, Egypt

H. M. Fathy · R. S. Abdel-Monem · K. A. Elsayed · O. F. Thabet · M. A. Adel · A. S. Abdel-Salam · A. M. Elshazly · E. M. Ahmed · M. M. Allam · S. S. Higazy · Y. M. Salah · B. N. Aziz · A. R. Ali · N. A. Soliman · F. A. Altamany
Natural and Medical Sciences Research Center, University of Nizwa, Nizwa, Oman

T. A. Mohamed
Dental Basic Sciences Department, College of Dentistry, University of Duhok, Duhok, Iraq

M. A. Abdel-Azeem
Pharmacognosy Department, Faculty of Pharmacy, Sinai University – Arish Branch, Arish, Egypt

L. Hajji-Hedfi
Regional Centre of Agricultural Research of Sidi Bouzid, CRRA, Sidi Bouzid, Tunisia

Research Laboratory of Agricultural Production Systems and Sustainable Development (LR03AGR02), Carthage University, Tunis, Tunisia

A. M. Abdel-Azeem
Botany and Microbiology Department, Faculty of Science, Suez Canal University, Ismailia, Egypt

Research Institute of University of the Bucharest (ICUB), University of Bucharest, Bucharest, Romania

The Centre for Mineral Biogeochemistry, The University of the Free State, Bloemfontein, Free State, South Africa

A. M. Abdel-Azeem, F. Degola (eds.), *Advances in Lichens and Non-Flowering Plants Fungal Endophytes Research*, Fungal Biology,
https://doi.org/10.1007/978-3-032-01506-8_12

Contents

1 Introduction

Gymnosperms represent a basal lineage of seed plants characterized by the exposure of their ovules or seeds, in contrast to the enclosed ovules within carpels observed in angiosperms. As one of the four principal divisions of terrestrial flora (along with bryophytes, ferns and fern allies, and angiosperms), gymnosperms exhibit certain shared characteristics with both ferns (e.g., the presence of archegonia in the female gametophyte and the utilization of spermatozoids are characteristic of the reproductive biology of these plants) and angiosperms (Stevenson 2013). Emerging in the mid-Devonian period, approximately 385 million years ago, gymnosperms have undergone an extensive evolutionary trajectory marked by numerous extinction events. These extinctions have resulted in substantial morphological divergence among the four extant gymnosperm groups: cycadophytes, ginkgophytes, conifers, and gnetophytes (Wang and Ran 2014; Yang et al. 2022a). A revised taxonomic classification of extant gymnosperms is proposed, integrating morphological and phylogenomic evidence. The system recognizes three classes: Ginkgoopsida, Pinopsida, and Cycadopsida. These are further subdivided into five subclasses: Ginkgoidae, Cycadidae, Pinidae, Gnetidae, and Cupressidae. A total of 8 orders are recognized within this classification: Cycadales, Ginkgoales, Cupressales, Araucariales, Ephedrales, Pinales, Welwitschiales, and Gnetales, and 13 families encompassing 86 genera (Table 1; Fig. 1). Conifers hold significant economic and ecological value, forming vast forests across both hemispheres. This group constitutes the most species-rich clade within extant gymnosperms, encompassing 7 families (Cupressaceae, Pinaceae, Araucariaceae, Sciadopityaceae, Podocarpaceae, Taxaceae, and Cephalotaxaceae), 72 genera, and approximately 702 recognized species (Yang et al. 2022a). A hemispheric-scale investigation of conifer evolutionary dynamics by Leslie et al. (2012), utilizing molecular phylogenetic analyses of

Table 1 Gymnosperm family diversity

Family	Genus	Species
Araucariaceae	3	40
Cephalotaxaceae	1	10
Cupressaceae	31	169
Cycadaceae	1	126
Ephedraceae	1	70
Ginkgoaceae	1	1
Gnetaceae	1	46
Pinaceae	11	272
Podocarpaceae	20	181
Sciadopityaceae	1	1
Taxaceae	5	29
Welwitschiaceae	1	1
Zamiaceae	9	255
Total	86	1201

489 species, determined that the major diversification of extant conifers occurred predominantly during the Neogene period, with earlier divergence events occurring in the Southern Hemisphere. Within the Northern Hemisphere's temperate and subtropical zones, *Pinaceae* and *Cupressaceae* are the dominant conifer families. Conversely, Southern Hemisphere conifer diversity is largely represented by *Araucariaceae, Podocarpaceae,* and *Callitroideae* subfamily of *Cupressaceae* (Ran et al. 2010). Phylogenetically, the *Podocarpaceae* are closest to the *Araucariaceae, Sciadopityaceae,* and *Taxaceae* (Leslie et al. 2018). *Podocarpaceae* holds considerable evolutionary and systematic importance due to its notable eco-physiological adaptations, which distinguish it from other conifer families. Consequently, this family offers valuable insights into evolutionary processes and responses to climate change (Kelch 2002).

Endophytes are ubiquitous in plants; nearly all plant species studied have been found to harbor them, and endophyte-free individuals within a species are rare (Zabalgogeazcoa 2008). Fossil evidence from plant stems and leaves indicates that endophytic associations, where microorganisms colonize plant tissues, may have originated over 400 million years ago (Kusari et al. 2012). Fungi represent the most observed endophytes. Endophytic fungi colonize the internal tissues of living plants, existing in a symbiotic relationship where they do not elicit disease symptoms or readily observable harm to the host. They establish mutualistic relationships with their plant hosts, at least for a portion of their life cycle. These fungi can inhabit a diverse range of plant tissues, including roots, stems, leaves, needles, branches, twigs, bark, fruit, seeds, scales, resin canals, meristems, petioles, and flowers, while some fungal endophytes exhibit specificity toward particular plant families (Abo Nouh 2019). The initial observation of fungal endophytes in conifers dates back to 1924 (Lewis 1924). Since their initial identification, these fungi have been found to colonize a broad taxonomic range of plant hosts, from herbaceous species and ferns to coniferous trees and deciduous (Johnson and Whitney 1989; Hata and

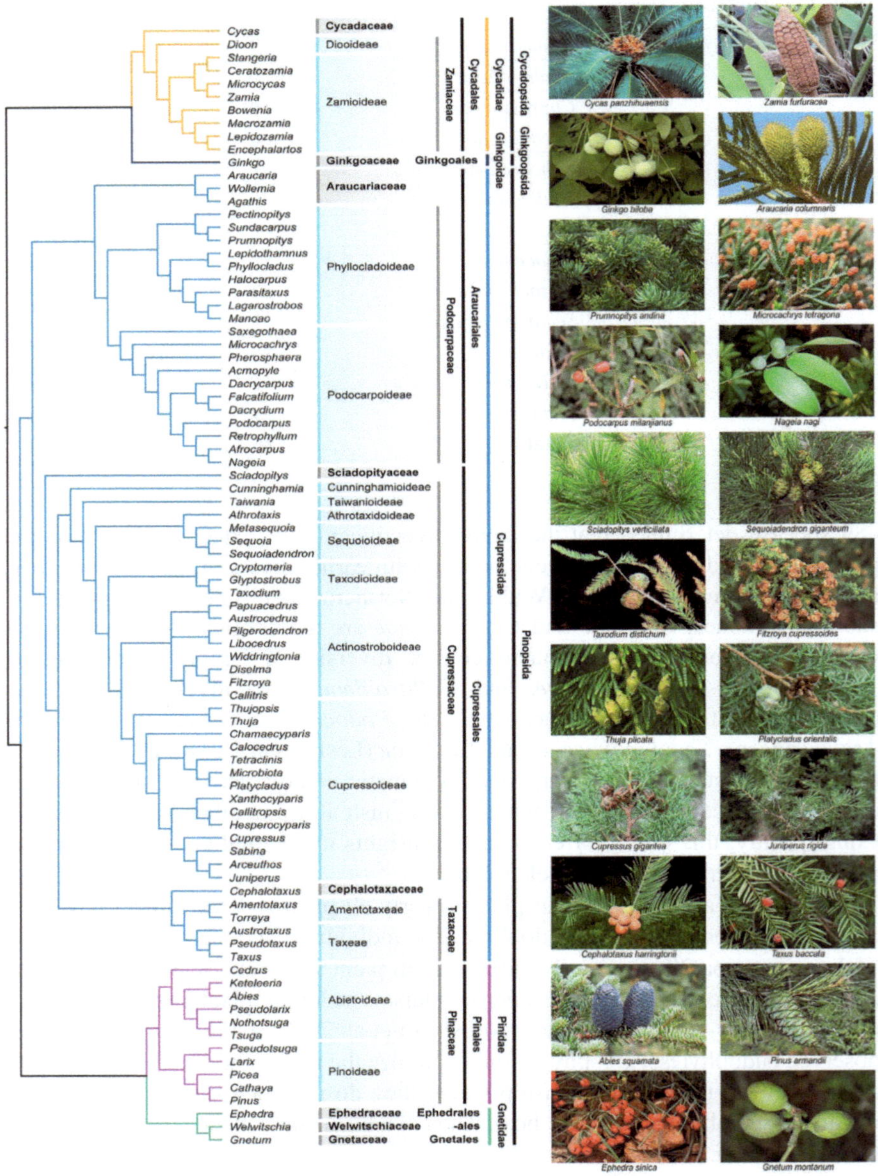

Fig. 1 Cladogram displaying phylogenetic relationships among extant gymnosperms according to Yang et al. (2022b)

Futai 1996; Arnold et al. 2001; Wei et al. 2007; Shi et al. 2010; Soltani and Moghaddam 2015; Ortiz et al. 2022; Vu et al. 2023). Contemporary studies on endophytic fungi predominantly examine their biodiversity, spatial distribution, and symbiotic relationships with host plants (Carroll 1988; Ganley et al. 2004;

Rodriguez and Redman 2008). *Cupressaceae* and *Pinaceae* species have been the subject of the most extensive investigations regarding endophytic fungal presence. Several studies have investigated the diversity and distribution of endophytes in coniferous trees, including *Picea mariana* (Johnson and Whitney 1992), *Picea abies* (Barklund and Kowalski 1996), *Abies balsamea* (Johnson and Whitney 1989), and various *Pinus* species (Hata et al. 1998; Deckert and Peterson 2000). While endophyte species diversity is considerable both within and among the three species, related endophyte species often dominate communities within host species of the same family. Diaporthales fungi are commonly dominant in broadleaved trees (e.g., *Aceraceae, Betulaceae, Fagaceae*), whereas Helotiales fungi frequently dominate needle- or scale-leaved trees (e.g., *Cupressaceae, Pinaceae*) (Schneider et al. 2004).

2 Biodiversity of Endophytic Fungi

Endophyte research is a rapidly advancing field, with ongoing developments in areas ranging from species identification and community structure analysis to understanding their ecological roles and the bioactive secondary metabolites they produce (Staniek et al. 2008). While there are about 270,000 known vascular plant species, the estimated number of endophytic fungal species is much higher, around 1.5 million. However, this is a broad estimate, with some studies suggesting a range from as low as 100,000 to as high as 9.9 million (Abo Nouh 2019). The potential diversity of fungal species, possibly exceeding that of plants, has prompted research to focus on plant-associated endophytic fungi. These studies aim to elucidate the factors contributing to this high fungal biodiversity (Arnold 2007). While Anton de Bary's initial observation of endophytic fungi in 1866 was groundbreaking, his initial interpretation of them as parasites within reproductive organs proved to be an oversimplification. The understanding of the complex and frequently mutualistic nature of these plant-fungus interactions developed significantly (de Bary 1866). Subsequent to de Bary's initial work, Carroll (1988), Petrini (1991), and Wilson (1995) researchers recognized endophytes as asymptomatic inhabitants of healthy plant tissues, such as twigs and leaves. These fungi live within various plant tissues, from leaves to roots, without producing overt signs of disease in the host plant. They contribute to a balanced relationship between the plant, the endophytes themselves, and the surrounding environment, at least during some phase of their life cycle. The ubiquity of endophytes is well-established; virtually every plant species examined has been found to host them, and endophyte-free individuals within a species are uncommon (Zabalgogeazcoa 2008). Fossil evidence from plant stems and leaves indicates that endophytic associations, where microorganisms colonize plant tissues, likely originated more than 400 million years ago (Kusari et al. 2012). Over the course of plant evolution, as plants transitioned from aquatic to terrestrial environments, they faced challenges such as higher atmospheric carbon dioxide levels, nutrient-poor soils, and fluctuating temperature and water availability. Simultaneously, the endophytes adapted to these evolving plant microenvironments, possibly through genetic changes, such as the incorporation of host plant DNA (Arora and Ramawat 2017). Although a diverse array of microorganisms (including fungi, bacteria, and actinobacteria) can function as endophytes, fungi represent the most frequently observed group. These fungal endophytes exhibit the capacity to colonize a broad spectrum of plant tissues, encompassing roots, leaves, stems, twigs, branches, bark, seeds, fruit, scales, meristems, petioles, resin canals, and even flowers (Abo Nouh 2019). The distribution and habitat of host plants are key factors influencing the ecological niche and biodiversity of their associated endophytic fungi (Saikkonen 2007). While endophytic fungi can exhibit species-specific symbiotic relationships with particular host plants, the nature of these relationships is highly variable (Deckert et al. 2001). Some endophytic fungi exhibit tissue specificity, colonizing particular plant organs such as leaves, petioles, or twigs. Other endophyte taxa, however, can systemically colonize the entire plant (Arnold 2007). Because plant distribution is largely determined by climatic conditions, Consequently, the presence of endophytic fungi is directly linked to the distribution patterns of their host plants, as well as factors like the plant's reproductive and vegetative organs, and both biotic and abiotic environmental influences (Koukol et al. 2012). Endophytic fungi dispersal occurs through two primary mechanisms: horizontal transmission and vertical transmission (Schulz and Boyle 2005). While horizontal and vertical transmission patterns influence the distribution of endophytic fungi within individual plants and contribute to broader geographical patterns, a complete understanding of endophyte distribution requires considering the geographical range of their host plants (Saikkonen et al. 2004). Endophyte communities at higher latitudes tend to exhibit lower species richness, while tropical endophyte communities, often characterized by a dominance of a few fungal classes, exhibit a high degree of species-level diversity (Lugtenberg et al. 2016). Ascomycetes are the dominant group of endophytic fungi, appearing to be ubiquitous across diverse ecosystems globally. They have been isolated from plants thriving in habitats ranging from hot deserts and Arctic tundra to mangroves, temperate and tropical forests, grasslands, savannas, croplands, and extreme environments such as Arctic, alpine, and xeric regions (Abo Nouh 2019). Endophytes in temperate regions often exhibit host specificity, whereas those in tropical regions tend to be less host specific. These fungi have a cosmopolitan distribution, having been isolated from plants in ecosystems ranging from temperate and tropical forests to harsh environments like deserts and Arctic tundra, as well as mangroves, grasslands, savannas, and croplands (Chhipa and Deshmukh 2019). Broadly, endophytic fungi can be categorized into two main groups based on evolutionary relationships, taxonomy, host preferences, and ecological roles: clavicipitaceous endophytes (C-endophytes), primarily associated with grasses, and non-clavicipitaceous endophytes (NC-endophytes), found in both vascular and nonvascular plants (Rodriguez et al. 2009; Jain and Pundir 2017). The classification of fungal endophytes into subclasses is based on several criteria, including host range, reproductive strategy, tissue specificity of colonization, transmission mode, nutritional mode, and the capacity to elicit disease symptoms in the host (Bamisile et al. 2018). Within the Clavicipitaceae family, a small group of phylogenetically related species, known as clavicipitaceous endophytes

(C-endophytes or Class 1 endophytes), includes genera such as *Balansia* and *Epichloë*. These endophytes are fastidious in culture and exhibit a limited host range, primarily restricted to certain cool- and warm-season grasses (Rodriguez et al. 2009). Characterized by high diversity, non-clavicipitaceous endophytes (NC-endophytes) are a polyphyletic group, mainly composed of ascomycetous fungi, and have been found in all major plant lineages and terrestrial ecosystems, including agricultural and diverse natural biomes from tropical to tundra zones (Rodriguez et al. 2009). NC-endophytes are further categorized into three functional classes (Class 2, 3, and 4) based on several key characteristics: how they colonize the host plant, how they are transmitted between generations, the level of diversity they exhibit within a single plant, and their specific ecological roles (Lugtenberg et al. 2016). Class 2 endophytes, a diverse group encompassing both Ascomycota and some Basidiomycota, are associated with above-ground tissues of a wide range of plants. These include nonvascular plants, seedless vascular plants, conifers, and both woody and herbaceous angiosperms, across diverse biomes from tropical forests to boreal and Arctic/Antarctic communities. Class 3 endophytes, characterized by high diversity, infect specific, localized areas, typically in aerial tissues such as leaves, and are found in both tropical trees and numerous species of vascular and nonvascular plants (Lugtenberg et al. 2016). Class 4 endophytes, also known as dark septate endophytes (DSE), are root-colonizing, facultative biotrophs. Their melanized and septate hyphae are characteristic, and they are frequently ascomycetes belonging to the Helotiales order. DSE have been identified in the roots of nearly six hundred plant species across a broad spectrum of terrestrial ecosystems, though they are particularly prevalent in polar and alpine environments (Abo Nouh 2019).

3 Historical Classification and Distribution of Coniferous Plants

Conifers constitute the majority of extant gymnosperm species, representing approximately 58.5% of the total. These plants, which include both trees and shrubs, are a dominant presence in roughly 39% of the world's forested areas, particularly within boreal and subtropical montane forests at higher elevations (Yang et al. 2022a). Traditional classifications of conifers, based primarily on morphological characteristics, often divide them into two groups. The first group, exemplified by families like *Pinaceae, Araucariaceae, and Cupressaceae*, is characterized by typical, cone-like female reproductive structures. The second group, including families such as *Podocarpaceae, Cephalotaxaceae*, and *Taxaceae*, possesses atypical female cones, which are often reduced and fleshy (Yang et al. 2022b). Li (1953), Keng (1975), and Cheng and Fu (1978) utilized this characteristic as a basis for conifer classification (Fig. 2). The evolutionary relationship between these two cone types (typical and atypical) has been a subject of debate, with differing views on which represents the more ancestral form. In the nineteenth century, the superficial similarity of their

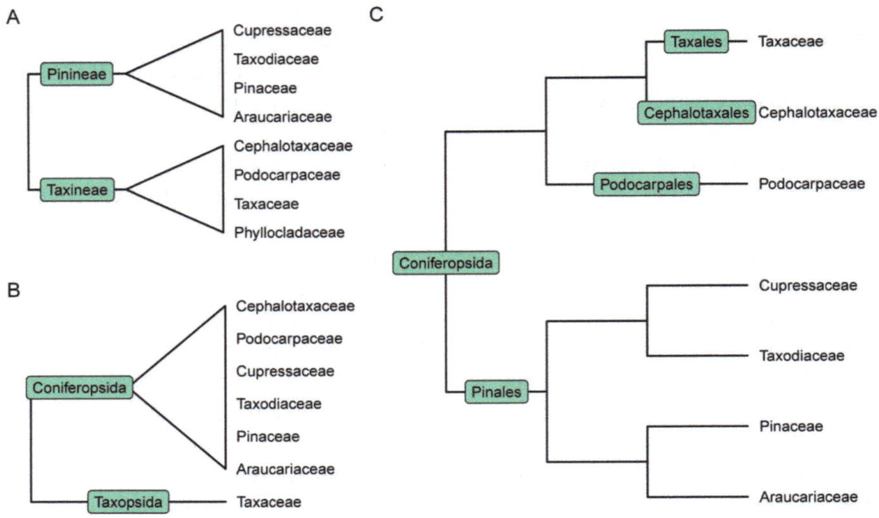

Fig. 2 Historical classifications of conifers. (**a**) Li (1953), (**b**) Keng (1975), and (**c**) Cheng and Fu (1978)

terminal ovulate organs led some botanists to classify Ginkgo with the taxads (families like *Taxaceae*). However, following the discovery of flagellated sperm in both *Ginkgo* and *cycads* by Japanese botanists, *Ginkgo* was recognized as distinct from the taxads, and both *Ginkgo* and *cycads* were subsequently considered to be more primitive gymnosperms (Hirase 1896). Florin (1948) hypothesis suggests that the unique female cone structure observed in the *Taxaceae* family arose independently from the reproductive structures found in *Lebachia*, rather than being a reduced form of the compound cone typical of other conifers. This distinct reproductive morphology led to the proposal that Taxaceae warrants a higher taxonomic classification, specifically the rank of Taxopsida. Contrary to Florin's view, Cheng and Fu's (1978) classification system proposed four conifer orders: Pinales, Podocarpales, Cephalotaxales, and Taxales. This classification implies a gradual evolutionary reduction in female cone structure, transitioning from the typical multiovulate cones found in families such as Araucariaceae, Pinaceae, Taxodiaceae, and Cupressaceae to the more modified cone structures observed in Podocarpaceae and Cephalotaxaceae, ultimately culminating in the single-ovulate structures of *Taxaceae*. Molecular phylogenetic/phylogenomic studies have challenged the hypotheses of Florin (1948) and Cheng and Fu (1978). The current understanding is that the reduced, fleshy female cones characteristic of the Podocarpaceae, Taxaceae, and Cephalotaxaceae families are not homologous structures derived from a common ancestor with typical cones, but rather represent independent evolutionary events in each lineage (Contreras et al. 2017). Phylogenomic analyses support the classification of conifers into three distinct subgroups. The first subgroup comprises solely the Pinaceae family (11 genera) within the order Pinales. The second subgroup includes two southern conifer families, Araucariaceae (3 genera) and Podocarpaceae (20 genera), both belonging to the order Araucariales. The third subgroup consists of Sciadopityaceae (1 genus), Cupressaceae (31 genera), Cephalotaxaceae (1 genus), and Taxaceae (5 genera), all within the order Cupressales. This classification also recognizes Pinidae, consisting solely of the Pinaceae family, and Cupressidae, encompassing all other conifer families (Yang et al. 2022b).

The monophyly of conifers has long been accepted. Early molecular phylogenetic studies, utilizing limited barcoding markers, provided support for this view, suggesting that conifers indeed represent a single, unified evolutionary lineage (Ran et al. 2010), While some earlier studies proposed that conifers are paraphyletic, with gnetophytes nested within them (Bowe et al. 2000), more recent phylogenomic analyses have confirmed that conifers are indeed paraphyletic, with gnetophytes forming a sister group to the Pinaceae (Ran et al. 2018). While some earlier classifications included ginkgophytes within the coniferophytes, current phylogenomic analyses of gymnosperms support a more restricted definition. Here, we define coniferophytes as comprising only conifers and gnetophytes, excluding Ginkgo. Conifers display a striking diversity in leaf morphology, varying from needle-like (e.g., *Picea*, *Pinus*, *Cedrus*) and linear (e.g., *Metasequoia*, *Larix*, *Sequoia*, *Abies*, *Pseudolarix*, *Taxus*, *Wollemia*) to lanceolate scale-like (e.g., *Platycladus*, *Sabina*), subulate (e.g., *Taiwania*, *Cryptomeria*), acicular (e.g., *Juniperus*), phyllode-like (e.g., *Phyllocladus*), elliptic (e.g., *Agathis*, *Nageia*), and (e.g., *Podocarpus*, *Cunninghamia*) (Farjon 2018). Representing the largest extant gymnosperm group, conifers comprise approximately 722 species with a cosmopolitan distribution across all six continents (Yang et al. 2022b). Northern Hemisphere conifers are predominantly cone-bearing, long-lived trees and shrubs, characteristic of mountainous and boreal ecosystems (Farjon 2010). In contrast, tropical and southern conifers exhibit considerably greater morphological and ecological diversity (Farjon 2008). Coniferous species and genera are unevenly distributed across the globe. Asia has the highest concentration, with 39 out of the world's 69 genera (approximately 57%) and around 200 to 250 species (about 35%). In comparison, Africa, Europe, and South America exhibit the lowest diversity, containing only 8 to 9 genera and 29 to 30 species. In contrast, Australasia and the South Pacific islands support a greater diversity, with 27 genera and approximately 150 species, while North and South America, including the West Indies, harbor 17 genera and more than 150 species. These latter regions represent an intermediate level of diversity between the highest and lowest extremes (Singh et al. 2018).

4 Biodiversity of Families of Coniferous Plants

4.1 Pinaceae

The Pinaceae family, belonging to the Pinales order, encompasses 11 genera and 272 species. This family represents the most significant of the seven conifer families and its members are found in a wide range of forest ecosystems, spanning boreal, arid, subalpine, temperate, and tropical zones (Nkongolo and Mehes-Smith 2012). Within the Pinaceae family, the *Pinus* genus holds considerable significance, comprising over 100 species. These species are primarily distributed across forests in the Northern Hemisphere, spanning Europe, Asia, North Africa, North America, and Central America. Commonly referred to as "pines," they are a dominant component of these forest ecosystems, these trees are ecologically vital, often representing the dominant arboreal species within numerous ecosystems across these regions (Nobis et al. 2012). Pine trees exhibit a global distribution, thriving in acidic and well-drained soils, and demonstrating a tolerance for poorly drained, waterlogged conditions. The *Pinus* genus is further classified into two subgenera based on needle vascular bundle number: *Pinus* subgenus *Pinus* and *Pinus* subgenus *Strobus*. Members of the former subgenus are characterized by hardwood and needle-like leaves arranged in fascicles, typically containing two or three needles, while the latter subgenus is distinguished by softwood and fascicles of five needles (Hernández-León et al. 2013).

4.2 *Araucariaceae*

The Araucariaceae family, within the Araucariales order, comprises 3 genera and 40 species. This family possesses an extensive fossil record, indicating a widespread distribution across both hemispheres during the Mesozoic era. Contemporary Araucariaceae diversity is represented by three genera: *Araucaria* and *Agathis*, both identified in the nineteenth century, and the recently discovered (approximately 15 years ago) monotypic genus *Wollemia*, endemic to New South Wales, Australia (Setoguchi et al. 1998). While *Araucaria* possesses the most extensive and diverse fossil record among the three extant araucarian genera, the taxonomic assignment of numerous remains, particularly impressions, to *Araucaria*, and even to the Araucariaceae family itself, remains questionable (Kershaw and Wagstaff 2001). The extant species of the conifer family Araucariaceae are predominantly distributed in the Southern Hemisphere. The majority of these species are endemic to Australia, New Zealand, New Guinea, or New Caledonia. Only two species, *Araucaria araucana* and *Araucaria angustifolia*, are endemic to South America (Liu et al. 2009).

4.3 *Podocarpaceae*

The Podocarpaceae family, belonging to the Araucariales order, comprises 20 genera and 181 species. As the second largest conifer family, its distribution is primarily concentrated in the Southern Hemisphere (Abdillahi et al. 2010). New Zealand exhibits the greatest generic diversity within Podocarpaceae (nine genera). Podocarp diversity is distributed across various regions, with the highest genus representation found in New Caledonia and Malesia (eight genera), followed by Australia (seven), South America (four), and both Africa and Asia (two each). Among the 20 recognized genera, three are endemic to Australia, while two are restricted to New Zealand. Several countries exhibit notable species richness within the Podocarpaceae family, with Indonesia hosting the highest number of species (51), followed by Papua New Guinea (43), Malaysia (39), and the Philippines (23). New Caledonia supports 20 species, while New Zealand harbors 19. China accounts for 17 species, Venezuela for 15, and Australia for 14. Brazil, Peru, and Bolivia each sustain 12 species, whereas Fiji contains 11. Ecuador supports nine species, while Madagascar, Thailand, Brunei Darussalam, and Colombia each contain eight species (Enright and Jaffré 2011). The extant Podocarpaceae species represent a diminished fraction of a previously much more diverse group. The family's diversity is currently centered in Australasia (encompassing New Caledonia, Tasmania, New Zealand, and Malesia), the South American Andes, Indo-China, and the Philippines. *Podocarpus* boasts the highest species diversity, encompassing approximately 120 species across approximately 70 countries. *Podocarpus* species are predominantly found in China, the islands of the southeastern Pacific Ocean, southeastern Africa, and South America (Brummitt et al. 2015).

4.4 Sciadopityaceae

The Sciadopityaceae family, within the Cupressales order, is monospecific, containing only one genus and one species: *Sciadopitys verticillata.* Commonly known as the Japanese umbrella-pine, this species is currently endemic to Japan (Farjon and Page 2003). *Sciadopitys verticillata* exhibits significant morphological and genetic divergence from all other extant conifers (Crisp and Cook 2011). This taxon occupies a sister position to a clade comprising the Cupressaceae, Taxaceae, and Cephalotaxaceae families. The Sciadopityaceae stem lineage likely originated in the Triassic, approximately 229.99 million years ago, according to molecular dating of gymnosperms (Ran et al. 2018). The fossil record indicates that Sciadopityaceae existed in Europe as far back as the Late Jurassic period (Farjon and Page 2003). Subsequent to this, diverse Sciadopityaceae fossils, including leaves, seed cones, cone scales, and wood, have been documented from regions spanning Europe to Japan and North America (Farjon 2008). The potential presence of Sciadopityaceae in the Jurassic period was indicated by the discovery of a fossil conifer wood with preserved pith, primary, and secondary xylem in western Liaoning Province, Northeast China, as reported by Jiang et al. (2012).

4.5 Cupressaceae

The Cupressaceae family, belonging to the Cupressales order, is comprised of 31 genera and 169 species, and is commonly known as the Cypress family. This family exhibits a near-global distribution. Beyond their substantial ecological roles, Cupressaceae members, like other conifers, hold significant economic value. They are utilized for timber production, landscaping, and ornamental purposes. Furthermore, they contribute to industries such as perfumery, beverage flavoring, and medicine (Gadek et al. 2000). The Cupressaceae family is organized into seven subfamilies: Callitroideae, Sequoioideae, Taxodioideae, Cupressoideae, Taiwanioideae, Cunninghamhioideae, and Athrotaxidoideae (Farjon et al. 2002). Thirty-one genera are distributed across these seven subfamilies. Genera within the Cupressoideae subfamily include *Chamaecyparis*, *Cupressus*, *Fokienia*, *Xanthocyparis*, *Thujopsis*, *Platycladus* (syn. *Thuja*), *Tetraclinis*, *Calocedrus*, *Juniperus*, and *Microbiota* (Dörken et al. 2017).

4.6 Cephalotaxaceae

The Cupressales order includes the Cephalotaxaceae family, which has only one genus and ten species and is not related to the Taxaceae. Gnetophytes form a monophyletic group and are sister to the Pinaceae family. Because of convergent evolution, the structure and fleshy nature of female cones are not considered reliable characters for defining clades above the family level (Yang et al. 2023). Cephalotaxaceae are exclusively found in the Old World, with a discontinuous distribution ranging from the Himalayas to Taiwan, Korea, and Japan, and southward to Penang. Their primary, albeit small, center of diversity is located in western China (Lang et al. 2011). The plum yews, belonging to the genus *Cephalotaxus*, constitute a small group of conifers native to eastern Asia, specifically north of the Indo-Chinese peninsula. The genus was originally included in Taxaceae by Pilger (1903), but Neger (1907) separated it into its own family, Cephalotaxaceae (Lang et al. 2013). *Cephalotaxus* species have a history of use both as ornamental garden plants and as a source of building and wood carving material. Currently, their primary importance lies in their medicinal applications. This is due to the presence of unique alkaloids, such as cephalotaxine esters, and terpenoids, compounds that are characteristic of this genus (Abdelkafi and Nay 2012).

4.7 Taxaceae

The Taxaceae family, within the Cupressales order, comprises 5 genera and 29 species. This yew family is characterized by slow-growing conifers primarily distributed throughout the Northern Hemisphere (Cope 1998). Of all the genera in the Taxaceae family, Taxus has the broadest geographic range, occurring in both the understory and canopy of diverse forest ecosystems, including those dominated by *Abies* spp., *Picea* spp., *Quercus semicarpifolia*, *Betula* spp., *Acer* spp., *Pinus* spp., *Rhododendron* spp., and *Betula alnoides* (Pant and Samant 2008). Key distinguishing features of the *Taxus* genus within the Taxaceae family include microsporophylls, a red aril, and variable foliage, contributing to its significant horticultural importance. *Taxus baccata* (English Yew), *Taxus brevifolia* (Pacific or Western Yew), *Taxus canadensis* (American Yew), and *Taxus cuspidata* (Japanese Yew) are examples of recognized species within the genus *Taxus* (Cope 1998). The Taxaceae family is generally recognized as comprising five genera: *Taxus*, *Pseudotaxus*, *Austrotaxus*, *Torreya*, and *Amentotaxus*. Indigenous tribes have historically used the bark of *Taxus brevifolia* for medicinal purposes, while its wood was employed by the people of British Columbia in crafting tools such as bows, digging sticks, needles, whaling harpoons, and wedges.

5 Biodiversity of Fungal Endophytes in Coniferous Plants

The symbiotic relationship between fungi and plants is a ubiquitous and often hidden phenomenon in natural ecosystems. Conidia and spore dispersal, facilitated by physical or mechanical vectors, is the typical mode of horizontal transmission for these microorganisms, spreading them to neighboring plants. While Ascomycota comprises the majority of endophytic fungi, some genera from Glomeromycota, Basidiomycota, and Zygomycota also occur as endophytes (Rodriguez et al. 2009). Numerous studies have investigated the composition of fungal endophyte communities in leaves, stems, and roots (or other specific tissues) (Guo et al. 2008; Qadri et al. 2013; Jin et al. 2013). Cupressaceae, Pinaceae, and Taxaceae are the three conifer families in which endophyte biology has received the most attention. Current studies of Taxaceae, and somewhat less so Araucariaceae, are mainly focused on the discovery and industrial applications of endophytes that produce taxol (Zhou et al. 2010). Bioactive endophytes with potential agricultural and biopharmaceutical applications have been derived from Pinaceae (Stierle and Stierle 2015). Cupressaceae is also increasingly recognized as a rich source of diverse endophytes with considerable biotechnological potential. The Cupressoideae subfamily has become a leading model for research in fungal endophyte biology and biotechnology (Soltani and Moghaddam 2015). Research efforts have largely focused on cultivable endophytic fungi within genera such as *Calocedrus*, *Chamaecyparis*, *Cupressus*, *Juniperus*, and *Platycladus* (syn. *Thuja*). This fungal community exhibits considerable taxonomic diversity within the Pezizomycotina subphylum of Ascomycota. Dothideomycetes and Sordariomycetes appear to be the dominant endophytic fungal classes colonizing Cupressoideae plants, based on current evidence. Eurotiomycetes, Leotiomycetes, and Pezizomycetes (within Pezizomycotina) are also frequently observed as endophytes (Soltani and Moghaddam 2015). Arizona and North

Carolina, USA, were the locations of the first study on fungal endophytism in Cupressus, Juniperus, and Thuja, which identified several associated fungal genera within healthy foliage. Four fungal classes were represented by the identified genera. These included Dothideomycetes (*Alternaria*, *Aureobasidium*, *Ascochyta*, *Botryosphaeria*, *Cladosporium*, *Kabatina*, *Guignardia*, *Monodictys*, *Phoma*, *Leptosphaerulina*, *Rhizosphaera*, *Phyllosticta*, *Stagonospora*, *Preussia*), Sordariomycetes (*Bartalinia*, *Nemania*, *Biscogniauxia*, *Diaporthe*, *Chaetomium*, *Lecythophora*, *Cordyceps*, *Pestalotia*, *Thielavia*, *Pestalotiopsis*, *Phomopsis*, *Xylaria*), Eurotiomycetes (*Paecilomyces*, *Penicillium*), and Pezizomycetes (*Morchella*, *Peziza*) (Hoffman and Arnold 2008). The study focused on cupressaceous genera cultivated in four distinct regions of Iran: Fars, Guilan, Hamedan, and Markazi Provinces identified a similar suite of associated foliar endophytes. Identified genera, categorized by class, included: Dothideomycetes (*Alternaria*, *Aureobasidium*, *Bipolaris*, *Cladosporium*, *Didymella*, *Embellisia*, *Leptosphaeria*, *Phoma*, *Pleospora*, *Pyrenochaeta*); Eurotiomycetes (*Aspergillus*, *Penicillium*, *Talaromyces*); Pezizomycetes (*Ascorhizoctonia*); and Sordariomycetes (*Coniochaeta*, *Cytospora*, *Fusarium*, *Thielavia*) (Moghaddam and Soltani 2014). Endophytic fungal strains were isolated from *Juniperus communis* samples collected in Hungary. In total, 14 genera were identified: *Alternaria*, *Aspergillus*, *Bipolaris*, *Cladosporium*, *Colletotrichum*, *Curvularia*, *Didymella*, *Fusarium*, *Penicillium*, *Pestalotiopsis*, *Phomopsis*, *Purpureocillium*, *Trichoderma*, and *Xylaria* (Vigneshwari et al. 2021). A total of 144 endophytic fungal isolates were cultured from *Juniperus procera* (Cupressaceae) twigs collected from the Taif region of Saudi Arabia. Of these, 52 demonstrated antimicrobial activity against at least one of the tested microorganisms (*Staphylococcus aureus*, *Klebsiella pneumoniae*, *Candida albicans*, and *Fusarium solani*). The most active isolates were identified as *Aspergillus fumigatus* (7 isolates), *Preussia* sp. (5), *Hypocrea lutea* (4), and *Penicillium oxalicum* (10) (Gherbawy and Elhariry 2014). Of the 58 isolates from *Juniperus communis* that were tested, all demonstrated activity against *Candida albicans*, *Bacillus subtilis*, and *Staphylococcus aureus*. Only a limited number of extracts, chiefly mycelial extracts, showed antifungal activity against the tested filamentous fungi. Specifically, *Trichoderma* and *Purpureocillium* mycelial extracts were effective against both fungi. *Fusarium culmorum* was more resistant than *Rhizoctonia solani* (Vigneshwari et al. 2021). It has been established that *Aspergillus fumigatus*, an endophytic fungus residing within *Juniperus communis*, exhibits antimicrobial efficacy through the production of deoxypodophyllotoxin against a range of pathogenic bacteria (Kusari et al. 2009). Twenty-six distinct fungal taxa were isolated as endophytes from *Juniperus procera* (Cupressaceae) twigs collected in the Taif region of Saudi Arabia. The most prevalent species were *Aspergillus fumigatus*, *Alternaria tenuissima*, *Preussia* sp., *Penicillium oxalicum*, *Peyronellaea sancta*, and *Peyronellaea eucalyptica* (Gherbawy and Elhariry 2014). Healthy *Chamaecyparis* foliage has been observed to exhibit a dominant endophytic association with several fungal genera. Within the Sordariomycetes class, these include *Glomerella*, *Harknessia*, *Xylaria*, *Microdochium*, *Coniochaeta*, *Phomopsis*, *Gelasinospora*, *Nodulisporium*, *Pestalotiopsis*, *Tubercularia*, and *Mycoleptodiscus*.

The Dothideomycetes class is represented by *Sporidesmium*, *Epicoccum*, *Diplodia*, *Phyllosticta*, *Cladosporium*, *Hormonema*, *Pleurophoma*, and *Alternaria* (Petrini 1991). Furthermore, Leotiomycetes within Pezizomycotina, including *Chloroscypha* and *Cryptosporiopsis*, are frequently identified as endophytes of both *Chamaecyparis* and *Thuja* species. *Leotiomycetes* sp. and *Lophodermium* (also Leotiomycetes) have been documented as *Juniperus* endophytes (Ellsworth et al. 2013). *Retinocyclus* (Lecanoromycetes, Pezizomycotina) is a commonly observed endophyte of *Juniperus* species. As observed in other plant-endophyte associations, specific fungal endophyte classes often dominate within particular plant lineages. Sordariomycetes are frequently dominant in Fagaceae (e.g., *Quercus* spp.), while Leotiomycetes are often prevalent in Pinaceae (e.g., *Pinus ponderosa*) (Arnold 2007). Fungal endophytes were isolated from the Mediterranean cypress, *Cupressus sempervirens*. Eleven fungal taxa, all belonging to the Ascomycota phylum, were identified as endophytes. These included *Alternaria multiformis*, *Didymella* sp., *Phoma* sp., *Phoma herbarum*, and *Pyrenochaeta* sp. (Dothideomycetes); *Penicillium brevicompactum* and *Talaromyces* sp. (Eurotiomycetes); *Ascorhizoctonia* sp. (Pezizomycetes); and *Thielavia microspora* and *Thielavia* spp. (Sordariomycetes) (Soltani and Moghaddam 2015). While numerous studies have investigated endophytic fungi associated with Pinaceae species, comparatively little research has focused on the mycoflora associated with *Cedrus atlantica* (Bensaci et al. 2015). Certain fungal endophytes exhibited tissue specificity within *Pinus* species. *Acremonium* spp., *Ceratobasidium* spp., *Isaria* spp., *Phlebia* spp., *Teichospora* spp., *Trichocomaceae* spp., and *Xylariomycetidae* spp. needle tissue exclusively yielded *Acremonium* spp., *Teichospora* spp., *Trichocomaceae* spp., *Isaria* spp., *Xylariomycetidae* spp., *Ceratobasidium* spp., and *Phlebia* spp. Stem tissue endophytes included *Epicoccum* spp., *Peyronellaea* spp., *Togninia* spp., *Ulocladium* spp., and *Venturia* spp. Root tissue harbored *Cadophora* spp., *Cylindrocarpon* spp., *Dothideomycetes* spp., *Ericoid* spp., *Eurotiales* spp., *Mycoleptodiscus* spp., *Periconia* spp., *Pholiota* spp., *Psathyrella* spp., and *Xylariomycetidae* spp. With the exception of *Cadophora* spp., *Dothideomycetes* spp., *Epicoccum* spp., and *Ulocladium* spp., these fungal endophytes represent first isolations from their respective tissues (Higgins et al. 2007). The fungal endophytes are influenced by a range of environmental factors, including climatic conditions, geographic location, edaphic (soil-related) factors, and host-specific factors such as plant genotypes (Wang and Guo 2007). *Cladosporium* spp. were the dominant fungal species across three *Pinus* species (R, D, and T), while *Umbelopsis* spp. predominated in *Pinus koraiensis* (K). Host specificity is a well-documented phenomenon among fungal endophytes. Among the four *Pinus* species examined, *P. thunbergii* exhibited the highest endophyte diversity, followed by *P. koraiensis*. These two species are known for their ability to thrive in challenging environments (Kim et al. 2013). Dominant endophytes included *Lophodermium* spp. in *Pinus densiflora* and *Pinus rigida* (57%–79% of isolates), *Beltrania* sp. in *P. thunbergii* (approximately 58%), and *Phyllosticta spinarum* in *P. koraiensis* (48%–65%). *Larix kaempferi* and *Juniperus rigida* did not show a clear pattern of dominance (Eo et al. 2013). *Xylaria* endophytes were isolated from the leaves and seeds of two tree species in Puerto Rico: *Casuarina equisetifolia* (Australian pine) and *Manilkara bidentata* (Ausubo). They were also isolated from the needles of Eastern white pine (*Pinus strobus*). It is known that some of these *Xylaria* endophytes associated with healthy trees can range from weak to strong parasites on woody plants (Richardson et al. 2014). Fourteen fungal species, classified into six genera, were isolated from seven *Pinus* species in Honduras. These included *Fusarium lateritium*, *F. pseudocircinatum*, *Pestalotiopsis pini*, *P. microspora*, *Xylaria grammica*, *Trichoderma atroviride*, and *Nigrospora oryzae* (Ortiz et al. 2022). The most diverse fungal communities, representing 13 genera (nearly half of all genera isolated), were found in *Cedrus deodara*, *Pinus roxburgii*, and *Abies pindrow*. Fermented broth extracts from several of these fungi demonstrated strong bioactivity against *Escherichia coli* and *Staphylococcus aureus*. The *Trichophaea abundans* extract (from a *Pinus* sp. endophyte) exhibited the lowest IC50 (18 μg/mL) (Qadri et al. 2013). Twenty-two endophytic fungal taxa were identified from *Cedrus atlantica* needles collected from three sites within Belezma National Park, Algeria. The complete list of isolated fungal endophytes included *Alternaria* sp., *Aspergillus* sp., *Lophodermium* sp., *Alternaria alternata*, *Gliocladium* sp., *Cladosporium cladosporioides*, *Lophodermium cedrinum*, *Phomopsis* sp., *Epicoccum* sp., *Arthrinium phaeospermum*, *Fusarium larvarum*, *Ulocladium* sp., *Aureobasidium pullulans*, *Drechslera* sp., *Paecilomyces* sp., *Curvularia* sp., *Phyllosticta* sp., *Trichoderma* sp., *Brachysporium* sp., *Pestalotiopsis* sp., and *Botryodiplodia* sp. (Bensaci et al. 2015). Endophytic fungi were isolated and identified from the leaves of coniferous trees on Bohyeon Mountain, Korea. Leaf samples from *Pinus densiflora* and *Pinus koraiensis* were collected from 11 locations. A total of 58 isolates were characterized using molecular and morphological methods. Four fungal endophyte species were identified from *P. densiflora*: *Lophodermium conigenum*, *Septoria pini-thunbergii*, *Leotiomycetes* sp., and *Polyporales* sp. *Pinus koraiensis* yielded two species: *Eurotiomycetes* sp. and *Rhytismataceae* sp. *L. conigenum* and *S. pini-thunbergii* were the most frequently isolated (Yoo and Eom 2012).

For much of the twentieth century, *Cephalotaxus* was typically considered the sole genus within the Cephalotaxaceae family. *Cephalotaxus* is an ancient genus, with a recognizable ancestor, *Thomasiocladus zamioides*, dating back to the mid-Jurassic period. The earliest *Cephalotaxus* species are known from the early Cretaceous (Shi et al. 2010). During the Tertiary period, *Cephalotaxus* enjoyed a wide distribution across much of the Northern Hemisphere, with fossil evidence found in Asia, Europe, and North America (Manchester et al. 2009). *Cephalotaxus* is currently found in Korea, China, Japan, Burma, Laos, Vietnam, and India, with its center of distribution in China, which encompasses parts of the native range of most species. Despite its use in leukemia treatment, the rare medicinal plant *Cephalotaxus mannii* (Cephalotaxaceae) has understudied fungal endophytes, which may represent a rich source of pharmaceutical compounds. The endophytic fungi of *C. mannii* from Vietnam, along with an evaluation of their biological potential, are investigated for the first time in this study. Six genera of endophytic fungi (*Aspergillus, Diaporthe, Fusarium, Megasporoporia, Penicillium, and Trichoderma*) were represented by the 18 isolates obtained. Notably, the ethyl acetate extracts of *Penicillium citrinum*

WDF8 and *Fusarium perseae* WDF12 demonstrated significant antibacterial activity against at least five of the tested bacterial species (Vu et al. 2023). Fifty-two isolates of endophytic fungi were collected from the bark of *Cephalotaxus mannii* (plum-yew) trees in northern Thailand and southern China. In total, 13 genera were represented among the 35 isolates (67.3%) identified: *Acremonium* sp., *Botrytis* sp., *Cephalosporium* sp., *Chalaropsis* sp., *Cladosporium* sp., *Fusarium* sp., *Geotrichum* sp., *Humicola* sp., *Monilia* sp., *Penicillium* sp., *Rhizoctonia* sp., *Spicaria* sp., and *Trichoderma* sp. While 17 strains (32.7%) could not be identified, the dominant endophytic fungal genera found in *Cephalotaxus mannii* trees from both northern Thailand and southern China were *Acremonium*, *Fusarium*, and *Monilia*. However, there were notable differences: *Trichoderma* and *Cladosporium* were unique to the southern China samples, while *Botrytis, Cephalosporium, Chalaropsis, Geotrichum, Humicola, Penicillium, Rhizoctonia, and Spicaria* were exclusively found in the northern Thailand samples. These results indicate a significant difference in the endophytic fungal communities associated with *C. mannii* trees from these two geographic locations (Saithong et al. 2010). Six fungal isolates were recovered from the roots of *C. mannii* trees in Ha Giang province, Vietnam. Notably, one isolate, WQF5, was identified as *Fusarium fujikuroi*. This particular isolate demonstrated the ability to produce paclitaxel and exhibited antibacterial activity against five different tested pathogens (Tung et al. 2022). This study explored the potential of endophytic fungi from *Cephalotaxus hainanensis* as sources of antimicrobial and antitumor agents. Around 265 endophytic fungal isolates were obtained from *C. hainanensis* and screened for their activity against a panel of microbes (tilapia, banana, rice, and rape pathogens) and human leukemia cell lines (K562, NB4, and HL-60). The findings revealed that 17.7% of the isolates exhibited antimicrobial activity against at least three of the tested microbes. The highest antimicrobial activity (15.8%) was observed against *Fusarium oxysporum* RKY102. In addition, 18.5% of the isolates demonstrated cytotoxicity against at least one of the tested tumor cell lines, with the highest activity (10.6%) observed against the K562 leukemia cell line. These results suggest that endophytic fungi from *C. hainanensis* represent a promising source of bioactive compounds with potential applications in medicine and agriculture (Liu et al. 2016). Several endophytic *Pestalotiopsis* species were isolated from *Taxus yunnanensis* (Taxaceae) in Kunming, China, including *P. neglecta*, *P. paeoniae*, and *P. subcuticularis*. A similar diversity of endophytic *Pestalotiopsis* species, including *P. neglecta*, *P. photiniae*, and *P. subcuticularis*, was found in *Taxus chinensis* var. *mairei* in Hangzhou, southern China (Wei et al. 2007). Analysis of asymptomatic leaf samples collected from nine *Taxus mairei* trees on Fu-Shan, Taiwan, yielded 40 endophytic fungal isolates, which were identified as *Colletotrichum gloeosporioides, Aspergillus nidulans*, *Fusarium solani*, and several isolates identified only to genus: *Colletotrichum* sp., *Phanerochaete* sp., and *Rhizoctonia* sp. *Colletotrichum* and the most frequently isolated endophytic fungi from *T. mairei* leaves belonged to the genus *Fusarium*. *Colletotrichum gloeosporioides* is reported here for the first time as an endophyte of *T. mairei* (Wang and Guo 2007).

Diverse endophytic *Pestalotiopsis* species were isolated from *Podocarpus macrophyllus* (Podocarpaceae) across three locations in southern China—Kunming: *P. heterocornis*, *P. kunmingensis*, *P. menezesiana*, *P. microspora*, *P. neglecta*, *P. olivacea*, *P. oxyanthi*, *P. paeoniae*, *P. photiniae*, *P. rhododendri*, and *P. yunnanensis*; Hangzhou: *P. clavispora*, *P. crassiuscula*, *P. heterocornis*, *P. neglecta*, *P. photiniae*, *P. rhododendri*, and *P. zonata*; and Nanning: *P. aquatica*, *P. clavispora*, *P. heterocornis*, *P. menezesiana*, *P. microspora*, *P. neglecta*, *P. photiniae*, *P. rhododendri*, and *P. theae*. A similar diversity of endophytic *Pestalotiopsis* species was found in *Podocarpus nagi* across the same locations—Kunming: *P. photiniae*, *P. mangifolia*, *P. neglecta*, and *P. olivacea*; Hangzhou: *P. neglecta*, *P. paeoniicola*, and *P. photiniae*; and Nanning: *P. neglecta* and *P. photiniae* (Wei et al. 2007). Research investigating the diversity and ecological interactions of endophytes associated with the endangered conifer *Araucaria angustifolia* identified 37 ascomycete fungal genera. These genera were classified into the classes Eurotiomycetes, Sordariomycetes, and Dothideomycetes, representing 11 orders and 13 families. The majority of isolates were classified within the Sordariomycetes class (40%), with the *Xylaria* genus being particularly prevalent (14%). Eurotiomycetes represented the least abundant class within the community (Groff et al. 2024).

6 Conclusion

The evolutionary history of fungal endophytes on conifers and angiosperms suggests a long period of coevolution. Fungi on conifers evolved into modern Helotialean fungi, while those on angiosperms developed into Diaporthalean fungi. This coevolution has been ongoing for over 300 million years. Both bitunicate ascomycetes (Dothideales, Pleosporales, and Mycosphaerellales) and Xylariales fungi can be dominant endophytes in both angiosperms and gymnosperms. The "Bitunicatae" likely diverged from the common ancestor of Helotiales and Diaporthales more than 300 million years ago, before the separation of gymnosperms and angiosperms. This predivergence likely explains why bitunicate ascomycetes are found as dominant endophytes in both groups. Conifer-associated endophytic fungi, having coevolved with their hosts, occupy specific ecological niches. Therefore, studying conifers and their endophytic fungi concurrently is crucial, especially in the context of climate change and deforestation. Beyond traditional macroscopic and microscopic methods, understanding biodiversity based on ecological principles is essential. A balanced approach to research on endophytic fungi, incorporating diverse perspectives, will provide deeper insights into their ecology and evolution.

References

Abdelkafi H, Nay B (2012) Natural products from *Cephalotaxus* sp.: chemical diversity and synthetic aspects. Nat Prod Rep 29:845–869. https://doi.org/10.1039/C2NP20037F

Abdillahi HS, Stafford GI, Finnie JF, Van Staden J (2010) Ethnobotany, phytochemistry and pharmacology of *Podocarpus sensu latissimo* (s.l.). S Afr J Bot 76:1–24. https://doi.org/10.1016/j.sajb.2009.09.002

Abo Nouh FA (2019) Endophytic fungi for sustainable agriculture. Microb Biosyst 4:31–44. https://doi.org/10.21608/MB.2019.38886

Arnold AE (2007) Understanding the diversity of foliar endophytic fungi: progress, challenges, and frontiers. Fungal Biol Rev 21:51–66. https://doi.org/10.1016/j.fbr.2007.05.003

Arnold AE, Maynard Z, Gilbert GS (2001) Fungal endophytes in dicotyledonous neotropical trees: patterns of abundance and diversity*. Mycol Res 105:1502–1507. https://doi.org/10.1017/S0953756201004956

Arora J, Ramawat KG (2017) An introduction to endophytes. In: Maheshwari DK (ed) Endophytes: biology and biotechnology. Springer, Cham, pp 1–23

Bamisile BS, Dash CK, Akutse KS et al (2018) Fungal endophytes: beyond herbivore management. Front Microbiol 9. https://doi.org/10.3389/fmicb.2018.00544

Barklund P, Kowalski T (1996) Endophytic fungi in branches of Norway spruce with particular reference to Tryblidiopsis pinastri. Can J Bot 74:673–678. https://doi.org/10.1139/b96-085

Bensaci O, Harzallah D, Kamel G (2015) Endophytic mycoflora of Cedrus atlantica: diversity patterns and determinism of the phytosanitary situation of atlas cedar forests in Belezma massif (Algeria). For Sci Technol 11:36–43. https://doi.org/10.1080/21580103.2014.957352

Bowe LM, Coat G, dePamphilis CW (2000) Phylogeny of seed plants based on all three genomic compartments: extant gymnosperms are monophyletic and Gnetales' closest relatives are conifers. Proc Natl Acad Sci 97:4092–4097. https://doi.org/10.1073/pnas.97.8.4092

Brummitt NA, Bachman SP, Griffiths-Lee J et al (2015) Green plants in the red: a baseline global assessment for the IUCN sampled red list index for plants. PLoS One 10:e0135152. https://doi.org/10.1371/journal.pone.0135152

Carroll G (1988) Fungal endophytes in stems and leaves: from latent pathogen to mutualistic symbiont. Ecology 69:2–9. https://doi.org/10.2307/1943154

Cheng W, Fu L (1978) Flora Reipublicae Popularis Sinicae, Volume 7, Gymnospermae. Sci Press, Beijing

Chhipa H, Deshmukh SK (2019) Fungal endophytes: rising tools in sustainable agriculture production. In: Mérillon J-M, Ramawat KG (eds) Bioactive molecules in food. Springer, Cham, pp 1–24

Contreras DL, Duijnstee IAP, Ranks S et al (2017) Evolution of dispersal strategies in conifers: functional divergence and convergence in the morphology of diaspores. Perspect Plant Ecol Evol Syst 24:93–117. https://doi.org/10.1016/j.ppees.2016.11.002

Cope EA (1998) Taxaceae: the genera and cultivated species. Bot Rev 64:291–322. https://doi.org/10.1007/BF02857621

Crisp MD, Cook LG (2011) Cenozoic extinctions account for the low diversity of extant gymnosperms compared with angiosperms. New Phytol 192:997–1009. https://doi.org/10.1111/j.1469-8137.2011.03862.x

de Bary A (1866) Morphologie und Physiologi der Plize, Flechten und Myxoceten. Verlag von Wilhelm Engelmann, Leipzig. Flecht Myxoceten Leipz Verl Von Wilhelm Engelmann

Deckert RJ, Melville LH, Peterson RL (2001) Structural features of a *Lophodermium* endophyte during the cryptic life-cycle phase in the foliage of *Pinus strobus*. Mycol Res 105:991–997. https://doi.org/10.1016/S0953-7562(08)61957-7

Deckert RJ, Peterson RL (2000) Distribution of foliar fungal endophytes of Pinus strobus between and within host trees. Can J For Res 30:1436–1442. https://doi.org/10.1139/x00-078

Dörken VM, Nimsch H, Jagel A (2017) Morphology, anatomy and morphogenesis of seed cones of *Cupressus vietnamensis* (Cupressaceae) and the taxonomic and systematic implications. Flora 230:47–56. https://doi.org/10.1016/j.flora.2017.03.008

Ellsworth KT, Clark TN, Gray CA, Johnson JA (2013) Isolation and bioassay screening of medicinal plant endophytes from eastern Canada. Can J Microbiol 59:761–765. https://doi.org/10.1139/cjm-2013-0639

Enright NJ, Jaffré T (2011) Ecology and distribution of the Malesian podocarps. In: Smithsonian contributions to botany, pp 57–77. https://doi.org/10.5479/si.0081024X.95.57

Eo J-K, Kim C-K, Lee H, Eom A-H (2013) Diversity of endophytic fungi isolated from Pinus densiflora and Larix kaempferi in Mt. Oser, Korea. Kor J Mycol 41:137. https://doi.org/10.4489/KJM.2013.41.3.137

Farjon A (2018) The Kew review: conifers of the world. Kew Bull 73:8. https://doi.org/10.1007/s12225-018-9738-5

Farjon A (2010) A handbook of the world's conifers. Brill

Farjon A (2008) A natural history of conifers. Timber Press, Portland, pp 46–1470. https://doi.org/10.5860/CHOICE.46-1470

Farjon A, Hiep NT, Harder DK et al (2002) A new genus and species in Cupressaceae (Coniferales) from northern Vietnam, Xanthocyparis vietnamensis. Novon 12:179–189. https://doi.org/10.2307/3392953

Farjon A, Page CN (2003) Diversity in strategies for conifer conservation, the action plan and future developments. Acta Hortic 615:397–403. https://doi.org/10.17660/ActaHortic.2003.615.44

Florin R (1948) On the morphology and relationships on the Taxaceae. Bot Gaz 110:31–39. https://doi.org/10.1086/335515

Gadek PA, Alpers DL, Heslewood MM, Quinn CJ (2000) Relationships within Cupressaceae sensu lato: a combined morphological and molecular approach. Am J Bot 87:1044–1057

Ganley RJ, Brunsfeld SJ, Newcombe G (2004) A community of unknown, endophytic fungi in western white pine. Proc Natl Acad Sci 101:10107–10112. https://doi.org/10.1073/pnas.0401513101

Gherbawy Y, Elhariry H (2014) Endophytic fungi associated with high-altitude Juniperus trees and their antimicrobial activities. Plant Biosyst Int J Deal Asp Plant Biol 11:1–10. https://doi.org/10.1080/11263504.2014.984011

Groff D, Marmentini J, Gaglioti A et al (2024) Endophytic fungi associated with Araucaria angustifolia (Bertol.) Kuntze. An Acad Bras Ciênc 96:10.1590/0001-3765202420230251

Guo L-D, Huang G-R, Wang Y (2008) Seasonal and tissue age influences on endophytic fungi of Pinus tabulaeformis (Pinaceae) in the Dongling Mountains, Beijing. J Integr Plant Biol 50:997–1003. https://doi.org/10.1111/j.1744-7909.2008.00394.x

Hata K, Futai K (1996) Variation in fungal endophyte populations in needles of the genus Pinus. Can J Bot 74:103–114. https://doi.org/10.1139/b96-015

Hata K, Futai K, Tsuda M (1998) Seasonal and needle age-dependent changes of the endophytic mycobiota in Pinus thunbergii and Pinus densiflora needles. Can J Bot 76:245–250. https://doi.org/10.1139/b97-177

Hernández-León S, Gernandt DS, Pérez de la Rosa JA, Jardón-Barbolla L (2013) Phylogenetic relationships and species delimitation in pinus section trifoliae inferrred from plastid DNA. PLoS One 8:e70501. https://doi.org/10.1371/journal.pone.0070501

Higgins KL, Arnold AE, Miadlikowska J et al (2007) Phylogenetic relationships, host affinity, and geographic structure of boreal and arctic endophytes from three major plant lineages. Mol Phylogenet Evol 42:543–555. https://doi.org/10.1016/j.ympev.2006.07.012

Hirase S (1896) Spermatozoid of Ginkgo biloba. Bot Mag 10:171–172

Hoffman MT, Arnold AE (2008) Geographic locality and host identity shape fungal endophyte communities in cupressaceous trees. Mycol Res 112:331–344. https://doi.org/10.1016/j.mycres.2007.10.014

Jain P, Pundir RK (2017) Potential role of endophytes in sustainable agriculture-recent developments and future prospects. In: Maheshwari DK (ed) Endophytes: biology and biotechnology, vol 1. Springer, Cham, pp 145–169

Jiang Z, Wang Y, Zheng S et al (2012) Occurrence of Sciadopitys-like fossil wood (Coniferales) in the Jurassic of western Liaoning and its evolutionary implications. Chin Sci Bull 57:569–572. https://doi.org/10.1007/s11434-011-4850-z

Jin H, Yan Z, Liu Q et al (2013) Diversity and dynamics of fungal endophytes in leaves, stems and roots of Stellera chamaejasme L. in northwestern China. Antonie Van Leeuwenhoek 104:949–963. https://doi.org/10.1007/s10482-013-0014-2

Johnson JA, Whitney NJ (1992) Isolation of fungal endophytes from black spruce (Picea mariana) dormant buds and needles from New Brunswick, Canada. Can J Bot 70:1754–1757. https://doi.org/10.1139/b92-217

Johnson JA, Whitney NJ (1989) A study of fungal endophytes of needles of balsam fir (Abies balsamea) and red spruce (Picea rubens) in New Brunswick, Canada, using culture and electron microscope techniques. Can J Bot 67:3513–3516. https://doi.org/10.1139/b89-431

Kelch DG (2002) Phylogenetic assessment of the monotypic genera Sundacarpus and Manoao (Coniferales: Podocarpaceae) utilising evidence from 18S rDNA sequences. Aust Syst Bot 15: 29–35. https://doi.org/10.1071/sb01002

Keng H (1975) A new scheme of classification of the conifers. Taxon 24:289–292. https://doi.org/10.2307/1218337

Kershaw P, Wagstaff B (2001) The southern conifer family Araucariaceae: history, status, and value for Paleoenvironmental reconstruction. GJB Annu Rev Ecol Syst 1605:397–414. https://doi.org/10.1146/annurev.ecolsys.32.081501.114059

Kim C-K, Eo J-K, Eom A-H (2013) Diversity and seasonal variation of Endophytic fungi isolated from three conifers in Mt. Taehwa. Korea Mycobiol 41:82–85. https://doi.org/10.5941/MYCO.2013.41.2.82

Koukol O, Kolařík M, Kolářová Z, Baldrian P (2012) Diversity of foliar endophytes in wind-fallen Picea abies trees. Fungal Divers 54:69–77. https://doi.org/10.1007/s13225-011-0112-2

Kusari S, Hertweck C, Spiteller M (2012) Chemical ecology of Endophytic fungi: origins of secondary metabolites. Chem Biol 19:792–798. https://doi.org/10.1016/j.chembiol.2012.06.004

Kusari S, Lamshöft M, Spiteller M (2009) Aspergillus fumigatus Fresenius, an endophytic fungus from Juniperus communis L. Horstmann as a novel source of the anticancer pro-drug deoxypodophyllotoxin. J Appl Microbiol 107:1019–1030. https://doi.org/10.1111/j.1365-2672.2009.04285.x

Lang X, Su J, Zhang Z, Lu S (2011) The taxonomic history and original literatures arrangements of living Cephalotaxus (Cephalotaxaceae). Am J Plant Sci 2:496–506. https://doi.org/10.4236/ajps.2011.23058

Lang X-D, Su J, Lu S-G, Zhang Z-J (2013) A taxonomic revision of the genus Cephalotaxus (Taxaceae). Phytotaxa 84:1–24. https://doi.org/10.11646/phytotaxa.84.1.1

Leslie AB, Beaulieu J, Holman G et al (2018) An overview of extant conifer evolution from the perspective of the fossil record. Am J Bot 105:1531–1544. https://doi.org/10.1002/ajb2.1143

Leslie AB, Beaulieu JM, Rai HS et al (2012) Hemisphere-scale differences in conifer evolutionary dynamics. Proc Natl Acad Sci USA 109:16217–16221. https://doi.org/10.1073/pnas.1213621109

Lewis FJ (1924) An Endotropic fungus in the Coniferæ. Nature 114:860–860. https://doi.org/10.1038/114860b0

Li H-L (1953) Present distribution and habitats of the conifers and Taxads. Evolution 7:245–261. https://doi.org/10.2307/2405735

Liu N, Zhu Y, Wei Z et al (2009) Phylogenetic relationships and divergence times of the family Araucariaceae based on the DNA sequences of eight genes. Chin Sci Bull 54:2648–2655. https://doi.org/10.1007/s11434-009-0373-2

Liu Y-H, Hu X-P, Li W et al (2016) Antimicrobial and antitumor activity and diversity of endophytic fungi from traditional Chinese medicinal plant Cephalotaxus hainanensis Li. Genet Mol Res GMR 15. https://doi.org/10.4238/gmr.15028633

Lugtenberg BJJ, Caradus JR, Johnson LJ (2016) Fungal endophytes for sustainable crop production. FEMS Microbiol Ecol 92:fiw194. https://doi.org/10.1093/femsec/fiw194

Manchester SR, Chen Z-D, Lu A-M, Uemura K (2009) Eastern Asian endemic seed plant genera and their paleogeographic history throughout the northern hemisphere. J Syst Evol 47:1–42. https://doi.org/10.1111/j.1759-6831.2009.00001.x

Moghaddam MS, Soltani J (2014) Bioactivity of endophytic Trichoderma fungal species from the plant family Cupressaceae. Ann Microbiol 64:753–761. https://doi.org/10.1007/s13213-013-0710-1

Neger F (1907) Cephalotaxaceae. In: Göschen, G.J. (ed.) Die Nadelhölzer (Koniferen) und übrigen Gymnospermen. Leipzig, pp. 23

Nkongolo KK, Mehes-Smith M (2012) Karyotype evolution in the Pinaceae: implication with molecular phylogeny. Genome 55:735–753. https://doi.org/10.1139/g2012-061

Nobis MP, Traiser C, Roth-Nebelsick A (2012) Latitudinal variation in morphological traits of the genus Pinus and its relation to environmental and phylogenetic signals. Plant Ecol Divers 5:1–11. https://doi.org/10.1080/17550874.2012.687501

Ortiz B, Enríquez L, Mejía K et al (2022) Molecular characterization of endophytic fungi from pine (Pinus oocarpa) in Honduras. Bionatura 7:1–5. https://doi.org/10.21931/RB/2022.07.03.13

Pant S, Samant SS (2008) Population ecology of the endangered Himalayan yew in Khokhan wildlife sanctuary of North Western Himalaya for conservation management. J Mt Sci 5:257–264. https://doi.org/10.1007/s11629-008-0078-z

Petrini O (1991) Fungal endophytes of tree leaves. In: Andrews JH, Hirano SS (eds) Microbial ecology of leaves. Springer, New York, pp 179–197

Pilger R (1903) Taxaceae. In: Engler A (ed) Das Pflanzenreich 5 (Heft 18), 4th edn. Wilhelm Engelmann, Leipzig, pp 99–105

Qadri M, Johri S, Shah BA et al (2013) Identification and bioactive potential of endophytic fungi isolated from selected plants of the Western Himalayas. Springerplus 2:8. https://doi.org/10.1186/2193-1801-2-8

Ran J-H, Gao H, Wang X-Q (2010) Fast evolution of the retroprocessed mitochondrial rps3 gene in conifer II and further evidence for the phylogeny of gymnosperms. Mol Phylogenet Evol 54: 136–149. https://doi.org/10.1016/j.ympev.2009.09.011

Ran J-H, Shen T-T, Wang M-M, Wang X-Q (2018) Phylogenomics resolves the deep phylogeny of seed plants and indicates partial convergent or homoplastic evolution between Gnetales and angiosperms. Proc R Soc B Biol Sci 285:20181012. https://doi.org/10.1098/rspb.2018.1012

Richardson SN, Walker AK, Nsiama TK et al (2014) Griseofulvin-producing *Xylaria* endophytes of *Pinus strobus* and *Vaccinium angustifolium*: evidence for a conifer-understory species endophyte ecology. Fungal Ecol 11:107–113. https://doi.org/10.1016/j.funeco.2014.05.004

Rodriguez R, Redman R (2008) More than 400 million years of evolution and some plants still can't make it on their own: plant stress tolerance via fungal symbiosis. J Exp Bot 59:1109–1114. https://doi.org/10.1093/jxb/erm342

Rodriguez RJ, White JF, Arnold AE, Redman RS (2009) Fungal endophytes: diversity and functional roles. New Phytol 182:314–330. https://doi.org/10.1111/j.1469-8137.2009.02773.x

Saikkonen K (2007) Forest structure and fungal endophytes. Fungal Biol Rev 21:67–74. https://doi.org/10.1016/j.fbr.2007.05.001

Saikkonen K, Wäli P, Helander M, Faeth SH (2004) Evolution of endophyte-plant symbioses. Trends Plant Sci 9:275–280. https://doi.org/10.1016/j.tplants.2004.04.005

Saithong P, Panthavee W, Stonsaovapak S, Congfa L (2010) Isolation and primary identification of endophytic fungi from Cephalotaxus mannii trees. Maejo Int J Sci Technol 4:446–453

Schneider H, Schuettpelz E, Pryer KM et al (2004) Ferns diversified in the shadow of angiosperms. Nature 428:553–557. https://doi.org/10.1038/nature02361

Schulz B, Boyle C (2005) The endophytic continuum. Mycol Res 109:661–686. https://doi.org/10.1017/s095375620500273x

Setoguchi H, Asakawa Osawa T, Pintaud JC et al (1998) Phylogenetic relationships within Araucariaceae based on rbcL gene sequences. Am J Bot 85:1507–1516

Shi G, Zhou Z, Xie Z (2010) A new *Cephalotaxus* and associated epiphyllous fungi from the Oligocene of Guangxi, South China. Rev Palaeobot Palynol 161:179–195. https://doi.org/10.1016/j.revpalbo.2010.04.002

Singh R, Biswas J, Bisht S (2018) Gymnosperm diversity of the Himalayan biodiversity hotspot. In: Plant diversity in the Himalaya hotspot region, Vol. I, M/s Bishen Singh Mahendra pal Singh, Dehra Dun. ISBN: 978-81-211-0946-8, pp 129–162

Soltani J, Moghaddam MS (2015) Fungal endophyte diversity and bioactivity in the Mediterranean cypress Cupressus sempervirens. Curr Microbiol 70:580–586. https://doi.org/10.1007/s00284-014-0753-y

Staniek A, Woerdenbag HJ, Kayser O (2008) Endophytes: exploiting biodiversity for the improvement of natural product-based drug discovery. J Plant Interact 3:75–93. https://doi.org/10.1080/17429140801886293

Stevenson D (2013) Chapter 5: gymnosperms. Ann Plant Rev 45:141e162

Stierle AA, Stierle DB (2015) Bioactive secondary metabolites produced by the fungal endophytes of conifers. Nat Prod Commun 10:1671–1682

Tung QN, Nguyen VTH, Chi LP et al (2022) Fusarium fujikuroi WQF5 ISOLATED FROM Cephalotaxus mannii Hook.f. AS A PRODUCER OF ANTIBACTERIAL AGENT AND PACLITAXEL. Acad J Biol 44:53–63. https://doi.org/10.15625/2615-9023/17577

Vigneshwari A, Škrbić BD, Kredics L et al (2021) Bioprospecting and biodiversity investigations of endophytic fungi isolated from Juniperus communis. Acta Biol Szeged 64:129–138. https://doi.org/10.14232/abs.2020.2.129-138

Vu THN, Quach NT, Le PC et al (2023) Bioprospecting Endophytic fungi isolated from Cephalotaxus mannii hook f. as prolific sources of antibacterial, anticancer, and antioxidant agents. Microbiology 92:284–292. https://doi.org/10.1134/S0026261722602834

Wang X-Q, Ran J-H (2014) Evolution and biogeography of gymnosperms. Mol Phylogenet Evol 75:24–40. https://doi.org/10.1016/j.ympev.2014.02.005

Wang Y, Guo L (2007) A comparative study of endophytic fungi in needles, bark, and xylem of Pinus tabulaeformis. Can J Bot 85:911–917. https://doi.org/10.1139/B07-084

Wei J-G, Xu T, Guo L, et al (2007) Endophytic Pestalotiopsis species associated with plants of Podocarpaceae, Theaceae and Taxaceae in southern China. Fungal Divers 24:55

Wilson D (1995) Endophyte: the evolution of a term, and clarification of its use and definition. Oikos 73:274–276. https://doi.org/10.2307/3545919

Yang JH, Oh S-Y, Kim W, Hur J-S (2022a) Endolichenic fungal community analysis by pure culture isolation and Metabarcoding: a case study of Parmotrema tinctorum. Mycobiology

Yang Y, Ferguson DK, Liu B et al (2022b) Recent advances on phylogenomics of gymnosperms and a new classification. Plant Divers 44:340–350. https://doi.org/10.1016/j.pld.2022.05.003

Yang Y, Yang Z, Ferguson DK, Shong J-Y (2023) An integrative view on the systematic position of the cupressophyte Cephalotaxus. Ecol Evol 13:e10273. https://doi.org/10.1002/ece3.10273

Yoo J-J, Eom A-H (2012) Molecular identification of Endophytic fungi isolated from needle leaves of conifers in Bohyeon Mountain, Korea. Mycobiology 40:231–235. https://doi.org/10.5941/MYCO.2012.40.4.231

Zabalgogeazcoa I (2008) Fungal endophytes and their interaction with plant pathogens: a review. Span J Agric Res 6:138–146. https://doi.org/10.5424/sjar/200806S1-382

Zhou X, Zhu H, Liu L et al (2010) A review: recent advances and future prospects of taxol-producing endophytic fungi. Appl Microbiol Biotechnol 86:1707–1717. https://doi.org/10.1007/s00253-010-2546-y

Temporal Variations in Endophytic Fungal Assemblages of *Ginkgo biloba*

Laith Khalil Tawfeeq Al-Ani, Lobna Hajji-Hedfi, Takwa Wannassi, Abdelhak Rhouma, Amira Khlif, Omaima Bargougui, and Samar Dali

Contents

L. K. T. Al-Ani (✉)
School of Biology Science, Universiti Sains Malaysia, Minden, Pulau Pinang, Malaysia

L. Hajji-Hedfi
Regional Centre of Agricultural Research of Sidi Bouzid, Gafsa, Sidi Bouzid, Tunisia

Research Laboratory of Agricultural Production Systems and Sustainable Development (LR03AGR02), Carthage University, Zaghouan, Tunisia

T. Wannassi · A. Rhouma · A. Khlif · O. Bargougui · S. Dali
Regional Centre of Agricultural Research of Sidi Bouzid, Gafsa, Sidi Bouzid, Tunisia

A. M. Abdel-Azeem, F. Degola (eds.), *Advances in Lichens and Non-Flowering Plants Fungal Endophytes Research*, Fungal Biology,
https://doi.org/10.1007/978-3-032-01506-8_13

1 Introduction

Ginkgo biloba, as a tree belonging to the Ginkgoaceae family, has a rich fossil history, going back 300 million years to the Permian period. During the Jurassic period, approximately 200 million years ago, this family reached its peak diversity and abundance. Many of its ancient members have gone extinct, while only one, *Ginkgo biloba* L., still survives today. This tree has demonstrated its hardness, having lived through the mass extinction events, such as those of the Cretaceous-Paleogene boundary, and the challenges of *Pleistocene glaciation. G. biloba*, therefore, is a living relic and a testimony to life's persistence in the face of radical environmental changes (Huh and Staba 1992; Singh et al. 2008; Weinmann et al. 2010; Mohanta 2012).

Ginkgo biloba is a living fossil, which is characterized by its longevity and medicinal values, and represents a diverse community of endophytic fungi in its tissues (Pawle and Singh 2014; Toghueo 2020; Tiwari and Bae 2022). These fungi live symptomlessly inside the tree host and are considered very important in plant health and growth, besides ecological interactions (Bao et al. 2022). For that purpose, the essential understanding of temporal dynamics of such fungal communities within *G. biloba* may elucidate complex relations between a host and its microbial inhabitants (Thongsandee et al. 2012).

The factors that could determine the temporal variation in endophytic fungal assemblages are varied and may range from seasonal changes and host developmental stages to environmental conditions (Cui et al. 2012). Seasonal variations in temperature, humidity, and photoperiod would influence fungal growth, reproduction, and dissemination directly (Islam et al. 2019). Changes in physiological and biochemical processes associated with the progress of *G. biloba* through its phenological cycle would change the nutritional status of the tree and provide favorable conditions for some fungal groups (Toghueo 2020). Other environmental factors, such as rainfall, wind, and soil type can also facilitate or impede the process of new fungal species introduction and establishment within the host (Deng et al. 2021).

Temporal variation in endophytic fungi *of G. biloba* can be highly useful in uncovering the ecology of these microorganisms (Islam et al. 2019). Characterization of seasonal succession may help in recognizing the putative roles of such dominant fungal species in plant defense, nutrient cycling, and overall plant health (Toghueo 2020). Further, their temporal stability and resilience may well reflect the potential of fungal communities to adapt to changing environmental conditions, with implications for the long-term health and survival of *G. biloba* (Deng et al. 2021; Tiwari and Bae 2022). This chapter book aims to delve into the intricate relationship between the living fossil, *G. biloba*, and its resident endophytic fungi. By investigating the temporal variations in these fungal assemblages, the research seeks to understand the dynamic interplay between the host and its microbial inhabitants.

2 Endophytic Fungi in Temporal Variations

Ginkgo biloba, a living fossil tree species, contains a complex assemblage of endophytic fungi that can fluctuate temporally due to seasonal variations, host phenology, and environmental factors. The temporal variation of endophytic fungi could affect their synthesis of bioactive compounds like secondary metabolites, which are pertinent for agricultural applications.

Furthermore, a few dominant fungal species are associated with ginkgo trees, like *Alternaria*, *Colletotrichum*, Cladosporium, *Fusarium*, *Pestalotiopsis*, *Peyronellaea*, and *Phoma*. Five common endophytic fungal taxa, including *Alternaria*, *Cladosporium*, *Fusarium*, and *Colletotrichum*, with their relative abundances changing over time (Kim et al. 1999; Toghueo 2020).

2.1 Host Phenology

The composition and diversity of endophytic fungi differed based on the *G. biloba* host organs. The endophytic fungi have organ preference or specificity to host plants that have been demonstrated by the frequent occurrences and dominance of some endophytic fungi in specific organs. Toghueo (2020) showed that *Phomopsis* sp. was much more frequent in twigs, whereas *Phyllosticta* sp. was most frequently isolated from both leaves and petioles. *Alternaria*, *Phyllosticta,* and *Rhizoctonia* were reported as the most dominant endophyte fungi taxa in *G. biloba* leaves (Du et al. 2009).

However, the life cycle of *G. biloba* and the endophytic fungal diversity are varied. For example, leaf development: juvenile leaves in spring may host distinct fungal populations relative to mature leaves in summer. During the leaf-emerging phase, no fungus was observed. *G. biloba* endophytic fungi in seed and bark tissues may exhibit reduced temporal variation relative to leaves, as these tissues demonstrate more stability throughout time.

3 Secondary Metabolites of Endophytic Fungi

Plants are associated with several endophytic microorganisms that live inside plant tissues with no detectable disease symptoms (Patchett and Newman 2021). Among these, endophytic fungi which divided into facultative and obligatory groups, based on how associated with their host plants (Gupta et al. 2020; Alam et al. 2021). These fungi produce a wide range of bioactive secondary metabolites with varied molecular structures (Mengistu 2020). Under stressful conditions, endophytic fungi may release various secondary metabolites in plant cells making them play a crucial role in enhancing plant resilience to environmental stresses, facilitating nutrient supply, and promoting growth and development (Singh et al. 2023). Extensive research was conducted toward the study of secondary metabolites and endophytic fungi-plant interactions, which demonstrate their important role in plant resilience. Among these plants, *G. biloba* hosts numerous endophytic fungi that produce a variety of physiologically active secondary metabolites (Toghueo 2020).

3.1 Identification of Secondary Metabolites of Endophytic Fungi

There are several procedures that are employed to identify secondary metabolites from endophytic fungi linked to *Ginkgo biloba*. Collect of samples from different parts of *G. biloba,* including root, stem, leaf, seed, and bark, were performed. Then, a conventional procedure of endophytic fungi isolation was conducted based on protocol availability in the laboratory, by sterilizing plant tissues, inoculating them on PDA medium at 28°C, endophytes were extracted from *Ginkgo biloba*, followed by a DNA extraction using a universal primer such as ITS5/ITS4 to amplify the conserved region of ITS (Toghueo 2020). Purified strains of endophytic fungi were cultivated then in a liquid fermenting medium to generate secondary metabolites. After the components in the crude extract were separated and analyzed using high-performance liquid chromatography (HPLC), the secondary metabolites were extracted, screened, and identified using mass spectrometer (MS) to assess the molecular weights and structures of the compounds (Li and Fitzloff 2002; Zou et al. 2021).

Flavonoids, terpenes, and phenolic compounds are the most important secondary metabolites of *G. biloba* with a focus on academic research from root exudates, leaves, and twigs (Guo et al. 2020; Wang et al. 2024). The leaves of this plant are among the most widely used dietary supplements, including pharmacologically active components among its diverse chemical elements, such as secondary metabolites (Ražná et al. 2019). Among secondary metabolites, ginkgolide is the most interesting one, classified as a terpenoid lactone, which is beneficial to human health (Wang et al. 2024). Taking into consideration those endophytes, most of them identified from the genera of *Fusarium, Aspergillus, Penicillium, Cladosporium, Chaetomium, and Xylaria* (Yuan et al. 2019).

3.2 Secondary Metabolites of **Chaetomium**

Among the largest group of endophytic fungi of *G. biloba*, *Chaetomium,* where *Chaetomium globosum is* considered the main endophyte of this genus. Many secondary metabolites detected of broth culture for Chaetomium that estimated 115 metabolites as chaetoglobosin (A,C, E, G, Vb) (Fig. 1). Some of polyketide

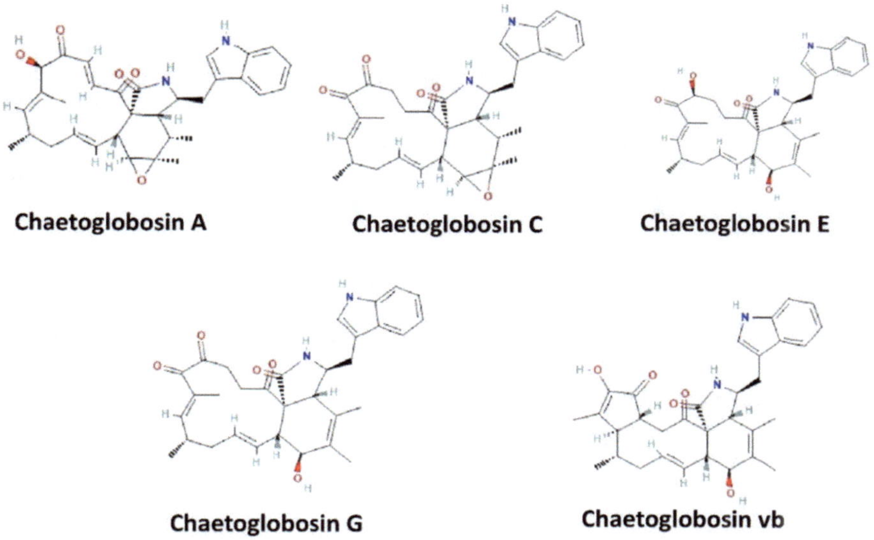

Fig. 1 Molecular structure of chaetoglobosin identified from *Chaetomium globosum.* (PubChem)

Fig. 2 Molecular structure of flavipin isolated from *C. globosum* associate with leaves of *G. biloba.* (PubChem)

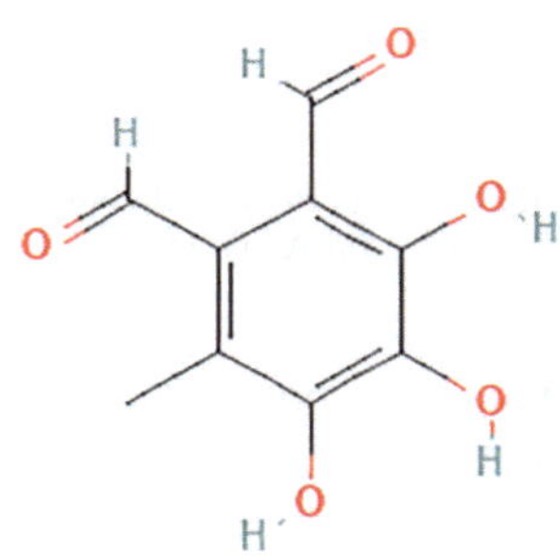

derivatives and around 80 chaetoglobosin are a cytotoxic substances targeting the filaments of pathogen by disrupting the cytoskeleton. In addition, ergosterol peroxide and chaetomugilin (A and D) reported as cytotoxicity compounds. This can contribute in biocontrol against fungal diseases (Xue et al. 2012; Li et al. 2013, Darwish et al. 2020). Three cytochalasan alkaloids, chaetoglobosins A, C, and D, were extracted from the fungal strain *C. globosum* no. 04, which is produced from *G. biloba*, and were found to have strong antifungal properties (Zhang et al. 2013).

Furthermore, strong antioxidant activity (As antagonistic compound against fungi and nematodes) is exhibited by the flavipin as a secondary metabolite from *C. globosum* associate with leaves of *G. biloba* (Ye et al. 2013) (Fig. 2).

One of the important secondary metabolites, the antifungal and antibacterial compounds of azaphilones isolated from *C. globosum,* and from other different species of *Chaetomium* display various biological activities (Xue et al. 2012;

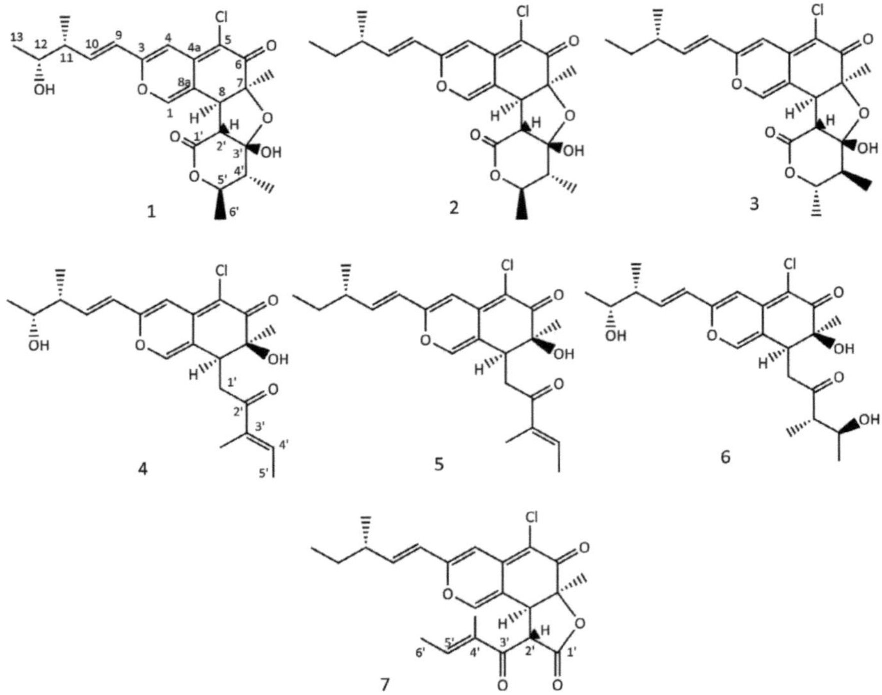

Fig. 3 Molecular structures of metabolites isolated from *Chaetomium globosum* TY1 chaetomugilin A (1), D (2), S (3), I (4), J (5), Q (6), and O (7). (Wang et al. 2017)

Wang et al. 2017). Among these azaphilones, seven, including chaetomugilin A (1), D (2), S (3), I (4), J (5), Q (6), and O (7), were identified from the endophytic *C. globosum* with the strain of TY1, all of these azaphilones were assessed for their ability to regulate plant growth with a positive demonstration on the seedling's impact on plant growth (Wang et al. 2017) (Fig. 3). *C. globosum* can produce two polyhydroxylated steroid metabolites (Qin et al. 2009).

3.3 Secondary Metabolites of Aspergillus

Aspergillus is the dominant genus of endophytic fungi found in *Ginkgo biloba*, isolated from various parts of the plant cultivated in different regions (Yuan et al. 2019). In recent research, flavonoid metabolites are studied from *Aspergillus*, and extracted from *G. biloba*; however, these kinds of metabolites directly from wild Ginkgo have the problems of a long growth cycle (Jiang et al. 2005). For that reason, using flavonoid biosynthesis pathway of *Aspergillus* sp. was predicted and reconstructed to identify secondary metabolites. Among identified flavonoid

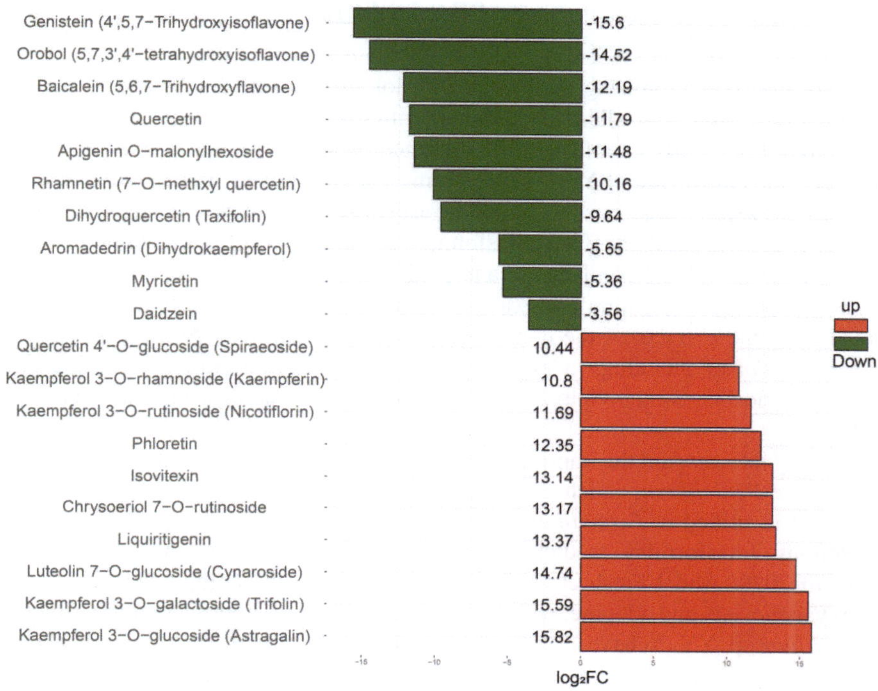

Fig. 4 The relative concentration difference of intracellular and extracellular flavonoid metabolites resulting from *Aspergillus* metabolites

metabolites, a total of 71 flavonoids were identified including 13 flavones, 4 isoflavones, 18 flavonols, 8 flavanones, 14 C-glycosyl flavonoids, 12 anthocyanins, and 2 procyanidins. According to concentration of compounds (Fig. 4), cyanin, afzelechin, dihydromyricetin, kaempferol 3-O-glucoside, di-O-methylquercetin, genistein, trifolin, and 4′-hydroxy-5,7-dimethoxyflavanone had higher total signal intensity. The results showed that genistein, orobol, baicalein, quercetin, apigenin O-malonylhexoside, rhamnetin, taxifolin, and other metabolites were much higher extracellularly than intracellularly. Meanwhile, astragalin, trifolin, cynaroside, liquiritigenin, chrysoeriol-7-O-rutinoside, isovitexin, phloretin, kaempferol-3-O-rhamnoside, kaempferin, spiraeoside, and other metabolites mainly accumulated intracellularly. Moreover, most of the flavonoids gathered intracellular contain glycosylation groups, such as astragalin, trifolin, cynaroside, and kaempferol, while almost all metabolites with much higher concentration extracellular than intracellular were flavonoid monomers with multiple hydroxyls (Wu et al. 2022; Zhao et al. 2008). It was reported on each of the following compounds' metabolites, 3-hydroxy-terphenyl, 4″-deoxycandidusin A, and 4″-deoxytripentin exhibit strong inhibitory activity against neuraminidase (Guo et al. 2012).

3.4 *Secondary Metabolites of* Penicillium

Among the most promising sources of physiologically active substances, such as hormones, mycotoxins, alkaloids, and antibiotics, is the genus *Penicillium.* Results from Wu et al. (2018) confirmed the presence of new compounds extracted from *Penicillium cataractum* SYPF 7131 strain, which showed antibacterial activity against *Staphylococcus aureus*, including isobenzofuranbutanoic acid, hydroxy-6-methoxy-5-methylisobenzofuran, rasfonin, and 7-hydroxy-2,5-dimethyl-chromen, penicimenolidyu B, alteryulactone, and penicimenolidying. Six very common secondary metabolites with an antioxidant and growth-promoting activities were isolated from the extracts of the endophytic fungus *Penicillium* sp. from the root of *Ginkgo biloba*, including adenosine, methyl β-D-ribofuranoside, adenine, 2′-deoxyadenosine, 3-methylpiperazine-2,5-dione, and 2′-deoxyuridine (Yuan et al. 2013).

Taking into account the importance of the secondary metabolites derived from the endophyte *Penicillium*, which including one of the important antibiotics, penicillin, was first identified in *Penicillium chrysogenum*, which effectively treated bacterial infections, transforming medicine (Peñalva et al. 1998; Kouipou Toghueo and Boyom 2020; Abdel-Fatah et al. 2021). Additionally, the compounds adenosine, deoxyadenosine, and adenine, which were isolated from the fermentation product of *Penicillium* sp. YY-20 have a strong scavenging capacity for DPPH free radical (Liu et al. 2008). We suggest, it would be necessary to separate the strains of *Penicillium* from the plant, cultivate them under ideal conditions, and then conduct fermentation to see if they generate penicillin or other secondary metabolites to investigate the possibility of extracting penicillin from Penicillium endophytes in *G. biloba.*

4 Application of Secondary Metabolites

Schwabe's discovery in Germany that *Ginkgo biloba* contains active compounds, such as ginkgo flavonoids and ginkgolides, led to its recognition for its benefits in the prevention and treatment of cardiovascular, cerebrovascular, and neurological diseases. This breakthrough sparked a growing interest in ginkgo research, with Germany and France taking the lead in developing ginkgo leaf-based products. In the 1970s, they introduced *Ginkgo biloba* preparations to treat cardiovascular diseases. Ginkgo extracts, such as EGb 761, have demonstrated effectiveness in treating noise-induced hearing loss (NIHL) in animal models by protecting the cochlea from hair cell damage through the reduction of reactive oxygen species (ROS). EGb 761 also shows potential in addressing secondary conditions, such as tinnitus (Konstantin et al. 2014). Furthermore, a *Ginkgo biloba* extract (GbE) has been used as a nontoxic agent for reducing and stabilizing graphene, creating a cytocompatible form of graphene oxide (Gb-rGO) that shows significant biocompatibility with cancer cells. This cost-effective and environmentally friendly method has potential for various biomedical applications, including tissue engineering, drug delivery, biosensing, and molecular imaging (Gurunathan et al. 2014).

Ginkgo preparations primarily include capsules, tablets, granules, and tea bags, with capsules and tablets being the most popular forms. Recently, new products such as shampoo, facial cleansers, and hair moisturizers have been introduced in cosmetics. Most ginkgo products are registered as health foods, while a few are classified as over-the-counter drugs. In many existing medicines, the basic quality requirements for Ginkgo biloba extracts include 24% total flavonoids and 6% ginkgolides. Some manufacturers set higher standards, incorporating ginkgolides A, B, C, J, and biloba lactone as quality indicators. These typically require ginkgolides A, B, C, and J content to be above 2.5% and biloba lactone content to exceed 2.6%. Some examples of secondary metabolites produced by endophytes associated with the *Ginkgo biloba* plant, along with their applications (Table 1).

Table 1 Secondary metabolites of endophytic fungi in *Ginkgo biloba*

Metabolites	Endophytes	Application	References
(7Z,11E)-7,11-Hexadecadien-1-yl acetate	*Chaetomium globosum* no. 16	Pesticide	Li et al. (2011)
(Z,Z)-9,12-Octadecadienoic acid	*Chaetomium globosum* no. 16	Prostaglandin biosynthesis and cell membrane formation	Tian (2013)
2,4,5-Trimethyl-1,3-dioxolane	*Chaetomium globosum* T16	Flavors	Yu (2010)
2′-O-Methyladenosine	*Chaetomium globosum*	Suppression of vaccinia virus replication	Li et al. (2014)
Acetaldehyde, diethyl acetal	*Chaetomium globosum* T16	Utilized in fruity, rum, and whiskey aromas	Yu (2010)
Alpha-Methylstyrene	*Chaetomium globosum*	Adhesive membranes and sealing compounds	Qin et al. (2009)
Anthranilic acid	*Chaetomium globosum* MX-0510	A vitamin that dissolves in water	Li et al. (2011)
Flavipin (1,2-benzenedicarboxaldehyde-3,4,5-trihydroxy-6-methyl)	*Chaetomium globosum* CDW7	Antioxidant fungicides	Xue et al. (2012)
Gliotoxin	*Chaetomium globosum* (NM0066)	A mycotoxin; an immunosuppressive compound; a protein farnesyltransferase blocker; a proteasome inhibitor; and an antifungal substance	Li et al. (2011)

(continued)

Table 1 (continued)

Metabolites	Endophytes	Application	References
Indole-3-acetic acid	*Chaetomium globosum*	Plant growth-stimulating hormone	Li et al. (2011)
Alternariol monomethyl ether	*Aspergillus* sp. YXf3	An antifungal agent	Zhang et al. (2015)
Aspergiloid I	*Aspergillus* sp. YXf3	Anticancer properties and suppression of plant pathogens	Yan et al. (2007)
Chlorflavonin	*Aspergillus* sp. (strain no. YXf3)	An antifungal agent	Yan et al. (2007)
Prenylterphenyllin D Prenylterphenyllin E 2′-O-Methylprenylterphenyllin	*Aspergillus* sp. IFB-YXS	Antibacterial properties and anti-plant pathogenic activities	Guo et al. (2014)
Alternariol monomethyl ether/alternariol-4-methyl ether	*Alternaria* no. 28	An antifungal agent	Guo et al. (2012)
Tenuazonic acid	*Alternaria* no. 28	An antibiotic with antiviral and antineoplastic properties; also serve as a mycotoxin	Guo et al. (2012)
3-Methylorsellinic acid	*Penicillium* no. 97	Antibacterial activity	Guo et al. (2012)
Fructigenine A	*Penicillium* no. 97	Inhibits the growth of leukemia cells	Liu et al. (2009)
Indole-3-acetic acid	*Penicillium* no. 97	Employed to prevent, eliminate, or reduce pests	Liu et al. (2009)
Pentadecanoic acid, methyl ester	*Xylaria* sp. YX-28	Pesticide	Ju et al. (2006)
Ferruginol	*Xylaria* sp. YX-28	An anticancer agent; antimicrobial agent; and protective substance	Ju et al. (2006)
Benzeneethanol/ phenylethyl alcohol	*Fusarium* sp. G1024	Local antiinfective agents; disinfectants; and pharmaceutical preservatives	Balkwill et al. (2006)
Rutin	*Fusarium oxysporum*	Functions as an antioxidant; antiallergic; antiinflammatory; antiproliferative; and anticancer properties	Zhao et al. (2017)
Quercetin	*Stemphylium* sp. *Actinomyces*	Antioxidants	Zhou et al. (2010) and Zhao et al. (2017)

(continued)

Table 1 (continued)

Metabolites	Endophytes	Application	References
Enniatin D	*Tuberculariaceae* sp. F1-3	Suppression of *Botrytis cinerea* spore germination	Kumaran et al. (2012)
Benzeneethanol/ phenylethyl alcohol	*Muscodor albus* strain GBA	Antibacterial agents and anti-oxidants. Antimicrobial agents	Kumaran et al. (2012)
Antibiotic U-62162	*Plantactinospora* sp. NEAU-gxj3	Restricted the growth of Gram-positive bacteria	Tang et al. (2009) and Guo et al. (2016)
Abscisic acid	*Phoma betae*	Plant growth enhancer	Kumaran et al. (2012)
Acetic acid, 2-methylpropyl ester	*Muscodor albus* strain GBA	A substance with antifungal properties	Deng et al. (2009)

5 Conclusions

Ginkgo biloba is a harbor for many fungal species. Endophyte fungi are diverse in different parts of *G. biloba,* such as root, stem, leaf, seed, and bark. The dominant fungi are in relation to the organ structure of the host plants. The composition and diversity of endophytic fungi differed based on the *G. biloba* host organs. The most endophytic fungi isolated from *G. biloba* including the genera of *Fusarium*, *Aspergillus*, *Penicillium*, *Cladosporium*, *Chaetomium*, and *Xylaria*. Endophytic fungi could produce many bioactive secondary metabolites with varied molecular structures. Flavonoids, terpenes, and phenolic compounds are the most important secondary metabolites of *G. biloba*. Finally, the abundance of endophytic fungi in *G. biloba* overlaps in producing the different secondary metabolites and could be a source of abrupt pathogens for other plants.

References

Abdel-Fatah SS, El-Batal AI, El-Sherbiny GM, Khalaf MA, El-Sayed AS (2021) Production, bioprocess optimization and γ-irradiation of *Penicillium polonicum*, as a new Taxol producing endophyte from Ginko biloba. Biotechnol Rep *30*:e00623. https://doi.org/10.1016/j.btre.2021.e00623

Alam B, Lǐ J, Gě Q, Khan MA, Gōng J, Mehmood S, Yuán Y, G□ng W (2021) Endophytic Fungi: from symbiosis to secondary metabolite communications or vice versa? Front Plant Sci 12: 791033. https://doi.org/10.3389/fpls.2021.791033

Balkwill DL, Fredrickson JK, Romine MF (2006) Sphingomonas and related genera. In: Dworkin M et al (eds) The prokaryotes, a handbook of the biology of bacteria. Springer, New York

Bao L, Sun B, Liu J, Zhang S, Xu N, Zhang X, Bohu T, Bai Z (2022) Leaf-associated epiphytic fungi of *Gingko biloba*, *Pinus bungeana* and Sabina chinensis exhibit delicate seasonal variations. Journal of Fungi 8(6):631

Cui Y, Yi D, Bai X, Sun B, Zhao Y, Zhang Y (2012) Ginkgolide B produced endophytic fungus (Fusarium oxysporum) isolated from *Ginkgo biloba*. Fitoterapia 83(5):913–920

Darwish AMG, Abdelmotilib NM, Abdel-Azeem AM, Abo Nahas HH, Mohesien MT (2020) Applications of *Chaetomium* functional metabolites with special reference to antioxidants. In: Abdel-Azeem AM (ed) Recent developments on genus *Chaetomium*, fungal biology. Springer, Cham, pp 227–240

Deng BW, Liu KH, Chen WQ, Ding XW, Xie XC (2009) *Fusarium solani*, Tax-3, a new endophytic taxol-producing fungus from *Taxus chinensis*. World J Microbiol Biotechnol 25: 139–143

Deng Y, Huang H, Lei F, Fu S, Zou K, Zhang S, Liu X, Jiang L, Liu H, Miao B, Liang Y (2021) Endophytic bacterial communities of *Ginkgo biloba* leaves during leaf developmental period. Front Microbiol 12:698703

Du SK, Chen SL, Lin D, Wu MQ and Wang MX (2009) Diversity of endophytic fungi in leaves of Ginkgo biloba. Mycosystema 26(4):504–511

Guo ZK, Yan T, Guo Y (2012) p-Terphenyl and diterpenoid metabolites from endophytic *Aspergillus* sp. YXf3. J Nat Prod 75:15–21

Guo ZK, Wang R, Huang W, Li XN, Jiang R, Tan RX, Ge HM (2014) Aspergiloid I, an unprecedented spirolactone norditerpenoid from the plant-derived endophytic fungus *Aspergillus* sp. YXf3. Beilstein J Org Chem 10:1701–1709

Guo XW, Zhang J, Li JS (2016) A new amide metabolite from endophytic *Plantactinospora* sp. NEAU-gxj3. Nat Prod Res Dev 28:481–485

Guo Y, Gao C, Wang M, Fu F, El-Kassaby YA, Wang T, Wang G (2020) Metabolome and transcriptome analyses reveal flavonoids biosynthesis differences in *Ginkgo biloba* associated with environmental conditions. Ind Crop Prod 158:112963. https://doi.org/10.1016/j.indcrop.2020.112963

Gupta S, Chaturvedi P, Kulkarni MG, Van Staden J (2020) A critical review on exploiting the pharmaceutical potential of plant endophytic fungi. Biotechnol Adv 39:107462. https://doi.org/10.1016/j.biotechadv.2019.107462

Gurunathan S, Han JW, Park JH, Eppakayala V, Kim JH (2014) *Ginkgo biloba*: a natural reducing agent for the synthesis of cytocompatible graphene. Int J Nanomedicine 9:363–377

Huh H, Staba EJ (1992) The botany and chemistry of *Ginkgo biloba* L. Int J Geogr Inf Syst 1:91–124

Islam MN, Choi J, Baek KH (2019) Control of foodborne pathogenic bacteria by endophytic bacteria isolated from *Ginkgo biloba* L. Foodborne Pathog Dis 16(10):469

Jiang HX, Wood KV, Morgan JA (2005) Metabolic engineering of the phenylpropanoid pathway in *Saccharomyces cerevisiae*. Appl Environ Microbiol 71:2962–2969. https://doi.org/10.1128/AEM.71.6.2962-2969.2005

Ju XY, Feng YJ, Chen FM (2006) Volatile constituents and their fibrinolytic activity of endophytic fungus *Fusarium* sp. GI024 from *Ginkgo biloba*. Microbiology 33:8–11

Kim SU, Strobel G, Ford E (1999) Screening of taxol-producing endophytic fungi from Ginkgo biloba and Taxus cuspidata in Korea. J Appl Biol Chem 42(2):97–99

Konstantin T, Sabine K, Sönke A, Holger S (2014) Protective efects of *Ginkgo biloba* extract EGb 761 against noise trauma-induced hearing loss and tinnitus development. Neural Plast 2014: 427298

Kouipou Toghueo RM, Boyom FF (2020) Endophytic *Penicillium* species and their agricultural, biotechnological, and pharmaceutical applications. *3*. Biotech 10(3):107. https://doi.org/10.1007/s13205-020-2081-1

Kumaran RS, Choi YK, Lee S (2012) Isolation of taxol, an anticancer drug produced by the endophytic fungus, *Phoma betae*. Afr J Biotechnol 11:950–960

Li W, Fitzloff JF (2002) Simultaneous determination of terpene lactones and flavonoid aglycones in *Ginkgo biloba* by high-performance liquid chromatography with evaporative light scattering detection. J Pharm Biomed Anal 30(1):67–75. https://doi.org/10.1016/S0731-7085(02)00201-7

Li HQ, Li XJ, Wang YL (2011) Antifungal metabolites from *Chaetomium globosum*, an endophytic fungus in *Ginkgo biloba*. Biochem Syst Ecol 39:876–879

Li X, Tian Y, Yang SX, Zhang YM, Qin JC (2013) Cytotoxic azaphilone alkaloids from *Chaetomium globosum* TY1. Bioorg Med Chem Lett 23:2945–2947. https://doi.org/10.1016/j.bmcl.2013.03.044

Li H, Xiao J, Gao YQ (2014) Chaetoglobosins from Chaetomium globosum, an endophytic fungus in *Ginkgo biloba*, and their phytotoxic and cytotoxic activities. J Agric Food Chem 62:3734–3741

Liu X, Dong M, Chen X, Jiang M, Lv X, Zhou J (2008) Antimicrobial activity of an endophytic *Xylaria* sp. YX-28 and identification of its antimicrobial compound 7-amino-4-methylcoumarin. Appl Microbiol Biotechnol 78:241–247

Liu JJ, Chen SJ, Gong HX (2009) An endophytic fungus producing orange pigment isolated. Prog Modern Biomed 9:246–250

Mengistu AA (2020) Endophytes: Colonization, behaviour, and their role in defense mechanism. Int J Microbiol 2020:1–8. https://doi.org/10.1155/2020/6927219

Mohanta TK (2012) Advances in *Ginkgo biloba* research: genomics and metabolomics perspectives. Afr J Biotechnol 11:15936–15944

Patchett A, Newman JA (2021) Comparison of plant metabolites in root exudates of *Lolium perenne* infected with different strains of the fungal endophyte *Epichlo festucae* var. *lolii*. J Fungi 7:148. https://doi.org/10.3390/jof7020148

Pawle G, Singh SK (2014) Antimicrobial, antioxidant activity and phytochemical analysis of an endophytic species of Nigrospora isolated from living fossil *Ginkgo biloba*. CREAM 4(1):1–9

Peñalva MA, Rowlands RT, Turner G (1998) The optimization of penicillin biosynthesis in fungi. Trends Biotechnol 16(11):483–489. https://doi.org/10.1016/s0167-7799(98)01229-3

Qin JC, Gao JM, Zhang YM (2009) Polyhydroxylated steroids from an endo- phytic fungus, *Chaetomium globosum* ZY-22 isolated from *Ginkgo biloba*. Steroids 74:786–790. https://doi.org/10.1016/j.steroids.2009.04.011

Ražná K, Sawinska Z, Ivanišová E, Vukovic N, Terentjeva M, Stričík M, Kowalczewski PŁ, Hlavačková L, Rovná K, Žiarovská J, Kačániová M (2019) Properties of *Ginkgo biloba* L.: antioxidant characterization, antimicrobial activities, and genomic MicroRNA based marker fingerprints. Int J Mol Sci 21(9):3087. https://doi.org/10.3390/ijms21093087

Singh B, Kaur P, Gopichand Singh RD, Ahuja PS (2008) Biology and chemistry of *Ginkgo biloba*. Fitoterapia 79(6):401–418

Singh DP, Maurya S, Yerasu SR, Bisen MS, Farag MA, Prabha R, Shukla R, Chaturvedi KK, Farooqi MS, Srivastava S, Rai A, Sarma BK, Rai N, Behera TK (2023) Metabolomics of early blight (Alternaria solani) susceptible tomato (Solanum lycopersicum) unfolds key biomarker metabolites and involved metabolic pathways. Sci Rep 13(1):1–17. https://doi.org/10.1038/s41598-023-48269-0

Tang AM, Jeewon R, Hyde KD (2009) A re-evaluation of the evolutionary rela- tionships within the Xylariaceae based on ribosomal and protein-coding gene sequences. Fungal Divers 34:127–155

Thongsandee W, Matsuda Y, Ito S (2012) Temporal variations in endophytic fungal assemblages of *Ginkgo biloba* L. J Forest Res 17(2):213–218

Tian Y (2013) Identification of three *Chaetomium* strains and the activity of secondary metabolites. Jilin University, Changchun

Tiwari P, Bae H (2022) Endophytic fungi: key insights, emerging prospects, and challenges in natural product drug discovery. Microorganisms 10(2):360

Toghueo RMK (2020) Endophytes from *Gingko biloba*: the current status. Phytochem Rev 19:743–759. https://doi.org/10.1007/s11101-020-09679-4

Wang D, Zhang Y, Li X, Pan H, Chang M, Zheng T, Sun J, Qiu D, Zhang M, Wei D, Qin J (2017) Potential allelopathic azaphilones produced by the endophytic *Chaetomium globosum* TY1 inhabited in *Ginkgo biloba* using the one strain-many compounds method. Nat Prod Res 31: 724–728. https://doi.org/10.1080/14786419.2016.1217208

Wang H, Zeng L, Ding HA (2024) Novel strategy for screening of secondary metabolites in *Ginkgo biloba* leaves by ultra-performance liquid chromatography–tandem information dependent acquisition mass spectrometry. J Anal Chem 79:330–338. https://doi.org/10.1134/S1061934824030146

Weinmann S, Roll S, Schwarzbach C, Vauth C, Willich SN (2010) Effects of *Ginkgo biloba* in dementia: systematic review and meta-analysis. BMC Geriatr 10:14

Wu Y-Y, Zhang T-Y, Zhang M-Y, Cheng J, Zhang Y-X (2018) An endophytic fungi of *Ginkgo biloba* L. produces antimicrobial metabolites as potential inhibitors of FtsZ of *Staphylococcus aureus*. Fitoterapia 128:*265–271*. https://doi.org/10.1016/j.fitote.2018.05.033

Wu X, Zou K, Liu X, Fu S, Zhang S, Duan Z, Zhou J, Liang Y (2022) The novel distribution of intracellular and extracellular flavonoids produced by *Aspergillus* sp. Gbtc 2, an endophytic fungus from *Ginkgo biloba* root. Front Microbiol 13:972294. https://doi.org/10.3389/fmicb.2022.972294

Xue M, Zhang Q, Gao JM, Li H, Tian JM, Pescitelli G (2012) Chaetoglobosin Vb from endophytic *Chaetomium globosum*: absolute configuration of chaetoglobosins. Chirality 24(8):668–674. https://doi.org/10.1002/chir.22068

Yan ZY, Luo J, Guo XH, Zeng QQ (2007) Screening of ginkgolides-producing endophytic fungi and optimal study on culture condition. Nat Prod Res Dev 19:554–558

Ye, Y., Xiao, Y.U., Ma, L., Li, H., Xie, Z., Wang, M., Ma, H., Tang, H. and Liu, J. (2013). Flavipin in Chaetomium globosum CDW7, an endophytic fungus from Ginkgo biloba, contributes to antioxidant activity. Applied microbiology and biotechnology, 97(16):7131–7139

Yu HS (2010) Study on the diversity and activity of endophytic fungi from Ginkgo. The Second Military Medical University, Chongqing

Yuan Y, Tian JM, Xiao J, Shao Q, Gao JM (2013) Bioactive metabolites isolated from *Penicillium* sp. YY-20, the endophytic fungus from *Ginkgo biloba. Nat Prod Res 28*(4):278–281. https://doi.org/10.1080/14786419.2013.850686

Yuan Z, Tian Y, He F et al (2019) Endophytes from *Ginkgo biloba* and their secondary metabolites. *Chin Med* **14**:51. https://doi.org/10.1186/s13020-019-0271-8

Zhang G, Zhang Y, Qin J, Qu X, Liu J, Li X, Pan H (2013) Antifungal metabolites produced by *Chaetomium globosum* No.04, an endophytic fungus isolated from *Ginkgo biloba*. Indian J Microbiol 53(2):175. https://doi.org/10.1007/s12088-013-0362-7

Zhang W, Wei W, Shi J (2015) Natural phenolic metabolites from endophytic *Aspergillus* sp. IFB-YXS with antimicrobial activity. Bioorg Med Chem Lett 25:2698–2701

Zhao W, Li L, Wang Z, Wang Z, Gao X, Yu M (2008) Isolation and product identification of a flavonid-producing endophytic fungus. J Microbiol 28:88–91

Zhao SS, Zhang YY, Yan W, Cao LL, Xiao Y, Ye YH (2017) *Chaetomium globosum* CDW7, a potential biological control strain and its antifungal metabolites. FEMS Microbiol Lett 364: fnw287

Zhao W, Li L, Wang Z, Wang Z, Gao X, Yu M. (2008). Isolation and product identification of a flavonid-producing endophytic fungus. J Microbiol;28:88–91

Zhou SL, Chen SL, Tan GH (2010) Antibacterial substances of endophytic fungus isolated from *Ginkgo biloba*. Nat Prod Res Dev 22:193–196

Zou K, Liu X, Hu Q, Zhang D, Fu S, Zhang S, Huang H, Lei F, Zhang G, Miao B, Meng D, Jiang L, Liu H, Yin H, Liang Y (2021) Root endophytes and *Ginkgo biloba* are likely to share and compensate secondary metabolic processes, and potentially exchange genetic information by LTR-RTs. Front Plant Sci 12:704985. https://doi.org/10.3389/fpls.2021.704985

Diversity of Endophytic Fungi of Common Yew (*Taxus baccata* L.)

Qudsia Firdous and Sheharyar Ahmed Khan

Contents

The original version of the chapter has been revised. A correction to this chapter can be found at https://doi.org/10.1007/978-3-032-01506-8_17

Q. Firdous (✉)
Department of Biosciences, Grand Asian University Sialkot, Sialkot, Pakistan
e-mail: firdous07@gaus.edu.pk

S. A. Khan
Institute of Plant Sciences, University of Bern, Bern, Switzerland

A. M. Abdel-Azeem, F. Degola (eds.), *Advances in Lichens and Non-Flowering Plants Fungal Endophytes Research*, Fungal Biology,
https://doi.org/10.1007/978-3-032-01506-8_14

1 Introduction

1.1 Brief Introduction to **Taxus baccata** *(Common Yew)*

Taxus baccata, or the common yew, is an important evergreen shrub noted for its aesthetic appeal and therapeutic qualities. Nonetheless, it is infamous for its toxicity, primarily attributed to taxine B, which can result in serious health complications if consumed. This dual character renders it a topic of interest in both horticulture and pharmacology. The subsequent sections offer a comprehensive introduction to *T. baccata*. It contains taxine B, which may induce seizures, arrhythmias, and cardiovascular collapse upon intake (Arens et al. 2016). The plant's toxicity requires vigilance, particularly in areas frequented by pets and youngsters. The common yew can flourish in several climates, with successful cultivation noted in Zhejiang, China, where new planting techniques have attained survival rates above 90% (Ningyu et al. 2015). The species has been cultivated in botanical gardens for over 90 years, demonstrating its durability and adaptability (Hunt 2010). *T. baccata* is a source of taxanes, chemicals utilized in oncological therapy, particularly contributing to the formulation of docetaxel, a significant chemotherapeutic drug (Le Roux and Guéritte 2017). *T. baccata* provides considerable advantages in landscaping and medicine, yet its toxicity presents hazards that require management. The equilibrium between benefit and risk is essential for its responsible implementation and utilization in diverse settings.

1.2 Importance of Studying Endophytic Fungi in **T. Baccata**

Investigating endophytic fungi in *T. baccata* is essential because they contribute to plant health and stimulation of secondary metabolite synthesis and have prospective pharmaceutical uses (Tripathi and Agnihotri 2024). These fungi form symbiotic associations with the host, enhancing development and stress resilience, which is especially crucial for medicinal plants like *T. baccata*, recognized for yielding valuable chemicals such as taxol (Adhikari et al. 2023). Endophytic fungi can augment the accumulation of secondary metabolites, such as taxanes, which are essential for the therapeutic activities of *T. baccata* (Liu et al. 2023a, b). Certain strains have demonstrated a substantial enhancement of taxol and analogous chemicals in plant tissues, suggesting their viability for biotechnological applications in pharmaceutical manufacture (Liu et al. 2023a, b). These fungi enhance nutrient absorption and bolster plant resilience against abiotic stresses, including drought and salinity, which are essential for the survival of *T. baccata* in adverse settings (Ebadi and Ebadi 2024; Zhang et al. 2023). The symbiotic interaction enhances the plant's health, enabling it to endure environmental stressors while preserving its therapeutic properties (Pandey et al. 2023). Endophytic fungi from *T. baccata*, including *Anthostomelloides leucospermi*, demonstrate antibacterial properties against diverse diseases, indicating their potential for the development of novel antimicrobial drugs (Zerroug et al. 2020). The extraction of bioactive chemicals from these fungi may result in innovative therapies for human ailments, underscoring their significance in pharmaceutical applications (Wijesekara et al. 2023). Although the advantages of researching endophytic fungi in *T. baccata* are considerable, it is crucial to evaluate the ecological equilibrium and the repercussions of resource overexploitation, which may jeopardize the sustainability of both the fungi and the host plant (Mishra et al. 2021).

1.3 Ecological and Pharmaceutical Significance of Endophytes

Endophytes, microorganisms that inhabit plant tissues, possess considerable ecological and medicinal significance (Ogbe et al. 2020). They promote plant health by improving growth, nutrient absorption, and stress resilience, while also providing bioactive chemicals with possible medicinal uses (Wijesekara et al. 2023). The subsequent sections detail their ecological functions and medicinal importance.

Ecological Importance

Symbiotic Associations: Endophytes participate in mutualistic relationships with host plants, enhancing growth and fortifying resistance to diseases and environmental stressors (Muhammad et al. 2024a, b).

Biodiversity: They demonstrate significant biological variety across numerous plant species, enhancing ecological stability and resilience (Pennekamp et al. 2018).

Sustainable Practices: Endophytes can diminish the necessity for artificial fertilizers and pesticides, thereby enhancing soil health and mitigating environmental impact (Muhammad et al. 2024a, b).

Pharmaceutical Importance
Endophytes generate an array of secondary metabolites, such as steroids, terpenoids, and phenols, with therapeutic characteristics (Hashem et al. 2023; Rai et al. 2021).

Drug Development: They are investigated as alternatives for the creation of new medications, especially in reaction to antibiotic resistance and emerging disorders (Brüssow 2024).

Innovative Applications: Endophytes are under investigation for their potential in agriculture and medicine, providing environmentally sustainable solutions for disease management and crop enhancement (Chaudhary et al. 2022). Despite the promising prospects of endophytes for sustainable agriculture and novel pharmaceutical development, challenges persist in comprehensively understanding their intricate interactions and optimizing their applications across diverse sectors (Verma et al. 2022).

1.4 Overview of Previous Research on Fungal Endophytes in Yew Trees

Fungal endophytes of yew trees, especially *Taxus wallichiana*, have been shown to have important roles in supporting plant growth and generating bioactive chemicals, such as the anticancer medication paclitaxel (Adhikari and Pandey 2024). Endophytes are increasingly acknowledged for their ability to improve the health and productivity of host plants, along with their pharmaceutical applications (Adeleke and Babalola 2021). A study identified seven fungal endophytes from *T. wallichiana*, showing their capacity to produce growth-promoting substances such as indole acetic acid and ammonia (Kaur and Saxena 2023). A separate study identified 13 distinct endophytic fungi from various genera, highlighting the diversity linked to Himalayan yew (Gauchan et al. 2021). Endophytic fungi play a significant role in the biosynthesis of paclitaxel, as studies demonstrate their ability to produce this important compound, providing an alternative to conventional extraction methods from yew bark (Sati et al. 2024). Fungal endophytes derived from yew positively affect the growth of other plants, including *Stevia rebaudiana*, by enhancing various physiological parameters without causing pathogenic effects (Lv et al. 2024). The emphasis on the advantages of yew endophytes necessitates an examination of their potential ecological impacts and the importance of sustainable practices in their use, particularly in light of the rarity of yew trees (Drożdżyński et al. 2024) (Fig. 1).

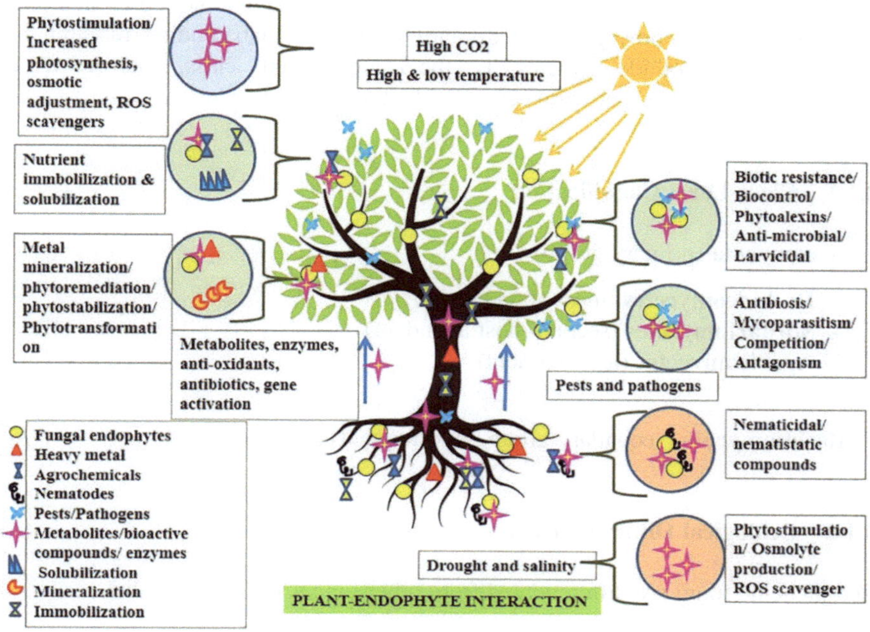

Fig. 1 The depiction of plant-endophyte dynamics amidst environmental challenges highlights essential processes and factors

2 Endophytic Fungi: An Overview

2.1 Definition and Classification of Endophytic Fungi

According to Baron and Rigobelo (2022), endophytic fungi are microorganisms that live asymptomatically inside plant tissues and establish mutually beneficial symbiotic relationships. These fungi are acknowledged for their various functions in improving plant health, generating bioactive compounds, and supporting ecological sustainability (Elhamouly et al. 2022). The classification of endophytic fungi relies on host plants, morphological traits, and molecular phylogenetic analysis, particularly emphasizing the internal transcribed spacer (ITS) region as a crucial molecular marker for precise identification (Rathnayaka et al. 2024). Endophytic fungi are symbiotic organisms that inhabit plant tissues without inducing disease (Alam et al. 2021). They enhance plant health by improving growth, nutrient uptake, and stress tolerance (Morales-Vargas et al. 2024). The classification relies on host plant associations (Healy et al. 2022), morphological characteristics, and molecular phylogeny, with the ITS region serving as a key marker (Pushpavathi and Krishnamurthy 2024). Endolichenic fungi represent a diverse assemblage of fungi that inhabit lichen thalli asymptomatically (Wethalawe et al. 2021). High biological diversity is exhibited across various plant groups (Wang et al. 2022). They are classified primarily within the phylum Ascomycota, with notable representation in the classes Dothideomycetes and Sordariomycetes. Metabarcoding research has revealed that Dothideomycetes is the predominant class, succeeded by Sordariomycetes and Agaricomycetes (Beck et al. 2023).

2.1.1 Class-Level Classification

- **Ascomycota** (primary group)
  - **Dothideomycetes** (most abundant)
  - **Sordariomycetes** (second most abundant)
  - **Agaricomycetes** (less frequent)

- **Basidiomycota** (secondary group, less frequent) (Fig. 2)

2.1.2 Ecological Differentiation

Endolichenic fungi are differentiated from other fungal symbionts of lichens:

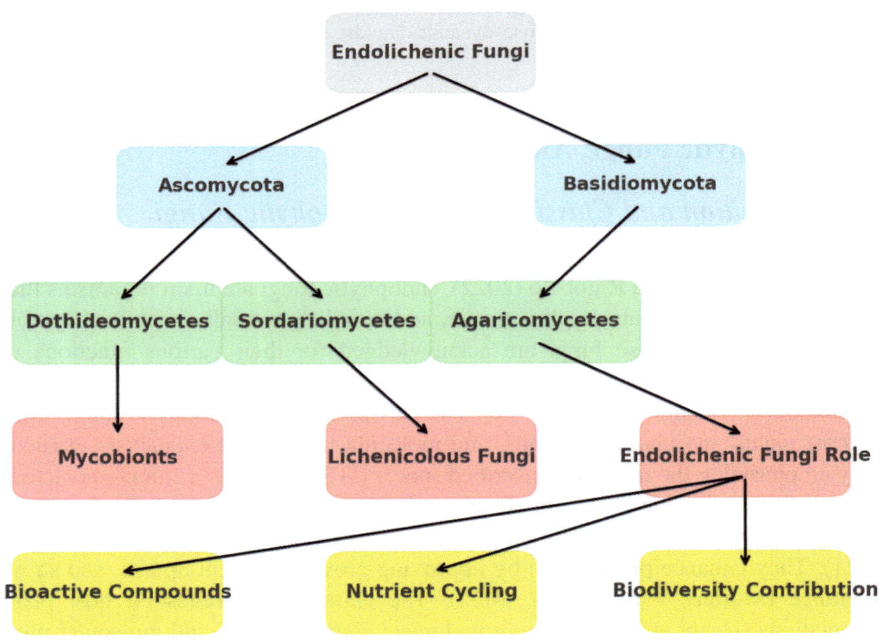

Fig. 2 Flowchart representation of the classification and roles of endolichenic fungi. It visually organizes their taxonomic structure (Ascomycota, Basidiomycota) and ecological roles, including their contributions to bioactive compounds, nutrient cycling, and biodiversity

Mycobionts Principal fungal associates in lichen symbiosis, constituting the primary structure of the lichen thallus (Biradar et al. 2023).

Lichenicolous Fungi Fungi that inhabit lichens, typically as parasites or commensals (Grube 2024).

Endolichenic Fungi Symbiotic fungi residing within the internal tissues of lichens without producing visible detriment.

2.1.3 Unique Characteristics

Endolichenic fungi demonstrate significant taxonomic diversity within Ascomycota (Wang et al. 2022). They enhance the microbiota of lichens and engage with their host in distinctive manners (Grimm et al. 2021). Molecular investigations affirm that they constitute a separate ecological group rather than mere incidental colonists (Yang et al. 2022). Endolichenic fungi remain a focus of ongoing research, as molecular techniques uncover new species and elucidate their ecological roles. Classification is crucial for comprehending their function and potential applications in biotechnology and environmental studies.

2.2 Roles of Endophytic Fungi in Plant Health, Growth, and Secondary Metabolite Production

Endophytic fungi significantly contribute to the improvement of plant health, growth, and the synthesis of secondary metabolites (Wijesekara and Xu 2023). These fungi form a symbiotic relationship with plants, enhancing various physiological and biochemical processes that are advantageous to both organisms (Mishra et al. 2024). This document delineates the primary functions of endophytic fungi within plant systems.

2.2.1 Promotion of Plant Growth

Endophytic fungi improve nutrient absorption, especially nitrogen and phosphorus, essential for plant development (Tang et al. 2022). Plant growth hormones, including auxins and gibberellins, are produced, stimulating root and shoot development (Kaur et al. 2024). Infected plants frequently demonstrate enhanced biomass and superior commercial yields, rendering them significant in agricultural contexts (Panno et al. 2021).

2.2.2 Stress Tolerance

These fungi enhance plant resilience to abiotic stresses, such as drought and salinity, by improving physiological functions (Morales-Vargas et al. 2024). They enhance plant defenses against biotic stresses, including pathogens and herbivores, by producing phytoalexins and other secondary metabolites (Saini et al. 2024a, b).

2.2.3 Secondary Metabolite Production

Endophytic fungi produce a significant array of bioactive secondary metabolites, applicable in both medicine and agriculture (Hashem et al. 2023). Prominent instances encompass anticancer agents such as taxol and camptothecin, which are sourced from particular endophytes (Nnaji 2025). These metabolites function as antimicrobial agents, offering a natural alternative to synthetic chemicals (Kaur et al. 2024). The benefits of endophytic fungi are substantial; however, their interactions with plants can differ significantly depending on environmental conditions and host species, resulting in variable outcomes in certain instances (Bard et al. 2024) (Table 1).

Table 1 Benefits of endophytic fungi across different categories

Aspect	Plant growth promotion	Stress tolerance	Metabolite production	References
Enhancement of nutrient uptake	Yes (nitrogen and phosphorus uptake)	No	No	Tang et al. (2022)
Production of plant hormones	Yes (auxins and gibberellins)	No	No	Kaur et al. (2024)
Increase in biomass and yield	Yes (increased biomass and yield)	No	No	Panno et al. (2021)
Abiotic stress resilience	No	Yes (drought and salinity resilience)	No	Morales-Vargas et al. (2024)
Biotic stress defense	No	Yes (defense against pathogens and herbivores)	No	Saini et al. (2024a, b)
Secondary metabolite production	No	No	Yes (bioactive compounds in medicine and agriculture)	Hashem et al. (2023)
Anticancer compound production	No	No	Yes (taxol and camptothecin)	Nnaji (2025)
Antimicrobial agent potential	No	No	Yes (natural antimicrobial alternative)	Kaur et al. (2024)

2.3 Symbiotic Interactions with Host Plants

Endophytic fungi inhabit the tissues of host plants, establishing either a mutualistic or commensal relationship. These fungi enhance the physiological functions of the plant without causing harm (Lu et al. 2021). They also enhance the absorption of vital nutrients, such as nitrogen and phosphorus, thereby improving plant nutrition and facilitating the mobilization of nutrients from the soil, enhancing their accessibility to the plant (Vishwakarma et al. 2021). Plant growth hormones like auxins and gibberellins, which promote root and shoot development, are produced by endophytic fungi. This promotes biomass accumulation and results in increased agricultural productivity (Baron and Rigobelo 2022). Endophytic fungi assist plants in adapting to environmental stresses such as drought and salinity. Modification of plant metabolic pathways enhances water retention and cellular stability (Bakhshi et al. 2023). These fungi generate secondary metabolites such as phytoalexins, which serve as antimicrobial agents, safeguarding the plant against pathogens and herbivores (Tiku 2020). Certain endophytes promote systemic resistance, enhancing the immune defense mechanisms of the plant. By altering the activity of antioxidant enzymes, endophytes improve plant life in the face of severe heat, heavy metal toxicity, and water scarcity (Kamran et al. 2022). They significantly enhance the overall resilience of plants in various environmental conditions. Through processes including nutrient exchange, growth hormone synthesis, and stress tolerance, endophytic fungi significantly improve the health and productivity of plants. Symbiotic interactions provide natural solutions for sustainable agriculture, disease resistance, and climate change adaptation in plants (Verma et al. 2021). Figure 3 depicts the symbiotic relationships between endophytic fungi and host plants, highlighting the various beneficial functions these fungi serve in promoting plant health (Akram et al. 2023).

3 Diversity and Distribution of Endophytic Fungi in Common Yew

3.1 Factors Influencing Fungal Diversity (Geographical Location, Environmental Conditions, Host Age, etc.)

The diversity of endophytic fungi is shaped by multiple factors, such as geographical location, environmental conditions, host characteristics, and seasonal variations. The interactions among these elements are complex, influencing the community structure and composition of endophytic fungi in various ecosystems (Qin et al. 2024). This study identifies the key factors that affect the diversity of endophytic fungi. The diversity of endophytic fungi is notably affected by geographical location. Research conducted on *Hippophae rhamnoides* in Northwest China demonstrated the

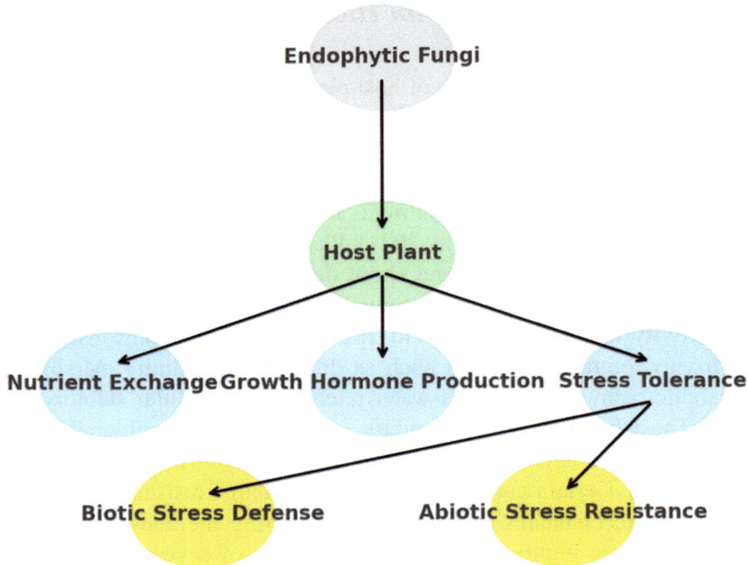

Fig. 3 Endophytic fungi's symbiotic interactions with host plants

presence of distinct fungal communities linked to particular local soil conditions. Geographic isolation in Eastern Asia and Eastern North America significantly affects endophyte diversity, with host identity and local environmental conditions exerting greater influence than historical biogeographic factors (Zhou et al. 2023). Environmental conditions, particularly climate and seasonality, are essential factors influencing endophytic diversity. Studies show that endophyte richness has a negative correlation with temperature seasonality, indicating that more stable climates foster greater diversity (Oita et al. 2021). Soil nutrients, including available phosphorus and nitrogen, exhibit a significant correlation with the composition of root-associated fungi, underscoring the importance of edaphic factors (Guo et al. 2024). Host characteristics, including the age and identity of the host plant, influence endophytic diversity. Older plants may host distinct fungal communities relative to younger plants, as demonstrated in research on *Huperzia serrata* (Shen et al. 2022). The host's specific tissues (roots, stems, and leaves) demonstrate differing levels of endophytic diversity, which are additionally affected by seasonal variations (Yang et al. 2022). Conversely, although these factors are important, some research indicates that phylogenetic divergence may have a lesser impact than previously assumed, suggesting that local environmental conditions and host identity play a more crucial role in shaping endophytic communities (Zhou et al. 2023).

3.2 *Different taxonomic groups of endophytic fungi found in T. baccata*

With several genera and species found in diverse investigations, the taxonomic richness of endophytic fungi linked to *Taxus baccata* is noteworthy. The fungi are essential for the ecology and potential pharmacological uses of *T. baccata*, especially in the production of taxol. This document presents the key findings concerning the taxonomic groups of endophytic fungi identified in *T. baccata*.

Percentages of the isolates of endophytic fungi identified in different taxonomic groups

Ascomycota	*Sterile fungi/Mycelia sterilia*	*Zygomycota*
The phylum accounted for 32% of the identified isolates, with significant genera such as *Cladosporium* and *Alternaria*, which were prevalent in multiple studies (Ashkezari and Fotouhifar 2017; Tayung and Jha 2010)	A significant 67.2% of the isolates were categorized as sterile mycelia, pointing to a considerable degree of undiscovered fungal variety (Ashkezari and Fotouhifar 2017)	Minimally represented at 0.8%, underscoring the dominance of Ascomycota and mycelia sterilia in *T. baccata* (Ashkezari and Fotouhifar 2017)

The diversity of endophytic fungi in *T. baccata* is considerable; nonetheless, many isolates are uncharacterized, indicating significant potential for the discovery of new species and their corresponding bioactive substances. This highlights the necessity for additional investigation into the ecological functions and uses of these fungi. *Fusarium* that often isolated and recognized for its antibacterial activities (Tayung and Jha 2010), *Gliocladium* that recognized as a taxol-producing endophyte, demonstrating potential for medicinal uses (Sreekanth et al. 2011), and *Fusarium redolens* that allegedly produces the highest taxol yield among isolated isolates (Garyali et al. 2014) are the most notable genera and species, respectively.

3.3 *Richness and Community Composition in Different Plant Tissues (Leaves, Bark, Roots, Seeds, etc.)*

The diversity and community structure of endophytic fungi exhibit considerable variation among distinct plant tissues, shaped by host species, climatic conditions, and seasonal fluctuations. Studies demonstrate that endophytic fungi exhibit greater diversity in roots and leaves compared with stems, with unique community structures identified in each tissue type. The diversity is typically greater in roots and leaves compared with stems, as demonstrated in studies of *Huperzia serrata* and several *Pinus* species (Shen et al. 2022; Rim et al. 2021). In *H. serrata*, 7005 operational taxonomic units (OTUs) were identified, revealing large variations in community composition among roots, stems, and leaves (Shen et al. 2022). Seasonal fluctuations influence endophytic diversity; specifically, elevated diversity was observed in summer for roots, but leaves exhibited greater diversity in winter (He et al. 2023). The identity of host species is essential in influencing endophytic communities, with certain taxa linked to specific plant types (He et al. 2023; Fan et al. 2020). These fungi enhance plant health by producing secondary compounds that increase tolerance to both biotic and abiotic stressors (Jha et al. 2023). Specific genera of endophytic fungi have been associated with enhanced synthesis of important chemicals, including huperzine A in *H. serrata* (Shen et al. 2022). Contrariwise, although endophytic fungi are mostly advantageous, their presence may occasionally result in harmful interactions, underscoring the intricacy of these partnerships within various plant tissues.

4 Isolation and Identification of Endophytic Fungi from *Taxus baccata*

4.1 Sampling Methods (Seasonal and Geographical Variations)

The extraction of endophytic fungi from *Taxus baccata* encompasses numerous critical procedures to guarantee the effective isolation and identification of fungal species. This is a comprehensive account of the employed methodologies:

4.1.1 Collection of Plant Material

Sample Types: Different components of the *Taxus baccata* plant, such as twigs, bark, roots, stems, and needles, are gathered for the extraction of endophytic fungi (Somjaipeng et al. 2015; Sharma et al. 2017; Ashkezari and Fotouhifar 2017).

Seasonal Variation: Sampling is frequently performed over various seasons to encompass the diversity of fungi linked to the plant. Research has been undertaken in summer, rainy, and winter seasons to evaluate the seasonal fluctuation in fungal diversity (Sharma et al. 2017; Tayung and Jha 2010).

4.1.2 Surface Sterilization

The objective of surface sterilization is to eliminate or deactivate microorganisms on the plant surface while preserving interior tissues and the residing endophytes. The plant material is generally subjected to washing with running water, subsequently immersed in ethanol or sodium hypochlorite solutions, and finally rinsed with sterile water to eliminate any remaining sterilizing chemicals (Somjaipeng et al. 2015; Mirjalili et al. 2012).

4.1.3 Isolation of Fungi

Aseptic Techniques: Following sterilization, the plant material is segmented into small pieces and inoculated onto nutrient media under aseptic conditions to facilitate the growth of endophytic fungi (Mattoo and Nonzom 2022).

Media Utilized: Frequently employed media consist of potato dextrose agar (PDA) and malt extract agar (MEA), which facilitate the proliferation of several fungal species (Black 2020).

4.1.4 Identification and Characterization

Morphological and Molecular Identification: Isolated fungi are first identified through their morphological traits. Molecular approaches, including the sequencing of internal transcribed spacer (ITS) regions, are utilized for accurate identification (Maria et al. 2024). The diversity of isolated fungi is evaluated by numerous indices, including Simpson and Shannon diversity indices, to comprehend the richness and evenness of fungal species across diverse samples (Safaie et al. 2024). The sampling of endophytic fungi from *Taxus baccata* entails a systematic methodology comprising the collection of plant material, surface sterilization, isolation in aseptic conditions, and identification by morphological and molecular approaches (Nagal and Panda 2022). This method enables researchers to investigate the varied fungal communities linked to *Taxus baccata* and their possible applications, including taxol synthesis.

4.2 Isolation Techniques (Surface Sterilization, Culture-Dependent vs. Culture-Independent Methods)

Endophytic fungi from *Taxus baccata*, or the English yew, are noteworthy for their potential to synthesize important chemicals such as taxol. The isolation of these fungi employs many procedures, notably emphasizing surface sterilization and both culture-dependent and culture-independent approaches.

Surface Sterilization Surface sterilization is an essential procedure in isolating endophytic fungi to guarantee that only internal bacteria are examined, hence excluding surface contamination. Common sterilants comprise sodium hypochlorite, ethanol, and mercuric chloride, utilized at diverse doses and exposure durations to enhance the elimination of epiphytic microorganisms while maintaining endophytic diversity (Yu et al. 2022). Sodium hypochlorite is commonly utilized; however, its content and exposure duration must be meticulously regulated to prevent impact on endophytes (Greenfield et al. 2015). An innovative technique has been established to augment the throughput of surface sterilization by concurrently processing numerous samples, thereby greatly improving the efficiency of endophyte isolation (Sahu et al. 2022).

Culture-Dependent Methods Culture-dependent procedures entail cultivating fungi on synthetic media for their isolation and identification. This method facilitates the morphological and molecular characterization of fungi, along with the preservation of fungal cultures for subsequent research (Somjaipeng et al. 2015; Sisti et al. 2023). Endophytic fungi from *Taxus baccata* have been isolated using potato dextrose agar, and their capacity to produce taxol has been validated by molecular approaches including PCR and sequencing (Mirjalili et al. 2012). This technique may fail to encompass the complete diversity of endophytic fungi, as numerous species are nonculturable (Prior et al. 2014).

Culture-Independent Methods Culture-independent techniques, including metagenomic analysis, offer a comprehensive insight into the variety and prevalence of endophytic fungi. These procedures entail the extraction of whole DNA from plant tissues and the application of molecular tools such as PCR and sequencing to detect fungi without culturing (Mirjalili et al. 2012). This method can uncover increased species diversity and is especially advantageous for investigating nonculturable fungi that may be overlooked by culture-dependent techniques (Prior et al. 2014). The isolation of endophytic fungi from *Taxus baccata* entails a combination of surface sterilization and both culture-dependent and culture-independent techniques (Mishra et al. 2021). Each technique possesses distinct advantages and limitations, and sometimes, a mix of methods is employed to attain a thorough understanding of the endophytic fungal community.

4.3 Molecular Identification (ITS Sequencing, Metagenomic Approaches, DNA Barcoding)

Molecular identification techniques, including ITS sequencing, metagenomic approaches, and DNA barcoding, are essential for the study of endophytic fungi, offering greater accuracy and comprehensiveness than conventional methods (Fig. 4).

ITS Sequencing
Internal Transcribed Spacer (ITS) regions: ITS sequencing is a prevalent technique for the identification of endophytic fungi. This process entails the amplification and sequencing of the ITS sections of rDNA, which exhibit significant variability and offer species-level resolution (Guo et al. 2000; Mollaei et al. 2019; Guo 2010). This technique has been employed to categorize fungi into distinct genera and species, including *Alternaria, Fusarium*, and *Trichoderma* (Guo 2010).

Phylogenetic Analysis: ITS sequences are compared with established sequences in databases to generate phylogenetic trees, facilitating the precise identification of fungal species (Mollaei et al. 2019).

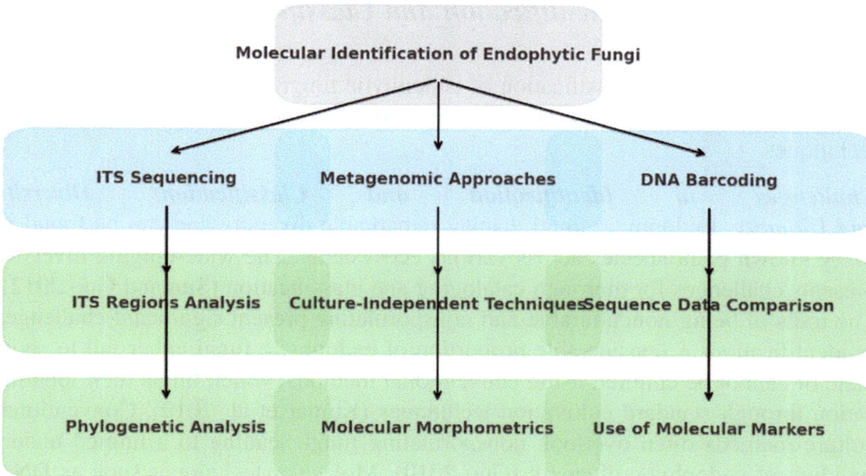

Fig. 4 Schematic representation of the molecular identification techniques used for studying endophytic fungi, highlighting ITS sequencing, metagenomic approaches, and DNA barcoding along with their specific methodologies

Metagenomic Approaches

Culture-Independent Techniques: Metagenomic techniques enable the identification of nonculturable fungi by extracting and analyzing whole DNA from plant tissues without the necessity for culture (Sisti et al. 2023). This strategy enables a more comprehensive understanding of fungal variety and abundance (Sun and Guo 2012; Sisti et al. 2023).

Molecular Morphometrics: This technique uses ITS2 sequence and secondary structure information to improve species delimitation, adding a third dimension to classic barcoding approaches (Guo et al. 2000).

DNA Barcoding

DNA barcoding entails the comparison of sequence data from fungal isolates against reference sequences available in public databases such as GenBank. Nevertheless, caution is warranted because of possible misidentifications within these databases (Ko et al. 2011). Molecular markers, specifically genes associated with biosynthetic pathways, can be utilized to screen for the production of bioactive compounds in endophytic fungi (Challinor et al. 2016). Molecular identification techniques, including ITS sequencing, metagenomic approaches, and DNA barcoding, are crucial for the precise identification and comprehension of endophytic fungi diversity (Rana et al. 2024). These methods address the limitations of conventional morphological identification by offering comprehensive genetic insights, though challenges like database inaccuracies persist (Yang et al. 2024).

4.4 Challenges in Identification and Classification

The identification and classification of endophytic fungi pose numerous challenges owing to their diversity and the constraints of both traditional and contemporary techniques.

Challenges in Identification and Classification: Diversity and Ubiquity Endophytic fungi display remarkable diversity and can be found in every known plant species across various ecosystems. The wide-ranging diversity presents challenges for thorough cataloging and classification (Sun and Guo 2012). The traits of being nonculturable and nonsporulating present significant challenges in identification. A considerable proportion of endophytic fungi either fail to sporulate or cannot be cultured using conventional methods, which limits their identification through standard cultivation techniques (Kumar et al. 2019). Conventional culture methods often overlook nonsporulating fungi, leading to a limited understanding of endophytic diversity (Guo 2010). Molecular techniques such as DNA sequencing, ITS sequencing, and DNA barcoding have improved identification processes, yet they still face various challenges. ITS sequencing is capable of identifying fungi at the generic level; however, it often falls short of accurately determining specific species within different genera (Doilom 2017). The ecological roles and interactions of endophytic fungi with host plants and other organisms add to the intricacy of their classification. The categorization of endophyte communities is intricate due to their reliance on context, shaped by the identity of the host plant and the geographic setting (Toghueo and Boyom 2019). The classification of many endophytic fungi remains ambiguous because of insufficient morphological and genetic information, which complicates their phylogenetic positioning (Tibpromma et al. 2018). The identification and classification of endophytic fungi encounter obstacles stemming from their extensive diversity, nonculturable traits, and the limitations of both traditional and molecular methodologies. The advancement of molecular methods in this field has been notable; nevertheless, significant challenges remain in achieving accurate species-level classification of these fungi. Understanding the ecological roles and interactions of endophytic fungi is crucial for improving classification systems and exploring their potential applications.

5 Functional Roles and Applications of Endophytic Fungi in *T. Baccata*

5.1 Biocontrol Potential and Plant Protection (Antifungal and Antibacterial Properties)

Endophytic fungi, which inhabit plant tissues without inflicting damage, have emerged as potential biocontrol agents in agriculture. These fungi provide a sustainable substitute for chemical pesticides by augmenting plant defense mechanisms and facilitating growth. This overview examines the biocontrol potential and plant protection capacities of endophytic fungi, specifically regarding *T. baccata*.

Mechanisms of Biological Control Antagonistic Activity Endophytic fungi safeguard host plants via methods including antibiosis, parasitism, and competition. They generate bioactive secondary metabolites that suppress the proliferation of phytopathogens, consequently diminishing plant illnesses (Segaran and Sathiavelu 2019).

Induced Resistance These fungi can activate the plant's immune system, improving its capacity to withstand both biotic and abiotic stressors. This is accomplished by the synthesis of phytohormones and other signaling molecules that regulate plant defense mechanisms (Yan et al. 2019; Fite et al. 2023; Fadiji and Babalola 2020).

Advantages in Plant Protection: Disease Mitigation Endophytic fungi have demonstrated efficacy in suppressing plant diseases by directly decreasing pathogen activity and indirectly developing systemic resistance in the host plant (Latz et al. 2018; Silva et al. 2019).

Pest Management Specific endophytic fungi, including *Trichoderma* and *Beauveria* species, have shown considerable efficacy in diminishing pest populations, such as insects and nematodes, through the production of insecticidal chemicals and the enhancement of plant resilience (Agbessenou et al. 2020).

Obstacles and Prospective Pathways A comprehensive understanding of the intricate interactions among endophytic fungi, host plants, and diseases is essential to enhance their application as biocontrol agents. This entails examining the particular conditions necessary for successful colonization and effectiveness (Fadiji and Babalola 2020). The potential of endophytic fungi is acknowledged; nonetheless, additional study is required to create commercial biocontrol agents. This entails evaluating beneficial strains, comprehending their mechanisms, and verifying their safety and performance across various agricultural contexts (Nawrot-Chorabik et al. 2021). Endophytic fungi offer a promising opportunity for sustainable agriculture, providing environmentally friendly methods for plant protection and pest management (Adeleke et al. 2024). Their capacity to augment plant resilience and diminish dependence on chemical inputs establishes them as essential partners in the pursuit of sustainable crop production.

5.2 Enhancement of Plant Growth and Stress Tolerance

Mechanisms of Enhancement

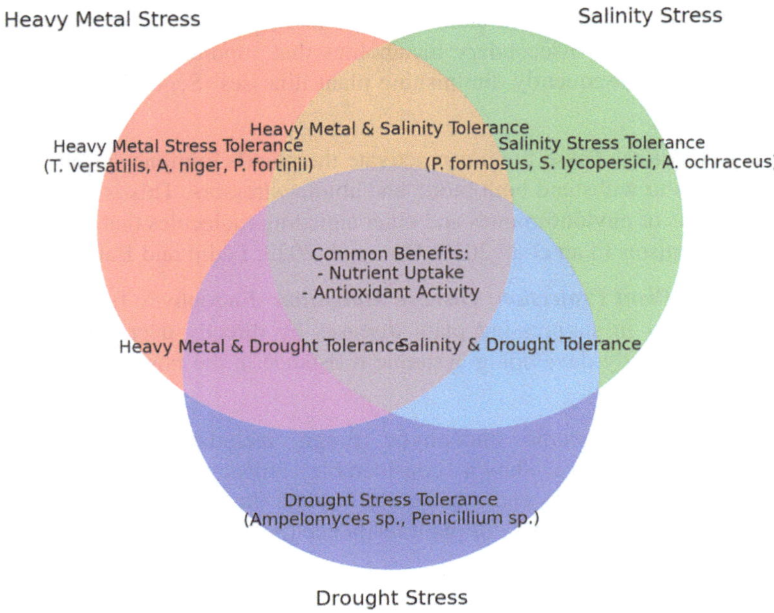

Heavy Metal Stress Tolerance Endophytic fungi, including *Talaromyces versatilis* and *Aspergillus niger*, have demonstrated the ability to promote plant development and diminish cadmium buildup in rose-scented geraniums by enhancing detoxification processes and antioxidant defenses. Likewise, root fungal endophytes such as *Phialocephala fortinii* enhance the growth of *Clethra barbinervis* by enhancing nutrient absorption and diminishing heavy metal levels in the roots.

Salinity Stress Tolerance Fungi such as *Paecilomyces formosus* and *Stemphylium lycopersici* enhance plant growth in saline environments by synthesizing phytohormones, including gibberellins and indole acetic acid, augmenting antioxidant activity, and preserving ionic equilibrium. Aspergillus ochraceus alleviates salt stress in barley by augmenting antioxidant capability and diminishing detrimental ion buildup.

Drought Stress Tolerance Endophytes like *Ampelomyces* sp. and *Penicillium* sp. enhance drought tolerance in tomatoes by improving growth and yield, even in stressful conditions. Endophytic fungi facilitate nutrient absorption, essential for plant growth in stressful environments. They also regulate hormonal balances, enhancing plant resilience.

Antioxidant and Metabolic Enhancements These fungi enhance antioxidant enzyme activity and metabolic processes, aiding plants in managing stress-induced damage.

5.3 Production of Bioactive Secondary Metabolites (E.G., Taxol and Other Anticancer Compounds)

These studies indicate that endophytic fungi are abundant sources of bioactive secondary metabolites exhibiting a range of biological activities, such as antimicrobial, anticancer, antioxidant, and other pharmacological properties, rendering them significant for prospective applications in medicine, agriculture, and biotechnology (Liu et al., 2023a, b). Endophytic fungi are a crucial source of bioactive secondary metabolites, utilized in medicine, agriculture, and biotechnology. These fungi inhabit plant tissues in a symbiotic relationship and generate chemicals that safeguard their host plants against many stressors and diseases (Muhammad et al. 2024a, b).

Essential Bioactive Constituents Fungi generate many secondary metabolites exhibiting antibacterial and anticancer activities. These encompass substances such as paclitaxel, camptothecin, and podophyllotoxin, which demonstrate efficacy against pathogenic microorganisms and cancer cells (Bano et al. 2016; Ancheeva et al. 2019; Rustamova et al. 2020). Endophytic fungi, especially marine varieties, generate compounds with potent antioxidant and anti-inflammatory properties, making them attractive for dermocosmetic uses (Calado et al. 2021).

Innovative Frameworks Endophytic fungi are recognized for synthesizing structurally distinctive metabolites, including alkaloids, peptides, terpenoids, and polyketides, which frequently have considerable biological activity (Ortega et al. 2021; Liu et al. 2023a, b).

Applications and Potential in Pharmaceuticals Metabolites derived from endophytic fungi are under investigation as prospective therapeutic candidates because of their unique modes of action and selectivity (Ancheeva et al. 2019; Rustamova et al. 2020). Fungi augment plant resistance to diseases and stress, rendering them significant for the development of biocontrol agents and the enhancement of crop resilience (Schulz et al. 2002).

Biotechnology Endophytic fungi are being studied for their capacity to sustainably generate plant-like compounds, providing an alternative to conventional plant extraction techniques (Torres-Mendoza et al. 2020). These studies indicate that host–fungal interactions and the coevolution of endophytic fungi entail intricate symbiotic relationships that vary from mutualistic to antagonistic, with endophytes promoting plant growth, nutrient absorption, stress resilience and affecting plant metabolic processes via secondary metabolites and biotransformation capabilities.

6 Ecological and Evolutionary Significance

6.1 Host–Fungal Interactions and Coevolution

Endophytic fungi constitute a diverse assemblage of fungi residing within plant tissues without inducing visible harm (Chaudhary et al. 2022). These fungi establish symbiotic interactions with their host plants, which may vary from mutualistic to hostile (Koski et al. 2024). The coevolution of endophytic fungus and their host plants has resulted in intricate interactions that affect plant physiology, ecology, and evolution (Magotra et al. 2024).

Evolutionary Origins and Diversity Endophytic fungi have independently evolved many times throughout the fungal kingdom, adapting to diverse ecological and host-specific constraints. This lifestyle is not ancestral to all fungi; nonetheless, its prevalence indicates that it has independently emerged in response to environmental stimuli (Nizamani et al. 2024). The diversity of endophytic fungi is extensive, with various species inhabiting numerous plant hosts, indicative of adaptations to particular host chemistries and environmental conditions (U'Ren et al. 2012).

Host–Fungal Interactions Endophytic fungi engage with their host plants along a spectrum of relationships, ranging from mutualism to hostility. The host's secondary chemistry influences these interactions, impacting the composition and assembly of endophytic communities (Christian et al. 2020). Fungi can influence host physiology, improving nutrient absorption, stress resilience, and disease resistance, therefore augmenting plant fitness (Sarkar et al. 2021).

Coevolution and Adaptation The coevolution of endophytic fungus and their host plants has resulted in specialized adaptations. Brassicaceae plants, lacking mycorrhizal relationships, have established distinctive interactions with root-endophytic fungi that facilitate phosphate transfer and enhance development. These interactions are governed by distinct secondary metabolite pathways that have developed in response to the host's characteristics. The capacity of endophytes to adjust to host-specific characteristics and environmental factors underscores their evolutionary plasticity (Schulz and Boyle 2005).

Ecological and Industrial Implications Endophytic fungi significantly influence plant ecology by affecting population dynamics, community diversity, and ecosystem services. They possess industrial potential owing to their capacity to generate bioactive secondary metabolites and enzymes, applicable in agriculture, medicine, and environmental management (Alam et al. 2021). The biotransformation capabilities of endophytes, evolved through coevolution, influence plant metabolic processes and present opportunities for biotechnological applications (Liu et al. 2021).

6.2 Influence of Environmental Factors on Fungal Communities

These studies indicate that environmental factors, including climate, elevation, soil type, and precipitation, substantially affect the composition and diversity of endophytic fungal communities, alongside influences from host species, plant organs, and geographic location (Moghaddam et al. 2021). The season can significantly influence community composition. Additional factors encompass plant organs, precipitation, and temperature, affecting endophytic composition. Conversely, canopy orientation induced minor alterations in fungal community composition, exhibiting unique impacts between the spring and autumn seasons (Gomes et al. 2018). Numerous studies have documented the influence of host, geographic, and environmental factors on endophyte communities; nevertheless, the direct impact of fungal growth rate on endophyte composition remains unexamined (Huang 2020).

6.3 Adaptations of Endophytes to Yew Tree Habitats

The adaptations of endophytes to yew tree environments, especially in species such as *Taxus wallichiana* and *Taxus cuspidata*, underscore their ecological significance and possible advantages. These endophytes have evolved in diverse ways to flourish in the distinctive habitats of yew trees, enhancing plant health and the synthesis of vital secondary metabolites (Zhang et al. 2024).

Diversity of Endophytes A considerable variety of endophytes is present in yew trees, with research identifying as many as 78 distinct species across 26 taxa (Yu et al. 2022). This diversity encompasses both fungal and bacterial endophytes, which can augment plant development and resilience.

Manufacture of Bioactive Compounds Endophytes in yew trees are recognized for synthesizing bioactive chemicals, including taxol, which is essential for their anticancer efficacy (Fischer and Schüffler 2018) (Schueffler and Anke 2011). Interactions among cohabiting fungi can markedly augment taxol synthesis, suggesting a complex ecology within the yew tree (Soliman and Raizada 2013).

Properties That Promote Growth Numerous endophytes extracted from yew trees demonstrate plant growth-promoting characteristics, including the synthesis of indole acetic acid and ammonia, which can enhance seed development in various plant species (Khan et al. 2022). Although the adaptations of endophytes to yew tree ecosystems are predominantly advantageous, it is crucial to acknowledge that certain endophytes may function as latent infections, potentially affecting the health of the host tree under specific circumstances (Schueffler and Anke 2011).

7 Challenges and Future Perspectives

7.1 *Current Limitations in Studying Endophytic Diversity*

The investigation of endophytic diversity encounters many contemporary constraints that impede thorough comprehension and utilization. These obstacles arise from methodological biases, ecological issues, and deficiencies in understanding host–endophyte interactions. The following are essential elements that demonstrate these constraints.

Methodological Difficulties
Surface Disinfection Challenges: Ineffective techniques may not eradicate epiphytic bacteria, resulting in erroneous evaluations of endophytic diversity (Reis et al. 2022).

Culture Medium Bias: The selection of culture conditions may preferentially support specific species, leading to an inadequate representation of actual diversity (Reis et al. 2022).

Identification Errors: Misidentification of fungal species might distort results and hinder the investigation of their potential uses (Reis et al. 2022).

Ecological and Environmental Variables
The variety of endophytes can be substantially affected by geographical location and environmental factors, confounding comparison analyses (Bamisile et al. 2021).

The Influence of Climate Change: Modernization and climate change jeopardize the existence of endophytic communities, complicating the examination of their biodiversity in natural environments (Packiam and Dhakshinamoorthy 2021).

Deficiencies in Knowledge
Genetic Influences: The impact of host genetic variables on endophyte interactions remains little understood, which is essential for crop enhancement techniques (Zhao et al. 2024).

Functional Diversity: Insufficient understanding of the functions of endophytes in plant health and stress response obstructs their utilization in agriculture (Zhao et al. 2024). Notwithstanding these constraints, continued study and progress in biotechnology may offer solutions to surmount these obstacles, hence improving the comprehension and application of endophytic diversity across multiple domains.

7.2 *Advances in Omics Technologies (Metagenomics, Metabolomics, and Proteomics)*

Progress in omics technologies, such as metagenomics, metabolomics, and proteomics, has markedly improved the comprehension of endophytic fungi and their interactions with host plants. These technologies enable the identification of bioactive chemicals and the clarification of intricate biological processes, resulting in prospective applications in agriculture and medicine.

Metagenomics Next-Generation Sequencing (NGS) This technology facilitates an extensive examination of endophytic microbial communities, uncovering taxonomic diversity and functional capabilities (Ali et al. 2021).

Whole Metagenome Analysis Facilitates the investigation of concealed microbial resources, offering insights into the genetic composition of endophytes and their functions in regulating plant growth (Soumya et al. 2024).

Metabolomics Chemical profiling employs advanced metabolomics systems, including liquid and gas chromatography combined with mass spectrometry, to detect and quantify secondary metabolites generated by endophytic fungi (Zakariyah et al. 2024).

Activation of Dormant Gene Clusters Methods such as cocultivation and epigenetic alterations are utilized to activate inactive biosynthetic gene clusters, hence increasing the synthesis of new metabolites (Zakariyah et al. 2024; Saini et al. 2024a, b).

Proteomics Deciphering Protein Interactions Proteomics aids in elucidating the molecular mechanisms of plant-fungal interactions, demonstrating the impact of endophytes on host plant physiology (Verma et al. 2023; Kaur et al. 2024). Although these advancements offer promising prospects for biotechnological applications, obstacles persist in comprehensively understanding the ecological functions and interactions of endophytic fungi, warranting additional research in this domain.

7.3 Potential for Sustainable Applications in Medicine, Agriculture, and Conservation

The potential of endophytic fungi for sustainable uses encompasses medicine, agriculture, and conservation, demonstrating their diverse advantages. These fungi, inhabiting plant tissues, promote plant development, facilitate nutrient absorption, and increase stress resilience, while also serving as biocontrol agents against diseases and pests (Ebadi and Ebadi 2024; Manathunga et al. 2024). Their capacity to generate bioactive substances renders them significant assets in pharmaceuticals and agriculture, fostering sustainable practices by diminishing dependence on synthetic chemicals (Semenzato and Fani 2024; Gutiérrez-Cárdenas et al. 2024). Moreover, endophytic fungi enhance soil health and biodiversity, which are crucial for sustainable agricultural systems (Rai et al. 2021).

Medical Applications Endophytic fungi serve as a source of bioactive substances possessing medicinal potential. They can be utilized to generate natural metabolites that replicate those derived from therapeutic plants (Semenzato and Fani 2024).

Agricultural Advantages They improve crop resilience to biotic and abiotic stressors, function as biopesticides, diminishing the necessity for chemical pesticides (Gutiérrez-Cárdenas et al. 2024) (Manathunga et al. 2024).

Conservation Initiatives They promote biodiversity by sustaining robust plant–microbe interactions. They assist in the conservation of endangered plant species by sustainable agriculture methods (Semenzato and Fani 2024). Despite the considerable advantages of endophytic fungi, obstacles such as legal constraints and the necessity for additional study on their application methods persist as vital for optimizing their potential in sustainable practices.

7.4 Need for Conservation of Taxus baccata *and Its Associated Microbiome*

1. *Environmental Significance*
   - *Taxus baccata* (European yew) is a pivotal species in forest ecosystems, offering shelter and sustenance for diverse wildlife (Thomas and Polwart 2003).
   - It is essential for preserving biodiversity by sustaining various plant, fungal, and microbial communities (Petr and Paul 2017).
   - It contributes to soil stabilization, erosion prevention, and the preservation of ecological equilibrium in temperate woods (Gugger and Peñalba 2021).
2. *Therapeutic and Economic Importance*
   - The bark and foliage of *Taxus baccata* encompass taxanes, crucial constituents for the synthesis of anticancer pharmaceuticals such as paclitaxel (Croteau 2006).
   - Conservation guarantees the sustainable availability of these essential medications (Cusido et al. 2014).
   - It possesses conventional therapeutic applications, encompassing remedies for respiratory ailments and cardiovascular conditions (Verma et al. 2019).
3. *Perils to Existence*
   - The pharmaceutical industry and timber extraction have caused a reduction in its population (McGraw et al. 2013).
   - Habitat damage resulting from deforestation and land-use alteration endangers natural populations.
   - Climate change and environmental stressors impact its regeneration and long-term viability (Hampe and Petit 2005).
4. *Significance of Its Associated Microbiome*
   - The endophytic and rhizospheric microbiome of *Taxus baccata* enhances plant health, stress resilience, and secondary metabolite synthesis (Zhou et al. 2023).
   - Microbial communities augment nutrient assimilation and confer resistance to diseases (Pirttilä et al. 2021).

- Specific bacteria play a vital role in taxane biosynthesis, rendering them essential for sustainable pharmaceutical manufacture (Caruso et al. 2018).

5. *Conservation Approaches*

- In situ conservation via the safeguarding of natural ecosystems and afforestation initiatives (Kaye and Kuykendall 2001).
- Ex situ conservation through the establishment of seed banks, botanical gardens, and tissue culture methodologies (Fay 2018).
- Advocating for sustainable harvesting methods to guarantee the enduring supply of therapeutic substances (Schippmann et al. 2002).

Investigation of microbiome-assisted conservation to utilize advantageous microorganisms for enhancing plant health and secondary metabolite synthesis (Köberl et al. 2019). Enhancing knowledge and enforcing legal safeguards to deter illicit logging and trafficking (Oldfield 2003). Preserving *Taxus baccata* and its microbiome is crucial for ecological stability, maintaining its medicinal value, protecting biodiversity, and safeguarding resources for future generations.

8 Conclusions

Taxus baccata, more frequently referred to as English yew, is a species of evergreen tree that is indigenous to a wide range of areas, including Ireland, Sweden, Morocco, Algeria, and northern Iran. According to Maleki et al. (2024), this species is of particular concern because of the loss of habitat that has occurred as a result of human activities, including forest management, which has led to a decrease in population. The investigation of *T. baccata* and the endophytic fungi that it harbors has led to the discovery of the significant ecological and medicinal roles that these organisms play. To improve plant health and create bioactive substances like taxol, it is vital to have a comprehensive understanding of the diversity and activity of these endophytes. While the problems that are now being faced in research continue to exist, recent developments in molecular techniques have opened up interesting options for investigating the diversity of endophytes. Not only does this have consequences for the protection of biodiversity, but it also has potential uses in the fields of medicine and agriculture. The continuation of research into these connections is necessary to make the most of the advantages offered by *T. baccata* and to guarantee the protection of its one-of-a-kind microbiome in the face of fluctuations in the environment.

References

Adeleke BS, Babalola OO (2021) Pharmacological potential of fungal endophytes associated with medicinal plants: a review. J Fungi 7(2):147

Adeleke BS, Olowe OM, Ayilara MS, Fasusi OA, Omotayo OP, Fadiji AE et al (2024) Biosynthesis of nanoparticles using microorganisms: a focus on endophytic fungi. Heliyon 10(21)

Adhikari P, Pandey A (2024) Phytochemicals, and antioxidants and antibacterial activities of Taxus wallichiana Zucc. Root associated endophytic fungi. Biologia:1–13

Adhikari P, Joshi K, Pandey A (2023) Taxus associated fungal endophytes: anticancerous to other biological activities. Fungal Biol Rev 45:100308

Agbessenou A, Akutse K, Yusuf A, Ekesi S, Subramanian S, Khamis F (2020) Endophytic fungi protect tomato and nightshade plants against Tuta absoluta (Lepidoptera: Gelechiidae) through a hidden friendship and cryptic battle. Sci Rep 10. https://doi.org/10.1038/s41598-020-78898-8

Akram S, Ahmed A, He P, He P, Liu Y, Wu Y et al (2023) Uniting the role of endophytic fungi against plant pathogens and their interaction. J Fungi 9(1):72

Alam B, Li J, Ge Q, Khan M, Gōng J, Mehmood S, Yuan Y, Gōng W (2021) Endophytic fungi: from symbiosis to secondary metabolite communications or vice versa? Front Plant Sci 12. https://doi.org/10.3389/fpls.2021.791033

Ali M, Ali Q, Sohail MA, Ashraf MF, Saleem MH, Hussain S, Zhou L (2021) Diversity and taxonomic distribution of endophytic bacterial community in the rice plant and its prospective. Int J Mol Sci 22(18):10165

Ancheeva E, Daletos G, Proksch P (2019) Bioactive secondary metabolites from endophytic fungi. Curr Med Chem. https://doi.org/10.2174/0929867326666190916144709

Arens AM, Anaebere TC, Horng H, Olson K (2016) Fatal Taxus baccata ingestion with perimortem serum taxine B quantification. Clin Toxicol 54(9):878–880

Ashkezari S, Fotouhifar K (2017) Diversity of endophytic fungi of common yew (Taxus baccata L.) in Iran. Mycol Prog 16:247–256. https://doi.org/10.1007/s11557-017-1274-4

Bakhshi S, Eshghi S, Banihashemi Z (2023) Application of candidate endophytic fungi isolated from extreme desert adapted trees to mitigate the adverse effects of drought stress on maize (Zea mays L.). Plant Physiol Biochem 202:107961

Bamisile BS, Siddiqui JA, Akutse KS, Ramos Aguila LC, Xu Y (2021) General limitations to endophytic entomopathogenic fungi use as plant growth promoters, pests and pathogens biocontrol agents. Plants 10(10):2119

Bano N, Rizvi I, Sharma N, Siddiqui M, Kalim M, Khan A, Akhtar S (2016). Production of bioactive secondary metabolites from endophytic fungi

Bard NW, Cronk QC, Davies TJ (2024) Fungal endophytes can modulate plant invasion. Biol Rev

Baron NC, Rigobelo EC (2022) Endophytic fungi: a tool for plant growth promotion and sustainable agriculture. Mycology 13(1):39–55

Beck A, Casanova-Katny A, Gerasimova J (2023) Metabarcoding of antarctic lichens from areas with different deglaciation times reveals a high diversity of lichen-associated communities. Genes 14(5):1019

Biradar S, Ramya R, Sankaranarayanan A (2023) Mycobionts interactions in lichen. In: Microbial symbionts. Academic Press, pp 215–233

Black WD (2020) A comparison of several media types and basic techniques used to assess outdoor airborne fungi in Melbourne, Australia. PLoS One 15(12):e0238901

Brüssow H (2024) The antibiotic resistance crisis and the development of new antibiotics. Microb Biotechnol 17(7):e14510

Calado M, Silva J, Alves C, Susano P, Santos D, Alves J, Martins A, Gaspar H, Pedrosa R, Campos M (2021) Marine endophytic fungi associated with Halopteris scoparia (Linnaeus) Sauvageau as producers of bioactive secondary metabolites with potential dermocosmetic application. PLoS One 16. https://doi.org/10.1371/journal.pone.0250954

Caruso R, Magon A, Baroni I, Dellafiore F, Arrigoni C, Pittella F, Ausili D (2018) Health literacy in type 2 diabetes patients: a systematic review of systematic reviews. Acta Diabetol 55(1):1–12

Challinor V, Singhal N, Verma J, Reddy M, Vasundhara M, Kumar A, Reddy S (2016) Molecular approaches to screen bioactive compounds from endophytic fungi. Front Microbiol 7. https://doi.org/10.3389/fmicb.2016.01774

Chaudhary P, Agri U, Chaudhary A, Kumar A, Kumar G (2022) Endophytes and their potential in biotic stress management and crop production. Front Microbiol 13:933017

Christian N, Sedio B, Flórez-Buitrago X, Ramírez-Camejo L, Rojas E, Mejía L, Palmedo S, Rose A, Schroeder J, Herre E (2020) Host affinity of endophytic fungi and the potential for reciprocal interactions involving host secondary chemistry. Am J Bot. https://doi.org/10.1002/ajb2.1436

Croteau A (2006) Détermination de la distribution spatiale et temporelle de la recharge à l'aquifère regional transfrontalier du bassin versant de la rivière Châteauguay, Québec et États-Unis (Doctoral dissertation, Université du Québec, Institut national de la recherche scientifique)

Cusido RM, Onrubia M, Sabater-Jara AB, Moyano E, Bonfill M, Goossens A Palazon J (2014) A rational approach to improving the biotechnological production of taxanes in plant cell cultures of Taxus spp. Biotechnol Adv 32(6):1157–1167

Doilom M (2017) Can ITS sequence data identify fungal endophytes from cultures? A case study from Rhizophora apiculata. Mycosphere 8:1869–1892. https://doi.org/10.5943/MYCOSPHERE/8/10/11

Dos Reis JBA, Lorenzi AS, do Vale HMM (2022) Methods used for the study of endophytic fungi: A review on methodologies and challenges, and associated tips. Arch Microbiol 204(11):675

Drożdżyński P, Rutkowska N, Rodziewicz M, Marchut-Mikołajczyk O (2024) Bioactive compounds produced by endophytic bacteria and their plant hosts – an insight into the world of chosen herbaceous ruderal plants in Central Europe. Molecules 29(18):4456

Ebadi M, Ebadi A (2024) Importance and applications of endophytic fungi. Intechopen

Elhamouly NA, Hewedy OA, Zaitoon A, Miraples A, Elshorbagy OT, Hussien S et al (2022) The hidden power of secondary metabolites in plant-fungi interactions and sustainable phytoremediation. Front Plant Sci 13:1044896

Fadiji A, Babalola O (2020) Elucidating mechanisms of endophytes used in plant protection and other bioactivities with multifunctional prospects. Front Bioeng Biotechnol 8. https://doi.org/10.3389/fbioe.2020.00467

Fan S, Miao L, Li H, Lin A, Song F, Zhang P (2020) Illumina-based analysis yields new insights into the diversity and composition of endophytic fungi in cultivated Huperzia serrata. PLoS One 15(11):e0242258. https://doi.org/10.1371/journal.pone.0242258

Fay MF (2018) Orchid conservation: how can we meet the challenges in the twenty-first century?. Bot Stud 59(1):16

Fischer J, Schüffler A (2018) Bioactive compounds from tree endophytes. Springer, Cham, pp 385–440. https://doi.org/10.1007/978-3-319-89833-9_16

Fite T, Kebede E, Tefera T, Bekeko Z (2023) Endophytic fungi: versatile partners for pest biocontrol, growth promotion, and climate change resilience in plants. Front Sustain Food Syst. https://doi.org/10.3389/fsufs.2023.1322861

Garyali S, Kumar A, Reddy MS (2014) Diversity and antimitotic activity of taxol-producing endophytic fungi isolated from Himalayan yew. Ann Microbiol 64:1413–1422. https://doi.org/10.1007/s13213-013-0786-7

Gauchan DP, Vélëz H, Acharya A, Östman JR, Lundén K, Elfstrand M, Garcia-Gil MR (2021) Annulohypoxylon sp. strain MUS1, an endophytic fungus isolated from Taxus wallichiana Zucc., produces taxol and other bioactive metabolites. 3 Biotech 11:1–16

Gomes T, Pereira JA, Benhadi J et al (2018) Endophytic and epiphytic phyllosphere fungal communities are shaped by different environmental factors in a mediterranean ecosystem. Microb Ecol 76:668–679. https://doi.org/10.1007/s00248-018-1161-9

Greenfield M, Pareja R, Ortiz V, Gómez-Jiménez M, Vega F, Parsa S (2015) A novel method to scale up fungal endophyte isolations. Biocontrol Sci Tech 25:1208–1212. https://doi.org/10.1080/09583157.2015.1033382

Grimm M, Grube M, Schiefelbein U, Zühlke D, Bernhardt J, Riedel K (2021) The lichens' microbiota, still a mystery? Front Microbiol 12:623839

Grube M (2024) Lichens. In: Fungal associations. Springer, Cham, pp 145–179
Guo L (2010) Molecular diversity and identification of endophytic fungi. pp 277–296. https://doi.org/10.1007/978-3-642-05042-8_13
Guo L, Hyde K, Liew E (2000) Identification of endophytic fungi from Livistona chinensis based on morphology and rDNA sequences. New Phytol 147(3):617–630. https://doi.org/10.1046/J.1469-8137.2000.00716.X
Guo S, Ye G, Liu W, Liu R, Liu Z, Ma Y (2024) Environmental factors drive the biogeographic pattern of hippophae rhamnoides root endophytic fungal diversity in the arid regions of Northwest China. J Fungi 10(10):679. https://doi.org/10.3390/jof10100679
Gutiérrez-Cárdenas OG, Cortéz-Madrigal H, Garzón Hidalgo A (2024) Use and impact of endophytic entomopathogenic fungi: their potential in the context of agricultural sustainability. Biotecnia 26:507–556. https://doi.org/10.18633/biotecnia.v26.1986
Hampe A, Petit RJ (2005) Conserving biodiversity under climate change: the rear edge matters. Ecol Lett 8(5):461–467
Hashem AH, Attia MS, Kandil EK, Fawzi MM, Abdelrahman AS, Khader MS et al (2023) Bioactive compounds and biomedical applications of endophytic fungi: a recent review. Microb Cell Factories 22(1):107
He C, Meng D, Li W, Li X, He X (2023) Dynamics of endophytic fungal communities associated with cultivated medicinal plants in farmland ecosystem. J Fungi 9(12):1165. https://doi.org/10.3390/jof9121165
Healy RA, Arnold AE, Bonito G, Huang YL, Lemmond B, Pfister DH, Smith ME (2022) Endophytism and endolichenism in Pezizomycetes: the exception or the rule? New Phytol 233(5)
Huang Y-L (2020) Effect of host, environment and fungal growth on fungal leaf endophyte communities in Taiwan. J Fungi 6(4):244. https://doi.org/10.3390/jof6040244
Hunt AMD (2010) The taxus national collection. University of Bath
Jha P, Kaur T, Chhabra I, Panja A, Paul S, Kumar V, Malik T (2023) Endophytic fungi: hidden treasure chest of antimicrobial metabolites interrelationship of endophytes and metabolites. Front Microbiol 14:1227830
Kamran M, Imran QM, Ahmed MB, Falak N, Khatoon A, Yun BW (2022) Endophyte-mediated stress tolerance in plants: a sustainable strategy to enhance resilience and assist crop improvement. Cells 11(20):3292
Kaur R, Saxena S (2023) Penicillium citrinum, a drought-tolerant endophytic fungus isolated from wheat (Triticum aestivum L.) leaves with plant growth-promoting abilities. Curr Microbiol 80(5):184
Kaur T, Negi R, Sharma B, Kaur S, Khan SS, Kour D, Singh S, Rustagi S, Shreaz S, Yadav N, Kumar M, Kumar A, Yadav AN (2024) Omics technologies for understanding the plant–fungal endophyte interactions: crop improvement for future security. J Appl Biol Biotechnol. https://doi.org/10.7324/jabb.2024.191389
Kaye TN, Kuykendall K (2001) Effects of scarification and cold stratification on seed germination of Lupinus sulphureus ssp. kincaidii. Seed Sci Technol 29(3):663–668
Khan SS, Verma V, Rasool S (2022) Isolation and characterisation of plant growth-promoting bacterial and fungal endophytes from Himalayan Yew (Taxus wallichiana)—an economically imperative plant of Himalayas. Sains Malaysiana 51(9):2789–2802. https://doi.org/10.17576/jsm-2022-5109-05
Köberl M, Erschen S, Etemadi M, White III RA, El-Arabi TF, Berg G (2019) Deciphering the microbiome shift during fermentation of medicinal plants. Sci Rep 9(1):13461
Ko T, Stephenson S, Bahkali A, Hyde K (2011) From morphology to molecular biology: can we use sequence data to identify fungal endophytes? Fungal Divers 50:113–120. https://doi.org/10.1007/s13225-011-0130-0
Koski TM, Zhang B, Wickham JD, Bushley KE, Blanchette RA, Kang L, Sun J (2024) Chemical interactions under the bark: bark-, ambrosia-, and wood-boring beetles and their microbial associates. In: Reviews in environmental science and bio/technology. Springer, pp 1–26

Kumar V, Soni R, Jain L, Dash B, Goel R (2019) Endophytic fungi: recent advances in identification and explorations. In: Advances in endophytic fungal research. https://doi.org/10.1007/978-3-030-03589-1_13

Latz M, Jensen B, Collinge D, Jørgensen H (2018) Endophytic fungi as biocontrol agents: elucidating mechanisms in disease suppression. Plant Ecol Divers 11:555–567. https://doi.org/10.1080/17550874.2018.1534146

Le Roux M, Guéritte F (2017) From the pacific yew (Taxus brevifolia) to the english yew (Taxus baccata): steps towards the discovery of docetaxel (Taxotere®). In: Le Roux M, Guéritte F (eds) Navelbine and taxotere histories of science, pp 151–212

Liu X, Zhou Z, Cui J, Wang M, Wang J (2021) Biotransformation ability of endophytic fungi: from species evolution to industrial applications. Appl Microbiol Biotechnol 105:7095–7113. https://doi.org/10.1007/s00253-021-11554-x

Liu Q, Li L, Chen Y, Wang S, Xue L, Meng W et al (2023a) Diversity of endophytic microbes in Taxus yunnanensis and their potential for plant growth promotion and taxane accumulation. Microorganisms 11(7):1645

Liu P, Tan Y, Yang J, Wang Y, Li Q, Sun B, Xing X, Sun D, Yang S, Ding G (2023b) Bioactive secondary metabolites from endophytic strains of Neocamarosporium betae collected from desert plants. Front Plant Sci 14. https://doi.org/10.3389/fpls.2023.1142212

Lu H, Wei T, Lou H, Shu X, Chen Q (2021) A critical review of communication mechanism within plant-endophytic fungi interactions to cope with biotic and abiotic stresses. J Fungi 7(9):719

Lv J, Yang S, Zhou W, Liu Z, Tan J, Wei M (2024) Microbial regulation of plant secondary metabolites: impact, mechanisms and prospects. Microbiol Res:127688

McGraw AP, Bagley J, Chen WS, Galayda C, Nickerson H, Armani A, Jaffe IZ (2013) Aldosterone increases early atherosclerosis and promotes plaque inflammation through a placental growth factordependent mechanism. J Am Heart Assoc 2(1):e000018

Magotra S, Negi NP, Kumar H (2024) Co-shaping and co-evolution of microbial biodiversity: study for identification of potential plant growth promoting microbes. In: Metabolomics, proteomics and gene editing approaches in biofertilizer industry: volume II, pp 261–281

Maleki K, Chmielarz P, Wawrzyniak MK, Maleki K, Maleki A, Soltani E (2024) Seed regeneration in taxus baccata: unveiling ecological restrictions and paving the way for future studies. Ecol Evol 14(11):e70534

Manathunga KK, Gunasekara NW, Meegahakumbura MK, Ratnaweera P, Faraj TK, Wanasinghe DN (2024) Exploring endophytic fungi as natural antagonists against fungal pathogens of food crops. J Fungi 10. https://doi.org/10.3390/jof10090606

Maria Z, Shamsi S, Hosen S, Bashar MA (2024) Morpho-molecular characterization of endophytic fungi from three traditional medicinal plants. Bangladesh J Bot 53(2):361–372

Mattoo AJ, Nonzom S (2022) Investigating diverse methods for inducing sporulation in endophytic fungi. Stud Fungi 7(1):1–10

Mirjalili M, Farzaneh M, Bonfill M, Rezadoost H, Ghassempour A (2012) Isolation and characterization of Stemphylium sedicola SBU-16 as a new endophytic taxol-producing fungus from Taxus baccata grown in Iran. FEMS Microbiol Lett 328(2):122–129. https://doi.org/10.1111/j.1574-6968.2011.02488.x

Mishra S, Sahu PK, Agarwal V, Singh N (2021) Exploiting endophytic microbes as micro-factories for plant secondary metabolite production. Appl Microbiol Biotechnol 105(18):6579–6596

Mishra S, Srivastava A, Singh A, Pandey GC, Srivastava G (2024) An overview of symbiotic and pathogenic interactions at the fungi-plant interface under environmental constraints. Front Fungal Biol 5:1363460

Moghaddam M, Safaie N, Tedersoo L, Hagh-Doust N (2021) Diversity, community composition, and bioactivity of cultivable fungal endophytes in saline and dry soils in deserts. Fungal Ecol 49: 101019. https://doi.org/10.1016/j.funeco.2020.101019

Mollaei S, Khanehbarndaz O, Gerami-Khashal Z, Ebadi M (2019) Molecular identification and phytochemical screening of endophytic fungi isolated from Lithospermum officinale L. roots: a new source of shikonin. Phytochemistry 168:112116. https://doi.org/10.1016/j.phytochem.2019.112116

Morales-Vargas AT, López-Ramírez V, Álvarez-Mejía C, Vázquez-Martínez J (2024) Endophytic fungi for crops adaptation to abiotic stresses. Microorganisms 12(7):1357

Muhammad M, Basit A, Ali K, Ahmad H, Li WJ, Khan A, Mohamed HI (2024a) A review on endophytic fungi: a potent reservoir of bioactive metabolites with special emphasis on blight disease management. Arch Microbiol 206(3):129

Muhammad M, Wahab A, Waheed A, Mohamed HI, Hakeem KR, Li L, Li WJ (2024b) Harnessing bacterial endophytes for environmental resilience and agricultural sustainability. J Environ Manag 368:122201

Nagal S, Panda BP (2022) Isolation, identification and characterization of β-carotene producing endophytic fungi-Penicillium citrinum from bark of Taxus baccata

Nawrot-Chorabik K, Marcol-Rumak N, Latowski D (2021) Investigation of the biocontrol potential of two ash endophytes against hymenoscyphus fraxineus using in vitro plant–fungus dual cultures. Forests. https://doi.org/10.3390/f12121750

Ningyu L, Yufang X, Junfeng W (2015) Method for planting Taxus baccata in Zhejiang, China. Patent

Nizamani M, Hu Y, Wei G, Wen J, Chang L, Chen Y, Nizamani M (2024) Endophytic fungi: tracing the evolutionary roots and exploring the diversity of plant-fungal symbioses. Curr Res Environ Appl Mycol. https://doi.org/10.5943/cream/14/1/1

Nnaji ND (2025) Anticancer and antimicrobial potential of forest endophytes. In: Forest Fungi. Academic Press, pp 293–310

Ogbe AA, Finnie JF, Van Staden J (2020) The role of endophytes in secondary metabolites accumulation in medicinal plants under abiotic stress. S Afr J Bot 134:126–134

Oita S, Ibáñez A, Lutzoni F et al (2021) Climate and seasonality drive the richness and composition of tropical fungal endophytes at a landscape scale. Commun Biol 4:313. https://doi.org/10.1038/s42003-021-01826-7

Oldfield S (Ed.). (2003) The trade in wildlife: regulation for conservation. Routledge

Ortega HE, Torres-Mendoza D, Caballero Z, Cubilla-Rios L (2021) Structurally uncommon secondary metabolites derived from endophytic fungi. J Fungi 7. https://doi.org/10.3390/jof7070570

Packiam KK, Dhakshinamoorthy M (2021) Endophytic fungi: a review of survival strategies that influence the biodiversity studies associated with biopotential products. In: Fungi bio-prospects in sustainable agriculture, environment and nano-technology, pp 1–15

Pandey P, Tripathi A, Dwivedi S, Lal K, Jhang T (2023) Deciphering the mechanisms, hormonal signaling, and potential applications of endophytic microbes to mediate stress tolerance in medicinal plants. Front Plant Sci 14:1250020

Panno S, Davino S, Caruso AG, Bertacca S, Crnogorac A, Mandić A et al (2021) A review of the most common and economically important diseases that undermine the cultivation of tomato crop in the mediterranean basin. Agronomy 11(11):2188

Pennekamp F, Pontarp M, Tabi A, Altermatt F, Alther R, Choffat Y et al (2018) Biodiversity increases and decreases ecosystem stability. Nature 563(7729):109–112

Pirttilä AM, Mohammad Parast Tabas H, Baruah N, Koskimäki JJ (2021) Biofertilizers and biocontrol agents for agriculture: How to identify and develop new potent microbial strains and traits. Microorganisms 9(4):817

Prior R, Görges K, Yurkov A, Begerow D (2014) New isolation method for endophytes based on enzyme digestion. Mycol Prog 13:849–856. https://doi.org/10.1007/s11557-014-0968-0

Pushpavathi D, Krishnamurthy YL (2024) Study on endolichenic fungal assemblage in Parmotrema and Heterodermia lichens of Shivamoga, Karnataka. Mol Biol Rep 51(1):549

Qin X, Xu J, An X, Yang J, Wang Y, Dou M et al (2024) Insight of endophytic fungi promoting the growth and development of woody plants. Crit Rev Biotechnol 44(1):78–99

Rai N, Keshri PK, Verma A, Kamble SC, Mishra P, Barik S, Singh SK, Gautam V (2021) Plant associated fungal endophytes as a source of natural bioactive compounds. Mycology 12(3): 139–159

Rana S, Kumar M, Maurya DK, Singh SK (2024) Fungal metagenomics: an emerging approach to determine the imperceptible facets of fungal diversity. In: Applied mycology for agriculture and foods. Apple Academic Press, pp 19–36

Rathnayaka AR, Tennakoon DS, Jones GE, Wanasinghe DN, Bhat DJ, Priyashantha AH et al (2024) Significance of precise documentation of hosts and geospatial data of fungal collections, with an emphasis on plant-associated fungi. N Z J Bot:1–28

Rim SO, Roy M, Jeon J, Montecillo JAV, Park S-C, Bae H (2021) Diversity and communities of fungal endophytes from four Pinus species in Korea. Forests 12(3):302. https://doi.org/10.3390/f12030302

Rustamova N, Bozorov K, Efferth T, Yili A (2020) Novel secondary metabolites from endophytic fungi: synthesis and biological properties. Phytochem Rev 19:425–448. https://doi.org/10.1007/s11101-020-09672-x

Safaie N, Salehi M, Felegari M, Farhadi S, Karimzadeh S, Asadi S et al (2024) Culture-based diversity of endophytic fungi of three species of Ferula grown in Iran. Front Microbiol 15: 1363158

Sahu P, Tilgam J, Mishra S, Hamid S, Gupta A, Jayalakshmi K, Verma S, Kharwar R (2022) Surface sterilization for isolation of endophytes: ensuring what (not) to grow. J Basic Microbiol 62:647–668. https://doi.org/10.1002/jobm.202100462

Saini N, Anmol A, Kumar S, Bakshi M, Dhiman Z (2024a) Exploring phenolic compounds as natural stress alleviators in plants-a comprehensive review. Physiol Mol Plant Pathol 133: 102383

Saini P, Verma A, Tiwari H, Mishra V, Gautam V (2024b) Omics-based approaches in studying fungal endophytes and their associated secondary metabolites. Springer, pp 209–227. https://doi.org/10.1007/978-3-031-49112-2_10

Sarkar S, Dey A, Kumar V, Batiha G, El-Esawi M, Tomczyk M, Ray P (2021) Fungal endophyte: an interactive endosymbiont with the capability of modulating host physiology in myriad ways. Front Plant Sci 12. https://doi.org/10.3389/fpls.2021.701800

Sati P, Sharma E, Dhyani P, Attri DC, Rana R, Kiyekbayeva L et al (2024) Paclitaxel and its semi-synthetic derivatives: comprehensive insights into chemical structure, mechanisms of action, and anticancer properties. Eur J Med Res 29(1):90

Schueffler A, Anke T (2011) Antimicrobial compounds from tree endophytes. Springer, Dordrecht, pp 265–294. https://doi.org/10.1007/978-94-007-1599-8_17

Schulz B, Boyle C (2005) The endophytic continuum. Mycol Res 109(Pt 6):661–686. https://doi.org/10.1017/S095375620500273X

Schulz B, Boyle C, Draeger S, Römmert A, Krohn K (2002) Endophytic fungi: a source of novel biologically active secondary metabolites * * paper presented at. Fungal Biol. https://doi.org/10.1017/S0953756202006342

Schippmann U, Leaman DJ, Cunningham AB (2002) Impact of cultivation and gathering of medicinal plants on biodiversity: global trends and issues. Biodiversity and the ecosystem approach in agriculture, forestry and fisheries

Segaran G, Sathiavelu M (2019) Fungal endophytes: A potent biocontrol agent and a bioactive metabolites reservoir. Biocatal Agric Biotechnol. https://doi.org/10.1016/J.BCAB.2019.101284

Semenzato G, Fani R (2024) Endophytic bacteria: a sustainable strategy for enhancing medicinal plant cultivation and preserving microbial diversity. Front Microbiol 15. https://doi.org/10.3389/fmicb.2024.1477465

Sharma A, Sagar A, Joshi M (2017) Diversity of fungi associated with Taxus baccata Linn. In different seasons. Indian Forester 143:380–384

Shen Z, Liu X, Yang J et al (2022) The temporal and spatial endophytic fungal community of Huperzia serrata: diversity and relevance to huperzine A production by the host. BMC Microbiol 22:281. https://doi.org/10.1186/s12866-022-02702-y

Silva N, Brooks S, Lumyong S, Hyde K (2019) Use of endophytes as biocontrol agents. Fungal Biol Rev. https://doi.org/10.1016/J.FBR.2018.10.001

Sisti L, Pena-Passos M, Mayer J (2023) Isolation, characterization, and total DNA extraction to identify endophytic fungi in mycoheterotrophic plants. J Vis Exp 195. https://doi.org/10.3791/65135

Soliman SSM, Raizada MN (2013) Interactions between co-habitating fungi elicit synthesis of taxol from an endophytic fungus in host taxus plants. Front Microbiol 4:3. https://doi.org/10.3389/FMICB.2013.00003

Somjaipeng S, Medina Á, Kwaśna H, Ortiz O, Magan N (2015) Isolation, identification, and ecology of growth and taxol production by an endophytic strain of Paraconiothyrium variabile from English yew trees (Taxus baccata). Fungal Biol 119(11):1022–1031. https://doi.org/10.1016/j.funbio.2015.07.007

Soumya P, Jayachandran K, Radhakrishnan EK (2024) Endophytic microbiome: an insight into the hidden world of microorganisms within plants. In: Plant-microbe interaction and stress management. Singapore, Springer, pp 265–287

Sreekanth D, Sushim GK, Syed A, Khan BM, Ahmad A (2011) Molecular and morphological characterization of a taxol-producing endophytic fungus, Gliocladium sp., from Taxus baccata. Mycobiology 39(3):151–157. https://doi.org/10.5941/MYCO.2011.39.3.151

Sun X, Guo L (2012) Endophytic fungal diversity: review of traditional and molecular techniques. Mycology 3:65–76. https://doi.org/10.1080/21501203.2012.656724

Tang MJ, Lu F, Yang Y, Sun K, Zhu Q, Xu FJ et al (2022) Benefits of endophytic fungus Phomopsis liquidambaris inoculation for improving mineral nutrition, quality, and yield of rice grains under low nitrogen and phosphorus condition. J Plant Growth Regul 41(6):2499–2513

Tayung K, Jha D (2010) Antimicrobial endophytic fungal assemblages inhabiting bark of Taxus baccata L. of Indo-Burma mega biodiversity hotspot. Indian J Microbiol 50:74–81. https://doi.org/10.1007/s12088-010-0056-3

Thomas PA, Polwart A (2003) Biological Flora of the British Isles, no. 229. Taxus baccata L. J Ecol 91(3):489–524

Tibpromma S, Hyde K, Bhat J, Mortimer P, Xu J, Promputtha I, Doilom M, Yang J, Tang A, Karunarathna S (2018) Identification of endophytic fungi from leaves of Pandanaceae based on their morphotypes and DNA sequence data from southern Thailand. MycoKeys:25–67. https://doi.org/10.3897/mycokeys.33.23670

Tiku AR (2020) Antimicrobial compounds (phytoanticipins and phytoalexins) and their role in plant defense. In: Co-evolution of secondary metabolites. Springer, Cham, pp 845–868

Toghueo R, Boyom F (2019) Endophytic fungi from terminalia species: a comprehensive review. J Fungi 5. https://doi.org/10.3390/jof5020043

Torres-Mendoza D, Ortega H, Cubilla-Rios L (2020) Patents on endophytic fungi related to secondary metabolites and biotransformation applications. J Fungi 6. https://doi.org/10.3390/jof6020058

Tripathi M, Agnihotri V (2024) Beyond the bark: endophytic fungal diversity in taxus spp. and their crucial role in medicinally relevant secondary metabolites. In: Ethnopharmacology and OMICS advances in medicinal plants volume 2: revealing the secrets of medicinal plants. Springer, Singapore, pp 117–135

U'Ren J, Lutzoni F, Miadlikowska J, Laetsch A, Arnold A (2012) Host and geographic structure of endophytic and endolichenic fungi at a continental scale. Am J Bot 99(5):898–914. https://doi.org/10.3732/ajb.1100459

Verma S, Lu S, Kenis PJ (2019) Co-electrolysis of CO_2 and glycerol as a pathway to carbon chemicals with improved technoeconomics due to low electricity consumption. Nat Energy 4(6):466–474

Verma SK, Sahu PK, Kumar K, Pal G, Gond SK, Kharwar RN, White JF (2021) Endophyte roles in nutrient acquisition, root system architecture development and oxidative stress tolerance. J Appl Microbiol 131(5):2161–2177

Verma A, Shameem N, Jatav HS, Sathyanarayana E, Parray JA, Poczai P, Sayyed RZ (2022) Fungal endophytes to combat biotic and abiotic stresses for climate-smart and sustainable agriculture. Front Plant Sci 13:953836

Verma V, Srivastava A, Garg SK, Singh V, Arora P (2023) Incorporating omics-based tools into endophytic fungal research. Biotechnol Notes. https://doi.org/10.1016/j.biotno.2023.12.006

Vishwakarma K, Kumar N, Shandilya C, Varma A (2021) Unravelling the role of endophytes in micronutrient uptake and enhanced crop productivity. In: Symbiotic soil microorganisms: biology and applications. Springer, Cham, pp 63–85

Wang Q, Li J, Yang J, Zou Y, Zhao XQ (2022) Diversity of endophytic bacterial and fungal microbiota associated with the medicinal lichen Usnea longissima at high altitudes. Front Microbiol 13:958917

Wethalawe AN, Alwis YV, Udukala DN, Paranagama PA (2021) Antimicrobial compounds isolated from endolichenic fungi: a review. Molecules 26(13):3901

Wijesekara T, Xu B (2023) Health-promoting effects of bioactive compounds from plant endophytic fungi. J Fungi 9(10):997

Yan L, Zhu J, Zhao X, Shi J, Jiang C, Shao D (2019) Beneficial effects of endophytic fungi colonization on plants. Appl Microbiol Biotechnol 103:3327–3340. https://doi.org/10.1007/s00253-019-09713-2

Yang JH, Oh SY, Kim W, Hur JS (2022) Endolichenic fungal community analysis by pure culture isolation and metabarcoding: a case study of Parmotrema tinctorum. Mycobiology 50(1):55–65

Yang B, Yang H, Liang J, Chen J, Wang C, Wang Y et al (2024) A review on the screening methods for the discovery of natural antimicrobial peptides. J Pharm Anal:101046

Yu Y, Chen Z, Xie H, Feng X, Wang Y, Xu P (2022) Overhauling the effect of surface sterilization on analysis of endophytes in tea plants. Front Plant Sci 13. https://doi.org/10.3389/fpls.2022.849658

Zakariyah RF, Ajijolakewu K, Ayodele AJ, Folami-A BI, Samuel EP, Otuoze SO, Abdulrauf LB, Ahmed RN (2024) Progress in endophytic fungi secondary metabolites: biosynthetic gene cluster reactivation and advances in metabolomics. Bull Natl Res Cent 48:1–18. https://doi.org/10.1186/s42269-024-01199-x

Zerroug A, Sadrati N, Demirel R, Bakli S, Harzallah D (2020) Antimicrobial activity of Anthostomelloides leucospermi TBT10, an endophytic fungus isolated from Taxus baccata stems against human pathogens. Curr Res Environ Appl Mycol

Zhang J, Lu J, Zhu Y, Shen X, Zhu B, Qin L (2023) Roles of endophytic fungi in medicinal plant TCM quality development and abiotic stress response. Chin Herb Med 16:204–213

Zhang H, Lin W, Ma R, Zang Y, Hou K, Xu Z et al (2024) Fungal endophytes of Taxus species and regulatory effect of two strains on taxol synthesis. BMC Microbiol 24(1):291

Zhao C, Onyino J, Gao X (2024) Current advances in the functional diversity and mechanisms underlying endophyte–plant interactions. Microorganisms 12(4):779

Zhou W, Shi W, Soltis PS, Soltis DE, Xiang QY (2023) Foliar endophyte diversity in eastern Asian-eastern north American disjunct tree species–influences of host identity, environment, phylogeny, and geographic isolation. Front Plant Sci 14:1274746

Biodiversity of Algicolous Fungi Related with Macroalgae (Seaweeds)

Hawraa F. H. Al-Abedi, Maha Abdel Kareem Taher, Raid D. Thanoon, Israa Ibrahim Khalil, Semaa F. H. Al-Abedi, Nawar Ali Jasim, Manal Muwaffaq Mohammed, and Suhail Gawdat Fadhil

Contents

H. F. H. Al-Abedi (✉) · R. D. Thanoon · S. F. H. Al-Abedi · M. M. Mohammed
Department of Biology, Collage of Education for Pure Sciences, University of Al-Hamdaniya, Nineveh, Iraq
e-mail: hawraafaisal@uohamdaniya.edu.iq

M. A. K. Taher
Laser and Photonics Research Center, University of Al-Hamdaniya, Nineveh, Iraq

I. I. Khalil
Department of Microbiology, College of Veterinary Medicine, University of Mosul, Mosul, Iraq

N. A. Jasim
Department of Pathology & Poultry Diseases, College of Veterinary Medicine, University of Tikrit, Tikrit, Iraq

S. G. Fadhil
Mosul Medical Technical Institute Northern Technical University (NTU), Mosul, Iraq

A. M. Abdel-Azeem, F. Degola (eds.), *Advances in Lichens and Non-Flowering Plants Fungal Endophytes Research*, Fungal Biology,
https://doi.org/10.1007/978-3-032-01506-8_15

1 Introduction

Marine fungi represent a rich source of bioactive minor metabolites with potential applications in medicine, as highlighted by studies such as those by Schulz et al. (2008) and Lavadia et al. (2017). Between 2013 and 2014, researchers discovered a total of 541 new components extracted from marine fungi, as reported by Blunt et al. (2016). These compounds exhibit a diverse range of biological activities, such as anticoagulant, antibacterial, antifungal, anti-inflammatory, antiplatelet, antimalarial, antituberculosis, and antiviral effects. One key group of fungi known for producing minor metabolites is macroalgal fungi, also referred to as algal fungi (AF). Bugni and Ireland's (2004) study revealed that 27% of newly discovered marine compounds originated from these fungi. They are known to produce significant pharmaceuticals and notable metabolites such as mycostatic macrocyclic dilactones, sesquiterpenes, cytotoxic dimeric diketopiperazines and lactones, and free radical scavenging epoxycyclohexenones.

In the Asia-Pacific region, particularly in the Philippines, current research on marine fungi has largely concentrated on nonmacroalgal sources, including mangroves, marine sediments, seagrasses, and seawater (Su et al. 2014; Notarte et al. 2017; Lavadia et al. 2017). Nonetheless, earlier studies have suggested a potential role for marine fungi in causing ice disease in cultivated seaweed (Solis et al. 2010). As the region and its subsurface are underexplored, additional research specifically focusing on Philippine algal fungi would be highly beneficial for the discovery of novel metabolites. The Philippine archipelago forms part of the Indo-Australian Archipelago (IAA), recognized as a biodiversity hotspot. The IAA serves as the largest hub of marine biodiversity and is notably abundant in benthic macroalgae, encompassing around 250 to 300 genera (Lohman et al. 2011; Cowman and Bellwood 2013). Within this context, the Philippines stands out, harboring over 820 species of macroalgae (Lavadia et al. 2017). However, despite the potential abundance of macroalgae hosts in the region, local research remains limited. The diversity of algal fungi in the Philippines is yet to be fully explored, making the country a promising candidate for isolating and examining lesser-utilized but bioactive microorganisms that produce novel metabolites. These groundbreaking metabolites could play a critical role in addressing the pressing issue of antimicrobial resistance posed by clinically significant pathogens. Bacterial cells employ various mechanisms that enable them to resist antibacterial agents, making infections harder to eliminate and treatments less effective. One such mechanism is the formation of biofilms, which render microorganisms embedded within them more resilient to adverse conditions compared with their planktonic counterparts. Biofilms are reinforced by extracellular polymeric substances (EPS), which aid in nutrient acquisition from their surroundings while providing protection against desiccation, ultraviolet radiation, and predation by protozoan grazers, as noted by Flemming and Wingender (2010).

The structural complexity of biofilms adds another layer of difficulty by interfering with the transport of antimicrobial agents. Factors such as limited antibiotic penetration and elevated multidrug efflux pump activity significantly compromise treatment effectiveness, as highlighted by Fux et al. (2005) and Leid (2009). Moreover, biofilms undermine the host immune response, impairing normal leukocyte activities. This includes reducing phagocytosis of bacteria within biofilms and suppressing the effector functions of leukocytes, further discussed by Leid (2009).

A significant contributor to biofilm-associated infections is *Staphylococcus aureus*, a bacterium capable of colonizing human skin and mucosal surfaces, which facilitates its invasion of medical devices, as described by Otto (2004). With the increasing occurrence of biofilm-related infections linked to staphylococci, various strategies have been suggested to address the issue. These include using topical antimicrobial ointments, administering antibiotics, applying bactericidal or bacteriostatic coatings to catheter lumens, and incorporating matrix-degrading enzymes, as explored by Smith (2005) and Chen et al. (2013). However, in severe cases, infection management may necessitate the removal of implants, resulting in greater trauma for the patient and increased treatment costs (Mah and O'Toole 2001). This chapter focuses on describing recent advancements in understanding the interactions between macroalgae and fungi. Specifically, it highlights algicolous fungi and explores the biological diversity associated with this group of fungi.

2 Estimation of Fungal Species Diversity

Fungi constitute a substantial portion of Earth's microbial diversity, as noted by Hibbett et al. (2007) and Ebead (2013). Estimates of total fungal diversity vary widely, ranging from as high as ten million to lowermillion to as low as 611,000 types (Mora et al. 2011). One prominent estimate suggests a total of 1.5 million species, attributed to Hawksworth's work. His estimate relied on the examination of fungal communities associated with plant ecosystems in Western Europe, which were then extrapolated globally using data on plant diversity worldwide. However, in 2001, Hawksworth himself acknowledged that this number was conservative and revised it upward to 2.7 million species (Hawksworth 2001). According to final edition of the *Dictionary of Fungi*, the number of described fungal types currently stands at 98,128 (Bass 2011). Regardless of the discrepancies in these estimates, it is widely accepted that over 94% of fungal species diversity remains uncharacterized (Hawksworth 2001).

3 Variety and Evolutionary Relationships of Fungi Associated with Algae and Sediments in the Pei Marine Habitats

Marine organisms, particularly macroalgae (commonly known as seaweeds), are an essential source of economically significant products, especially chemicals used in the food and pharmaceutical industries. Research by Bugni and Ireland (2004) shows that between 1977 and 1987, algae contributed approximately 35% of newly discovered bioactive compounds, surpassing sponges at 29% and cnidarians at 22%. However, since 1995, research interest in discovering new products of seaweeds has declined, shifting attention toward marine microorganisms (Kelecom 2002). The earliest documented marine fungus was an isolate from marine algae reported by Sutherland in 1916. This fungus, identification as a member of the genus *Dendryphiella* and originally named *Cercospora salina*, was noted for its saprophytic relationship with various seaweeds.

Over the last 60 years, research on marine fungi has predominantly focused on lignicolous species, while fungi associated with algae (algicolous fungi) remain comparatively underexplored (Kohlmeyer and Volkman-Kohlmeyer 2003). An early attempt to catalog marine fungal diversity was made in 1944 by Barghoon and Linder (as cited by Ebead 2013), who reported 25 species living on submerged wood in New England and California. Later, in 1979, Kohlmeyer and Volkmann-Kohlmeyer published Marine Mycology, which documented 209 filamentous fungal species. This number grew to 321 species by 1991 and expanded further to 537 species across 326 genera by 2011. This included 431 species from the phylum Ascomycota in 256 genera, 12 species from Basidiomycota distributed across 9 genera, and 94 anamorphic ascomycetes in 61 genera without specific family assignments (Jones 2011). Additionally, about 20% of these fungi were associated with algae, either as parasites or symbionts (Kohlmeyer and Volkman-Kohlmeyer 2003). By the early 2000s, Hyde and Pointing (2000) reported a total of 444 identified marine fungal species worldwide.

Significant research efforts have centered on elucidating the ecological roles of marine fungi within marine ecosystems. These organisms function as pathogens, decomposers aiding organic matter breakdown, or symbionts within habitats such as algal communities and mangroves (Schulz et al. 2008). A notable model is the mutualistic relationship between the brown alga *Ascophyllum nodosum* and its ascomycete partner *Mycophycias ascophylli*, forming what Garbary and Gautam (1989, as cited by Ebead 2013) described as a "symbiotum" or complex organism. Interactions between marine fungi and algae, whether epiphytic or endophytic, have drawn considerable attention due to their potential in natural product drug discovery (Suryanarayanan et al. 2010). The review conducted in 2004 highlighted marine algae as the second most important source of marine-derived fungi isolated for chemical research (Bugni and Ireland 2004). The study by Rateb and Ebel (2011) observed that algae had become the leading source of marine fungi studied for chemical compounds between 2006 and 2010. One striking model is the marine algicolous fungus *Drechslera dematioidea*, which produces various bioactive components, including ten newly identified sesquiterpenes (Osterhage et al. 2002).

The global distribution of resources significantly influences research advancements in marine microbial natural products. Among the leading seaweed-producing nations, Canada stands out due to the extensive diversity of algae found in its eastern coastal regions. The Maritime Provinces, Nova Scotia (NS), Prince Edward Island (PEI), and New Brunswick (NB), encompass three distinct coastal ecosystems: the southern Gulf of St. Lawrence, the Atlantic Coastal, and the Bay of Fundy. PEI's south shore intertidal zone is particularly rich in biodiversity, shaped by semidiurnal tides with average amplitudes of 1.83 meters (Jones et al. 2012). Seasonal environmental variability is pronounced, with water temperatures fluctuating from 22 to 25 °C in summer to 1.8 °C in winter and salinity levels ranging between 30‰ in winter and 26–27‰ in summer. These dynamic conditions drive seaweed populations to adapt to a range of challenging environmental factors (Osterhage et al. 2002).

Evidence suggests that endophytic fungi can provide vital support to their host plants by enduring extreme environmental stresses, enhancing heat tolerance, moisture retention, and disease resistance. Despite significant efforts devoted to studying fungal diversity in tropical regions, there remains a noticeable gap in research focusing on marine fungi associated with temperate algae (Osterhage et al. 2002).

This study marks the first exploration of fungal diversity linked to algae along the south shore of PEI. It provides key insights into the distribution and prevalence of marine-derived fungi in this underexplored area and serves as an essential baseline for comparisons with other Canadian coastal regions. As PEI represents an untapped reservoir for drug discovery, this investigation also lays a solid foundation for future efforts to build a comprehensive library of fungal extracts. These extracts will be subjected to biological testing to uncover new bioactive natural products, thereby paving the way for advancements in natural drug discovery and innovation (Osterhage et al. 2002).

4 Macroalgae

Macroalgae, widely known as seaweeds, are multicellular, macroscopic autotrophs divided into three main groups based on the coloration of their thallus: Rhodophyta (red algae), Phaeophyta (brown algae), and Chlorophyta (green algae). Much like terrestrial plants, they are key contributors to marine ecosystems, supporting benthic animal communities as primary producers in the aquatic food web. Moreover, macroalgae act as ecological bioindicators of water quality and play a vital role in bioremediation efforts (Lourenço-Lopes et al. 2020). The commercial farming of seaweed involves two main approaches: harvesting naturally occurring wild varieties or cultivating them in onshore and nearshore setups. However, onshore cultivation often competes with other coastal activities, such as aquaculture, recreation, and compatible uses like wind energy generation. Direct harvesting, on the other hand, can lead to the overexploitation of wild macroalgae species. To address these challenges, offshore cultivation in both sheltered and exposed coastal areas has been explored, particularly for applications such as biofuel production and feed supply (Duarte et al. 2020). The success of such farming depends on multiple factors, including selecting appropriate macroalgae species, identifying optimal cultivation sites, considering environmental factors like light, temperature, nutrients, and current flows, as well as the ability to withstand high waves in offshore waters (Kiers et al. 2010). With a global production value of approximately USD 6 billion, seaweed farming has become a cornerstone of the ocean economy. Applications range from utilizing molecules to whole plants in diverse industries, such as food, fuel, feed, fertilizers, pharmaceuticals, nutraceuticals, and cosmetics (Choudhary et al. 2021). Out of the 200 species cultivated for commercial purposes, about 10 are intensely farmed. Among these species, *Saccharina japonica* represents 33% of total production, followed by *Kappaphycus alvarezii* at 17%. Other notable species include *Undaria pinnatifida*, *Porphyra* spp., *Sargassum fusiforme*, *Eucheuma, Gracilaria* spp., *Enteromorpha clathrata*, *Monsotroma nitidum*, and *Caulerpa* spp. (Choudhary et al. 2021). Many macroalgal species act as ecosystem engineers or foundational species, making them integral to restoration and sustainability efforts. These macroalgae play a vital role in marine ecosystems by providing direct or indirect food sources, facilitating attachment, growth, or reproduction for various organisms, offering shelter from predators, and modifying physical conditions to benefit marine invertebrates, vertebrates, seabirds, and mammals. Despite their ecological importance, research on the biodiversity associated with marine macroalgae is significantly less developed compared with studies on terrestrial plant ecosystems (Choudhary et al. 2021).

Macroalgae exhibit a range of morphological and structural traits that are closely linked to habitat colonization and utilization by both infaunal and epifaunal organisms (Marzinelli et al. 2016). However, these traits remain underexplored, particularly regarding their influence on biodiversity dynamics and the potential shifts driven by ocean warming. Investigating these traits at both individual and population levels is essential for advancing scientific knowledge and supporting conservation efforts Christensen-Dalsgaard et al. 2017 and Gibbons and Quijón 2023). This becomes especially critical for macroalgal species (Tano et al. 2016) that are under threat or experiencing persistent decline (Best et al. 2014). Demonstrating their contributions to biodiversity and identifying the traits that sustain high biodiversity levels could strengthen arguments for their protection and restoration (El-Khaled et al. 2022; Norderhaug et al. 2005).

To address these gaps, this minireview focused on examining the role of macroalgae in supporting biodiversity and evaluating specific traits of individual species (Chemello and Milazzo 2002) or populations that influence different dimensions of biodiversity. The assessment encompassed multiple metrics, including species richness, diversity indices, evenness, biomass, organism abundance, and other relevant parameters, as illustrated in Fig. 1 (Tummon Flynn et al. 2019).

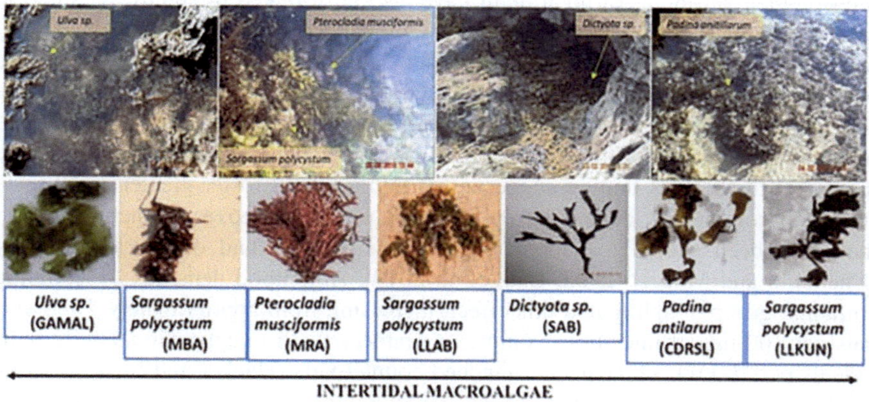

Fig. 1 Photographs of macroalgal samples collected in situ, alongside their respective designation codes (underlined), were obtained from various locations. The codes and their descriptions are as follows: LLKUN (Leaf-like sample from Kunkeshwar), MRA (Malwan Red Algae), MBA (Malwan Brown Algae), SAB (Subtidal algae from Anjuna Beach), GAMAL (Green Algae from Malwan), CDRSL (Cabo-de-Rama Sample), and LLAB (Leaf-like sample from Anjuna Beach). (Kumar et al. 2022)

5 The Effect of Individual Seaweed Features

Several studies have identified varied relationships between algal structural complexity and associated community structures, with some reporting strong positive correlations involving features like holdfast diameters and frond length and negative correlations linked to attributes like the number of stipes (Velasco-Charpntier et al. 2021). However, a few studies found that algal complexity has minimal impact on invertebrate species richness (Dijkstra et al. 2017) or abundance (Lenzo et al. 2023), while others note that these relationships can fluctuate over time. For example, Da Rocha et al. (2006) observed that structurally different macroalgae could host similar assemblages, yet complication still exerts an effect on dominant invertebrate groups. Seasonal changes in macroalgal morphology, driven by life cycle variation (Mancuso et al. 2023), have led researchers to investigate how these dynamics affect associated epifaunal communities. Specific morphological changes—such as variations in fractal dimensions through vegetative cycles—have been linked to shifts in gastropod populations (Mancuso et al. 2023) and altered seasonal peaks in ostracod densities, depending on algal complexity (Hull 1997, cited by Gibbons and Quijón 2023). For instance, ostracod densities peaked in February on the highly complex *Ceramium nodulosum* but peaked in December on the simpler *Chondrus crispus* (Hull, 1997, cited by Gibbons and Quijón 2023). Interestingly, seasonal biomass variation in other seaweeds, such as *Sargassum* spp., showed no correlation with fluctuations in their associated epifaunal assemblages (Gibbons and Quijón 2023). In other studies, predictors such as fractal area combined with biomass emerged as the strongest determinants of epifaunal compositions (Gallardo et al. 2021). The physiological attributes of algal thalli have also been linked to significant shifts in invertebrate biodiversity and density. A comparison by Gallardo et al. (2021) revealed a 20–50% rise in biodiversity for species with harder thalli compared with those with softer ones. Similarly, Taylor and Cole (1994, cited by Gibbons and Quijón 2023) identified an inverse relationship between thallus structure and invertebrate densities, with a nearly 300% increase in density from the widest to the narrowest thalli. This finding suggests that thinner thalli foster greater small-scale structural complexity, enhancing epifaunal densities. Moreover, branching within thalli has been shown to support higher taxa richness and overall biodiversity (Chemello and Milazzo 2002; Lutz et al. 2019). Amphipod densities—a common epifaunal group—highlighted this effect, increasing from approximately 2 individuals per 100 mL of interstitial volume on leathery thalli to approximately 22 individuals per 100 mL on coarsely branched counterparts (Hacker and Steneck 1990, cited by Gibbons and Quijón 2023). Structural complexity is particularly well demonstrated in red coralline algae like rhodoliths, which form porous habitats that enhance physical complexity and support significantly higher biodiversity compared with bare sediments or even certain kelp beds (Neves and Costa 2022; Schoenrock et al. 2018; Cerqueira Veras et al. 2020). Factors such as habitat depth (shallow tropical versus deep polar regions) and rhodolith bed density further contribute to species richness (Cerqueira Veras et al. 2020). However, contrasting studies have identified weaker associations between rhodolith morphology and species assemblages (Harvey and Bird 2008) or emphasized host morphology over habitat form in shaping cryptofaunal assemblages on crustose coralline algae (Glanz 2021). The holdfasts of macroalgae also significantly influence invertebrate diversity. Richness, species count, and diversity levels all showed notable increases with heightened holdfast structural complexity (Hauser et al. 2006; Lutz et al. 2019). For instance, Hauser et al. (2006) documented nearly 170%, 250%, and 260% increases in species richness, diversity, and abundance, respectively, when comparing the simplest to the most complex artificial kelp holdfasts. Similarly, Lutz et al. (2019) found that the invasive green alga (*Codium fragile* spp. *fragile*), noted for its more intricate holdfast structure compared with native congeners (*C. fragile*), supported double the species richness. Nonetheless, some studies have reported exceptions, finding no substantial differences in invertebrate taxa richness (Schmidt and Scheibling 2006).

6 Marine Macro-algicolous Endophytes

Mangrove forests are highly productive coastal ecosystems that support an extensive variety of microorganisms, including bacteria, algae, protozoa, and fungi. Among these, manglicolous fungi play a significant role (Sahoo and Dhal 2009; Lagunes et al. 2023). Marine fungi are categorized into four primary groups: arenicolous fungi, manglicolous fungi, algicolous fungi, and foliicolous fungi. In Goa, 62 species of manglicolous fungi have been documented. The prevalent species include

Halocyphina villosa, *Massarina vellatospora*, *Verruculina enalia*, *Lophiostoma mangroei*, *Lulworthia gradispora*, and *Cirrenalia pygmea*.

Out of eight mangrove species studied, *Rhizophora mucronata* supported the highest number of fungal species with 52 identified, followed closely by *Rhizophora apiculata* with 52 species, *Kandelia candel* with 50 species, *Avicennia marina* with 48 species, *Avicennia officinalis* with 45 species, *Sonneratia alba* with 42 species, *Aegiceras corniculatum* with 28 species, and *Excoecaria agallocha* with 9 species. While approximately 15 mangrove species are known, further studies on the remaining mangrove species could potentially uncover new fungal species (Chinnaraj 1993; Untawale 1999).

The distribution of manglicolous fungi appears to be closely associated with specific mangrove species and their substrates. However, there is still much to explore regarding the influence of hypersaline environments on fungi, particularly those thriving in the mangrove rhizosphere (Sen et al. 2022). Fungi are essential contributors to marine microbial communities, playing a crucial role in nutrient cycling (Tisthammer et al. 2016). They are divided into obligate marine fungi and facultative marine fungi. Obligate marine fungi are specialized to reproduce exclusively in aquatic environments, while facultative marine fungi can grow in both aquatic and terrestrial settings (Raghukumar 2017a, b). When the precise adaptation status of marine fungi is unclear, they are referred to as marine-derived fungi (Raghukumar 2017a, b).

Marine fungi function as saprobes, parasites, or symbionts and can colonize various substrates, such as sponges, corals, mangroves, seagrasses, and algae (Jones et al. 2015). Although fungi associated with algae remain largely understudied, they play vital ecological roles and show potential for industrial applications. For instance, algae-associated fungi have been found to produce numerous bioactive secondary metabolites (Miao et al. 2019).

7 Marine Fungal Diversity: Present Status and Future Perspectives

The diversity of fungi in marine habitats is influenced by the methodologies employed for their study. Traditional techniques, such as processing water and soil samples on artificial media in Petri dishes, typically yield results comparable to those found in terrestrial environments. In contrast, direct examination of decaying plant materials often highlights the presence of litter fungi, predominantly ascomycetes. With advancements in molecular techniques and the identification of common soil fungi in deep-sea environments, it has become evident that earlier definitions of marine fungi were overly restrictive. Consequently, the scope and criteria for defining marine fungi have been expanded. As of 2009, 530 marine fungal species had been documented. However, with the broadened definition, this number rose to 1112 species within 472 genera by 2015. This revised classification now includes marine-derived fungi under the umbrella of marine fungi. Currently, the number of reported marine fungal species has reached 1206. Within this group, the Halosphaeriales order of Ascomycota stands out as the most species-rich. Though taxonomically diverse, marine fungi can be identified as a relatively defined group in physiological or ecological terms. Molecular sequencing studies have revealed a wide range of forms and lineages in marine environments, including chytrids, filamentous hyphal structures, and multicellular forms. From an ecological perspective, marine fungi play important roles as both saprophytes and parasites. Additionally, they are a valuable source of bioactive compounds with therapeutic potential, demonstrating antimicrobial, antioxidant, and anticancer properties, as well as applications in treating various other disease conditions (Sarma 2019).

8 Fungal Endophytes Found in Seaweeds and Their Potential Applications in Agriculture

Seaweeds are macroscopic benthic algae and rank among the most productive organisms in ecosystems. The marine ecosystem hosts the second-largest diversity of seaweed-associated endophytic fungi. These fungal endophytes, a group of microorganisms that inhabit the internal tissues of plants, have recently drawn significant attention from researchers due to their largely untapped diversity (Noorjahan et al. 2021). Endophytic fungi play a crucial role in helping their host plants withstand external biotic and abiotic stresses, contributing to their survival by activating defense mechanisms that trigger the production of secondary metabolites (Noorjahan et al. 2021). While numerous studies have focused on secondary metabolites derived from endophytes associated with terrestrial and aquatic plants, the endophytes of marine macroalgae remain an underexplored group. These fungi hold immense potential as sources of bioactive and chemically novel compounds, with promising applications across medicine, agriculture, and industry (Noorjahan et al. 2021). The growing interest in isolating new beneficial fungal strains from seaweeds is driven by advancements in technologies for developing effective formulations, alongside an increasing global demand for biofertilizers to enhance sustainable agricultural practices. Nevertheless, limited research has been conducted to understand the specific roles of endophytic fungi derived from seaweeds (Noorjahan et al. 2021).

9 Biodiversity of Algicolous Fungi Associated with Macroalgae

Marine habitats, which dominate the planet's surface, remain largely unexplored from a biological perspective. Among the diverse organisms present, fungi have been observed across various marine environments, often in association with macroalgae. Macroalgae, a prominent group in coastal marine ecosystems, are considered the second-largest reservoir of marine fungi after corals (Bugni and Ireland 2004; Vicente et al. 2021). Their role in nutrient cycling is vital for maintaining the stability and functionality of coastal ecosystems (Vicente et al. 2021).

Algicolous fungal communities associated with macroalgae exhibit a range of ecological roles, including parasitic, saprobic, and endophytic behaviors (Raghukumar, 2017a, b). These fungi commonly belong to phyla such as Ascomycota, Basidiomycota, Chytridiomycota, Labyrinthulomycota, and certain fungal-like taxa within Stramenopiles (Jones et al. 2019). Over recent decades, significant strides have been made to uncover the diversity of algicolous fungi; nonetheless, many taxa are identified only at the genus level (Jones et al. 2019). The advent of high-throughput sequencing technologies has further uncovered the vast diversity of fungi in marine habitats, including those associated with macroalgae.

Despite their prevalence in well-lit coastal regions and their apparent richness in fungal diversity, studies on the mycobiota of macroalgae remain scarce. Notable efforts include research by Gonçalves et al. (2020), which documented 46 fungal taxa primarily associated with *Fucus* and *Ulva* species along Portugal's coast; Garzoli et al. (2018), which described 134 taxa on 20 thalli of the brown alga *Padina pavonica*; and Gnavi et al. (2017), which identified 64 fungal taxa in the green alga *Flabellia petiolata* from Italy's coast. The composition of fungal communities within macroalgae is influenced by both biotic and abiotic factors (Raghukumar 2012), including host-derived antifungal compounds, salinity variations, water levels, and air exposure (Sridhar 2017).

Although algicolous fungi remain poorly studied in terms of their biopotential, emerging research highlights their promising metabolic capabilities. Various studies have identified antimicrobial and cytotoxic activities in these fungi, revealing their capacity to produce bioactive natural compounds with potential applications in pharmaceuticals and agriculture (Furbino et al. 2014; Raghukumar 2012). This underscores the need for further exploration of their untapped potential. (Figs. 2 and 3).

A recent study by Vicente et al. (2021) forms part of a comprehensive survey aimed at exploring marine fungal diversity within Portuguese marine environments, encompassing both coastal and estuarine habitats. By employing culture-dependent techniques combined with a polyphasic approach that integrates morphological and molecular studies, the research aimed to address the existing knowledge gap concerning the diversity of algicolous fungi in the Ria de Aveiro estuary. The

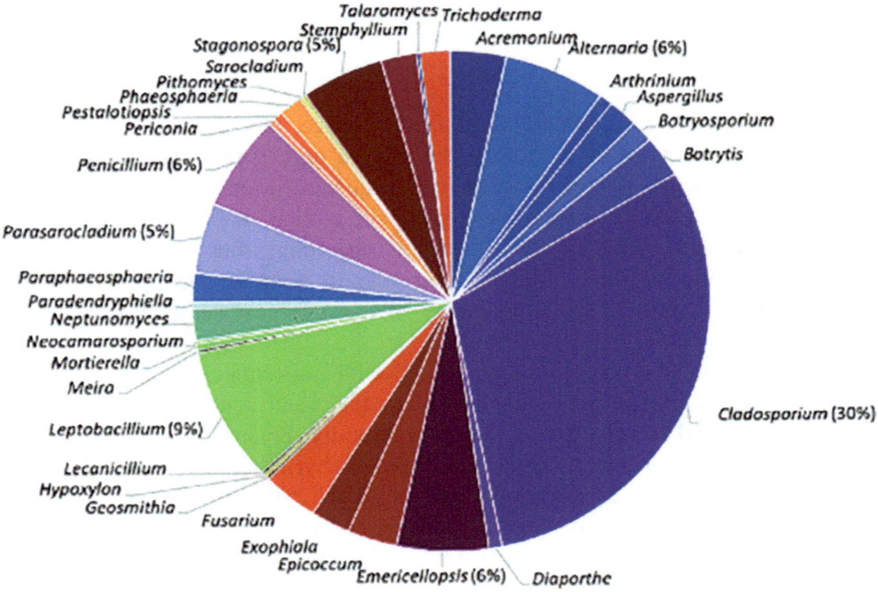

Fig. 2 Overall fungal diversity, at the genus level, from macroalgae samples collected at the Ria de Aveiro estuary. The abundance percentage of each genus above 5% is shown. (Vicente et al. 2021)

Fig. 3 Graphical marine fungi associated with macroalgae. (Patyshakuliyeva et al. 2020)

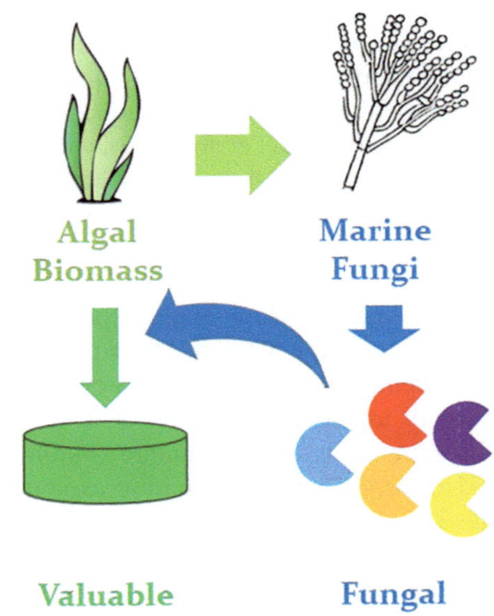

three predominant fungal classes, Dothideomycetes, Sordariomycetes, and Eurotiomycetes, have been previously observed in marine habitats through earlier studies on Mediterranean seaweeds (Gnavi et al. 2017) and Portuguese coastal beach ecosystems (Gonçalves et al. 2019). Meanwhile, the least abundant class identified was Leotiomycetes, represented solely by species of the genus *Botrytis*, which appeared in multiple algae samples. These species were reported earlier in marine environments by Costello et al. (2021), while Gonçalves et al. (2020) later described their isolation from seawater, algae, and driftwood samples. Additionally, several genera were minimally represented in the study, each with only a single isolate or species documented. These include *Geosmithia*, *Hypoxylon*, *Lecanicillium*, *Meira*, *Neocamarosporium*, *Pithomyces*, *Sarocladium*, and *Talaromyces*, as outlined in Vicente et al. (2021). Within the subkingdom Dikarya, two isolates were identified as belonging to the genus *Mortierella*, classified under the phylum Mortierellomycota. Species of *Mortierella* are typically associated with the decomposition of soil organic matter (Hyde et al. 2017).

Nonetheless, certain species have also been documented in marine environments, particularly in association with red and green algae (Ogaki et al. 2019; Costello et al. 2021). The presence of fungi in marine ecosystems was initially reported in the late nineteenth century, making marine mycology a relatively nascent field of study (Shearer et al. 2007). Fungal species in marine habitats can be categorized into three groups: "residents" or "obligate marine" fungi, which are specifically adapted to complete their life cycles within aquatic systems and are not found in nonaquatic environments; "transients," which are introduced into water bodies incidentally, often through wind or water movement; and "facultative" marine species, which are capable of thriving in marine, freshwater, or terrestrial habitats (Shearer et al. 2007; Goecke and Imhoff 2016).

Macroalgae are believed to host the highest diversity of aquatic fungi, closely followed by sponges and mangroves as significant habitats (Rateb and Ebel 2011). Comprehensive overviews of algicolous fungi have been presented by Kohlmeyer and Volkmann-Kohlmeyer (2003) and Zuccaro and Mitchell (2006). Particularly noteworthy are the enduring symbiotic relationships between systemic marine fungi and macroalgae, often termed mycophycobioses, where the algal component predominates. These associations are regarded as early evolutionary stages of lichenization (Goecke and Imhoff 2016). In this regard, the brown macroalgae *Ascophyllum nodosum* and *Pelvetia canaliculata* exhibit significant interactions with their obligate fungal symbiont *Mycophycias ascophylli* (Ascomycetes) (Xu et al. 2008). Studies suggest that *M. ascophylli* may have a protective role, aiding the development of algal zygotes while also preventing desiccation. Although both *A. nodosum* and *M. ascophylli* can be cultivated independently, their growth rates are considerably reduced in isolation. Moreover, *A. nodosum* and *P. canaliculata* are not found in their natural habitats without the presence of this fungal partner (Toxopeus et al. 2011). The ecological importance of enhanced growth rates may lie in their function as an effective escape mechanism against herbivory, as observed by Deckert and Garbary (2005). Among the fungi frequently detected, the Ascomycota are the most prevalent, consistent with results from cultivation-based studies. Other fungal groups, such as Basidiomycetes, Chytridiomycota, and Zygomycota, are also found in algal samples (Gleason et al. 2011), although these groups are rarely identified through molecular analyses. Phylogenetic analysis of *Fucus serratus* along Germany's Atlantic coast has highlighted the presence of four primary ascomycete orders: Halosphaeriales, Hypocreales, Lulworthiales, and Pleosporales (Zuccaro et al. 2003). Within these, species like *Lindra*, *Lulworthia*, *Engyodontium*, the *Sigmoidea/Corollospora* complex, and *Emericellopsis/Acremonium* types are predominant (Zuccaro et al. 2008). While a growing body of research recognizes fungi as significant contributors to marine plant pathology, there is a lack of understanding about the underlying mechanisms of pathogenesis. Most investigations into these fungi have been grounded in taxonomic perspectives, yielding limited insights into host-parasite interactions, pathogenicity, predisposition factors, and epidemiology (Goecke and Imhoff 2016). Moreover, studies on asymptomatic fungal endosymbionts remain sparse (Suryanarayanan 2012). The vast, largely unexplored diversity of fungi, particularly those associated with macrophytes, presents both a significant opportunity and a complex challenge. While considerable insights have been gained regarding the roles fungi play in shaping and sustaining terrestrial ecosystems, such as forests, their contributions to the dynamics of marine ecosystems dominated by macrophytes remain poorly understood. Drawing from the extensive diversity of both fungal species and their hosts, along with established evidence of symbiotic relationships with terrestrial algae (as seen in lichens) and land plants (through mycorrhizal symbiosis), it is likely that numerous intricate and specific associations between fungi and macrophytes await discovery (Wainwright et al. 2019). Fungi derived from algae are known to associate with a wide range of algal groups, including brown algae such as *Agarum clathratum*, *Fucus* spp., *Laminaria* spp., and *Sargassum* spp.; green algae such as *Ulva* spp., *Enteromorpha* spp., and *Flabellia* spp.; as well as red algae such as *Chondrus* spp., *Dilsea* spp., and *Ceramium* spp. Among the most commonly identified fungal groups associated with algae are members of the Ascomycota, encompassing a broad spectrum of genera. Examples include *Aspergillus*, *Alternaria*, *Phoma*, *Acremonium*, *Penicillium*, *Spathulospora*, *Trichoderma*, *Cladosporium*, *Emericellopsis*, *Retrosium*, *Pontogenia*, and *Sigmoidea*. Many genera within this group include terrestrial species capable of producing enzymes that break down plant biomass, converting lignocellulosic material into various high-value products such as biofuels and biochemicals (Mäkelä et al. 2014). For example, enzymes derived from terrestrial species of *Trichoderma* (Peterson and Nevalainen 2012) and *Aspergillus* (De Vries and Visser 2001) have been commercially utilized in various industrial applications. Nevertheless, the production of plant-based biofuels and other high-value products faces several disadvantages, such as competition with agriculture for cropland, limited harvesting cycles (occurring only two to four times a year), and the need for optimal growth conditions that often involve energy-intensive processing techniques (Wei et al. 2013).

In contrast, seaweeds offer distinct advantages for biomass production. They do not require freshwater or arable land and can thrive in saltwater or nutrient-rich wastewater systems (Wei et al. 2013). Seaweeds are fast-growing organisms with high photosynthetic efficiency, resulting in significantly greater biomass yields compared with terrestrial plants (Wei et al. 2013). Additionally, macroalgae have a low lignin content, which makes them ideal for enzymatic downstream processing. This reduces the need for energy-intensive and costly pretreatment methods typically required for plant biomass (John et al. 2011). Remarkably, nearly one-third of all identified marine fungal species are associated with algae (Balabanova et al. 2018), suggesting that their enzymes are likely well suited to breaking down the polymers found in macroalgal cell walls.

Given the distinct roles carried out by various algal structures, such as the holdfast, vesicles, and leaf-like laminae (referred to here as leaves) in supporting and promoting growth, we propose that each structure harbors a unique fungal community. For instance, the grazing-resistant holdfast, as noted by Löffler et al. (2018), does not participate in nutrient absorption but instead serves to anchor the algae and maintain contact with the substrate. In contrast, the leaves, often observed with bite marks (Low 2015), are primarily responsible for photosynthesis, while gas-filled vesicles provide buoyancy. Since fungi play numerous essential roles within their hosts, the differing functions of these structures may drive the selection of distinct fungal communities (Lee et al. 2019).

10 Conclusion

The relationship between seaweeds and the diversity of their associated organisms remains significantly underexplored compared with the extensive research on the ecological roles of terrestrial plants. This gap is worrisome, particularly in light of mounting evidence indicating widespread declines in macroalgae and localized extinctions caused by ocean warming and climate change. Macroalgae play a crucial role in supporting a diverse range of associated organisms, and this review emphasizes their broad ecological significance by distinguishing traits specific to individual seaweed species from those linked to larger populations or habitats. Studies conducted at both individual and population scales consistently reveal positive correlations between morphological attributes, such as structural complexity (e.g., fractal dimensions), and higher biodiversity among invertebrates as well as greater abundance across faunal groups. Nonetheless, these relationships often exhibit complexity and variability, highlighting the dependence of macroalgae's ecological functions, both current and future, on species- and habitat-specific dynamics. As climate change continues to threaten seaweed through displacement or localized losses, identifying these critical traits could become essential in shaping effective conservation and restoration efforts.

References

Balabanova L, Slepchenko L, Son O, Tekutyeva L (2018) Biotechnology potential of marine fungi degrading plant and algae polymeric substrates. Front Microbiol 9:15–27

Bass RF (2011) Stochastic processes, vol 33. Cambridge University Press, Cambridge

Best RJ, Chaudoin AL, Bracken MES, Graham MH, Stachowicz JJ (2014) Plant-animal diversity relationships in a rocky intertidal system depend on invertebrate body size and algal cover. Ecology 95:1308–1322

Blunt JW, Copp BR, Keyzers RA, Munro MH et al (2016) Marine natural products. Nat Prod Rep 33:382–431

Bugni TS, Ireland CM (2004) Marine-derived fungi: a chemically and biologically diverse group of microorganisms. Nat Prod Rep 21:143–163

Cerqueira Veras P, Pierozzi-Jr I, Lino JB, Amado-Filho GM, Senna AR, Santos CSG et al (2020) Drivers of biodiversity associated with rhodolith beds from euphotic and mesophotic zones: insights for management and conservation. Perspect Ecol Conserv 18:37–43

Chemello R, Milazzo M (2002) Effect of algal architecture on associated fauna: some evidence from phytal molluscs. Mar Biol 140:981–990

Chen M, Yu Q, Sun H (2013) Novel strategies for the prevention and treatment of biofilm related infections. Int J Mol Sci 14:18488–18501

Chinnaraj A (1993). Studies on fungi associated of the Indian subcontinent, Ph D. thesis submitted to Goa University, p 147

Choudhary P, Venkata SG, Monika Khade M, Savant S, Musale A, Raja KKG, Chelliah MS, Dasgupta S (2021) Empowering blue economy: from underrated ecosystem to sustainable industry. J Environ Manag 291(1):112697

Christensen-Dalsgaard S, Mattisson J, Bekkby T, Gundersen H, May R, Rinde E et al (2017) Habitat selection of foraging chick-rearing European shags in contrasting marine environments. Mar Biol 164:1–12

Costello MJ, Emblow C, White RJ (2021) European register of marine species a check-list of the marine species in Europe and a bibliography of guides to their identification. Patrimoines Naturels

Cowman PF, Bellwood DR (2013) The historical biogeography of coral reef fishes: global patterns of origination and dispersal. J Biogeogr 40:209–224

Da Rocha CMC, Venekev V, Bezerra TNC, Souza JRB (2006) Phytal marine nematode assemblages and their relation with the macrophytes structural complexity in a Brazilian tropical rocky beach. Hydrobiologia 553:219–230

De Vries RP, Visser J (2001) Aspergillus enzymes involved in degradation of plant cell wall polysaccharides. Microbiol Mol Biol Rev 65:497–522

Deckert RJ, Garbary DJ (2005) *Ascophyllum* and its symbionts. VI. Microscopic characterization of the symbiotum. Algae 20:225–232

Dijkstra JA, Harris LG, Mello K, Litterer A, Wells C, Ware C (2017) Invasive seaweeds transform habitat structure and increase biodiversity of associated species. J Ecol 105:1668–1678

Duarte RCS, Mota ELS, Dias TLP (2020) Algal complexity positively affects the abundance, richness and diversity of molluscan assemblages of a semiarid hypersaline mangrove. Aquat Ecol 54:1001–1013

Ebead GA (2013) Assessment of the diversity of algicolous fungi from prince edward island, new species discovery and antimicrobial screening of selected isolates. A thesis submitted to the graduate faculty in partial fulfillment of the requirements for the Degree of Doctor of Philosophy. Department of Biomedical Sciences Faculty of Veterinary Medicine University of Prince Edward Island, p 9

El-Khaled YC, Daraghmeh N, Tilstra A, Roth F, Huettel M, Rossbach FI et al (2022) Fleshy red algae mats act as temporary reservoirs for sessile invertebrate biodiversity. Commun Biol 5:579

Flemming HC, Wingender J (2010) The biofilm matrix. Nat Rev Microbiol 8:623–633

Furbino LE, Godinho VM, Santiago IF, Pellizari FM, Alves TMA et al (2014) Diversity patterns, ecology and biological activities of fungal communities associated with the endemic macroalgae across the Antarctic Peninsula. Microb Ecol 67:775–787
Fux CA, Costerton JW, Stewart PS, Stoodley P (2005) Survival strategies of infectious biofilms. Trends Microbiol 13:34–40
Gallardo D, Oliva F, Ballesteros M (2021) Marine invertebrate epibionts on photophilic seaweeds: importance of algal architecture. Mar Biodivers 51:16
Garbary DJ and Gautam A (1989) The Ascophyllum, Polysiphonia, Mycosphaerella Symbiosis. I. Population Ecology of Mycosphaerella from Nova Scotia. Botanica Marina 32:181–186
Garzoli L, Poli A, Prigione V, Gnavi G, Varese GC (2018) Peacock's tail with a fungal cocktail: first assessment of the mycobiota associated with the brown alga Padina pavonica. Fungal Ecol 35:87–97
Gibbons EG, Quijón PA (2023) Macroalgal features and their influence on associated biodiversity: implications for conservation and restoration. Front Mar Sci 10:1304000
Glanz JS (2021) Relationship between a branch-forming crustose coralline algae, associated small motile invertebrates, and water flow. California State University, Northridge
Gleason FH, Küpper FC, Amon JP, Picard K, Gachon CMM, Marano AV, Sime-Ngando T, Lilje O (2011) Zoosporic true fungi in marine ecosystems: a review. Mar Freshw Res 62:383–393
Gnavi G, Garzoli L, Poli A, Prigione V, Burgaud G et al (2017) The culturable mycobiota of Flabellia petiolata: first survey of marine fungi associated to a Mediterranean green alga. PLoS One 12:e0175941
Goecke F, Imhoff JF (2016) Microbial biodiversity associated with marine macroalgae and seagrasses. In: Marine macrophytes as foundation species. CRC Press/Taylor & Francis Group Editors: Emil Olausson
Gonçalves MFM, Santos L, Silva BMV, Abreu AC, Vicente TFL et al (2019) Biodiversity of Penicillium species from marine environments in Portugal and description of Penicillium lusitanum sp. nov., a novel species isolated from sea water. Int J Syst Evol Microbiol 69: 3014–3021
Gonçalves MFM, Esteves AC, Alves A (2020) Revealing the hidden diversity of marine fungi in Portugal with the description of two novel species, Neoascochyta fuci sp. nov. and Paraconiothyrium salinum sp. nov. Int J Syst Evol Microbiol 70:5337–5354
Hacker SD, Steneck RS (1990) Habitat architecture and the abundance and body-size dependent habitat selection of a phytal amphipod. Ecology 71:2269–2285
Harvey AS, Bird FL (2008) Community structure of a rhodolith bed from cold-temperature waters (southern Australia). Aust J Bot 56:437–450
Hauser A, Attrill MJ, Cotton PA (2006) Effects of habitat complexity on the diversity and abundance of macrofauna colonising artificial kelp holdfasts. Mar Ecol Prog Ser 325:93–100
Hawksworth DL (2001) The magnitude of fungal diversity: the 1.5 million species estimate revisited. Mycol Res 105(12):1422–1432
Hibbett DS, Binder M, Bischoff JF, Blackwell M, Cannon PF, Eriksson OE, Huhndorf S, James T, Kirk PM, Lücking R, Lumbsch HT (2007) A higher-level phylogenetic classification of the fungi. Mycol Res 111(Pt 5):509–547
Hull SL (1997) Variable effects of a kelp foundation species on rocky intertidal diversity and species interactions in Central California. Mar Ecol Prog Ser 161:71–82
Hyde KD, Pointing SB (2000) Marine mycology: a practical approach. Fungal Divers Res Ser 1:1–377
Hyde KD, Norphanphoun C, Abreu VP, Bazzicalupo A, Thilini Chethana KW et al (2017) Fungal diversity notes 603–708: taxonomic and phylogenetic notes on genera and species. Fungal Divers 87:1–235
John RP, Anisha G, Nampoothiri KM, Pandey A (2011) Micro and macroalgal biomass: a renewable source for bioethanol. Bioresour Technol 102:186–193
Jones E (2011) Fifty years of marine mycology. Fungal Divers 50:73–112

Jones E, Pang K, Stanley S (2012) Fungi from marine algae. In: Jones E, Pang K (eds) Marine fungi and fungal-like organisms. Walter de Gruyter, Berlin/Boston, pp 329–344

Jones EBG, Suetrong S, Sakayaroj J, Bahkali AH, Abdel-Wahab MA, Boekhout T, Pang KL (2015) Classification of marine Ascomycota, Basidiomycota, Blastocladiomycota and Chytridiomycota. Fungal Divers 73:1–72

Jones EBG, Pang K-L, Abdel-Wahab MA, Scholz B, Hyde KD et al (2019) An online resource for marine fungi. Fungal Divers 96:347–433

Kelecom A (2002) Secondary metabolites from marine microorganisms. Acad Bras Cienc 74(1): 151–170

Kiers E, Palmer TM, Ives AR, Bruno JF, Bronstein JL (2010) Mutualisms in a changing world: an evolutionary perspective. Ecol Lett 13:1459–1474

Kohlmeyer J, Volkmann-Kohlmeyer B (2003) Marine ascomycetes from algae and animal hosts. Bot Mar 46:285–306

Kumar P, Verma A, Sundharam SS, Ojha AK, Krishnamurthi S (2022) Exploring diversity and polymer degrading potential of epiphytic bacteria isolated from marine macroalgae. Microorganisms 10(12):2513

Lagunes I, Lumbreras-Martínez H, Espinoza C, Padrón JM, López-Portillo J, Trigos A (2023) Diversity and bioactivity of sediment-associated fungi from a mangrove forest in Mexico with different conservation conditions. Lat Am J Aquat Res 51(3):379–387

Lavadia MGB, Dagama NHA, dela Cruz TEE (2017) Diversity and biofilm inhibition activities of algicolous fungi collected from two remote islands of the Philippine archipelago. Curr Res Environ Appl Mycol 7(4):309–321

Lee S, Park MS, Lee H, Kim JJ, Eimes JA, Lim YW (2019) Fungal diversity and enzyme activity associated with the macroalgae, Agarum clathratum. Mycobiology 47:50–58

Leid J (2009) Bacterial biofilms resist key host defenses. Microbe 4:66–70

Lenzo D, Colangelo MA, Pasteris A, Rindi F, Pistocchi R, Pezzolesi L (2023) Understanding the role of macroalgal complexity and allelochemicals production in invasive and non-invasive macroalgae in the North-Western Adriatic Sea: effect on the associated communities. Water 15: 1697

Loffler Z et al (2018) Holdfasts of Sargassum swartzii are resistant to herbivory and resilient to damage. Coral Reefs 37:1075–1084

Lohman DJ, de Bruy M, Page T, von Rintelen K et al (2011) Biogeography of the Indo–Australian archipelago. Annu Rev Ecol Evol Syst 42:205–226

Lourenço-Lopes C, Fraga-Corral M, Jimenez-Lopez C, Pereira AG, Garcia-Oliveira P, Carpena M, Prieto MA, Simal-Gandara J (2020) Metabolites from macroalgae and its applications in the cosmetic industry: a circular economy approach. Resources 9(9):101

Low JKY (2015) Sargassum on Singapore's reefs. PhD thesis, National University of Singapore. Singapore

Lutz ML, Minchinton TE, Davis AR (2019) Differences in architecture between native and non-indigenous macroalgae influence associations with epifauna. J Exp Mar Biol Ecol 514: 76–86

Mah TC, O'Toole GA (2001) Mechanisms of biofilm resistance to antimicrobial agents. Trends Microbiol 9:34–39

Mäkelä MR, Donofrio N, de Vries RP (2014) Plant biomass degradation by fungi. Fungal Genet Biol 72:2–9

Mancuso FP, Milazzo M, Sarà G, Chemello R (2023) Bi- and threedimensional fractal analysis of the brown seaweed Gongolaria montagnei and their relationship with gastropod molluscs assemblage. Mar Pollut Bull 186:114396

Marzinelli EM, Leong MR, Campbell AH, Steinberg PD, Verges A (2016) Does restoration of a habitat-forming seaweed restore associated faunal diversity? J Soc Ecol Restoration 24:81–90

Miao F, Zuo J, Liu X, Ji N (2019) Algicidal activities of secondary metabolites of marine macroalgal-derived endophytic fungi. J Oceanol Limnol 37:112–121

Mora C, Tittensor DP, Adl S, Simpson AGB, Worm B (2011) How many species are there on Earth and in the ocean? PLoS Biol 9(8):e1001127

Neves SB, Costa KG (2022) Diversity of benthic fauna of rhodoliths and sediments deposited on sandstone reefs in Southeast Brazil. Ocean Coast Res 70:22010

Noorjahan A, Aiyamperumal B, Anantharaman P (2021) Chapter 3: fungal endophytes from seaweeds and bio-potential applications in agriculture. In: Fungi bio-prospects in sustainable agriculture, environment and nano-technology, volume 1: fungal diversity of sustainable agriculture. Elsevier, pp 83–95

Norderhaug KM, Christie H, Fosså JH, Fredriksen S (2005) Fish macrofauna interactions in a kelp (Laminaria hyperborea) forest. J Mar Biol Assoc UK 85:1279–1286

Notarte KI, Nakao Y, Yaguchi T, Bungihan M et al (2017) Trypanocidal activity, cytotoxicity and histone modifications induced by malformin A1 isolated from the marine-derived fungus aspergillus tubingensis IFM 63452. Mycosphere 8:111–120

Ogaki MB, de Paula MT, Ruas D, Pellizzari FM, GarcíaLaviña CX (2019) Marine fungi associated with antarctic macroalgae. In: The ecological role of micro-organisms in the antarctic environment. Springer, Cham, pp 239–255

Osterhage C, König GM, Höller U, Wright AD (2002) Rare sesquiterpenes from the algicolous fungus Drechslera dematioidea. J Nat Prod 65:306–313

Otto M (2004) Quorum-sensing control in staphylococci—a target for antimicrobial drug therapy. FEMS Microbiol Lett 241:135–141

Patyshakuliyeva A, Falkoski DL, Wiebenga A, Timmermans K, de Vries RP (2020) Macroalgae derived fungi have high abilities to degrade algal polymers. Microorganisms 8:52

Peterson R, Nevalainen H (2012) Trichoderma reesei RUT-C30–thirty years of strain improvement. Microbiology 158:58–68

Raghukumar C (2012) Biology of marine fungi, vol 53. Springer

Raghukumar S (2017a) The marine environment and the role of fungi. In: Fungi in coastal and oceanic marine ecosystems: marine fungi. Springer, Cham, pp 17–38

Raghukumar S (2017b) Fungi in coastal and oceanic marine ecosystems. Springer, New York, p 378

Rateb ME, Ebel R (2011) Secondary metabolites of fungi from marine habitats. Nat Prod Rep 28: 290–344

Sahoo K, Dhal NK (2009) Potential microbial diversity in mangrove ecosystems: a review. Indian J Mar Sci 38:249–256

Sarma VV (2019) Marine fungal diversity: present status and future perspectives. In: Microbial diversity in ecosystem sustainability and biotechnological applications. Springer

Schmidt AL, Scheibling RE (2006) A comparison of epifauna and epiphytes on native kelps (Laminaria species) and an invasive alga (Codium fragile spp. tomentosoides) Nova Scotia Canada. Bot Mar 49:1–16

Schoenrock KM, Vad J, Muth A, Pearce DM, Rea BR, Schofield JE et al (2018) Biodiversity of kelp forests and coralline algae habitats in Southwestern Greenland. Diversity 10:1–20

Schulz B, Draeger S, Cruz TE, Rheinheimer J, Siems K, Loesgen S, Bitzer J, Schloerke O, Zeeck A, Kock I (2008) Screening strategies for obtaining novel, biologically active, fungal secondary metabolites from marine habitats. Bot Mar 51:219–234

Sen K, Sen B, Wang G (2022) Diversity, abundance and ecological roles of planktonic fungi in marine environments. J Fungi 8:491

Shearer CA, Descals E, Kohlmeyer B, Kohlmeyer J, Marvanova L, Padgett D, Porter D, Raja HA, Schmit JP, Thorton HA, Voglymayr H (2007) Fungal biodiversity in aquatic habitats. Biodivers Conserv 16:49–67

Smith A (2005) Biofilms and antibiotic therapy: is there a role for combating bacterial resistance by the use of novel drug delivery systems? Adv Drug Deliv Rev 57:1539–1550

Solis MJ, Draeger S, dela Cruz TEE. (2010) Marine-derived fungi from Kappaphycus alvarezii and K. Striatum as potential causative agents of ice-ice disease in farmed seaweeds. Bot Mar 53: 587–594

Sridhar KR (2017) Marine filamentous fungi: diversity, distribution and bioprospecting. Dev Fungal Biol Appl Mycol:59–73

Su GS, Dueñas K, Roderno K, Sison MA et al (2014) Distribution and diversity of marine fungi in Manila Bay, Philippines. Annu Res Rev Biol 4:4166–4173

Suryanarayanan TS (2012) Fungal endosymbionts of seaweeds. In: Raghukumar C (ed) Biology of marine fungi, Progress in molecular and subcellular biology, vol 53. Springer, Berlin, pp 53–69

Suryanarayanan TS, Venkatachalam A, Thirunavukkarasu N, Ravishankar JP, Doble M, Geetha V (2010) Internal mycobiota of marine macroalgae from the Tamilnadu coast: distribution, diversity and biotechnological potential. Bot Mar 53:457–468

Tano S, Eggertsen M, Wikstrom SA, Berkstrom C, Buriyo AS, Halling C (2016) Tropical seaweed beds are important habitats for mobile invertebrate epifauna. Estuar Coast Shelf Sci 183:1–12

Taylor RB, Cole RG (1994) Mobile epifauna on subtidal brown seaweeds in northeastern New Zealand. Mar Ecol Prog Ser 115:271–282

Tisthammer KH, Cobian GM, Amend AS (2016) Global biogeography of marine fungi is shaped by the environment. Fungal Ecol 19:39–46

Toxopeus J, Kozera CJ, O'Leary SJB, Garbary DJ (2011) A reclassifi cation of Mycophycias ascophylli (Ascomycota) based on nuclear large ribosomal subunit DNA sequences. Bot Mar 54:325–334

Tummon Flynn P, Lynn KD, Cairns D, Quijón PA (2019) The role of the non-indigenous green crab (Carcinus maenas) in the decline of a unique strain of Irish moss (Chondrus crispus): direct and indirect effects. ICES J Mar Sci 76:2338–2348

Untawale AG (1999) Status and strategies for marine biodiversity of Goa. National Institute Oceanography, Goa, p 403004

Velasco-Charpentier C, Pizarro-Mora F, Navarro NP, Valdivia N (2021) Disentangling the links between habitat complexity and biodiversity in a kelpdominated subantarctic community. Ecol Evol 11:1214–1224

Vicente TFL, Gonçalves MFM, Brandão C, Fidalgo C, Alves A (2021) Diversity of fungi associated with macroalgae from an estuarine environment and description of Cladosporium rubrum sp. nov. and Hypoxylon aveirense sp. nov. Int J Syst Evol Microbiol 71:004630

Wainwright BJ, Bauman AG, Zahn GL, Todd PA, Huang D (2019) Characterization of fungal biodiversity and communities associated with the reef macroalga Sargassum ilicifolium reveals fungal community differentiation according to geographic locality and algal structure. Mar Biodivers 49:2601–2608

Wei N, Quarterman J, Jin YS (2013) Marine macroalgae: an untapped resource for producing fuels and chemicals. Trends Biotechnol 31:70–77

Xu H, Deckert RJ, Garbary DJ (2008) *Ascophyllum* and its symbionts. X. Ultrastructure of the interaction between *a. nodosum* (Phaeophyceae) and *Mycophycias ascophylli* (Ascomycetes). Botany 86:185–193

Zuccaro A, Mitchell JI (2006) Fungal communities of seaweeds. In: Dighton J, White JF, Oudemans P (eds) The fungal community. CRC/Taylor and Francis, New York, pp 533–579

Zuccaro A, Schulz B, Mitchell JI (2003) Molecular detection of Ascomycetes associated with Fucusserratus. Mycol Res 107:1451–1466

Zuccaro A, Schoch CL, Draeger S, Spatafora JW, Kohlmeyer J, Mitchell JI (2008) Detection and identifi cation of fungi intimately associated with the brown seaweed Fucus serratus. Appl Environ Microbiol 74:931–941

Effect of Host, Environment, and Climate Change on Fungal Endophytic Communities Appealing

Lobna Hajji-Hedfi, Takwa Wannassi, Rym Jaouadi, Okon Godwin Okon, Parwiz Niazi, and Laith Khalil Tawfeeq Al-Ani

Contents

L. Hajji-Hedfi (✉)
Regional Centre of Agricultural Research of Sidi Bouzid, CRRA, Sidi Bouzid, Tunisia

Laboratory of Agriculture Production Systems and Sustainable Development (LR03AGR02), Higher School of Agriculture of Mograne (ESAM), University of Carthage, Zaghouane, Tunisia
e-mail: lobna.hajji@iresa.agrinet.tn

T. Wannassi
Regional Centre of Agricultural Research of Sidi Bouzid, CRRA, Sidi Bouzid, Tunisia

R. Jaouadi
Laboratory of Agriculture Production Systems and Sustainable Development (LR03AGR02), Higher School of Agriculture of Mograne (ESAM), University of Carthage, Zaghouane, Tunisia

O. G. Okon
Department of Botany, Akwa Ibom State University, Akwa Ibom, Nigeria

P. Niazi
Faculty of Education, Department of Biology, Kandahar University, Kandahar, Afghanistan

L. K. T. Al-Ani
School of Biology Science, Universiti Sains Malaysia, Minden, Malaysia

A. M. Abdel-Azeem, F. Degola (eds.), *Advances in Lichens and Non-Flowering Plants Fungal Endophytes Research*, Fungal Biology,
https://doi.org/10.1007/978-3-032-01506-8_16

1 Introduction

Fungal endophytes represent a diverse and ubiquitous group of microorganisms that inhabit the internal tissues of plants without causing visible symptoms of disease (Arnold et al. 2000; Wen et al. 2022). These fungi form symbiotic relationships with their host plants, significantly contributing to the health and fitness of these plants. They enhance growth, improve stress tolerance, and provide protection against pathogens, making them vital components of plant ecosystems (Lodge et al. 1996; Rodriguez et al. 2009). Recent studies have further elucidated the ecological and functional roles of fungal endophytes, emphasizing their importance in mitigating abiotic and biotic stresses, such as drought, salinity, and pathogen attacks (Ripa et al. 2019; Compant et al. 2021).

The composition and structure of fungal endophytic communities are influenced by a variety of factors, with the host plant playing a central role in shaping these communities. Host-related factors, including plant species, tissue type, and physiological characteristics, play a crucial role in determining which fungal taxa colonize and thrive within the plant (Porras-Alfaro and Bayman 2011). Advances in high-throughput sequencing and multiomics approaches have revealed that host plants actively modulate their endophytic communities through the secretion of secondary metabolites and the regulation of immune responses (Hiruma et al. 2016; Carrión et al. 2019).

The influence of the host plant on fungal endophytic communities is not merely a result of ecological interactions but also reflects deeper evolutionary dynamics. Each plant species possesses unique biochemical and physical traits that create distinct microhabitats, which in turn favor the colonization of specific fungal species (Berg et al. 2014). Recent research has highlighted the role of coevolutionary processes in shaping these interactions, with studies demonstrating that plants and their fungal endophytes have developed mutually beneficial adaptations over time (Almario et al. 2017; Selosse et al. 2018). Consequently, the diversity, abundance, and functional potential of fungal endophytic communities can vary significantly between different host plants (Zhang et al. 2016). These variations highlight the intricate and coevolved relationships between plants and their fungal partners, which are increasingly recognized as key drivers of ecosystem functioning (Trivedi et al. 2020).

Understanding the intricate interplay between host identity, environmental factors, and climate change on endophytic communities is crucial for predicting and mitigating the impacts of global change on plant health and ecosystem function. As climate change continues to alter environmental conditions, endophytic communities may undergo significant shifts, potentially impacting plant productivity, disease resistance, and overall ecosystem stability. This chapter emphasizes the diversity and community structure of fungal endophytes, as well as their ecological functions. Additionally, since fungal endophytes interact with their ecosystem, the effects of environment, plant host, and climate change on the communities overview was described.

2 Fungal Endophytic Communities: Diversity, Community Structure, and Ecological Function

2.1 Diversity of Fungal Endophytes (FE)

Without causing any obvious symptoms of disease and processing extensive beneficial functional diversity, endophytic fungi are a diverse group of microorganisms associated with plants in all their structures, including fruits, flowers, stems, roots, and leaves, that colonize the internal tissues of healthy host plants (Wen et al. 2022).

Regarding the diversity of fungal communities, the number of fungi was initially estimated to be approximately 100,000 (Bisby and Ainsworth 1943). However, Martin (1951) later increased that estimate to nearly 250,000, and then, places them at about 1.5 million (Wu et al. 2019). More recent studies estimate the diversity of fungi to range between 2 and 11 million estimated species, yet the number of described taxa stands at 150,600 (Hawksworth and Lücking 2017; Lücking et al. 2021; Baldrian et al. 2021; Phukhamsakda et al. 2022). The number of species estimated from metabarcoding sequence data is the highest, estimated to be 13.2 million species (Wu et al. 2019). Among all identified fungi, endophytic fungi represent a significant portion of fungal diversity. The precise diversity of these endophytes is still unknown, although their identification and categorization have increased significantly due to developments in molecular and sequencing technologies. Only 7% of the 1.5 million fungal species are represented by the fungi that have been described (Hawksworth 2004). As new screening methods are developed, scientists continue to discover new species, providing insight into their ecological functions and potential applications in biotechnology, agriculture, and medicine. The research of EF is therefore becoming increasingly crucial to understanding plant-microbe interactions and leveraging their benefits for sustainable growth (Yeh and Kirschner 2019).

In general, two major groups of endophytic fungi have been previously recognized, reflecting differences in taxonomy, plant hosts, and ecological functions, which are the Clavicipitaceous endophytes and the non-clavicipitaceous endophytes. This classification is based on their host range, ecological roles, and evolutionary relationships (Rodriguez et al. 2009). Data analysis from recent reference sequences obtained from the NCBI database (Madden et al. 1996) and from collected research were used to construct a phylogeny to identify the main groups of endophytic fungi as mentioned earlier as Ascomycota, Basidiomycota, and Mucoromycota.

Figure 1 shows the data collected from 2014 to 2017 mentioning EF with cytotoxicity effects, it revealed that the main phyla from EF are Ascomycota, Basidiomycota, and Mucoromycota, comprising 97%, 2%, and 1%, respectively (Rana et al. 2019; Aamir et al. 2020) (Fig. 1a), which was mainly concentrated in four classes: Sordariomycetes (40%), Dothideomycetes (31%), Eurotiomycetes (24%), Pezizomycetes (1%), and others belonging to Agaricomycetes (2%) (Figs. 2 and 3). Similarly, data from Rana et al. (2019) showed that the main groups of endophyte fungi were Ascomycota (89%), Basidiomycota (9%), Mucoromycota (2%) (Fig. 1b). However, Yadav et al. (2022) reported slightly different proportions and mentioned that endophytes fungi grouped in the main phylum as Ascomycota (96%), Basidiomycota (3%) and Mucoromycota (3%) (Fig. 1c). The difference in statistical variation between studies can be attributed to differences in screening techniques, such as metabarcoding (Li et al. 2019). Globally, based on data collected from published papers between 2014 and 2017, 109 strains of endophytic fungi were isolated from plants; the main identified genera were five, including *Aspergillus*, *Penicillium*, *Pestalotiopsis*, *Fusarium*, *Phomopsis*, and *Alternaria*. Indeed, it was found that *Aspergillus*, *Fusarium*, *Penicillium*, *Talaromyces*, and *Trichoderma* were the dominant genera in wetland sediments along the Yangtze River (Wu et al. 2013).

In other similar studies, the main endophytic fungal communities were revealed to consist of genera such as *Aspergillus*, *Chaetomium*, *Fusarium*, *Gaeumannomyces*, *Metarhizium*, *Microsphaeropsis*, *Paecilomyces*, *Penicillium*, *Piriformospora*, *Talaromyces*, *Trichoderma*, *Verticillium*, and *Xylaria* (Yadav et al. 2022). One of the largest families of Ascomycota, the Xylariaceae, has at least 85 accepted genera and 1343 described species (U'Ren and Arnold 2016).

Data published on the composition of fungal communities in terrestrial environments, such as soil and plant-associated habitats, were collected and validated to study the diversity of EF. This strategy led to the construction of the *GlobalFungi* database, including over 110 million sequences showing variation in the nuclear ribosomal internal transcribed spacers (ITS) and contained in 178 original studies (Větrovský et al. 2020). The approach consisted of collecting samples; then the geographic location of each sample was recorded and released using GPS coordinates. All the detected fungi were isolated, purified, and subjected to amplicon sequencing. High-throughput sequencing (HTS) has enabled integrative microbiome studies, and the identified sequences were assigned to biomes according to the environment ontology (Sun and Guo 2012; Yeh and Kirschner 2019).

The composition of fungal endophytic communities varies across plant hosts, geographic regions, and ecosystems (Arnold 2007). These fungi contribute to plant fitness through:

Enhanced Nutrient Uptake: Arbuscular mycorrhizal fungi (AMF) improve plant phosphorus and nitrogen acquisition (Smith and Read 2008).

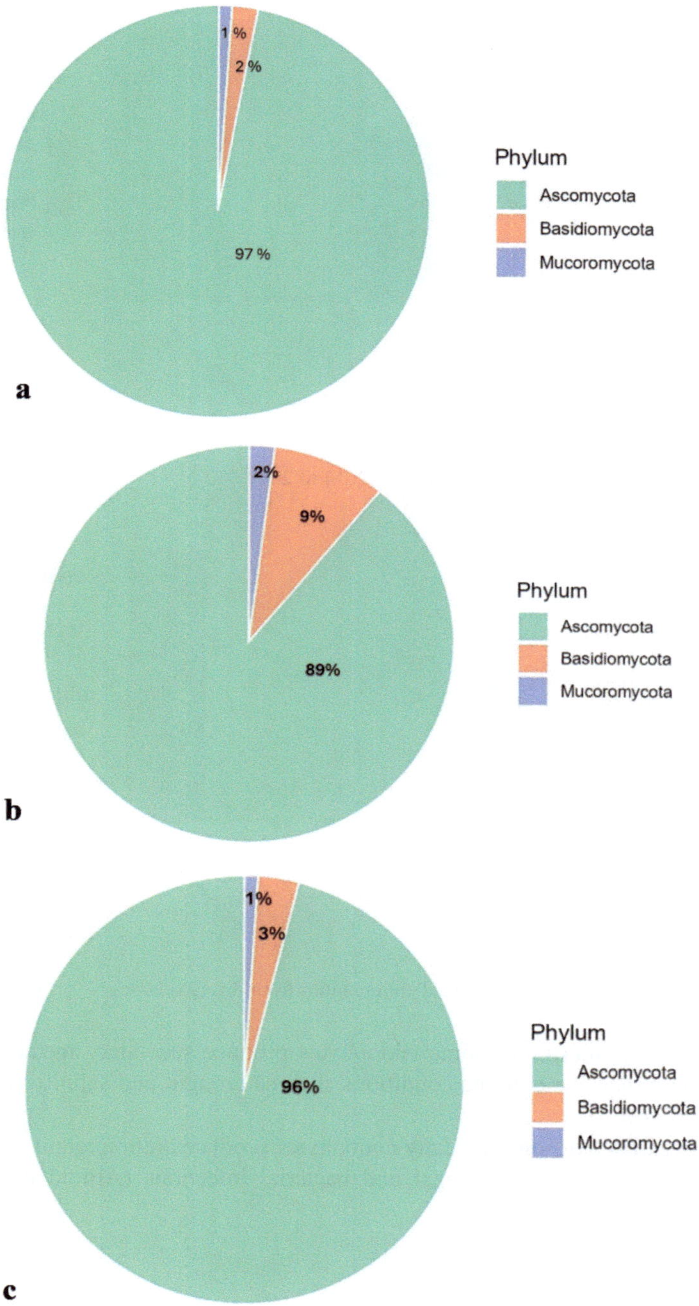

Fig. 1 Main phyla of endophytic fungi isolated from 2014 to 2017. Figure generated using R 4.2.1, with ggplot package. (Wickham 2016)

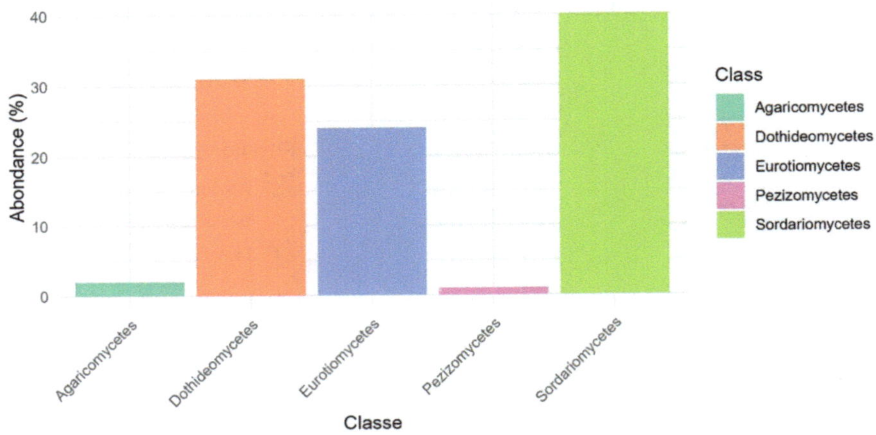

Fig. 2 Main class of Ascomycota phyla from 2014 to 2017

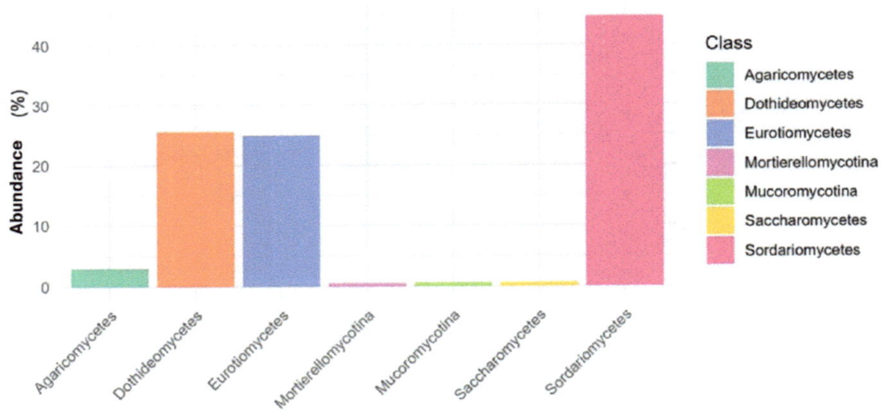

Fig. 3 Abundance of endophytic fungal communities from diverse classes

Abiotic Stress Tolerance: Some endophytes produce secondary metabolites that help plants survive extreme conditions such as drought and salinity (Rodriguez et al. 2009).

Defense Against Pathogens: Many endophytes produce antimicrobial compounds that protect plants from fungal and bacterial infections (Arnold et al. 2000) (Figs. 1, 2 and 3).

2.2 Ecological Functions of Fungal Endophytes

Studying the ecology, evolution, and importance of fungal endophytes is an important prospect given the serious number of fungi investigated capable of forming endophytic associations and their unique characteristics relative to other plant-associated microbes (Arnold 2007). The diverse biological activities of endophytic fungi stem from the secondary metabolites they produce. EF are microorganisms that live inside plant tissues and produce a wide range of secondary metabolites, such as polysaccharides, alkaloids, flavonoids, phenols, and terpenes. These substances are crucial for both the fungi's ecological processes and their interactions with their host plants. Because of their diverse biological activities, the secondary metabolites obtained from endophytic fungi are useful for several purposes. They exhibit antibacterial, anticancer, and antimalarial qualities that have been studied and show promise for use in medical and pharmaceutical research. Furthermore, these metabolites have nematicidal qualities that can be used in agriculture to manage nematode pests and support phytoremediation, which helps detoxify contaminated areas (Duggal et al. 2025).

Furthermore, Fungal Endophytes (EF) that has been isolated from a variety of plant sources has been identified in recent decades as important sources of natural products for agronomy, industry, and biomedical development. These fungi also produce extracellular enzymes that serve as defense mechanisms against pathogenic organisms and are necessary for obtaining nutrients from the host. Bezerra et al. (2012) isolated 44 isolates of fungal endophytes from *Opuntia ficus-indica* and assessed their ability to synthesize hydrolytic enzymes such as cellulases, pectinases, proteases, and xylanases.

2.2.1 Plant Growth Promoting

Isolated EF identified as *Aspergillus niger*, *Aspergillus* sp., *A. versicolor*, *Chaetomium globosum*, *Fusarium fusarioides*, *F. moniliforme*, *F. oxysporum*, *F. semitectum*, *F. solani*, *Mucor hiemalis*, *Mucor* sp., and *Trichoderma pseudokoningii* from the roots of tomato were further screened for different plant growth-promoting attributes. These isolates produced ammonia, four produced HCN, seven produced siderophores, and all showed the ability to solubilize phosphorus. It was discovered that *Fusarium fusarioides* produced the most indole acetic acid (Baron and Rigobelo 2021). As a spray application for enhancing maize growth and yield, the endophyte *Colletotrichum tofieldiae* Ct0861 significantly improved plant growth, and yield increases reached 21% to 22% compared with the control group not treated with the endophyte fungi. This strain of EF, Ct0861, also significantly minimized the prevalence of *Aspergillus* spp. in maize grains. The reduction was manifested by a lower aflatoxin level, demonstrating the biocontrol potential of Ct0861 (Marín et al. 2020). Additionally, EF produce phytohormones such as auxins, gibberellins, and cytokinins, which regulate plant growth and development

(Hassan and Mathesius 2012). By creating secondary metabolites and triggering the plant's defense mechanisms, endophytic fungi also increase a plant's tolerance to biotic and abiotic stressors. For instance, recent research has found that salinity stress has less impact on rice plants when associated with plant growth-promoting endophytic fungi, such as *Aspergillus welwitschiae* Ocstreb1 (AwOcstreb1), *Talaromyces adpressus*, and *Talaromyces argentinensis*, which were identified from the wild halophytic rice *Oryza coarctata* (Airin et al. 2023). The growth enhancement associated with S127 contributes to improved tolerance to abiotic stress. Recent studies revealed that the fungal strain S127 of the genus *Leptobacillium*, functioning as a tomato (*Solanum lycopersicum*) endophyte, promoted tomato growth by enhancing chlorophyll and lycopene content in fruits (Vicedo et al. 2025). In other words, a significant increase in shoot biomass was demonstrated under drought conditions, with *Penicillium chrysogenum*, *Penicillium minioluteum*, *Serendipita indica* increase in biomass in parts of tomato plants under drought conditions when inoculated with these fungal endophytes (González et al. 2025).

Increased nutrient uptake and phytohormone levels in the plant's soil are considered a direct beneficial plant–endophytic fungi interaction. These factors are directly linked to increased biomass production, expanded root system development, plant height, weight, reproduction, and yield. These advantages allow them to contribute to plant growth promotion as part of biofertilizers (Amir et al. 2020). Endophyte fungi that promote plant growth include *Trichoderma* (Tseng et al. 2020), *Piriformospora indica, Colletotrichum tofieldiae* enhance phytohormone production as auxin (Svoboda et al. 2021), *Aspergillus* sp. generates phytohormones and secondary metabolites that promote plant development such as *Aspergillus chevalieri* and *Aspergillus egyptiacus* (Daigham et al. 2024), *Penicillium* sp. stimulates plant development by stimulating the synthesis of phytohormones (gibberellins, auxin, cytokinin, and siderophores) like the example of *Penicillium menonorum* which is detected as a novel endophyte to promote growth and Nutrient management in cucumber plants (Babu et al. 2015).

One species of *Aspergillus* that revealed a promising result regarding plant growth, *Aspergillus flavus* CHS1, an endophytic fungus isolated from *Chenopodium album* roots, was examined by Asaf et al. (2018). The results showed notable improvements in a number of growth metrics, such as higher biomass production, improved root and shoot elongation, and increased chlorophyll content. The researchers also investigated whether the fungus could reduce the negative effects of salt stress on soybean development. According to the findings, *Aspergillus flavus* CHS1 has potential as a bioinoculant to improve agricultural yield and resilience in stressful environmental settings, such as high salinity.

2.2.2 Plant Disease Control

Endophytic fungi produce bioactive compounds as secondary metabolites, which may be used in agriculture for plant protection from phytopathogenic diseases.

Research studies focused on identifying and investigating beneficial bioactive compounds from endophytic fungi that have antibacterial and insecticidal properties and have provided promising results for disease control (Aravinthraju et al. 2024; Chandra et al. 2024). When examining compounds with antagonistic activity, the potential of *Aspergillus versicolor* to produce ten bioactive compounds with strong antifungal, antibacterial, nematicidal, and antioxidant properties has been cited. Among these compounds, there are effective antagonists against *Macrophomina phaseolina,* which include chondrillasterol, 1,2-benzedicarboxylic acid, diisooctyl ester, decane (16.63%), 9,12-octadecadienoic acid (Z,Z)-, stigmasterol, undecane, cis-1-chloro-9-octadecene, benzene, 1,3,5-trimethyl, and hexadecanoic acid, 2-hydroxy-1-(hydroxymethyl) ethyl ester (Javaid et al. 2023). Among endophyte fungi, *Trichoderma* spp., which showed a great result on antagonistic activity against pathogen of plant, with the ability to produce antimicrobial chemicals as secondary metabolites and growth-inhibiting enzymes are common characteristics of these incredibly successful endophytes, have garnered particular attention in this field of study among endophytes (Manathunga et al. 2024; Natsiopoulos et al. 2024). Among these, it was identified that *Trichoderma gamsii* were found to be antagonistic against the plant pathogenic fungus, *Rhizoctonia solani* in soil (Anees et al. 2019), as well as it was mentioned by Risoli et al. (2023), as biocontrol agent of *Fusarium* head blight with a relevant defense gene expression in wheat.

These endophytic fungi stop the growth of pathogens by enclosing their appressorium-like hyphae around the host and secreting several lytic enzymes. Prominent among these is *Trichoderma harzianum*, which is considered one of the widely used biological control agents in agriculture. *Trichoderma* was chosen as an antagonist for a variety of reasons, including its high capacity for reproduction, ability to survive in adverse environments, ability to modify the rhizosphere, extreme aggression against plant pathogenic fungi, capacity to stimulate plant growth, and potential to improve nutrient utilization efficiency (Yao et al. 2023).

Trichoderma is highly recommended as a biocontrol agent because of several characteristics, including its capacity for reproduction, adaptation to stressful environments, strong antagonistic activity against plant pathogenic fungi, ability to promote plant growth, and potential to improve nutrient utilization efficiency (Zin and Badaluddin 2020). In the agricultural industry, *Trichoderma* includes over 50 authorized product formulations globally as biofungicides (Waghunde et al. 2016) (Table 1).

Additionally, research studies revealed that *Penicillium* species are known for their antagonistic properties by producing bioactive secondary metabolites. They have been studied for their ability to manage various phytopathogenic fungal infections and their antiinsect properties. These species demonstrated efficacy against *Cercospora beticola* using the enzymes production as pectinase and cellulase (Mansour et al. 2018). In another study, it was revealed that the filtrate of *Penicillium* strains showed high antifungal effects on mycelial growth of *Fusarium oxysporum, Fusarium solani, Macrophomina phaseolina, Aspergillus japonicus,* and *Cladosporium cladosporioides*. These species secrete a diverse array of bioactive metabolites, including indole-3-acetic acid (IAA), siderophores, lipase, hydrocyanic

Table 1 *Trichoderma* species with promising antagonistic activities against plant pathogens

Trichoderma species	Antagonistic activities	Pathogens	References
Trichoderma harzianum	Mycoparasitism, antibiosis	*Fusarium oxysporum, Rhizoctonia solani, Colletotrichum gloeosporioides, Diaporthe* spp., *Neofusicoccum parvum, Phomopsis perseae*	Estela et al. (2022)
Trichoderma viride	Enzymatic degradation	*Fusarium, Aspergillus, Conocybe, Naganishia,* and *Monocillium*	Gao et al. (2023)
Trichoderma asperellum	Induced systemic resistance	*Phytophthora infestans, Sclerotinia* spp.	Pacheco et al. (2016)
Trichoderma atroviride	Mycoparasitism, oxidative stress induction, reduction in mycotoxin production	*Rhizoctonia solani, Fusarium graminearum*	Kullnig et al. (2000), Yörük et al. (2022)
Trichoderma koningii	Enzymatic degradation, competition	*Alternaria alternata, Colletotrichum* spp.	Alwadai (2022)
Trichoderma virens	Antibiosis, induced systemic resistance	*Pythium ultimum,* Fusarium spp.	Howell (2006)
Trichoderma longibrachiatum	Mycoparasitism, disruption of cell membranes by phenolic compounds	*Aspergillus flavus, Alternaria alternata, Fusarium oxysporum, Sclerotinia sclerotiorum*	Isabel et al. (2024), Li et al. (2024)

acid, protease, and β-1,3-glucanase, which contribute to disease suppression and enhanced iron absorption in plants (Chen et al. 2022). In addition to *Penicillium*, the genus *Muscodor* has emerged as a highly efficient endophytic fungus, renowned for synthesizing volatile organic compounds (VOCs) with potent antimicrobial properties, making it a promising candidate for mycofumigation (González et al. 2025). *Muscodor* species are widely distributed in tropical gramineous plants, and numerous strains have demonstrated the ability to inhibit or kill a variety of phytopathogens. For example, *Muscodor albus*, isolated from *Cinnamomum zeylanicum*, has been identified as an effective mycofumigant, producing a mixture of VOCs that exhibit broad-spectrum antifungal activity. Among these VOCs, 1-butanol and 3-methylacetate esters have been identified as the most effective inhibitory compounds, impacting pathogens such as *Fusarium oxysporum*, *Pythium ultimum, Rhizoctonia solani*, and *Ustilago hordei*. Furthermore, the *Muscodor albus* strain GBA, isolated from *Ginkgo biloba*, has shown exceptional efficacy in inhibiting *Botrytis cinerea*, *Phytophthora cinnamomi, Pythium ultimum*, and *Sclerotinia sclerotiorum*, achieving 100% mortality upon exposure. These findings underscore the potential of *Muscodor* species as sustainable alternatives for managing plant diseases in agriculture (Strobel et al. 2001).

3 Effect of Hosts on Fungal Endophytic Communities

The host plant acts as a selective force, influencing the composition of its associated endophytic community. Plant species, and even different cultivars or genotypes within a species, exhibit varying degrees of susceptibility or resistance to fungal colonization (Arnold 2007). This host specificity is driven by a range of factors, including:

Plant defense mechanisms: Plants possess a variety of defense mechanisms, such as physical barriers (e.g., thick cell walls), chemical defenses (e.g., production of antimicrobial compounds), and immune responses (e.g., hypersensitive response), that can restrict fungal colonization (Schulz et al. 2017).

Nutrient availability: The availability of nutrients within plant tissues can influence the growth and survival of endophytes. Certain plant species or genotypes may provide more favorable nutrient conditions for specific fungal taxa.

Plant signaling molecules: Plants produce a variety of signaling molecules that can attract or repel specific fungal taxa. These molecules can influence the recruitment and establishment of endophytes within plant tissues.

3.1 Host Genetics

The genetic composition of the host plant is a fundamental factor in shaping fungal endophytic communities, primarily through the regulation of secondary metabolites, defense compounds, and other biochemical traits. These genetic factors influence the production of compounds such as terpenoids, alkaloids, phenolics, and flavonoids, which act as chemical signals that either attract or repel specific fungal taxa. For example, grasses (*Poaceae*) produce alkaloids like lolines and peramine, which selectively promote the growth of mutualistic endophytes such as *Epichloë* species while inhibiting pathogenic fungi (Clay and Schardl 2002). Similarly, plants in the *Hypericum* genus produce hypericin and hyperforin, which shape the composition of their endophytic communities by selecting fungi capable of tolerating or metabolizing these bioactive compounds (Kusari et al. 2019).

Genetic variation within a host species also plays a significant role in determining endophyte colonization patterns. For instance, in maize (*Zea mays*), certain genotypes exhibit higher endophytic diversity due to variations in root exudates, such as benzoxazinoids, which influence microbial recruitment and colonization (Neal et al. 2012). Additionally, host genes involved in immune responses, such as those encoding pathogenesis-related proteins and receptor-like kinases, modulate the plant's ability to recognize and respond to microbial invaders, thereby influencing endophytic community assembly (Pieterse et al. 2014; Hardoim et al. 2015). For example, the overexpression of PR proteins in transgenic plants has been shown to reduce fungal colonization, highlighting the role of host immunity in regulating endophyte interactions (Van Loon et al. 2006). Understanding the genetic mechanisms underlying these interactions can enable the creation of crops with tailored microbiomes, enhancing agricultural productivity and bolstering resilience against the challenges posed by climate change.

3.2 Host Physiology

The physiological state of the host plant, encompassing factors such as age, health, and developmental stage, plays a critical role in shaping fungal endophytic communities. Younger plants typically harbor fewer endophytes due to underdeveloped tissues and limited nutrient availability, whereas mature plants offer more diverse microhabitats and resources, facilitating greater microbial colonization (Hardoim et al. 2015).

Environmental stressors, such as drought, salinity, or pathogen infection, can significantly alter the host's physiology, creating opportunities for stress-tolerant endophytes to thrive. For example, drought-stressed maize (*Zea mays*) plants often showed an increase in endophytes that produce osmoprotectants like proline and trehalose, which help the host retain water and sustain cellular functions under stress conditions. Similarly, plants exposed to high salinity often harbor endophytes that produce antioxidants and stress-alleviating compounds, helping the host cope with ionic and osmotic stress (Rodriguez et al. 2009). Diseased or nutrient-deficient plants may also host endophytes that mitigate these challenges by producing growth-promoting compounds, such as auxins and cytokinins, or by enhancing nutrient uptake through mechanisms like phosphate solubilization and nitrogen fixation (Hardoim et al. 2015; Rodriguez et al. 2009). For example, endophytic fungi in wheat (*Triticum aestivum*) have been shown to improve phosphorus uptake in nutrient-poor soils, promoting plant growth and resilience (Wakelin et al. 2007).

Additionally, the host's immune responses, such as the production of reactive oxygen species (ROS) and pathogenesis-related (PR) proteins, can influence endophyte colonization by creating a selective environment for compatible fungi (Pieterse et al. 2014). Thus, the physiological condition of the host acts as a dynamic and multifaceted factor that continuously shapes the diversity, composition, and functional roles of fungal endophytic communities.

3.3 Host Chemistry

The chemical composition of the host plant, including its secondary metabolites and structural compounds, is a key determinant in shaping fungal endophytic communities. Plants produce a diverse array of secondary metabolites, such as phenolics, terpenoids, alkaloids, and flavonoids, which serve as chemical signals that either attract or repel specific fungal taxa. These compounds create a selective environment that influences the colonization and persistence of endophytes. For example, plants with high lignin content, such as woody species, often host fungi like *Xylaria* and *Phomopsis*, which possess lignin-degrading enzymes that enable them to break down complex plant cell walls (Carroll 1988; Rodriguez et al. 2009). Similarly, grasses (*Poaceae*) produce alkaloids like lolines and ergot alkaloids, which selectively promote the growth of mutualistic endophytes such as *Epichloë* species while inhibiting pathogenic fungi (Clay and Schardl 2002).

Medicinal plants, such as *Hypericum perforatum*, produce bioactive compounds like hypericin and hyperforin, which attract unique endophytes capable of tolerating or metabolizing these chemicals. These endophytes often exhibit pharmaceutical potential, producing secondary metabolites with antimicrobial, anticancer, or antioxidant properties (Kusari et al. 2019). Root exudates, which vary significantly among plant species, also play a critical role in recruiting soil-derived endophytes. These exudates, composed of sugars, amino acids, organic acids, and other compounds, create a chemical gradient in the rhizosphere that influences microbial colonization. For example, maize (*Zea mays*) releases benzoxazinoids in its root exudates, which attract beneficial endophytes capable of enhancing plant growth and stress tolerance (Neal et al. 2012). Similarly, legumes secrete flavonoids that not only facilitate symbiotic relationships with nitrogen-fixing bacteria but also influence the assembly of fungal endophytic communities (Hassan and Mathesius 2012). Additionally, the production of volatile organic compounds by the host can influence endophyte interactions by mediating fungal communication and behavior (Schulz-Bohm et al. 2017).

3.4 Host Environmental Adaptations

Host plants adapted to specific environmental conditions, such as drought, salinity, nutrient-poor soils, extreme temperatures, or heavy metal contamination, significantly shape the composition and functional traits of their fungal endophytic communities. These adaptations create selective pressures that favor endophytes capable of enhancing the host's ability to survive and thrive under stress. In arid environments, endophytes like *Epichloë* species produce osmoprotectants (e.g., proline, trehalose) and improve root architecture to enhance drought tolerance (Rodriguez et al. 2009; Giauque and Hawkes 2013). In saline conditions, endophytes mitigate salt stress by producing antioxidants, osmoprotectants, and enzymes like superoxide dismutase (SOD), as observed in halophytic plants such as *Salicornia* and *Suaeda* (Rodriguez et al. 2009). In nutrient-poor soils, endophytes like *Trichoderma* and *Fusarium* solubilize phosphate, while others, such as *Penicillium*, enhance nitrogen uptake by improving root development and nutrient mobilization. These mechanisms support plant growth in low-fertility ecosystems (Wakelin et al. 2007; Puri et al. 2016). In extreme temperatures, endophytes from Antarctic plants (*Colobanthus quitensis*, *Deschampsia antarctica*) produce antifreeze and heat shock proteins, while those in desert plants like *Agave* produce heat-stable enzymes to mitigate thermal stress (Marquez et al. 2007). In heavy metal-contaminated environments, endophytes sequester metals like cadmium and arsenic and produce compounds such as metallothioneins to reduce toxicity (Chen et al. 2011). These interactions highlight the critical role of endophytes in enhancing host resilience to environmental challenges. By associating with fungi that improve stress tolerance, nutrient acquisition, and metal detoxification, host plants can thrive in harsh conditions. This dynamic interplay underscores the importance of microbial symbiosis in plant adaptation and survival, offering potential applications in sustainable agriculture and ecosystem restoration.

4 Effect of Climate Change on Fungal Endophytic Communities

Climate change is a major driver of ecosystem transformations, affecting species distributions, ecological interactions, and biodiversity patterns globally. Among the many ecological relationships influenced by these changes, plant–endophytic fungal interactions remain underexplored. Fungal endophytes, which are microorganisms that colonize plant tissues without causing disease, contribute significantly to plant health by enhancing stress tolerance, nutrient acquisition, and defense mechanisms (Rodriguez et al. 2009).

Climate change, through rising temperatures, altered precipitation patterns, increased atmospheric CO_2 levels, and habitat disturbances, is reshaping fungal endophyte diversity, composition, and ecological roles. Increased temperatures can favor the growth of certain thermophilic endophytes, while negatively impacting the growth and survival of others (Jumpponen and Trappe 1998). Drought stress can lead to shifts in community composition toward drought-tolerant species, while negatively impacting the growth and survival of water-dependent species (Rodriguez et al. 2009). Altered precipitation patterns can affect the dispersal and colonization of endophytes, influencing their distribution and abundance across landscapes (Jumpponen and Trappe 1998). This section examines the multiple dimensions through which climate change affects fungal endophytic communities (Compant and Sessitsch 2010).

4.1 Temperature Shifts and Fungal Endophyte Distribution

Global temperature rise is one of the most profound consequences of climate change, with direct and indirect effects on fungal endophytes. As ectothermic organisms, fungi are highly sensitive to temperature variations, influencing their growth rates, colonization patterns, and symbiotic efficiency. Studies on *Colobanthus quitensis* in Antarctica showed that simulated warming altered the communities of bacterial and fungal endophytes, with microbial richness inside warming chambers significantly lower than in open spaces (Perazzolli et al. 2022).

Additionally, the mutualistic interactions between fungal endophytes and plants are similarly impacted by temperature variations. The response of plant biomass to arbuscular mycorrhizal (AM) fungal inoculation varied with temperature, according to a meta-analysis looking at how climate change affects AM fungi. In particular, plant growth responses were more favorable at higher temperatures, indicating that warming may intensify the advantages that AM fungi offer (Duarte and Maherali 2022). Furthermore, symbiosis between plants and fungi may be disrupted by temperature rises. According to research, warming may change the natural state of these interactions, which could have adverse or even reduced benefits for the host plants. For instance, it has been demonstrated that during symbiotic relationships, high temperatures inhibit the ethylene and jasmonic acid signaling pathways in plants, which can impact defense and growth promotion mechanisms (Chen et al. 2022).

4.1.1 Thermal Tolerance and Community Shifts

Rising temperatures can lead to shifts in fungal community composition by favoring thermophilic (heat-tolerant) fungi while reducing the prevalence of cold-adapted species. In Arctic and alpine environments, where cold-adapted endophytes have evolved to thrive in low temperatures, warming trends may lead to species displacement or extinction. Studies on *Helotiales* endophytes from tundra plants suggest that increasing temperatures are altering fungal diversity and activity, impacting plant resilience in these regions (Geml et al. 2015).

4.1.2 Fungal Response to Heat Stress

Certain fungal endophytes exhibit adaptive responses to heat stress by altering their secondary metabolite production and colonization dynamics. For instance, endophytes in *Neotyphodium* species associated with grass have been observed to produce increased levels of alkaloids under heat stress, providing enhanced protection against herbivores and environmental stressors. However, these metabolic shifts can also disrupt existing ecological balances by modifying trophic interactions (Rodriguez et al. 2009).

4.1.3 Geographic Range Expansion

As temperatures rise, fungal species capable of withstanding warmer conditions may expand their geographic range into higher latitudes and altitudes. This shift can lead to novel plant–endophyte associations, some of which may be beneficial, while others may introduce competition or pathogenicity risks (Faticov et al. 2021).

4.2 Changes in Precipitation Patterns and Drought Stress

Climate change is intensifying the frequency and severity of droughts in many regions, significantly impacting fungal endophyte communities and their relationships with host plants.

Drought conditions impose selective pressures on fungal endophytes, leading to a shift in community composition favoring drought-resistant species. Studies on endophytes in desert plants show that dark septate endophytes (DSE), such as *Fusarium* and *Cladosporium* species, enhance host drought tolerance by modulating water retention and osmolyte production (Rodriguez et al. 2009). Changes in precipitation patterns also influence the dispersal mechanisms of fungal spores. Increased rainfall in some regions can enhance fungal dispersal via water-mediated transport, while prolonged droughts may restrict endophyte transmission by limiting sporulation and infection opportunities (Faticov et al. 2021).

Arbuscular mycorrhizal fungi (AMF), which facilitate water and nutrient uptake, are particularly susceptible to changes in soil moisture. Decreased rainfall can limit AMF colonization, reducing plant resilience to drought stress (Sendek et al. 2019). Conversely, nonmycorrhizal fungal endophytes in xerophytic (drought-adapted) plants may become more dominant, altering community structure and functionality.

4.3 Elevated CO_2 Levels and Endophyte–Plant Symbiosis

Rising atmospheric CO_2 concentrations have complex effects on plant physiology, which in turn influence fungal endophyte interactions. Increased CO_2 levels can lead to changes in plant physiology, such as enhanced photosynthesis and altered carbon-nitrogen ratios, which affect endophyte colonization and metabolic interactions (Compant and Sessitsch 2010). Studies have shown that plants grown under elevated CO_2 conditions may exhibit increased root exudation, altering fungal recruitment in the rhizosphere (Wang et al. 2017). Endophytes produce secondary metabolites such as alkaloids, flavonoids, and terpenoids, which play crucial roles in plant defense. Changes in CO_2 levels have been linked to shifts in the production of these compounds, potentially influencing fungal-mediated plant resistance against herbivores and pathogens (Arnold 2007). Some studies suggest that elevated CO_2 may strengthen endophyte–plant mutualisms by increasing the availability of carbohydrates for fungal symbionts (Compant and Sessitsch 2010). This could promote fungal colonization and enhance stress resistance in plants, but the long-term ecological consequences remain uncertain.

4.4 Habitat Loss and Biodiversity Decline

Climate change, combined with land-use changes such as deforestation and urbanization, is contributing to habitat loss, which has direct implications for fungal endophyte diversity.

Many fungal endophytes exhibit high host specificity, meaning the loss of particular plant species could lead to the extinction of associated fungal taxa. Tropical forests, which harbor some of the highest fungal endophyte diversity, are particularly vulnerable to deforestation-driven biodiversity loss (Arnold et al. 2000).

Long-standing coevolutionary relationships between plants and their endophytes may be disrupted by habitat fragmentation and shifting climate conditions. For instance, fungal species that rely on specific environmental cues for sporulation may fail to complete their life cycles under changing climate conditions. Disturbances to fungal communities may facilitate the invasion of opportunistic or pathogenic fungi, potentially leading to increased plant disease incidence (Kumar and Ria 2024).

5 Conclusions

Fungal endophytes play an essential role in plant ecosystems, influencing plant adaptation, stress tolerance, and overall ecosystem stability. Their presence within plant tissues facilitates nutrient acquisition, enhances resistance to pathogens, and enables plants to withstand harsh environmental conditions. However, the composition and function of fungal endophytic communities are not static; they are shaped by a dynamic interplay of host identity, environmental factors, and climate change.

Host specificity remains a key determinant of endophyte diversity, as different plant species and genotypes exhibit varying levels of susceptibility to fungal colonization. This specificity is driven by multiple factors, including plant defense mechanisms, nutrient availability, and signaling molecules that regulate fungal recruitment. Meanwhile, environmental conditions such as temperature, soil composition, and moisture availability further influence endophyte diversity and function. These factors collectively shape the structure of endophytic communities, determining which fungal species thrive in specific plant hosts and ecological niches.

Climate change introduces additional complexity to these interactions by altering temperature regimes, precipitation patterns, and atmospheric CO_2 levels. Rising temperatures may favor thermophilic endophytes while negatively impacting cold-adapted species, leading to shifts in fungal community composition. Similarly, drought stress can cause an increase in drought-tolerant endophytes while suppressing water-dependent species. Changes in precipitation patterns affect fungal dispersal, potentially influencing the colonization success and geographic distribution of endophytes. Elevated CO_2 levels also have significant consequences, as they can modify plant physiology and alter the metabolic interactions between plants and their fungal symbionts. The impact of climate change on fungal endophytes extends beyond individual plant-fungal interactions. Since endophytes play a vital role in ecosystem processes such as carbon cycling, nutrient dynamics, and disease suppression, disruptions to their communities could have cascading effects on plant biodiversity and ecosystem functionality. Habitat loss due to deforestation and land-use changes further threatens endophyte diversity, potentially leading to the extinction of host-specific fungal species and the invasion of opportunistic or pathogenic fungi. Given these challenges, future research must focus on understanding the molecular mechanisms underlying endophyte–plant interactions and identifying strategies to harness endophytes for agricultural and conservation purposes. This includes exploring the potential of endophytes in promoting stress tolerance in crops, enhancing plant resilience in changing climates, and restoring degraded ecosystems. Additionally, global monitoring of fungal endophyte communities will be essential in predicting long-term ecological shifts and mitigating the impacts of environmental changes.

In conclusion, fungal endophytes are indispensable allies in plant evolution and ecosystem sustainability. Their interactions with host plants are complex and dynamic, shaped by an intricate web of biological and environmental factors. As climate change continues to reshape ecosystems, understanding and conserving fungal endophyte diversity will be crucial for maintaining plant health, agricultural productivity, and global biodiversity.

References

Airin AA, Arafat MI, Begum RA et al (2023) Plant growth-promoting endophytic fungi of the wild halophytic rice. Oryza coarctata. Ann Microbiol 73:36. https://doi.org/10.1186/s13213-023-01738-3

Almario J, Jeena G, Wunder J, Langen G (2017) Root-associated fungal microbiota of nonmycorrhizal Arabis alpina and its contribution to plant phosphorus nutrition. Proc Natl Acad Sci 117:31549–31559

Alwadai AS et al (2022) The isolation and characterization of antagonist Trichoderma spp. from the soil of Abha, Saudi Arabia. Molecules 27(8):2525. https://doi.org/10.3390/molecules27082525

Anees M, Abid M, Chohan S, Jamil M, Ahmed N, Zhang L, Rha ES (2019) In situ impact of the antagonistic fungal strain, Trichoderma gamsii T30 on the plant pathogenic fungus, Rhizoctonia solani in soil. Pol J Microbiol 68(2):211. https://doi.org/10.33073/pjm-2019-021

Aravinthraju K, Shanthi M, Murugan M, Srinivasan R, Maxwell LA, Manikanda Boopathi N, Anandham R (2024) Endophytic entomopathogenic fungi: their role in enhancing plant resistance, managing insect pests, and synergy with management routines. J Fungi 10(12):865. https://doi.org/10.3390/jof10120865

Arnold AE (2007) Understanding the diversity of foliar endophytic fungi: progress, challenges, and frontiers. Fungal Biol Rev 21(2–3):51–66. https://doi.org/10.1016/j.fbr.2007.05.003

Arnold AE, Maynard Z, Gilbert GS, Coley PD, Kursar TA (2000) Are tropical fungal endophytes hyperdiverse? Ecol Lett 3:267–274

Asaf S, Hamayun M, Khan AL, Waqas M, Khan MA, Jan R, Lee I-J, Hussain A (2018) Salt tolerance of Glycine max. L induced by endophytic fungus aspergillus flavus CSH1, via regulating its endogenous hormones and antioxidative system. Plant Physiol Biochem 128:13–23

Babu AG, Kim SW, Yadav DR, Hyum U, Adhikari M, Lee YS (2015) Penicillium menonorum: a novel fungus to promote growth and nutrient management in cucumber plants. Mycobiology 43(1):49. https://doi.org/10.5941/MYCO.2015.43.1.49

Baldrian P, Větrovský T, Lepinay C, Kohout P (2021) High-throughput sequencing view on the magnitude of global fungal diversity. Fungal Divers 19:1–9

Baron NC, Rigobelo EC (2021) Endophytic fungi: a tool for plant growth promotion and sustainable agriculture. Mycology 13(1):39. https://doi.org/10.1080/21501203.2021.1945699

Berg G, Grube M, Schloter M, Smalla K (2014) The plant microbiome and its importance for plant and human health. Front Microbiol 5:491

Bezerra JDP, Santos MGS, Svedese VM et al (2012) Richness of endophytic fungi isolated from Opuntia ficus-indica Mill. (Cactaceae) and preliminary screening for enzyme production. World J Microbiol Biotechnol 28:1989–1995. https://doi.org/10.1007/s11274-011-1001-2

Bisby G, Ainsworth G (1943) The numbers of fungi. Trans Br Mycol Soc 26(1–2):16–19. https://doi.org/10.1016/S0007-1536(43)80005-X

Carrión VJ, Perez-Jaramillo J, Cordovez V, Tracanna V, de Hollander M, Ruiz-Buck D, Mendes LW, Van Ijcken WFJ, Gomez-Exposito R, Elsayed S, Mohanraju P, Arifah A, van der Oost J, Paulson JN, Mendes R, Van Wezel GP, Medema MH, Raaijmakers JM (2019) Pathogen-induced activation of disease-suppressive functions in the endophytic root microbiome. Science 366:606–612

Carroll G (1988) Fungal endophytes in stems and leaves: from latent pathogen to mutualistic symbiont. Ecology 69:2–9

Chandra H, Yadav A, Prasad R, Kalra SJS, Singh A, Bhardwaj N, Gupta KK (2024) Fungal endophytes from medicinal plants acting as natural therapeutic reservoir. The Microbe 3: 100073. https://doi.org/10.1016/j.microb.2024.100073

Chen J, Hu KX, Hou XQ, Guo SX (2011) Endophytic fungi assemblages from 10 dendrobium medicinal plants (Orchidaceae). World J Microbiol Biotechnol 27:1009–1016

Chen J, Yin Q, Zhu M, Xia X, Han J (2022) High ambient temperature regulated the plant systemic response to the beneficial endophytic fungus serendipita indica. Front Plant Sci 13:844572. https://doi.org/10.3389/fpls.2022.844572

Clay K, Schardl C (2002) Evolutionary origins and ecological consequences of endophyte symbiosis with grasses. Am Nat 160:S99–S127

Compant S, Sessitsch A (2010) Climate change effects on beneficial plant–microorganism interactions. FEMS Microbiol Ecol 73(2):197–214. https://doi.org/10.1111/j.1574-6941.2010.00900.x

Compant S, Samad A, Faist H, Sessitsch A (2021) A review on the plant microbiome: ecology, functions, and emerging trends in microbial application. J Adv Res 19:29–37

Daigham GE, Mahfouz AY, Abdelaziz AM et al (2024) Protective role of plant growth-promoting fungi aspergillus chevalieri OP593083 and aspergillus egyptiacus OP593080 as biocontrol approach against Alternaria leaf spot disease of Vicia faba plant. Biomass Conv Bioref 14: 23073–23089. https://doi.org/10.1007/s13399-023-04510-4

Duarte AG, Maherali H (2022) A meta-analysis of the effects of climate change on the mutualism between plants and arbuscular mycorrhizal fungi. Ecol Evol 12(1):e8518. https://doi.org/10.1002/ece3.8518. PMID: 35127032; PMCID: PMC8796888

Duggal P, Kumar A, Bhambhu MK, Kumar V, Kumar P, Mann SS, Saini S. Occurrence, distribution and management of cereal cyst nematode: Global impact to cutting edge solutions. Physiol Mol Plant Pathol. 2025;138:102675. https://doi.org/10.1016/j.pmpp.2025.102675

Estela M, Lizette C, Antonio J, Angel M, Thalia A, Guadalupe M, Pierre J, Javier P (2022) Isolation and characterization of Trichoderma spp. for antagonistic activity against avocado (Persea americana mill) fruit pathogens. Horticulturae 8(8):714. https://doi.org/10.3390/horticulturae8080714

Faticov M, Ahmed A, Tomas R, Corinne V, Peter Hambäck FGB, Björn DL, Ayco JMT (2021) Climate warming dominates over plant genotype in shaping the seasonal trajectory of foliar fungal communities on oak. New Phytol 231(5):1770–1783. Accessed February 27, 2025. https://doi.org/10.1111/nph.17434

Gao P, Qi K, Han Y, Ma L, Zhang B, Zhang Y, Guan X, Qi J (2023) Effect of Trichoderma viride on rhizosphere microbial communities and biocontrol of soybean root rot. Front Microbiol 14: 1204688. https://doi.org/10.3389/fmicb.2023.1204688

Geml J, Morgado LN, Semenova TA, Welker JM, Walker MD, Smets E (2015) Long-term warming alters richness and composition of taxonomic and functional groups of arctic fungi. FEMS Microbiol Ecol 91(8):fiv095. https://doi.org/10.1093/femsec/fiv095

Giauque H, Hawkes CV (2013) Climate affects symbiotic fungal endophyte diversity and performance. Am J Bot 100:1435–1444

González Ortega-Villaizán A, King E, Patel MK et al (2025) Identification of a drought stress response module in tomato plants commonly induced by fungal endophytes that confer increased drought tolerance. Plant Mol Biol 115:7. https://doi.org/10.1007/s11103-024-01532-y

Hardoim PR, van Overbeek LS, Berg G, Pirttilä AM, Compant S, Campisano A, Döring M, Sessitsch A (2015) The hidden world within plants: ecological and evolutionary considerations for defining functioning of microbial endophytes. Microbiol Mol Biol Rev 79:293–320

Hassan S, Mathesius U (2012) The role of flavonoids in root-rhizosphere signaling: opportunities and challenges for improving plant-microbe interactions. J Exp Bot 63:3429–3444

Hawksworth DL (2004) Fungal diversity and its implications for genetic resource collections. Stud Mycol 50:9–18

Hawksworth DL, Lücking R (2017) Fungal diversity revisited: 2.2 to 3.8 million species. Microbiol Spectr 5:79–95

Hiruma K, Gerlach N, Sacristán S, Nakano RT, Hacquard S, Kracher B, Schulze-Lefert P (2016) Root endophyte Colletotrichum tofieldiae confers plant fitness benefits that are phosphate status dependent. Cell 165:464–474

Howell CR (2006) Understanding the mechanisms employed by trichoderma virens to effect biological control of cotton diseases. Phytopathology 96(2):178–180. https://doi.org/10.1094/PHYTO-96-0178

Isabel A, Ernesto M, Rosa A (2024) Phenolic compounds synthesized by trichoderma longibrachiatum native to semi-arid areas show antifungal activity against phytopathogenic fungi of horticultural interest. Microbiol Res 15(3):1425–1440. https://doi.org/10.3390/microbiolres15030096

Javaid A, Ali A, Haider Khan I, Ferdosi M FH. Leaves of Chenopodium album as source of natural fungicides against Sclerotium rolfsii. Arab J Chem. 2023;16(5):104677. https://doi.org/10.1016/j.arabjc.2023.104677

Jumpponen A, Trappe JM (1998) Dark-septate endophytes: a review. New Phytol 140(3):295–310

Kullnig C, Mach RL, Lorito M, Kubicek CP (2000) EnzymeDiffusion from Trichoderma atroviride (=T. harzianum P1) to Rhizoctonia solani is a prerequisite for triggering of Trichoderma ech42 gene expression before mycoparasitic contact. Appl Environ Microbiol 66. https://doi.org/10.1128/AEM.66.5.2232-2234.2000

Kumar D, Ria M (2024) Climate change and plant pathogens: understanding dynamics, risks and mitigation strategies. Plant Pathol 74(1):59–68. https://doi.org/10.1111/ppa.14033

Kusari S, Singh S, Jayabaskaran C (2019) Biotechnological potential of plant-associated endophytic fungi: Hope versus hype. Trends Biotechnol 34:603–612

Li P, Jeewon R, Aruna B, Li H, Lin F, Wang H (2019) Metabarcoding reveals differences in fungal communities between unflooded versus tidal flat soil in coastal saline ecosystem. Sci Total Environ 690:911–922. https://doi.org/10.1016/j.scitotenv.2019.06.473

Li E, Zhu N, Zhang S, Xu B, Liu L, Zhang A (2024) Efficacy of Trichoderma longibrachiatum SC5 fermentation filtrate in inhibiting the sclerotinia sclerotiorum growth and development in sunflower. Int J Mol Sci 26(1):201. https://doi.org/10.3390/ijms26010201

Lodge DJ, Fisher PJ, Sutton BC (1996) Endophytic fungi of Manilkara bidentata leaves in Puerto Rico. Mycologia 88:733–738

Lücking R, Aime MC, Robbertse B, Miller AN, Aoki T, Ariyawansa HA, Cardinali G, Crous PW, Druzhinina IS, Geiser DM, Hawksworth DL, Hyde KD, Irinyi L, Jeewon R, Johnston PR, Kirk PM, Malosso E, May TW, Meyer W, Nilsson HR, Öpik M, Robert V, Stadler M, Thines M, Vu D, Yurkov AM, Zhang N, Schoch CL (2021) Fungal taxonomy and sequence-based nomenclature. Nat Microbiol 6:540–548

Madden TL, Tatusov RL, Zhang J (1996) Applications for network BLAST server. Methods Enzymol 266:131–141. https://doi.org/10.1016/s0076-6879(96)66011-x

Manathunga KK, Gunasekara NW, Meegahakumbura MK, Ratnaweera PB, Faraj TK, Wanasinghe DN (2024) Exploring endophytic fungi as natural antagonists against fungal pathogens of food crops. J Fungi 10(9):606

Mansour M, Rafeek MI, Kamal AM (2018) Evaluation of certain penicillium frequentans isolates against cercospora leaf spot disease of sugar beet. Egypt J Biol Pest Control 28(1):1–7. https://doi.org/10.1186/s41938-018-0053-0

Marín P, Sánchez R, Arribas C, Kruse J, Brunner F, Sacristán S (2020) Mutualistic fungal endophyte colletotrichum tofieldiae ct0861 colonizes and increases growth and yield of maize and tomato plants. Agronomy 10(10):1493. https://doi.org/10.3390/agronomy10101493

Marquez LM, Redman RS, Rodriguez RJ, Roossinck MJ (2007) A virus in a fungus in a plant: three-way symbiosis required for thermal tolerance. Science 315:513–515

Martin GW (1951) The numbers of fungi. Proc Iowa Acad Sci 58(1):175–178. https://scholarworks.uni.edu/pias/vol58/iss1/18

Natsiopoulos D, Topalidou E, Mantzoukas S, Eliopoulos PA (2024) Endophytic Trichoderma: potential and prospects for plant health management. Pathogens 13(7):548. https://doi.org/10.3390/pathogens13070548. PMID: 39057775; PMCID: PMC11279820

Neal AL, Ahmad S, Gordon-Weeks R, Ton J (2012) Benzoxazinoids in root exudates of maize attract pseudomonas putida to the rhizosphere. PLoS One 7:e35498

Pacheco KR et al (2016) Efficacy of Trichoderma asperellum, T. harzianum, T. longibrachiatum and T. reesei against Sclerotium rolfsii = Eficácia de Trichoderma asperellum, T. harzianum, T. longibrachiatum E T. eesei contra Sclerotium rolfsii. Biosci J 32:412–421

Perazzolli M, Vicelli B, Antonielli LO, Longa CM, Bozza E, Bertini L, Caruso C, Pertot I (2022) Simulated global warming affects endophytic bacterial and fungal communities of Antarctic pearlwort leaves and some bacterial isolates support plant growth at low temperatures. Sci Rep 12:18839. https://doi.org/10.1038/s41598-022-23582-2

Phukhamsakda C, Nilsson RH, Bhunjun CS et al (2022) The numbers of fungi: contributions from traditional taxonomic studies and challenges of metabarcoding. Fungal Divers 114:327–386. https://doi.org/10.1007/s13225-022-00502-3

Pieterse CMJ, Zamioudis C, Berendsen RL, Weller DM, Van Wees SCM, Bakker PAHM (2014) Induced systemic resistance by beneficial microbes. Annu Rev Phytopathol 52:347–375

Porras-Alfaro A, Bayman P (2011) Hidden fungi, emergent properties: endophytes and microbiomes. Annu Rev Phytopathol 49:291–315

Puri A, Padda KP, Chanway CP (2016) Nitrogen fixation by endophytic bacteria in agricultural crops: recent advances. Nitrogen Agric 36:1–20

Rana KL, Kour D, Sheikh I, Dhiman A, Yadav N, Yadav AN et al (2019) Endophytic fungi: biodiversity, ecological significance, and potential industrial applications. In: Nonparametric statistics. Springer, pp 1–62. https://doi.org/10.1007/978-3-030-10480-1_1

Ripa FA, Cao WD, Tong S, Sun JG (2019) Assessment of plant growth promoting and abiotic stress tolerance properties of wheat endophytic fungi. Microorganisms 7:672

Risoli S, Petrucci A, Vicente I, Sarrocco S (2023) Trichoderma gamsii T6085, a biocontrol agent of Fusarium head blight, modulates biocontrol-relevant defence genes expression in wheat. Plant Pathol 72(8):1442–1452. https://doi.org/10.1111/ppa.13773

Rodriguez RJ, White JF, Arnold AE, Redman RS (2009) Fungal endophytes: diversity and functional roles. New Phytol 182:314–330. https://doi.org/10.1111/j.1469-8137.2009.02773.x

Schulz-Bohm K, Geisen S, Wubs ERJ, Song C, de Boer W, Garbeva P (2017) The prey's scent volatile organic compound mediated interactions between soil bacteria and their protist predators. ISME J 11:817–820

Selosse MA, Schneider-Maunoury L, Martos F (2018) Time to re-think fungal ecology? Fungal ecological niches are often prejudged. New Phytol 210:334–337

Sendek A, Karakoç C, Wagg C, et al. Drought modulates interactions between arbuscular mycorrhizal fungal diversity and barley genotype diversity. Sci Rep. 2019;9:9650. https://doi.org/10.1038/s41598-019-45702-1

Smith SE, Read DJ (2008) Mycorrhizal symbiosis, 3rd edn. Academic Press

Strobel GA, Sears J, Dirkse E, Markworth C (2001) Volatile antimicrobials from Muscodor albus, a novel endophytic fungus. Microbiology 147(11):2943–2950. https://doi.org/10.1099/00221287-147-11-2943

Sun X, Guo LD (2012) Endophytic fungal diversity: review of traditional and molecular techniques. Mycology 3(1):65–76. https://doi.org/10.1080/21501203.2012.656724

Svoboda T, Thon MR, Strauss J (2021) The role of plant hormones in the interaction of colletotrichum species with their host plants. Int J Mol Sci 22(22):12454. https://doi.org/10.3390/ijms222212454

Trivedi P, Leach JE, Tringe SG, Singh BK (2020) Plant-microbiome interactions: from community assembly to plant health. Nat Rev Microbiol 18:607–621

Tseng Y, Rouina H, Groten K, Rajani P, Furch AC, Reichelt M, Baldwin IT, Nataraja KN, Uma Shaanker R, Oelmüller R (2020) An endophytic trichoderma strain promotes growth of its hosts and defends against pathogen attack. Front Plant Sci 11:573670. https://doi.org/10.3389/fpls.2020.573670

U'Ren JM, Arnold AE (2016) Diversity, taxonomic composition, and functional aspects of fungal communities in living, senesced, and fallen leaves at five sites across North America. PeerJ 4:e2768. https://doi.org/10.7717/peerj.2768

Van Loon LC, Rep M, Pieterse CMJ (2006) Significance of inducible defense-related proteins in infected plants. Annu Rev Phytopathol 44:135–162

Větrovský T, Morais D, Kohout P, Lepinay C et al (2020) GlobalFungi, a global database of fungal occurrences from high-throughput-sequencing metabarcoding studies. Sci Data 7(1):1–14. https://doi.org/10.1038/s41597-020-0567-7

Vicedo B, Papadopoulou KK, Camañes G, Llorens E (2025) Isolation and characterization of a new Leptobacillium species promoting tomato plant growth. Sci Rep 15(1):1–14. https://doi.org/10.1038/s41598-024-84951-7

Waghunde RR, Shelake RM, Sabalpara AN (2016) Trichoderma: a significant fungus for agriculture and environment. Afr J Agric Res 11(22):1952–1965

Wakelin SA, Warren RA, Harvey PR, Ryder MH (2007) Phosphate solubilization by Penicillium spp. closely associated with wheat roots. Biol Fertil Soils 44:1027–1036

Wang Z, Li T, Wen X, Liu Y, Han J, Liao Y and DeBruyn JM (2017) Fungal Communities in Rhizosphere Soil under Conservation Tillage Shift in Response to Plant Growth. Front. Microbiol. 8:1301. https://doi.org/10.3389/fmicb.2017.01301

Wen J, Okyere SK, Wang S, Wang J, Xie L, Ran Y, Hu Y (2022) Endophytic fungi: an effective alternative source of plant-derived bioactive compounds for pharmacological studies. J Fungi 8(2):205. https://doi.org/10.3390/jof8020205

Wickham H (2016) Ggplot2: elegant graphics for data analysis, 2nd edn. Springer

Wu B, Tian J, Bai et al (2013) The biogeography of fungal communities in wetland sediments along the Changjiang River and other sites in China. ISMEJ 7:1299–1309. https://doi.org/10.1038/ismej.2013.29

Wu B, Hussain M, Zhang W, Stadler M, Liu X, Xiang M (2019) Current insights into fungal species diversity and perspective on naming the environmental DNA sequences of fungi. Mycology 10:127–140

Yadav A, Kour D, Kaur T et al (2022) Endophytic fungal communities and their biotechnological implications for agro-environmental sustainability. Folia Microbiol 67:203–232. https://doi.org/10.1007/s12223-021-00939-0
Yao X, Guo H, Zhang K, Zhao M, Ruan J, Chen J (2023) Trichoderma and its role in biological control of plant fungal and nematode disease. Front Microbiol 3(14):1160551. https://doi.org/10.3389/fmicb.2023.1160551
Yeh H, Kirschner R (2019) Diversity of endophytic fungi of the coastal plant vitex rotundifolia in Taiwan. Microbes Environ 34(1):59. https://doi.org/10.1264/jsme2.ME18075
Yörük E, Özkale E, Sefer Ö et al (2022) Trichoderma atroviride triggers alterations at epigenetics, transcripts, oxidative stress, and apoptosis levels on Fusarium graminearum. J Plant Pathol 104:1039–1047
Zhang Q, Li HQ, Zong SC, Gao JM, Zhang AL (2016) Chemical and bioactive diversities of the genus Chaetomium secondary metabolites. Mini-Rev Med Chem 12:127–148
Zin NA, Badaluddin NA (2020) Biological functions of Trichoderma spp. for agriculture applications. Ann Agric Sci 65(2):168–178. https://doi.org/10.1016/j.aoas.2020.09.003

Correction to: Diversity of Endophytic Fungi of Common Yew (*Taxus baccata* L.)

Qudsia Firdous and Sheharyar Ahmed Khan

Correction to:
Chapter 14 in: A. M. Abdel-Azeem, F. Degola (eds.), *Advances in Lichens and Non-Flowering Plants Fungal Endophytes Research*, Fungal Biology,
https://doi.org/10.1007/978-3-032-01506-8_14

The name of the author has been corrected from *Sheharyar Ahmad Khan* to *Sheharyar Ahmed Khan*. The originally published version has now been updated.

The updated version of this chapter can be found at
https://doi.org/10.1007/978-3-032-01506-8_14

A. M. Abdel-Azeem, F. Degola (eds.), *Advances in Lichens and Non-Flowering Plants Fungal Endophytes Research*, Fungal Biology,
https://doi.org/10.1007/978-3-032-01506-8_17

Index

A. M. Abdel-Azeem, F. Degola (eds.), *Advances in Lichens and Non-Flowering Plants Fungal Endophytes Research*, Fungal Biology,
https://doi.org/10.1007/978-3-032-01506-8

MIX
Papier aus verantwortungsvollen Quellen
Paper from responsible sources
FSC® C105338

If you have any concerns about our products,
you can contact us on
ProductSafety@springernature.com

In case Publisher is established outside the EU,
the EU authorized representative is:
Springer Nature Customer Service Center GmbH
Europaplatz 3, 69115 Heidelberg, Germany

Printed by Libri Plureos GmbH
in Hamburg, Germany